D1082778

**HEAT TRANSFER IN COUNTERFLOW,
PARALLEL FLOW AND CROSS FLOW**

**McGRAW-HILL
BOOK COMPANY**

New York
St. Louis
San Francisco
Auckland
Bogotá
Hamburg
Johannesburg
London
Madrid
Mexico
Montreal
New Delhi
Panama
Paris
São Paulo
Singapore
Sydney
Tokyo
Toronto

HELMUTH HAUSEN
Translated from the German by M. S. Sayer
Translation edited by A. J. Willmott
University of York

Heat Transfer in Counterflow, Parallel Flow and Cross Flow

Library of Congress Cataloging in Publication Data

Hausen, Helmuth.
    Heat transfer in counterflow, parallel flow and cross flow.
    Translation of: Wärmeübertragung im Gegenstrom,
    Gleichstrom, und Kreuzstrom.   2., neubearb. Aufl. 1976.
    Includes bibliographies and index.
    1. Heat—Transmission.   I. Title.
    QC320.H2813   536′.2   81-11772

    ISBN 0-07-027215-8   AACR2

British Library Cataloguing in Publication Data

Hausen, Helmuth
    Heat transfer in counterflow, parallel flow and cross flow.
    1. Heat—Transmission—Mathematics
    2. Numerical calculations
    I. Title   II. Willmott, A. J.
    III. Wärmeübertragung im Gegenstrom,
    Gleichstrom und Kreuzstrom. *English*
    563′.2′015117   QC320.2

    ISBN 0-07-027215-8

1234567890   HAL/HAL   898765432

ISBN 0-07-027215-8

Printed and bound in the United States of America

CONTENTS

PART TWO

Heat Transfer and Pressure Drop, Particularly in Tubes and Ducts

PART THREE

Recuperators

Appendices

FOREWORD

In preparing an English translation of the German version of the second edition of this book, we have taken into particular consideration two specific matters. Firstly, we were asked by Professor Hausen to produce as close a translation as possible of the German text. Secondly, the agreement between the publishers of this translation and of the second edition in German (Springer-Verlag), required we produce a book with as few alterations as possible so that in no sense could our translation be thought of as a Third Edition of the book.

We have endeavoured to satisfy the wishes of the author and of Springer-Verlag to the best of our ability.

The most important difference between the German edition and this translation is that this book is divided into nineteen numbered chapters, the chapters being grouped together into four parts. Part One is essentially an introduction; Part Two deals with general heat transfer theory as applicable to heat exchangers. Parts Three and Four form the main body of the text, the first dealing with recuperators, the second with regenerators. The chapters themselves and their titles correspond exactly to the divisions into sections in the German edition. We were asked by McGraw-Hill to renumber all the equations and the figures so that, for example, the fourth equation in Chapter Nine is numbered (9-4).

In no sense have we attempted to liberally paraphrase the German text and if this translation has a strong Germanic flavour, this reflects the way Professor Hausen wrote his book in the first place. Thus the English reader will find that the references to equations and figures elsewhere in the text are presented in a very thorough and exhaustive way.

The first edition of the book was published in 1950 in German, but has been out of print for several years. It was written at a time when digital computers were not available for recuperator and regenerator calculations. This reflects itself in the second edition where, for example, graphical methods for solving equations are still mentioned. Even so, this does not detract from the importance of the book. Many more modern methods of calculation are included.

We have placed editorial footnotes in the text where comment or further explanation seemed to us to be necessary. On the other hand, we have endeavoured to keep these footnotes to a minimum.

We must express our thanks to our families, in particular Joan Willmott who has borne our involvement in the preparation of this translation with great patience. Fiona Sands has been a tower of strength in preparing over 600 pages of typescript from our notes. We shall always appreciate this. We must say thank you to T. E. Blackall of the Maidenhead offices of McGraw-Hill who has been both kind and efficient in his dealing with those minor difficulties which arose as this work progressed.

Finally, we have greatly appreciated the help we have received from Professor Hausen. It has been a great pleasure to have been associated with this book.

October 1980

M. S. Sayer
Pickering
North Yorkshire

A. J. Willmott
University of York

PREFACE TO THE ENGLISH EDITION

The purpose of this book is to describe the processes which occur in continuously operating recuperative heat exchangers and in regularly reversed regenerators. Its purpose is also to show how these processes can be modelled mathematically. Methods of calculation of the temperature distribution in such heat exchangers and of the quantities of heat transferred in these devices are derived, wherever possible, from exact theories. These methods have arisen out of industrial requirements and have proved, on the whole, to be practicable.

An early part of the book presents those portions of heat transfer theory which are of particular relevance to heat transfer in heat exchangers. This should enable the heat transfer coefficients which may be required to be easily determined or at least estimated. This section does not claim to be complete.

I should like to express my thanks to Dr. Willmott and to Mrs. Sayer who have collaborated to translate my book into English. Mrs. Sayer has prepared an English translation of the German text from a linguistic point of view, while Dr. Willmott has edited this English translation and has been responsible for the correct representation in English of the scientific meaning of the original text.

Dr. Willmott is a scientist who has since about 1960 been concerned mainly with regenerators. In particular he was one of the first to calculate the performance of regenerators with time varying flowrates. He has also computed the response of a regenerator to a sudden change in the gas entrance temperature or in gas flowrate and has calculated the time to re-establish cyclic equilibrium following such changes. Finally he has examined the detailed processes which take place inside a regenerator at a reversal.

Mrs. Sayer is a technical translator working in the field of engineering and the physical sciences. She first worked with Dr. Willmott in the mid 1970's upon a translation of the classical paper by Anzelius which deals with the heating of a porous solid by a flowing medium.

H. Hausen

PREFACE TO THE SECOND EDITION

The first edition of this book, which appeared in 1950, has already been out of print for more than 15 years. During this period I have been asked time and again, both at home and abroad, when a new edition would appear and it has been my firm intention to prepare the second edition as soon as possible. However numerous other commitments prevented me from doing this before the end of 1972. During the 26 years since the appearance of the first edition an extraordinarily large number of research papers have been published and I have had to considerably rewrite the text. This particularly applies to Part Two of the book which deals with the principles of heat transfer and with the pressure drops in the various types of tubes and ducts. It also applies, to a lesser extent, to Parts Three and Four: here the theoretical models of the processes occurring in heat exchangers are treated together with the development, from these models, of methods of heat exchanger calculation. Much of the knowledge necessary for this was gained before the Second World War. The new and latest research results are considered in all sections, as far as possible; nevertheless, because of their overwhelming numbers this has not always been possible. The presentation of many details has been avoided due to the appearance in 1974 of the second edition of the VDI Heat Atlas (VDI Wärmeatlas) to which reference should be made at various places in this book.*

Because of its unquestionable advantages I have been happy to switch over to the international system of units. In particular this makes consideration of the application of the similarity principle to heat transfer both easier and clearer. In the choice of symbols I have endeavoured to comply with current norms, although I have not been completely successful in this. Further, I have been unable to use the VDI Heat Atlas notation for distinguishing various temperatures, which notation uses bars above a symbol and indices; I apologise to readers who wish to use the

---

* Editor's note: At this juncture, it is to be hoped that the possibly necessary reference to the VDI Wärmeatlas, published in German, will not deter the reader of this English translation of Hausen's book.

Heat Atlas and my book at the same time. I have explained the difficulties which prevented my using the VDI Heat Atlas notation in toto in the explanatory note attached to the list of most frequently used symbols.

While I am aware of many imperfections in the present revised edition I nevertheless hope that it will be received in as friendly a manner, and prove to be as useful, as was the first edition.

Hannover, July 1976

York, October 1978

INDEX OF THE MOST FREQUENTLY USED SYMBOLS

1. Lengths and Surfaces

x or l	Length of a section of a heat exchanger, calculated from the entry position of one of the gases
L	Total length of a heat exchanger
d, d_i, d_a	Diameter of a tube
d_{gl}	Equivalent diameter from Eq. (2-14), p. 21
D_i	Internal diameter of a casing
δ	Thickness of a tube wall or a plane wall
f	Heating surface area, relative to the length x of the exchanger
F	Total heating surface area of a heat exchanger
s	Thickness of a layer of radiating gas
a, b, l, x, y	In finned tubes, see Fig. 8-23, p. 249

2. Time

t, t'	Time in s, in regenerators usually calculated from the beginning or middle of a period
T, T'	Length of a regenerator's hot or cold period

3. Flow Velocity

w	Average flow velocity in a cross-section
$\dot{m}$	Mass flow

4. Physical Properties

ρ	Density
η	Dynamic viscosity

$v = \eta/\rho$	Kinematic viscosity
λ	Thermal conductivity of a fluid
λ_s	Thermal conductivity of a fixed material (Index s: solid or heat-storing mass)
$a = \lambda/\rho c_p$	Diffusivity of a fluid

5. Temperatures

θ, θ'	Temperatures of the fluid, usually gas or liquid
θ_1, θ'_1	Entry temperatures
θ_2, θ'_2	Exit temperatures
$\Delta\theta = \theta - \theta'$	Temperature difference between two materials at a specific position x in the heat exchanger; in evaporation temperature difference between the surface of the heating surface area and the boiling liquid
$\tau = \Delta\theta_a/\Delta\theta_b$	Relation of this temperature difference to the end of the heat exchanger
Θ	Temperature of the wall or the heat-storing mass
$\bar{\theta}, \bar{\theta'}, \bar{\Theta}$	Chronological average value of the temperatures at a specific position x
Θ_0, Θ'_0	Surface temperature of the wall or the heat-storing element
Θ_m	Spatial average value of the temperature of the wall or the heat-storing mass at a specific point x
$\theta_M, \theta_M, \Theta_M, \Delta\theta_M$	Average value in the total heat exchanger

6. Quantities of Heat and Thermal Capacities

$\dot{q}$	Quantities of heat which are transferred in unit time in a section of the heat exchanger of length x; thermal flow density (heating surface area load) in evaporation (p. 78) and radiation (p. 110)
$\dot{Q}$	Quantities of heat which are exchanged in unit time in the complete heat exchanger
c_p, c'_p	Specific thermal capacities of the fluids
c_s	Specific thermal capacity of the heat-storing mass of a regenerator
C, C'	Thermal capacities of the quantities flowing through a regenerator in unit time
C_s	Thermal capacity of a regenerator heat-storing mass
r	Evaporation enthalpy of the unit of mass

7. Heat and Mass Transfer

α, α'	Heat transfer coefficients of fluids
$\bar{\alpha}$	Heat transfer coefficient at the average temperature Θ_m of the heat-storing mass according to Eq. (13-3), p. 346
k	Heat transmission coefficient

k_0 Heat transmission coefficient, relative to the zero eigen function of a regenerator, according to Eq. (11-4), p. 307

σ Evaporation coefficient, see p. 468

8. Thermal Radiation

$E, A, C, E_s, C_s, \lambda, A_\lambda, \varepsilon_g, \varepsilon_w, \overline{\varepsilon_g}, \overline{\varepsilon_w}, \alpha$ see Sec. 3-1 to Sec. 3-3

9. Dimensionless Quantities

Re, Pr, Pe, Gr	Dimensionless groups of flow and heat transfer, see pp. 19, 20, 37
ψ or ζ	Dimensionless groups of pressure drops, see pp. 119, 126
ξ, η, Λ, Π	Dimensionless length and time in regenerators, see pp. 353, 354
η	Recuperator efficiency
η^*	Efficiency function, see p. 164
η_{Reg}	Regenerator efficiency, see pp. 382, 383
η_R	Fin efficiency, see p. 251

NOTE

In my choice of symbols I have endeavoured, as far as possible, to comply with today's norms, and therefore, contrary to the first edition, I have carried out some alterations; in particular I have designated time with t instead of τ. Nevertheless I have not been completely successful in complying with these standards. As previously I can see an advantage in distinguishing partial quantities including lengths and surfaces by small letters and in using capital letters for total quantities; this is visually more clear than using indices. In designating time it is possible to use t or T as both comply with the norm. On the other hand it did not seem possible to replace f for a partial surface area and F for the total surface area with the letters a and A, the latter now being the sole standard symbol, because a is used to denote the thermal diffusivity. I have therefore retained f and F. In order to further emphasize the difference between the temperatures of the fluids and those of the solid walls or the heat storing elements of a regenerator, I have denoted the former with θ and θ' and the latter with Θ; however Θ is only admissible in the ISO standards but is no longer employed in the DIN-norms (German Industry Norms). Although I have in no way wished to undervalue the importance and achievement of standard notations, in my endeavours to comply with these I have gained the impression that occasionally these can be too restrictive and that more alternative letters should be allowed.

In particular I regret that I have been unable to comply with the system used in the VDI-Wärmeatlas (VDI-Heat Atlas) in my choice of indices and dashes which had to be applied to the letter θ for the temperatures. There, indices differentiate between the fluids but dashes (superfix) serve to characterize the entry and exit positions into and out of the heat exchanger. However in this book, particularly in the treatment of regenerators, numerous positions within a heat exchanger have to be considered and these positions can only be differentiated satisfactorily by means of indices. This explains the chosen method of designation which nevertheless can be used to advantage for example, in practice, when a gas enters, or is removed from, an average position in a heat exchanger.

I have conformed with the present day standard of using K or °C to denote unit temperature. Although I recognize the difficulty of another general adjustment of standards I should like to make the following observation. I feel it is most unsatisfactory to always use two different letters for one and the same unit temperature. Since the centigrade scale was first introduced 1°C has represented unit temperature and one had to decide as with other units, to free it from a pre-determined, fixed zero point. This was, in part, already the case when °C was also used for temperature differences. The possibility exists of course of denoting unit temperature which is independent of a zero point with C and denoting unit temperature which is linked to the customary zero point with °C.

H. Hausen

PART
ONE

INTRODUCTION

1-1 PURPOSE AND TECHNICAL IMPORTANCE OF HEAT EXCHANGERS

Heat exchangers are employed to transfer heat or "cold" from one fluid to another which has a lower or higher initial temperature. For the most part, heat is exchanged between liquids or gases. In principle heat can be transferred between more than two mediums in a heat exchanger.[1]

The numerous applications of heat exchangers in industry are based upon the following consideration. Combustion processes and many other chemical reactions take place at temperatures which are much higher than ambient. The gaseous or liquid products frequently will leave a process at a relatively high temperature and therefore will still contain large quantities of heat which should not go unused in an economical operation. Conversely it is often required to preheat a material before it is processed. The best way to utilize the heat contained in the exit materials is to transfer it to the entry materials using a heat exchanger. In chemical reactions, for example in the combustion in industrial furnaces, such a preheating, usually of the combustion air, will be often indispensable since the heat of reaction alone will not be sufficient to maintain the necessary high temperatures. Such recovery of the heat contained in the exit materials from a process always increases the economy of operation. One of the oldest and best known operations of this type is the preheating of the air and the fuel gases in the regenerator of a Siemens–Martin furnace; heat is retrieved from the waste combustion gases.

The use of heat exchangers is, however, in no way restricted to the cases

---

[1] For more detailed considerations of the term "heat exchanger" see Sec. 1-4.

described. In Cowper stoves for blast furnaces, for example, it is not hot exhaust gases but the heat from blast furnace gas, burnt specifically for this purpose, which serves to heat the air for the furnace.

Of particular importance is the exchange of heat between gases at various pressures in low temperature technology, for example in the condensation and separation of air. Low temperatures of around $-200°C$ and sometimes lower can neither be reached nor maintained with known methods for reducing temperature if the "cold" contained in the return gas is not transferred to the fresh, compressed inlet air in a heat exchanger.

1-2 CLASSIFICATION AND OPERATION OF HEAT EXCHANGERS

The fluids in heat exchangers do not, in general, come into direct contact with each other; if they did they would mix and the pressure gradient which generally exists between the fluids could not be maintained. Cases in which, for example, two immiscible liquids, or a liquid and a gas, or even a gas and a moving solid material exchange heat by direct contact, are relatively infrequent, except when mass transfer takes place as in rectification or evaporative cooling. Disregarding these cases then, heat transfer from one material to another requires there to be a heat conducting partition or a heat storing mass. Partitions have the dual task of guiding the fluids into *spatially divided channels* and at the same time transferring the heat over the shortest possible path. Fluids pass *simultaneously* and *continuously* through heat exchangers which have partitions. In the terminology of the iron and steel industry heat exchangers of this type are called "*recuperators*". The term recuperator also includes, however, the special case already mentioned, in which heat is exchanged between two immiscible fluids or even between a fluid and pieces of solid material. In this case the surface of the fluid or the surface of the solid material assumes the role of a partition.

Heat exchangers which contain a heat storing mass are called "*regenerators*". The heat storing mass, which is usually formed of checkerwork or is porous, has numerous, more or less continuous ducts passing through it and the walls of these ducts offer a large heat transfer surface to the fluids passing through them. Regenerators are *reversed* at fixed, usually regular, time intervals. Both fluids thus flow *alternately through the same cross-section* of each regenerator. The passages of the fluids, between which heat is exchanged, are *separated in time* but not spatially. The heat-storing mass removes heat or "cold" from the fluid passing through it, and after the reversal, releases it to the other fluid. At least two regenerators are needed for uninterrupted operation so that continuously and simultaneously heat can be removed from one fluid in one regenerator and can be released to the second fluid in the other regenerator. Regenerators with a rotating heat storing mass are an example of this; see Fig. 11-3.

Both recuperators and regenerators can be operated in *parallel flow*, *counterflow* or even *cross flow* mode. In parallel flow both fluids flow in the same direction through the heat exchanger. In recuperators parallel flow only allows the temperatures of the two fluids to approach a common average value. On the other hand with counterflow, in which the fluids flow in opposite directions through the heat exchanger, at least one of the two fluids can reach, theoretically, the entry temperature of the other. Consequently the heat transfer performance of counterflow is superior to parallel flow. In regenerators also, it represents the most favourable mode of operation and is used almost exclusively.

Finally, *cross flow* consists of the two fluids flowing at right angles, or approximately at right angles, to one another. Until now it has only been possible to use cross flow operation in recuperators although basically there is no reason why it could not be employed in regenerators. Given large enough heating surface areas and adequate heat transfer coefficients pure cross flow is superior in operation to parallel flow but none the less is inferior to counterflow. However cross flow has the particular advantage that a higher rate of heat transfer occurs when the gas passes over the outside surface of the tubes in a direction perpendicular to, instead of parallel to, the axis of such tubes. On this basis cross flow is often combined favourably with counterflow, for example in so called "mixed switching" or in the cross-counterflow modes of operation in low temperature technology (see Sec. 8-2).

It is not unusual for the heat transfer between two fluids to be accompanied by changes in state in one or both fluids, for example, liquids evaporating or gases and steam condensing. Conventional evaporators and condensers can thus be regarded as heat exchangers, as can those heat exchangers which operate as an evaporator on one side and as a condenser on the opposite side of the partition wall through which heat is transferred. However as evaporators and condensers represent an area of specialisation frequently considered elsewhere, problems in this area will be mentioned only briefly here. Further, it must be noted that numerous individual components, in liquid or solid form, can become separated from the fluid and are deposited on the walls of the heat exchanger; this usually occurs in low temperature technology and indeed is usually an unwelcome side effect. The reverse case can be of practical importance, for example in evaporative cooling where the evaporation of a liquid to which heat is transferred in the exchanger, takes place; however this will not be considered further in this book.

As has been pointed out solid substances can also take the role of a fluid in a heat exchanger. An example is provided by the ancient process of preheating pottery before firing; this is achieved by passing the pots through an oven heated by the waste gases from the kiln which flow over the pots moving in the opposite direction. Historically processes of this type represent the first uses of counterflow. Another example is the pusher furnace which is frequently used in the iron and steel industry; here steel slabs are slowly advanced through a hot gas. A recent suggestion is to transfer the heat contained in one gas to a moving granular solid material or "fluidised bed" which in turn releases the heat to another gas in a second heat exchanger; see for example [G2, N205]. The solid particles can be guided to flow either against, or across, an existing current of gas. This usually applies to recuperators, in which one of the fluids is stationary.

1-3 BASIS OF HEAT EXCHANGER CALCULATIONS AND STRUCTURE

Research into the principles of heat transfer has provided a secure basis for heat exchanger calculations. Fundamental to these are, above all, the works of Nusselt [N4 to 11], who has applied the similarity principle to heat transfer, enabling numerous test results relating to heat transfer to be presented in a clear, practical and simple form. Other experimental results show the need to take into account the influence of thermal radiation, the effect of which can be very considerable at high temperatures.

The principles of heat transfer and the drop in pressure combine quite clearly and form the basis for experimental calculations for heat exchangers. These principles will be described in Part Two of this book. These can be dealt with relatively briefly as more precise details can be obtained from other publications [1 to 19, esp. 1, 2 and 3].

Once the heat transfer coefficient is determined, a separate calculation is still necessary in order to obtain the thermal efficiency of a given heat exchanger or to estimate the necessary dimensions for given operating conditions and to determine the optimal design. The necessary calculation procedures can be derived by purely theoretical methods. The tracing of the temperature distribution in heat exchangers forms the basis for this. In recuperators this temperature profile usually only depends on the longitudinal coordinates. On the other hand, in regenerators chronological temperature changes also play an important part. From this emerges the idea that regenerators are essentially much more complicated than recuperators. In spite of this, however, in almost all cases, simple calculation procedures are indicated.

In the Preface it was mentioned that the main task of this book is to bring together the theories which secure the foundations of the methods of calculation and which are very numerous, reflecting the many forms of heat exchanger. These will appear in Parts Three and Four.

1-4 NOMENCLATURE

Objections can be raised to the term "heat exchanger", because heat is only transferred from one material to another in one direction, and no second quantity of heat exists, moving in the opposite direction, with which this heat can be said to be exchanged. However it is quite possible to conceive from physical, and particularly kinetic considerations, of a partial exchange of energy between one heat flux and another, moving in opposite directions, as witnessed by heat transfer by radiation.

But other reasons can be suggested for using the term "exchange". One perhaps arises from the idea of an exchange of heat with "cold", but this is nevertheless physically incorrect. Another reason might be that two mediums of the same heat capacity exchange their temperatures, indeed, in ideal circumstances, each medium takes as its exit temperature, the inlet temperature of the other medium. In

Germany the most appropriate expression "Wärmeübertrager" has been proposed, which translated into English would be "heat transferer" or "heat transmitter". However this expression has not been adopted despite various recommendations that it should be used. Moreover the terms "heat exchange" and "heat exchanger" are used internationally, so that whatever misgivings there might be, these terms cannot really be avoided.

In Germany regenerators have been occasionally called "Wärmespeicher" or "Kältespeicher" which mean "heat storage unit" or "cold storage unit". However these terms can be misleading. They suggest the primary function of the regenerator is to accumulate waste or surplus heat with a view to its recovery at a later time when there is an energy shortage. While this might be the aim in certain applications, the term "regenerative heat exchanger" is probably more appropriate in most other circumstances since it rightly suggests that the storage of heat is not an end in itself, but is a means to an end.

1-5 BIBLIOGRAPHY

Foreword

Bibliographies are to be found at the end of each large section.

A list of books and papers on the general area of heat transfer follows in Sec. 1-6. The more specialized lists further on in the book are tabulated in the alphabetical order of the surnames of the authors together with a reference number. The bibliographies on the various subjects begin with the following reference numbers:

1: Heat transfer in tubes and ducts (p. 42)
41: Heat transfer in cross flow, in finned tubes and in packed beds (p. 69)
61: Heat transfer in condensation and evaporation (p. 85)
101: Radiative heat transfer (p. 111)
151: Pressure drop (p. 129)
201: Recuperators (p. 279)
301: Regenerators (p. 481)

1-6 LITERATURE IN THE GENERAL AREA OF HEAT TRANSFER

Books

1. VDI-Wärmeatlas. Düsseldorf: VDI-Verlag 1953; Supplements 1956–1963; 2nd ed. 1974.
2. Gröber, H.; Erk, S.: Die Grundgesetze der Wärmeübertragung. 1st ed. by Gröber, 1921; 3rd ed. revised by U. Grigull. Berlin, Göttingen, Heidelberg: Springer 1955; Neudruck 1961.
3. Eckert, E.: Einführung in den Wärme- und Stoffaustausch. 1st ed. 1949; 3rd ed. Berlin, Heidelberg, New York: Springer 1966.
4. Eckert, E. R. G.; Drake, R. M.: Heat and Mass Transfer. New York, Toronto, London 1959.
5. Schack, A.: Der industrielle Wärmeübergang. 7th ed. Düsseldorf: Verlag Stahleisen 1969.
6. Jakob, M.: Heat Transfer. New York: John Wiley, vol. I 1949, vol. II 1957.

7. McAdams, W. H.: Heat Transfer. 6th ed. London: Chapman & Hall, vol. I 1958; vol. II 1959 (first ed. 1933 and 1942).
8. Fishenden, M.; Saunders, O.: An Introduction to Heat Transfer. Oxford: Clarendon Press 1950.
9. Giedt, W. H.: Heat Transfer. Toronto, New York, London: Nordstrand Comp. 1958.
10. Bird, R. B.; Steward, W. E.; Leighfoot, E. N.: Transport Phenomena. New York: John Wiley 1960.
11. Petuchov, B. S.: Experimentelle Untersuchung der Wärmeübertragung. Berlin: Verlag Technik 1958.
12. Chapman, A. J.: Heat Transfer, 2nd ed. New York: The Macmillan Comp. 1967.
13. Luikov, A. V.; Mikhailov, Ju. A.: Theory of Energy and Mass Transfer. Revised English Edition. Oxford, London, Edinburgh, New York, Paris, Frankfurt: Pergamon Press 1965.
14. Ibele, W.: Modern Developments in Heat Transfer (14 Papers by various Authors). New York and London: Academic Press 1963.
15. Haase, R.: Thermodynamik der irreversiblen Prozesse. Darmstadt: D. Steinkopf 1963.
16. Kays, W. M.: Convective Heat and Mass Transfer. New York: McGraw-Hill 1966.
17. Eckert, E. R. G.; Irvine, T. (Editors): Progress in Heat and Mass Transfer. Oxford, London, Edinburgh, New York, Paris, Frankfurt: Pergamon 1971.
18. Rohsenow, W. M.; Hartnett, J. P.: Handbook of Heat Transfer. New York: McGraw-Hill 1973.
19. Dwyer, O. E. (Editor): Progress in Heat and Mass Transfer. Oxford: Pergamon 1973.

Presentation of the theory of heat transfer in extracts from books and collected works

20. Landolt-Börnstein: Zahlenwerte und Funktionen aus Physik, Chemie, Astronomie, Geophysik und Technik, 6th ed. Berlin, Göttingen, Heidelberg; in particular vol. II, 5a, 1969 (Viscosity) and vol. IV, 4: Wärmetechnik, Part a, 1967: Thermodynamic entropy (including density); Part b, 1972 Transport phenomena (Thermal conductivity) and Heat transfer.
21. Schmidt, E.: Einführung in die technische Thermodynamik. 9th ed. Berlin, Göttingen, Heidelberg: Springer 1962, pp. 347–408.
22. Nesselmann, K.: Angewandte Thermodynamik. Berlin, Göttingen, Heidelberg: Springer 1950, pp. 254–301.
23. Hofmann, E.: Wärme- und Stoffübertragung. Handbuch der Kältetechnik edited by R. Plank, vol. III. Berlin, Göttingen, Heidelberg: Springer 1959, pp. 187–463.
24. Grassmann, P.: Physikalische Grundlagen der Chemie-Ingenieur-Technik. Aarau und Frankfurt: Verlag Sauerländer 1961, Chapter 9, pp. 593–698: Impuls-, Wärme- und Stoffaustausch.
25. Hütte, Des Ingenieurs Taschenbuch, Bd. I, 28th ed. Berlin: Ernst and Sohn 1955, pp. 491-506.
26. Haselden, G. G.: Cryogenic Fundamentals. London and New York 1971, pp. 17–197; see, in particular, pp. 92–197.

PERIODICALS

Special periodicals

Heat and Mass Transfer; Journal of Heat Transfer; Int. Journal of Heat and Mass Transfer; Révue Génerale de Thermique.

Numerous other periodicals also publish essays on heat transfer. The following are only mentioned by way of example:
Brennstoff–Wärme–Kraft; Stahl und Eisen; Chemie-Ingenieur-Technik; Verfahrenstechnik, Mainz; Kältetechnik–Klimatisierung.

In addition many magazines publish regular reviews of new research papers, for example:
Eckert, E., et al.: first reported in Mechan. Engng., then in 83 (1961) 7, 34–42 and 8, 56, 57; since 1964 in Int. J. Heat Mass Transfer, e.g. 17 (1974) 615–624. Fortschritte der Verfahrenstechnik. Weinheim: Verlag Chemie, e.g. 7 (1967) 347–394, Review of the years 1964 and 1965; 8 (1969) 328–389, Review of the years 1966 and 1967.
The report of the 5th International Heat Transfer Conference in Tokyo 1974 contains comprehensive material.

HEAT TRANSFER AND PRESSURE DROP PARTICULARLY IN TUBES AND DUCTS

HEAT TRANSFER BY HEAT CONDUCTION AND CONVECTION TOGETHER WITH CONDENSATION AND EVAPORATION

2-1 THE CONCEPT AND IMPORTANCE OF HEAT TRANSFER COEFFICIENTS

The principles of heat transfer and pressure drop in tubes and ducts will be briefly discussed in this section.[1] An introductory consideration will serve to refresh the reader's memory of the concepts of both heat transfer coefficients and heat transmission coefficients.

In recuperators, which are the most common type of heat exchanger construction and through which gases or liquids flow continuously and steadily, the heat is transferred through plane or curved walls. Primarily it is necessary to consider the exchange of heat between two gases (or liquids) through a plane wall and in so doing simply to assume that the gas temperatures are chronologically constant. Until it is close to the wall gas I (Fig. 2-1 left) has an overall constant temperature θ and gas II (right), an overall constant temperature θ'. The temperature of the wall is indicated by Θ, the temperatures at the two wall surfaces by Θ_0 and Θ'_0.[2]

---

[1] Further details can be obtained from the listed books on heat transfer.

[2] Here, and in what follows, temperatures will be indicated by θ and Θ and time by t. The upper case letter Θ is used to represent the temperature of the solid wall or the heat storing mass whereas, to provide greater clarification, the fluid temperatures are represented using the lower case θ, and are namely θ and θ'. This distinction, using upper and lower case letters, between solid and gas temperatures, obviates the need for indices, and in particular avoids a possible multiple indexing of symbols in later discussions. This arrangement is abandoned only in Secs. 3-1, 3-2 and 3-3 where T (but no other letter) is introduced to represent absolute temperature: for the gas the absolute temperature is T_g, for the wall T_w.

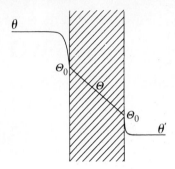

Figure 2-1 Heat transfer through a plane wall.

If $\theta > \theta'$, heat flows perpendicularly through the wall from gas I to gas II. At equilibrium a linear temperature drop occurs in the wall if the thermal conductivity of the material of which the wall is constructed is constant. However a steep temperature drop occurs in each of the gases within the thin boundary layer lying directly adjacent to each surface. As a consequence the surface temperature Θ_0 and Θ'_0 of the wall differ from the temperatures θ and θ' of the main bodies of the gases, see Fig. 2-1. The quantity of heat $\dot{Q}$, passing from gas I through the left hand surface area F of the wall in unit time, is proportional to the temperature difference $\theta - \Theta_0$ and the magnitude F of the surface area.[3] This is calculated using Eq. (2-1),

$$\dot{Q} = \alpha F(\theta - \Theta_0), \tag{2-1}$$

in which α denotes the *heat transfer coefficient*. The heat transfer coefficient can thus be regarded as the quantity of heat passing through a unit surface area in unit time for a unit temperature difference, 1 K, between gas and wall surface. In S.I. units α is measured in $J/(m^2 \, s \, K) = W/(m^2 K)$.

The rate of heat transfer, $\dot{Q}'$, through the opposite surface of the wall to gas II (Fig. 2-1) with heat transfer coefficient α' is obtained from Eq. (2-2) which takes the same form as Eq. (2-1),

$$\dot{Q}' = \alpha' F(\Theta'_0 - \theta') \tag{2-2}$$

The quantity of heat flowing through the wall itself in unit time is provided by Eq. (2-3), where λ_s is the thermal conductivity of the wall material, δ is the thickness of the wall and F, as previously specified, is the area of the surface on one side of the wall

$$\dot{Q}_w = \frac{\lambda_s}{\delta} F(\Theta_0 - \Theta'_0) \tag{2-3}$$

The units for thermal conductivity λ_s are $J/(m \, s \, K) = W/(m \, K)$. The value of λ_s

---

[3] All values relating to unit time should be designated by a dot above the relevant letter. They can then be understood to be differential coefficients in time, the values of which rarely alter chronologically. Thus for example $\dot{Q} = dQ/dt$. This particular notation will only be waived in the case of the heat capacities C and C' of quantities of gas flowing through a heat exchanger in unit time.

varies from one material to another and is dependent, in general, upon the local temperature of the material of which the wall is made.

At equilibrium, the rate $\dot{Q}_w$ is the same as the rates of heat $\dot{Q}$ or $\dot{Q}'$ passing through the surfaces of the wall. Equations (2-1), (2-2) and (2-3) can thus be solved simultaneously for the total temperature difference between the two gases, thereby eliminating Θ_0 and Θ'_0 and yielding Eq. (2-4),

$$\theta - \theta' = \frac{\dot{Q}}{F}\left(\frac{1}{\alpha} + \frac{\delta}{\lambda_s} + \frac{1}{\alpha'}\right) \tag{2-4}$$

Heat transfer through a wall can thus be expressed by Eq. (2-5) where the coefficient k is called the *heat transmission coefficient.**

$$\dot{Q} = kF(\theta - \theta') \tag{2-5}$$

The heat transmission coefficient can hereafter be defined as the quantity of heat which is transferred through a unit area of a wall in unit time for a unit temperature difference, 1 K, between the gases or liquids involved. Like the heat transfer coefficient, the units for the heat transmission coefficient are $J/(m^2\,s\,K) = W/(m^2 K)$.

By comparing Eq. (2-4) with Eq. (2-5) and eliminating $\theta - \theta'$, Eq. (2-6) is obtained.

$$\frac{1}{k} = \frac{1}{\alpha} + \frac{\delta}{\lambda_s} + \frac{1}{\alpha'} \tag{2-6}$$

In this way the heat transmission coefficient can easily be calculated from the heat transfer coefficients α and α', the thickness δ, and the thermal conductivity λ_s of the wall. The lumped heat transfer coefficient for curved walls, in particular the walls of tubes, can be established by similar considerations; how this can be done is shown in Sec. 5-3 in Part Three.

From these considerations, it must be noted however that the temperatures of the fluids flowing along the wall of a heat exchanger will vary as a result of the transfer of heat taking place. Thus in heat exchangers Eqs. (2-1), (2-2), (2-3), and (2-5) refer only to an infinitely small surface area. The heat transfer rate through a larger area can be obtained, fundamentally, by integration over this area (see Parts Three and Four). It therefore turns out to be of some relief that the heat transfer coefficients, and with them the heat transmission coefficient, in general vary only slightly along the length of the heat exchanger, at least as long as heat radiation plays no great part. Frequently the values of α, α' and k can more or less thus be considered as constant.

The thermal conductivity λ_s, which is required to determine the heat transmission coefficient, for example from Eq. (2-6) can almost always be obtained from

---

* Editor's note: the quantity k is commonly referred to in English as "the lumped heat transfer coefficient" or the "overall heat transfer coefficient". Hausen used the word "*Wärmedurchgangskoeffizient*", a possible translation of which is "heat transmission coefficient". All three expressions are acceptable; all that is required is that k should be carefully distinguished from the surface heat transfer coefficients α and α'.

tables of values of the measured thermal conductivities of numerous materials. On the other hand an exact knowledge of the complex principles of heat transfer is required to determine the heat transfer coefficients α and α'; above all these depend upon the velocities of flow and the physical properties of the liquids or gases, and, usually, only in the case where small dimensions are involved, upon the condition of the wall surface. These principles will be discussed below.

2-2 THE BASIC PROCESSES OF HEAT TRANSFER IN TUBES AND DUCTS

A gas or a liquid flows through a tube. While doing so the temperature of the fluid differs from that of the tube wall. The heat transfer which then takes place between the gas or liquid on the one hand and the tube wall on the other depends in general on the following three processes:

(1) Heat conduction in the fluid in the direction of the temperature drop.
(2) Convection, that is heat transported by fluid flow, indeed usually by means of a disorderly mixing motion.
(3) Thermal radiation.

The convection component in heat transfer depends to a great extent on the state of flow which can be laminar or turbulent.

In the pure *laminar region of flow* which encompasses small values of average fluid velocity, the flow can be imagined to be constructed of very thin concentric layers which slide past each other, exactly or approximately in an axial direction. With this type of flow, heat will only be transferred, apart from radiation, by *thermal conduction* in the fluid, specifically at right angles to the direction of flow. One can only really talk of convection in this case when the fluid carries away the heat extracted by conduction and thereby contributes to the maintenance of the temperature drop necessary for continued heat transfer.

In *turbulent flow* there appears the mixing motion, previously mentioned; this considerably enhances heat transfer in the direction perpendicular to the main fluid flow in a manner yet to be described. There is a certain velocity, the so called critical velocity, at which this mixing motion first manifests itself. The mixing motion influences heat transfer in two ways which can be explained clearly in the following way. Prandtl suggested that there remains, directly at the wall surface, a very thin laminar boundary layer of which the particles nearest to the wall stick to its surface. Through this boundary layer heat is transferred solely by conduction, that is if radiation is once again ignored. Beyond this boundary layer, the heat transfer is facilitated by the turbulent mixing motion which conveys the heat very quickly, further into the fluid. The agitated fluid particles, called "turbulence eddies" receive heat from the boundary layer or from other particles which themselves have obtained heat already from the boundary layer and give it up at positions remote from this layer. This latter process usually represents a major component in the heat transfer effect by convection under turbulent conditions.[4]

The strong influence of the speed of flow on heat transfer with turbulent flow can be understood on the basis of this concept. With an increasing average speed of flow the boundary layer not only becomes thinner but the intensity of the turbulent mixing motion also increases. The thinner the boundary layer, the more rapidly heat is conducted through it [see Eq. (2-3)]. At the same time the transport of heat is augmented by the increased mixing motion. The result is a strong increase in the heat transfer coefficient with increasing speed.

The transition from laminar to turbulent flow takes place as fluid velocity increases at the critical Reynolds number Re_{cr}.[5] Its lowest value $Re_{cr} = 2320$ occurs under agitated, greatly disturbed inlet conditions, while in other cases larger Re_{cr} values occur, frequently of between 3000 and 4000.

Heat transfer with laminar or turbulent flow is basically described by the differential equations for viscous liquid flow and heat conduction in fluids. A rigorous solution of these equations has, however, only been found, to date, for special cases, in particular for the so-called "fully developed laminar" flow such as appears in a tube at subcritical velocities at a sufficient distance from the inlet; this will be discussed further in Sec. 2-7. Heat transfer with turbulent flow is, on the other hand, so complicated that at first a rigorous solution might appear impossible. Our knowledge of heat transfer and similarly of the drop in pressure in turbulent flow is in essence due to the very large number of experimental investigations the results of which can be reproduced by means of relatively simple equations, thanks to the similarity principles; this has already been mentioned in Sec. 1-3. It is thus possible to calculate fairly accurately the heat transfer in turbulent flow for measured flow conditions (see Sec. 2-6).

2-3 MEASUREMENT OF HEAT TRANSFER COEFFICIENT

The apparatus for measuring the heat transfer coefficient is always a small heat exchanger which is as simple as possible and so arranged that the physical process to be observed can take place undisturbed and be examined as precisely as possible. Figure 2-2 shows the basic idea of an apparatus for measuring the heat transfer coefficient.

The fluid whose heat transfer coefficient is to be determined, e.g., a previously heated gas, generally flows in the inner tube from left to right. Heat will be extracted from the gas by means of a cooling fluid flowing through the outer duct. The entry temperature θ_1, the exit temperature θ_2, and the gas flowrate will be observed as

---

[4] On closer inspection the boundary layer is not sharply divided from the turbulent zone. There exists rather a gradual transition whereby the strength of the turbulence increases steadily with its distance from the tube wall.

[5] See Sec. 2-4 for the significance of the Reynolds number and Sec. 4-1 for further details of the transfer from laminar to turbulent flow.

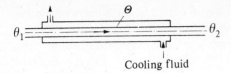

Figure 2-2 Basic idea of the equipment for measuring the heat transfer coefficient.

will the tube wall temperature Θ in almost all exact measurements of α. The temperature Θ_0 of the surface can be obtained by minimal extrapolation from Θ which can, for example, be measured by means of a thermocouple inserted in the tube wall. The rate of heat transfer $\dot{Q}$ from the gas to the tube wall is determined from the drop in temperature together with the flowrate and specific heat of the gas. This rate of heat transfer can, however, also be determined from corresponding measurements of the cooling fluid. Finally if the heating surface area F is known and the average value of the temperature difference $\theta - \Theta_0$ between the gas and the tube surface is determined, by a method which will be described later (see Part Three, Sec. 5-6) the heat transfer coefficients, which are then the only unknowns, can be obtained from Eq. (2-1).

The gas or liquid whose heat transfer coefficients are to be determined can flow in the annulus between the inner and outer tube instead of through the inside tube. Furthermore, the heating and cooling fluids can be interchanged; either or both fluids can be replaced by an evaporating liquid or a condensing gas. Heat transfer in cross flow can also be measured in smooth or ribbed tubes using the same principles, employing a suitably modified apparatus. Finally it is possible to measure the heat transfer in the various forms of the packing in regenerators, particularly in scaled down regenerators; see for example H. Glaser [G42].

Observations of average size and large heat exchangers, as are built for practical use, yield results which approach those obtained in similar laboratory experiments. Such observations generally enable the operational behaviour of such heat exchangers to be examined. Such measurements should not be underrated when it comes to checking and confirming the heat transfer coefficients determined by laboratory methods.

It is possible to conclude from the results of measurements made on practically operating heat exchangers that these measurements almost always confirm the heat exchanger theories developed below; further, these results also demonstrate that the average values for *heat transfer coefficients* and *drops in pressure* determined on *laboratory apparatus* can be *transferred* in their entirety to the *largest working plant, thanks to the similarity principle.*

2-4 THE SIMILARITY PRINCIPLE AND ITS APPLICATION TO HEAT TRANSFER

A very complex picture initially emerges from measurements of the heat transfer coefficients due to the numerous factors which influence heat transfer. This situation has been rationalized and clarified by the similarity principle. Furthermore the similarity principle has enabled the form of the empirical equations which reproduce experimental data, to be determined. In addition the *similarity principle* provides the basis for the employment of the heat transfer coefficient *values, often only measured between 0°C and 100°C and at roughly atmospheric pressure,* for very high and very low temperature conditions and at very high pressures. The similarity principle also enables the results of measurements carried out on only a few materials, such as air, water and oils, to be considered as valid for any other material. This extrapolation is based on the assumption that, in addition to temperatures and pressures, there is a sufficiently precise knowledge of the physical properties of the gases or liquids, such as heat conduction, specific heat capacity, viscosity and the like.

Nusselt [N4, 6] first applied the similarity principle to heat transfer following Reynolds [R5] who had already previously recognized its importance in the representation of the laws of pressure drop.[6]

The similarity principle is based on the following fundamental concepts. Geometrically similar figures, e.g., similar triangles, can, as is known, be mapped onto each other so that all straight lines present in such a figure can be lengthened or shortened in the same proportions. A logical extension of this idea comes in physical processes of the same type which are analogous to each other, provided they can be mapped onto one another numerically simply by appropriate alterations to the scale in which the physical dimensions involved are measured.

In general, however, with diverse physical dimensions the scales have to be altered differently. Of course these alterations are not completely independent of each other but are bound by a similarity condition. This condition exists in a simpler form for geometric similarity. Geometrical shapes are of course similar to each other when non-dimensional quantities like the ratio of one side to another or angles have the same values in the cases which are to be compared. Correspondingly the similarity of two physical processes presupposes that certain non-dimensional quantities, called dimensionless groups, are of the same size in both processes. These dimensionless groups can all be formed fundamentally of quantities participating in such a process by means of suitable multiplications and divisions. In spite of the equality of the dimensionless groups the individual physical quantities in both cases can be very different.

Under these conditions, test results gained in an individual case can be transferred by means of a simple calculation to all other processes which are physically similar to the test case.

The most general deduction from the similarity principle is, however, that with its help it is possible to bring the physical relationship existing between the two non-similar cases in a process into a basic and most easily understandable form. It is here that the similarity principle plays its most important part in the evaluation of a series of experiments. While the result of a single measurement only encompasses the cases which are similar to each other, attempts can be made in general, by means of a series of experiments, to establish a relationship between

---

[6] The first very general formulation of the similarity principle originates from Helmholz [H14].

non-similar cases. In so doing, therefore, the problem is one of determining the dependence of a physical quantity, e.g., the heat transfer coefficient, on other quantities which influence it such as speed of flow, heat conduction, tube diameter and the like. The similarity principle in its most general form leads to the important assertion that such a dependence can be represented by a relationship between those dimensionless groups which are, in the similarity sense, governing factors for the process in question. As the number of dimensionless groups is smaller than the original physical quantities, the relationship existing between the dimensionless groups can be more easily determined than the relationship existing between the original quantities. The relationship mentioned however, is not determined solely by the similarity principle; it is usually determined, on the contrary, by experiments. It can also be calculated in cases amenable to mathematical solution.

The dimensionless groups can be derived either from the governing differential equations for the process in question or by the general methods of dimensional analysis.[7]

Application of the Similarity Principle to Heat Transfer

In order to apply the similarity principle to heat transfer, the physical quantities, and hence the dimensionless groups which occur, must be determined. Thus heat radiation will be ignored and heat transfer only by conduction and convection will be considered.

As a consequence of the conduction of heat which takes place, essentially in the boundary layer, the heat transfer coefficient α is dependent on the thermal conductivity λ of the fluid; however the viscosity η is also a determining factor as this influences the thickness of the boundary layer. Convection causes α to be a function of the average speed of flow ω, the density ρ, and the specific heat c_p of the fluid. Furthermore the diameter d of the tubes or the equivalent measurement for the ducts through which the fluid flows has a significant influence on the pattern of fluid movement and therefore on the value of α. On the other hand the length L of the tube or duct is of lesser importance although it should never be disregarded.

The average speed of flow ω is introduced because the true fluid velocity which has its greatest value in the middle of the tube, decreases towards the wall until it is zero at the wall itself. Thus this average value should be taken as a basis of the calculation of the volume of fluid flowing through the tube in unit time, divided by the free flow cross-sectional area of the tube.

The quantities mentioned are grouped together in the first column of Table 2-1. The second column shows their dimensions expressed in terms of the four basic dimensions: length (L), time (Z), mass (M) and temperature (T). In doing this it must be taken into account that heat is given the same dimensions as work, namely ML^2/Z^2.[8] Intentionally only the dimensions and not the units will be established, in

---

[7] The similarity principle and the procedure for determining the dimensionless groups are discussed in almost every book on heat transfer. The very comprehensive literature on similarity includes the following publications: [B9, B17, C4, H14, N6, N7].

[8] Instead of the dimension of the temperature, the dimension of the heat quantity divided by the temperature can be introduced because in α, λ and c_p, these two variables only occur in this combination.

Table 2-1 Physical quantities which determine the heat transfer between a fluid and a tube wall

Quantity	Dimension
Heat transfer coefficient α	$\dfrac{M}{Z^3 T}$
Thermal conductivity λ	$\dfrac{ML}{Z^3 T}$
Speed of flow ω	$\dfrac{L}{Z}$
Density ρ	$\dfrac{M}{L^3}$
Specific heat c_p	$\dfrac{L^2}{Z^2 T}$
Viscosity η	$\dfrac{M}{LZ}$
Tube diameter d	L
Tube length L	L

order to show that these considerations are independent of the system of units chosen.[9]

It can easily be recognized that subsequently four dimensionless expressions can be selected from the physical quantities chosen.[10]

$$\alpha d/\lambda, \; \rho\omega d/\eta, \; c_p\eta/\lambda, \; L/d \tag{2-7}$$

Other dimensionless groups could be formed, e.g., $\rho c_p\omega d/\lambda$. This dimensionless group, however, would signify nothing new since it is not independent of the previous dimensionless groups, i.e., it results through the multiplication of the second and third expressions of (2-7). The first three dimensionless groups of (2-7) are known as

$$
\begin{aligned}
&\text{Nusselt number:} && Nu = \alpha d/\lambda, \\
&\text{Reynolds number:} && Re = \omega d/v, \\
&\text{Prandtl number:} && Pr = v/a.
\end{aligned}
\tag{2-8}
$$

---

[9] Incidentally it might be pointed out that in verifying the correctness of equations, this is often only a matter of checking the dimensions, although in practice it is the agreement of the units on both sides of the equation which has greater meaning.

[10] The number of dimensionless groups which are independent of each other (in fact 4) is, from Table 2-1, equivalent to the number of physical quantities (8) reduced by the number of base dimensions (4). This conforms to a general principle which is derived from dimensional analysis. See P. W. Bridgman and H. Holl [B17].

Here the thermal diffusivity $a = \lambda/(\rho c_p)$ and the kinematic viscosity $v = \eta/\rho$ are introduced as abbreviations.

These dimensionless groups[11] can be clearly interpreted as ratios of the following geometrical or physical quantities. λ/α is equal to the thickness of a laminar boundary layer of thermal conductivity λ, through which as much heat is transferred by conductivity with a linear temperature gradient, as is calculated using instead the heat transfer coefficient α, but with the same temperature difference. It follows that the Nusselt number, Nu, is equal to the ratio of the internal diameter of the tube to the thickness of this boundary layer.

The Reynolds number can be considered as the ratio of the force of inertia to the frictional force. For a small section of fluid of volume V and velocity u, which varies with time and position, the inertia (accelerating force) amounts to $V\rho \partial u/\partial t$ and the frictional force working on a surface area F amounts to $F\eta \partial u/\partial y$, in which y is the distance from the surface. In these expressions V/F has the dimensions of length, $(\partial u/\partial t)/(\partial u/\partial y)$ has the dimensions of velocity. If d is inserted for the length and w for the velocity, the Reynolds number immediately results from division of the inertia by the frictional force.

In the Prandtl number Pr, v is a measurement of the speed of the momentum exchange occurring between the molecules which gives rise to the internal friction: a is a corresponding measurement of the molecular exchange of energy which facilitates the conduction of heat. Pr can thus be taken as the ratio of the velocities to these two exchange processes.

Formerly the dimensionless group $\rho c_p wd/\lambda$, which has already been mentioned, i.e.,

the Peclet number: $Pe = wd/a = Re \cdot Pr$ (2-9)

would have been used together with Re and Pr. Today it still proves itself to be useful in the laminar domain. On the other hand in the turbulent domain it is being used less and less, principally for reasons of uniformity as the Reynolds number Re is, above all, the authoritative dimensionless group for pressure drop (see Sec. 4-2).

According to the similarity principle the laws governing heat transfer must be represented by a relationship between the four dimensionless groups (2-7). Thus, from Eq. (2-8)

$$Nu = f[Re, Pr, L/d]$$ (2-10)

or even, using the Peclet number (2-9)

$$Nu = F[Pe, Pr, L/d]$$ (2-11)

in which f and F are arbitrary functions of the independent variables in brackets. As has been mentioned, in many cases these functions are determined by experiment. Nusselt [N6] has suggested "power functions" for the reproduction of experimental

---

[11] For further dimensionless groups which are not, in general, very important for forced flow in tubes and ducts, see, for example, Gröber, Erk, Grigul: "The principles of heat transfer" [2] 3rd edition p. 419. Reference will only be made to Grashof dimensionless groups which deal with free convection. Its influence on heat transfer in tubes will be dealt with in Sec. 2-8.

results; subsequent measurements have proved that these are also useful over an amazingly wide area. Thus Eq. (2-10) and (2-11) are frequently used in the form:

$$Nu = \text{const. } Re^{m_1} Pr^{m_2} (L/d)^{m_3} \tag{2-12}$$

or

$$Nu = \text{const. } Pe^{n_1} Pr^{n_2} (L/d)^{n_3} \tag{2-13}$$

in which the values of the constants and the fixed exponents m_1, m_2, n_1 etc., are obtained from measurements.

With laminar flow exponents m_1 and n_1 of Eqs. (2-12) and (2-13) lie somewhere between zero and $\frac{2}{3}$, with turbulent flow they lie nearer to unity. It follows from this that in the laminar domain the heat transfer coefficient varies less with velocity than it does in the turbulent domain where, as a consequence of the influences of convection described in Sec. 2-2, α increases almost proportionally with velocity.

In the preceding equations d stands for the internal diameter of the tube when the material under consideration flows through the inside of the tube. When it is flowing in the outside space surrounding an assembly of tubes or through any other arbitrary cross-section, any other specific dimension can take the place of d. In almost all cases it has proved expedient to use, instead of d, the equivalent ("hydraulic") diameter

$$d_{gl} = 4F_q/U \tag{2-14}$$

in which F_q is the surface area and U the total length of the perimeter of the flow cross-section. The basis of Eq. (2-14) is the idea that for the flow conditions and thus the conditions for heat transfer depend essentially on the ratio of the flow cross-section to perimeter length. Equation (2-14) is, moreover, so constructed that d_{gl} changes over to the circle diameter, d for the circular cross-section.

Frequently only parts of the duct walls are heated or cooled. For example in the outer area surrounding an assembly of tubes, as a rule only the tube walls of the assembly, and not the casing, take part in heat transfer. In such cases, as Nusselt [N6, 7] has already shown, part of the test measurement is more readily reproduced if only the heated or cooled part U^* of the cross-section perimeter is considered and the *thermal* diameter formed

$$d_{th} = 4F_q/U^* \tag{2-15}$$

which is then introduced into the dimensionless groups.

A theoretical examination of heat transfer in slits that are heated one side only [H9, 10] indicates that the thermal diameter according to Eq. (2-15) is preferred for small values of Pr up to about 4, while for larger values of Pr the equivalent diameter according to Eq. (2-14) is preferred. From the same publication, a more precise heat transmission coefficient α^* can be obtained for only partially heated or cooled duct walls; whereas α is primarily determined with the help of Eq. (2-14), α^* is calculated from the equation

$$\frac{\alpha^*}{\alpha} = 1 - \frac{0.75}{1+Pr}\left(1 - \frac{U^*}{U}\right) \tag{2-16}$$

For the formulation of the heat transfer equation, Lorenz [L6] has suggested that in place of Nu the expression

$$\frac{Nu}{RePr} = \frac{\alpha}{w\rho c_p} = \frac{\theta_1 - \theta_2}{(\theta - \Theta_0)_M}\frac{d}{4L} \, (= \text{St}) \qquad (2\text{-}17)$$

be set up, in which $\theta_1 - \theta_2$ is the temperature variation of the fluid brought about by the exchange of heat, $(\theta - \Theta_0)_M$ is the average difference between the temperature θ of this fluid and the surface temperature Θ_0 of the tube wall. This expression will be termed the Stanton number, abbreviated as St.* It has the advantage of clarity because it is determined by means of the ratio of the two temperature differences and the ratio of the tube diameter to the tube length.

However there is no advantage in using the dimensionless group St for practical heat exchanger calculations since here it is desirable to become familiar with the heat transfer coefficients which are best determined by means of Nu.

2-5 EQUATIONS FOR HEAT TRANSFER IN TUBES AND DUCTS WITH TURBULENT FLOW

Of the numerous equations which have become available for the heat transfer coefficients in tubes or ducts only a few will be mentioned; these are experimentally well founded and hitherto, have proved to be useful. Additional relationships can be found in the literature, especially in the appropriate text books [2 to 26]. The VDI Heat Atlas [1] contains numerous diagrams from which the heat transfer coefficient can be determined.

Heat Transfer Coefficients in Turbulent Flowing Gas and Steam

Nusselt [N7] has outlined the following equation for heat transfer with turbulent flow of gases in tubes:

$$Nu = 0.0362(d/L)^{0.054}Pe^{0.786} \qquad (2\text{-}18)$$

This equation is valid, as a close approximation, for superheated steam. It has been found to be true up to now, for all practically occurring values of Re, at least for $Re > 10,000$. However, Eq. (2-18) is only valid in the region of $Pr = 1$ when the underlying measurements have been carried out with air and other gases at atmospheric pressure and when Pr was slightly less than 1. With higher gas pressures, however, values of Pr up to 5 can occur.

According to the theories and measurements applicable to liquids, the effect of Pr up to $Pr = 10$ has been successfully reproduced employing the power $Pr^{0.45}$. Furthermore the power $(d/L)^{0.054}$ in Eq. (2-18) proves to be correct in representing the influence of d/L only when average values of L/d are used. Below about $L/d = 8$

---

* Editor's note: Hausen acknowledges at this point that the abbreviation St is taken from the scientific literature in English.

a strong dependence on L/d is to be expected whereas above $L/d = 200$, the expression $(d/L)^{0.054}$ can be approximated by a constant value. These requirements can be satisfied by introducing $0.663\,[1+(d/L)^{2/3}]$ in place of $(d/L)^{0.054}$ in Eq. (2-18) [H7].

Taking these two principles into account the following heat transfer equation for gases and steam is obtained from Eq. (2-18)

$$Nu = 0.024[1 + (d/L)^{2/3}]\,Re^{0.786}Pr^{0.45} \qquad (2\text{-}19)$$

which is valid for Pr values of between 0.7 and 10.

In order to calculate the value of Nu and hence α from such an equation or from one of the other equations of general form (2-10) or (2-11) (these will be discussed in part later), L, d, w and the material values which generally depend on pressure and temperature, are inserted first of all into the dimensionless groups using Eq. (2-8). The heat transfer equation, for example, Eq. (2-19), then gives the value of Nu from which α can finally be obtained employing the first Eq. (2-8) directly. Numerous material values can be found in the VDI Heat Atlas [1]. According to Nusselt these should be based on the average boundary layer temperature which can be taken with a sufficient accuracy to be equivalent to the mean value of the average gas temperature and the wall temperature.[12]

Figure 2-3 shows the variation of the heat transfer coefficient for air at 0°C and 1.0133 bar with velocity of flow for various tube diameters d, with $L = 6\,\text{m}$,

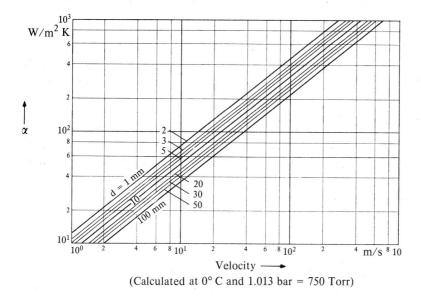

(Calculated at 0° C and 1.013 bar = 750 Torr)

Figure 2-3 Heat transfer coefficient of the air with turbulent flow according to Nusselt's equation.

[12] See W. Nusselt [N7] for the fixing of the reference temperature for the medium values with high temperature differences between gas and wall.

calculated using Eq. (2-19). The reduced velocity of flow $w_0 = \rho/\rho_0 w$ under the standard conditions of 0°C and 760 torr = 1.0133 bar is shown as abscissa. In so doing the density ρ_0 refers to these standard conditions, while ρ represents the real density. The introduction of this standard velocity has the advantage that the diagram is also valid for higher gas pressures of up to about 10 bar.

Figure 2-3 has an accuracy of 3 per cent between $L = 2$ and $L = 20$ m. It can be seen from Fig. 2-3 and also from Eqs. (2-18) and (2-19) that α *does not increase directly in proportion to the speed and decreases slowly with increasing tube diameters.*

As the values of λ, η and c_p in general vary with temperature within the dimensionless groups, according to Eqs. (2-18) and (2-19), the heat transfer coefficient is also dependent on temperature. This temperature dependence is shown in Fig. 2-4, in which the ratio of the heat transfer coefficient α of any gas at an arbitrary temperature to the heat transfer coefficient $\alpha_{0,air}$ of air at 0°C and at the same fluid velocity w_0 is shown as a function of the temperature.[13] Thus for an arbitrary gas, given the values of w_0, d and L, α can be determined by first taking $\alpha_{0,air}$ for example from Fig. 2-3, and then multiplying this value by the ratio $\alpha/\alpha_{0,air}$ taken from Fig. 2-4.

The *effect of pressure on heat transfer* is caused, according to Eq. (2-19), solely by the pressure dependence of λ, η and c_p when calculated with the standard velocity w_0. The magnitude of the pressure dependence of λ, η and c_p however is about the same as that of the influence of deviations from the ideal gas condition so that at low pressure, the effect of pressure plays hardly any part. Therefore Figs. 2-3 and

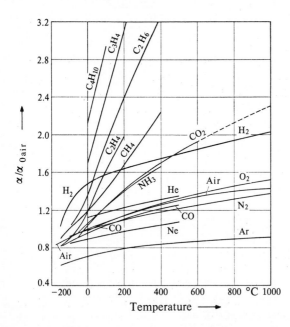

Figure 2-4 Heat transfer coefficient α of gases in relation to the heat transfer coefficient $\alpha_{0,air}$ of air at 0°C.

[13] The values were recently calculated by the author in 1975 employing Eq. (2-19).

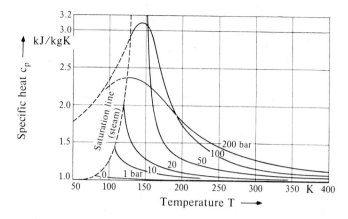

Figure 2-5 Specific heat c_p of air.

2-4 which are primarily designed for atmospheric pressure only are also valid, and provide good approximations for moderately high pressures of up to about 10 bar.

However at considerably higher pressures the influence of pressure upon λ, η and c_p must also be considered, the appropriate values of λ, η and c_p for the pressure in question are inserted into the heat transfer equation. λ, η and c_p all vary in approximately the same way with temperature and pressure; Fig. 2-5 displays in particular the temperature and pressure dependence of specific heat c_p. Figure 2-6

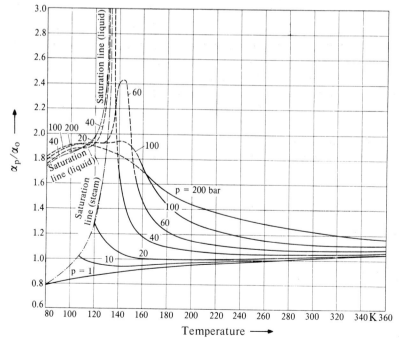

Figure 2-6 Heat transfer coefficient for air at various pressures. α_p is the heat transfer coefficient at pressure p and at arbitrary temperature; α_0 is the heat transfer coefficient at 1 bar and 0°C.

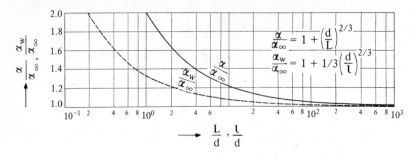

Figure 2-7 Dependence of the heat transfer coefficient of a gas in turbulent flow on the tube length. α is the average heat transfer coefficient in the tube of length L; α_w is the actual heat transfer coefficient at distance l from the entrance; α_∞ is the heat transfer coefficient at a large distance from the entrance.

shows how the heat transfer coefficient α of air flowing through tubes, influenced by these variations, is dependent on pressure and temperature, according to Eq. (2-19); α has been calculated using a previously developed approximation.

Finally, in Eq. (2-19) the factor $1 + (d/L)^{2/3}$ gives expression to the fact that the heat transfer coefficient α, at given fluid velocity w or w_0 and given tube diameter d, decreases with increasing tube length L. The dependence of the heat transfer coefficient on L/d as determined by this factor, is shown in Fig. 2-7 by a solid line. With the very diverse tube lengths L and diameters d which occur in practice, L/d as a rule lies roughly between 50 and 5000. It follows that $0.024\,[1 + (d/L)^{2/3}]$ in Eq. (2-19) lies between 0.026 and 0.024, so that this expression can be replaced quite satisfactorily by the average value 0.025; this approximation has already been used as a basis for Fig. 2-3. The values of the heat transfer coefficient discussed so far are always the average values for a heated or cooled tube of length L. However the actual heat transfer coefficient α_w at a particular position in the tube, which is at distance l from the entrance of the gas into a heated or cooled tube, can be calculated from Eq. (2-19) simply by substituting $1 + \frac{1}{3}(d/L)^{2/3}$ for $1 + (d/L)^{2/3}$. In doing this it is implied that the relationship

$$\alpha = \frac{1}{L_0} \int_0^L \alpha_w dl \qquad (2\text{-}20)$$

exists between α and α_w; this is intelligible but only approximately correct.[14] The lower broken line in Fig. 2-7 shows the actual heat transfer coefficient determined in this way which is dependent on the distance l from the entrance. In general only the average value α of the heat transfer coefficient is required.

According to the findings of G. Grass [G8], α_w can be treated as proportiona to $1 + c_0 d/l$ with reasonable accuracy, where c_0 represents a constant which is to be determined experimentally. If this is modified slightly into the form $1 + c_0 d/(l+1)$,

---

[14] From Sec. 5-13 only the integration of k is exact in that k alone is dependent on the longitudinal coordinates l or f.

integration using Eq. (2-20) leads to numerical values which agree very well with the expression $1 + (d/l)^{2/3}$ if $c_0 = 0.85$ is used.

Equation (2-19) is valid for gases which flow in *smooth tubes or ducts*. That somewhat higher heat transfer coefficients for gases in *rough tubes and ducts* are to be expected, will be discussed below in Sec. 2-9.

Finally it should be pointed out that Eqs. (2-18) and (2-19) as well as the corresponding exponential equations for liquids which are yet to be discussed, give high values for $Re < 10{,}000$ when contrasted with a large number of experimental results. An equation which is better adapted to such results in the area of all Reynolds numbers up to $Re = 2320$ [Eq. 2-27)] will be further discussed below.

Heat Transfer Coefficients for Turbulently Flowing Liquids

On average the heat transfer coefficients for liquids are considerably higher than those for gases. Apart from this, liquids differ from gases primarily because the Prandtl number has a considerably higher value [see Eq. (2-8)]. While Pr has an approximate value of 1, for gases at atmospheric pressure, for water Pr lies between 2 and 10 and for viscous oil the values of Pr can reach 300 and more, sometimes even as high as 5000. With liquids it is thus particularly important that the heat transfer equation gives expression to the dependence upon the Prandtl number.

A further important difference lies in the fact that viscosity η varies much more strongly with liquids than with gases. While the viscosity η for gases increases relatively slowly with respect to temperature, with liquids it decreases very rapidly with increasing temperature (see Table 2-2). With turbulently flowing liquids this has the effect that the thickness of the boundary layer, through which heat is transferred principally by conduction, as explained in Sec. 2-2 is also dependent on the direction of the flow of heat. For the same average liquid temperature, the temperature distribution in the boundary layer and hence the prevailing viscosity is determined by whether the liquid is being heated or cooled by the wall. In the first case the boundary layer is thinner, because of the lower viscosity, than it is in the second. This therefore results in remarkably differing heat transfer coefficients and indeed for the same average liquid temperature the heat transfer coefficients for heating are higher than those for cooling.

Nusselt has already demonstrated [N7, 8] that exponential formulae in the form of Eq. (2-12) are also applicable to a wide range of liquids. Later Kraussold [K17] established the following equations for heat transfer by liquids in tubes by

Table 2-2 Viscosity of water and air in kg/(m s) ($= 10$ poise)

Temperature		0°C	20°C	60°C	100°C	200°C
Water	$10^6 \eta =$	1798	1003	462	278	133
Air	$10^6 \eta =$	17.1	18.2	20.0	21.8	25.1

bringing together all previously known experimental results [e.g. E1, S15, 19, 24, 26] as well as by utilizing measurements specific to viscous oils [K15]:

$$Nu = 0.032Re^{0.8}Pr^{0.37}(d/L)^{0.054} \qquad \text{for heating a fluid} \qquad (2\text{-}21)$$

$$Nu = 0.032Re^{0.8}Pr^{0.3}(d/L)^{0.054} \qquad \text{for cooling a fluid} \qquad (2\text{-}22)$$

The physical properties occurring in the dimensionless groups of these equations should always be related to the average liquid temperature. In the most practically important area, in which L/d usually lies somewhere between 100 and 400, $0.032 \cdot (d/L)^{0.054}$ can be replaced reasonably accurately by 0.024 in Eqs. (2-21) and (2-22). The accuracy of these equations is greatest when the difference between the tube wall temperature and the average liquid temperature is about 30°.

With $Pr = 100$ the results of the Kraussold equations closely agree with the practical knowledge that a heat transmission coefficient for heating is about 38 per cent higher than that for cooling. However the fact that the two equations are not interchangeable when there are very small temperature differences is contrary to what might be expected from a theoretical point of view. Two significantly different methods have been suggested in order to obtain a continuous transition between the heat transfer coefficients for heating and cooling.

On the one hand, Kaye and Furnas [K4], proposed Eq. (2-23) which represents a small variation of an equation originating from Eckert [E2]:

$$\frac{\alpha}{\alpha_{\text{isotherm}}} = 4\sqrt{\frac{v_w}{v_{fl}}} \qquad (2\text{-}23)$$

where α is the actual heat transfer coefficient for heating or cooling, α_{isotherm} is the limiting value of α for isothermic flow and v_w and v_{fl} are the kinematic viscosities at the wall temperature and the average liquid temperature respectively.

Here it is assumed all the remaining physical properties in the heat transfer equation relate to the average temperature of the boundary layer, and thus to a certain extent, to the arithmetic mean of the wall temperature and the average liquid temperature.

Sieder and Tate [S17] on the other hand suggested Eq. (2-24) and introduced the values η_{fl} and η_w of the dynamic viscosity.

$$\frac{\alpha}{\alpha_{\text{isotherm}}} = \left(\frac{\eta_{fl}}{\eta_w}\right)^{0.14} \qquad (2\text{-}24)$$

where the collected physical properties in the heat transfer equation, with the exception of η_w, are related to the average liquid temperature. This is more computationally convenient. In the light of this suggestion the factors $Pr^{0.3}$ and $Pr^{0.37}$ can be replaced in the Kraussold equations by $0.68 \cdot Pr^{0.42} \cdot (\eta_{fl}/\eta_w)^{0.14}$.

Further possibilities consist not only in selecting other exponents in Eqs. (2-23) and (2-24) but also in introducing the ratio of the corresponding values of Pr in place of the ratio of physical properties.

Various test results, such as those by Malina and Sparrow [M1] suggest that in the case of $\eta_{fl} > \eta_w$, i.e., in the heating of the liquid, the exponent chosen for

Eq. (2-24) will have to be smaller than 0.14; Gnielinski [G3] has recently suggested the value of 0.11 for this. However the influence of the viscosity ratio is stronger in the case of liquids which are being cooled. Experimental values by Hackl and Gröll [H1] with $\eta_w > \eta_{fl}$ were reproduced by the author [H12] using the following equation, which is valid for η_w/η_{fl} up to 1000.

$$\frac{Nu}{Nu_{\text{isotherm}}} = 0.645 \cdot \left(\frac{\eta_w}{\eta_{fl}}\right)^{-0.3} + 0.355 \qquad (2\text{-}25)$$

Meanwhile, since these test results are not all consistent, the factor $(\eta_w/\eta_{fl})^{0.14}$ should continue to be used in the knowledge that later it will most probably have to be replaced by a more precise expression.

In the third edition of his book Eckert [3] has emphasized that, with gases at least, one can dispense with the factor $(\eta_{fl}/\eta_w)^{0.14}$, or another function substituted for it, if all the physical properties relate to the average boundary layer temperature instead of to the average fluid temperature; as a consequence only the speed of flow will relate to the average fluid temperature.

Influence of the Prandtl Number on the Heat Transfer Coefficient

The dependence on the Prandtl number, which is very important for liquids, is expressed in Eqs. (2-21) and (2-22) by the power $Pr^{0.37}$ or $Pr^{0.3}$, which can be taken on average to be $P^{1/3}$. Such a dependence is reproduced by a straight line in the log–log graph in Fig. 2-8. On the other hand slightly non-linear curves arise for this dependence on Pr from the theories of Prandtl [P9], Hofmann [H21], Kármán [K1], Reichardt [R2, 3] and others which are based on the analogy between heat transfer and the transfer of momentum in fluids; these are discussed in Sec. 2-6. The same follows from a newer equation by Petukhov [P5] which is based upon the same considerations. The dependence on Pr arising from Petukhov's equation is shown in Fig. 2-8. Up to now, it has not been possible to confirm experimentally the

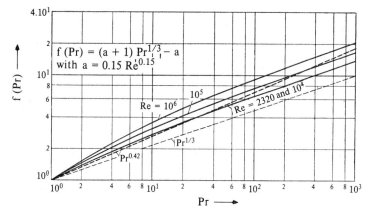

Figure 2-8 Heat transfer in tubes. Dependence of the Nu number on Pr according to Petukhov's equation and the powers $Pr^{1/3}$ and $Pr^{0.42}$.

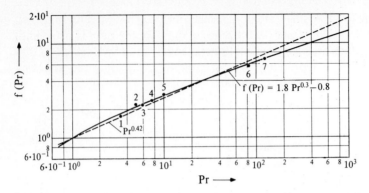

Figure 2-9 Heat transfer in tubes with turbulent flow. Dependence on Pr. Average of the test values of 1. Hufschmidt and Burck; 2. Eagle and Ferguson; 3. Ivanovski; 4. Allen and Eckert; 5. Stones; 6. Kraussold; 7. Hufschmidt and Burck.

slight influence of Re expressed here. Rather, the author [H13] has established, by the evaluation of old and new test measurements on liquids and gases, particularly [A4, E1, I1, H23, 24, K16, 18, S31], that the dependence on Pr can be reproduced by means of a single gently curved line, as may be seen in Fig. 2-9. It satisfies the empirical equation

$$f(Pr) = 1.8 Pr^{0.3} - 0.8 \qquad (2\text{-}26)$$

The power $Pr^{0.42}$, which has been widely used in recent times, resembles Petukhov's curve in Fig. 2-8 and the empirically based curve in Fig. 2-9; the form $Pr^{0.42}$ has been suggested by Friend and Metzner [F1] on the basis of particular results from test measurements. The dependence on Pr derived from these results is shown as a broken line in Figs. 2-8 and 2-9.

Heat Transfer Equation for Turbulent Flow, Valid Over a Very Wide Region

Kraussold himself indicated that his Eqs. (2-21) and (2-22) would give very high value for liquids between the smallest critical Reynolds number $Re_{kr} = 2320$ and $Re = 10{,}000$. He demonstrated this by comparison with test values obtained by Kiley and Mangsen [S16] [Fig. 2 in K17]. A more thorough survey of the relationships is found in the Sieder and Tates graph [S17] in Fig. 2-10 where

$$\frac{Nu}{Pr^{1/3}(\eta_{fl}/\eta_w)^{0.14}}$$

is shown to be dependent on Re.

If the slight splitting up of the curve with L/d is disregarded, it is evident that the dependence upon Re, especially between $Re = 2320$ and $Re = 10^4$, follows a curved line. In contrast, the representation of Nusselt's Eqs. (2-18) and (2-19) and Kraussold's Eqs. (2-21) and (2-22) employing the logarithmic scales used here, would yield a straight line.

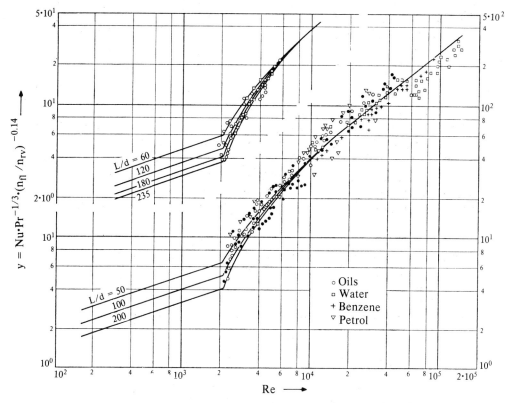

Figure 2-10 Dependence of the heat transfer coefficient upon Re according to the representation of Sieder and Tate.

In 1943 the author [H7] demonstrated that a curve of this type could be reproduced with an expression of the type

$$Re^m - K$$

where m and K are constants which are to be established by experiment. While the original values of Sieder and Tate can be best reproduced by $m = 2/3$ [H7], the values which were later calculated by the author using the factor $Pr^{0.42}$ are best reproduced with $m = 0.75$ [H11]; more recent data obtained by Hufschmidt and Burck [H23, 24], especially for the range between $Re = 10^5$ and $Re = 10^6$, indicate the need to raise the value of m to 0.8. Although this has already been suggested by Nusselt's Eq. (2-19) and Kraussold's Eq. (2-21), Gnielinski [G3] alludes to this yet again. The author [H13] has generated the heat transfer Eq. (2-27) for gases and liquids by bringing together as many as possible of the most recent experimental results, and at the same time utilizing the form of the dependence upon Pr expressed by Eq. (2-26).

$$Nu = 0.0235(Re^{0.8} - 230)(1.8Pr^{0.3} - 0.8)\left[1 + \left(\frac{d}{L}\right)^{2/3}\right]\left(\frac{\eta_{fl}}{\eta_w}\right)^{0.14} \qquad (2\text{-}27)$$

The fact that according to Eq. (2-27), Nu sharply decreases with diminishing Reynolds number for Re less than 10^4 has been proved by numerous heat transfer measurements, as mentioned previously. Some series of measurements, particularly with gases, working with higher values of Nu, appear to fit closer to the slope obtained from Nusselt's equation. Different experimental results between $Re = 2320$ and 10^4 do not always agree; the reason for this could be that this "Reynolds-zone" represents a transitory region between laminar and turbulent flow; the state of flow can therefore be very different depending on the inlet conditions, the roughness of the tube and the physical properties of the fluid.

The high viscosity makes the transition to turbulence difficult for liquids; this explains why the transition is represented, in Fig. 2-10 for example, by a curved line as opposed to a linear relationship. In general, it is probably safer to calculate the Nu values using Eq. (2-27) which are possibly too small in some cases, than to assume higher values without a secure basis for them.

For small values of Re of only slightly greater than 2320, there exists the difficulty that according to Eq. (2-27) α calculated using a heat transfer equation for laminar flow can be greater than that for turbulent flow. Sieder and Tate [S17] have recognized this difficulty previously. They endeavoured to eliminate it by splitting the curve between $Re = 2320$ and $Re = 10^4$; a single curve is replaced by a set of curves, each curve corresponding to a different value of d/L (see Fig. 2-10). As the author has shown [H8, page 35], this splitting can be expressed numerically by replacing the constant 125 in the 1943 equation [H7] or, the corresponding constant 230 in Eq. (2-27) by a simple function of d/L. Schlünder [S4] and Gnielinski [G3] alluded to this difficulty recently; they established that the Pohlhausen and Kroujiline's [P8] equation for laminar flow, namely

$$Nu = 0.664 Pr^{1/3}(Re\, d/L)^{1/2} \tag{2-28}$$

may yield higher values of Nu, for Re greater than 2320, than the 1959 equation for turbulent flow [H11] and thus also Eq. (2-27). However this observation relates only to short tubes where L/d is usually less than 10. Such short tubes rarely play an important part in heat exchangers; in general therefore, it is rarely necessary to modify Eq. (2-27). Nevertheless this apparent contradiction between Eqs. (2-28) and (2-27) can be eliminated by replacing the constant -230 in Eq. (2-27) by $-266 + 510\sqrt{d/L}$. In special cases it might be useful to know that, with very short tubes, and under certain circumstances, Eq. (2-27) is no longer valid if Eq. (2-28) gives higher values for Nu.

Reineke suggests in a yet unpublished study conducted in 1970 at the Thermodynamics Institute of the Technical University of Hannover, that the factor $(\eta_{fl}/\eta_w)^{0.14}$ should be retained, even for gases, in Eq. (2-27), or in any earlier relevant equation, but with a much higher exponent, for example 0.25. The dynamic viscosity of gases increases with temperature whereas for liquids, it decreases with temperature; as a consequence the factor containing η_{fl}/η_w has the opposite effect upon heat transfer in the case of gases, compared with liquids. At a given average gas temperature the heat transfer coefficient for heating a gas is less than that for cooling that gas; however this difference is not very important for gases.

It has been mentioned already that Eckert [3] has demonstrated that the factor $(\eta_{fl}/\eta_w)^{0.14}$ can be neglected, particularly in the case of gases, if all the physical properties refer to the average boundary layer temperature and thus only the fluid velocity relates to the average liquid or gas temperature. Reineke has investigated this idea in the study mentioned above, and has determined that the results of Humble *et al.* [H26] obtained from a wide range of experimental values and those of Deissler [D2] are in very close agreement, both with each other and with the equation drawn up by the author [H11] in 1959, and thus also with Eq. (2-27), when the physical properties are related to the average boundary layer temperature or, more precisely, to the mean value of the average liquid temperature and the wall temperature.

2-6 THEORETICALLY BASED ESTIMATES FOR HEAT TRANSFER WITH TURBULENT FLOW

The exponent relationships discussed this far have proved to be most fruitful but, apart from the similarity principle, they are purely experimental. For a deeper insight into the physical relationships, it is useful to note that Prandtl [P9] has derived a theoretically based approximate equation for heat transfer with turbulent flow in tubes. Much earlier than this, Reynolds [R5] conceived of the idea that the turbulent mechanism for heat transfer is the same as that for the transfer of momentum; this determines the velocity distribution over the tube cross-section and the drop in pressure in the direction of flow. However Prandtl was the first to develop successfully these ideas; he took the concept mentioned in Sec. 2-2 that, pure laminar flow prevails in a thin boundary layer at the tube wall while, on the other hand, a turbulent mixing motion dominates the residual zone of flow, usually called the "turbulent core". In this way he arrived at the equation

$$Nu = \frac{0.0395 Re^{3/4} \cdot Pr}{1 + \phi(Pr - 1)} \tag{2-29}$$

where ϕ denotes the ratio of the fluid velocity at the surface separating the boundary layer and the turbulent core, to the average speed of flow in the total cross section of the tube. Prandtl [P10] developed Eq. (2-30) relating ϕ to Reynolds number Re from the assumption, which is supported by experiment, that the fluid velocity in the turbulent core is proportional to the seventh root of the distance to the tube wall; Prandtl further assumed that the fluid temperature has a similar profile to the fluid velocity distribution in the tube.

$$\phi = 2.26 (Re/2)^{-1/8} \tag{2-30}$$

Working with liquids and taking a different value for ϕ, Bosch, Hofmann and others [K24] were able to obtain a satisfactory agreement, at least in part, with the Prandtl equation, using experimental data for turbulent flow. Hofmann formed the equation [H21]

$$\phi = 1.5 Re^{-1/8} Pr^{-1/6} \tag{2-31}$$

and established a common reference temperature θ_*, for determining the physical properties which appear in the dimensionless groups in Eqs. (2-29) and (2-31). This reference temperature can be calculated from the average liquid temperature θ_m and the temperature Θ_0 of the surface of the tube wall using the empirical equation

$$\theta_* = \theta_m - \frac{0.1\,Pr + 40}{Pr + 72}(\theta_m - \Theta_0) \tag{2-32}$$

These Eqs. (2-29) and (2-31) are applied in a similar manner to both heating and cooling. From these starting points Hofmann was able to reproduce, most satisfactorily, the heat transfer coefficients measured for liquids in tubes, measurements obtained for the most part prior to 1973.

The method described by Prandtl was developed in later theoretical works, among which those of Kármán [K1], Hofmann [H22], Reichardt [R2] and Petukhov [P3, 4, 5] are prominent. In these works particular consideration was paid to the fact that a gradual transition in reality takes place between the laminar flow in close proximity to the wall and the pronounced flow in the turbulent core. This fact is taken into account by inserting an intervening intermediate layer between the purely laminar boundary layer and the turbulent core. These improved calculations were in even better agreement with experimental data than were the earlier methods.

In order to supplement the above analogical considerations it must be mentioned that Steimle [S25] has derived a remarkable *connection between heat transfer and pressure drop* in fully developed *turbulent* flow, by considering the thickness of the hydrodynamic and thermal boundary layer, and by using empirical equations for heat transfer and velocity distribution in the cross-section of flow. With the quantity $\dot{m}$ flowing through the tube or duct in unit time, thermal diffusivity a, and tube or duct length L, Steimle formed the following new turbulent core values

$$Nu_{\dot{m}} = \frac{\alpha}{\lambda} \cdot \frac{\dot{m}}{\eta} \tag{2-33}$$

and the dimensionless group for flow

$$SK = \frac{\Delta p}{L} \frac{\dot{m}^3}{a \cdot \eta^4} \tag{2-34}$$

The *connection* mentioned is expressed by the following simple equation

$$Nu_{\dot{m}} = \text{const} \cdot (SK)^{0.37} \tag{2-35}$$

in which the constant varies from one case to another. By comparison with numerous experimental results Steimle was able to show that Eq. (2-35) is approximately correct, not only for straight tubes but also for tubes with turbulence inducing components and for tube coils. It is also a good approximation for heat transfer and pressure drops for cross flow in assemblies of smooth and ribbed tubes and it is applicable to regenerator heat-storing masses.

2-7 HEAT TRANSFER COEFFICIENTS OF LAMINAR FLOWING GASES AND LIQUIDS˙

Laminar flow occurs more frequently in liquids than in gases. This lies principally in the fact that for the same velocity of flow, the greater the density of the liquid, the larger the drop in pressure; as a consequence, liquids usually flow more slowly than gases. Taking into consideration that the kinematic viscosity $v = \eta/\rho$ is of the same order of magnitude for both liquids and gases, Eq. (2-8) gives lower values of Re for liquids. Thus it is not unusual in the case of liquids for the value of Re to fall below the critical value ($Re_{kr} \geqslant 2320$).

As has already been mentioned in Sec. 2-2, purely theoretical calculations for the heat transfer coefficients of liquids and gases have successfully been carried out for fully developed laminar flow in a straight tube of circular cross-section. This type of flow, in which the fluid velocity is parabolically distributed across the tube cross-section, first appears at a specific distance from the entrance that is, after covering an initial distance (see Sec. 4-1). Graz [G7] and later, independently, Nusselt [N5], found an exact solution to the differential equation for heat transfer, for the cases where the liquid or gas has already flowed through the hydrodynamic entry length in which laminar flow is established, before entering the heated or cooled part of the tube and where the wall temperature of this tube is constant. Their theory was completed by Léréque's equations [L2], and by Gröber's calculations [2 (1st edition 1921, p. 181)]. Thus, under the conditions specified, the true spatial heat transfer coefficients α_w for various distances l from the entrance into the heated or cooled tube is determined, as also is the average value α in a tube of length L. The values of both heat transfer coefficients resemble each other, as is shown in Fig. 2-7 for turbulent flow.

In Fig. 2-11 the values of the Nusselt number relating to the average heat

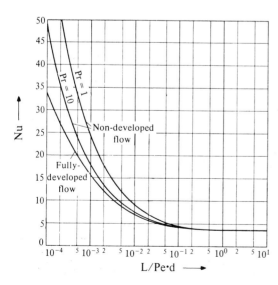

Figure 2-11 Heat transfer with laminar flow in tubes. Lower curve: Nu for hydrodynamically fully-developed flow using Eq.(2-37); Upper curve: Nu for non-developed flow, i.e., in a hydrodynamic entry length, from Eq. (2-37).

transfer coefficient α are shown as dependent on $L/Pe\,d$, where L is the tube length, and Pe is the Peclet number in Eq. (2-9). The bottom line in the graph is in very close agreement with the results of Nusselt's theory. It is calculated as follows from the heat transfer equation for fully developed laminar flow, derived empirically by the author [H7],

$$Nu = 3.66 + \left[\frac{0.0668 \dfrac{Pe \cdot d}{L}}{1 + 0.045 \dfrac{Pe \cdot d^{2/3}}{L}} \right] \left(\frac{\eta_{fl}}{\eta_w} \right)^{0.14} \tag{2-36}$$

by inserting $\eta_{fl}/\eta_w = 1$. The factor $(\eta_{fl}/\eta_w)^{0.14}$ corresponds to the difference between a heating and a cooling liquid; this approach of Seider and Tate [S17] has been mentioned already on p. 30. In Eq. (2-36) all physical properties, with the exception of η_w are to be related to the average liquid temperature; η_w is related to the wall temperature.

The constant 3.66 in Eq. (2-36) represents a limiting value towards which Nu tends in very long tubes. In practice this is seldom attained and further it only applies for a constant wall temperature. On the other hand, according to Eagle and Ferguson [E1], the limiting value amounts to 4.36 for a wall temperature which varies linearly with the direction of flow.[15]

An even better agreement with one of Lévêque's theoretical equations can be obtained for very short tubes [H11] by slight alterations to the constants and exponents in Eq. (2-36); however this is done at the expense of a deterioration in the correlation with important practical results. Baehr and Hicken [B4, Eqs. (2) and (28)] suggest a simple approximation which gives almost the same values as Eq. (2-36).

There exist numerous experimental results relating to heat transfer in laminar flowing liquids. Some of these results, among which are the early test results collected by Kraussold [K15, 18] and Böhm [B12, 13] for castor oil and glycol, confirm Nusselt's theory and therefore, also, Eq. (2-36). However most experimental results are higher than might be predicted by Nusselt, for example [D5, S17]. It is obviously of prime importance that, for a large part of each test, the flow into the entry of the heated tube was not yet hydrodynamically fully developed, so that the fluid had still to cover the thermal as well as the hydrodynamic entry length.

Stephan [S27] has examined theoretically the processes taking place in these frequently encountered conditions. First of all he calculated, as precisely as possible, the variations in the velocity distribution in consecutive cross-sections of the entry length. Then Stephan calculated the corresponding temperature distribution. The values of the average heat transfer coefficients α obtained in this way are in close agreement with more recent experimental results and could be

---

[15] For the limiting value in other cases, see Gröber, Erk, Grigul [2] 3rd edition, p. 184.

replicated using the following empirical equation

$$Nu = 3.66 + \frac{0.0677(Pr\,Re\,d/L)^{1.33}}{1 + 0.1Pr(Re\,d/L)^{0.83}}$$ (2-37)

In addition to the dependence on $Pe\,d/L = Pr\,Re\,d/L$, this equation, in contrast to Eq. (2-36), also introduces a particular, though only slight, dependence on Pr. The Pohlhausen equation mentioned, Eq. (2-28), was developed 38 years previously.

In Fig. 2-11 the values of Nu from Eq. (2-37) for $Pr = 1$ and $Pr = 10$ are represented by the two upper curves. The differences compared with the lower curve for flow conditions which have been fully developed throughout, turn out, practically, to be only very slight, because they only occur in a short section of the tube and quickly fade away. Stephan [S27] himself indicated that with $Pr = 10$ the differences can be ignored.

2-8 THE INFLUENCE OF FREE CONVECTION ON HEAT TRANSFER

With laminar flow and in the transition region between laminar and turbulent flow, heat transfer is often considerably increased, by means of superimposed free convection. For this reason free convection, and its influence on heat transfer with forced convection, will be briefly discussed.

Free convection which takes place without any forced flow whatsoever, particularly at hot vertical wall surfaces or on the outside of hot horizontal tubes, is usually an upward flow, which is caused by the buoyancy of that part of the gas or liquid which has been heated by the surface. Correspondingly a downward flow occurs at a cold wall. The Grashof number

$$Gr = d^3 g\beta(\Theta_O - \theta)/v^2$$ (2-38)

is determined as a dimensionless quantity relating to free convection at a horizontal tube, in which d is the outside diameter of the tube, g is the acceleration due to gravity, β is the coefficient of thermal expansion of the gas or liquid, Θ_O is the temperature of the outer tube surface, θ is the temperature of the gas or liquid outside the narrow area of free convection and v is the kinematic viscosity of the gas or liquid. The dimensionless group Gr also determines the free convection at a vertical wall. In this case however the height of the wall should replace d in both Eq. (2-38) and the expression for Nu determined by means of the Eq. (2-40) to (2-42) following.

In the English literature the dimensionless group

$$\text{Rayleigh number } Ra = Gr \cdot Pr$$ (2-39)

is widely used in this connection.

For heat transfer with free convection the similarity equation takes the form

$$Nu = f(Gr, Pr)$$ (2-40)

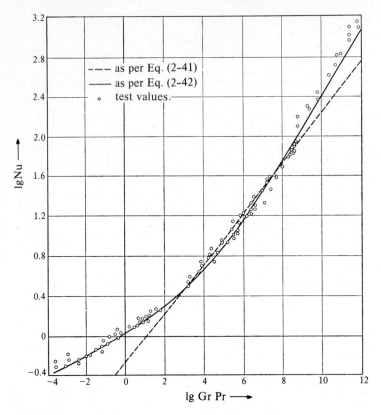

Figure 2-12 Heat transfer by free convection. Superimposition of free and forced convection in tubes.

The heat output of a heated horizontal tube or a heated vertical wall with free convection to an otherwise motionless gas or liquid can be calculated using:

$$Nu = 0.53(Gr \cdot Pr)^{1/4} \tag{2-41}$$

The coefficient 0.53, selected by McAdams [7], is reasonably accurate, provided $Gr \cdot Pr$ lies between 10^3 and 10^9, and this is usually the case in practice.

Figure 2-12 shows the results of numerous measurements of heat transfer with free convection over a very wide range of $Gr \cdot Pr$. Above $Gr \cdot Pr = 10^9$ turbulence occurs in the boundary layer in which free convection prevails; it is because of this that the corresponding experimental data points are relatively highly positioned in Fig. 2-12. The broken straight line reproduces Eq. (2-41). The equation [H11]

$$Nu = 0.11(Gr \cdot Pr)^{1/3} + (Gr \cdot Pr)^{0.1} \tag{2-42}$$

is reasonably accurate, between $Gr \cdot Pr = 10^{-7}$ and $Gr \cdot Pr = 10^{12}$ and is shown as a solid line in Fig. 2-12.

It is necessary to consider the superimposition of free convection on laminar or slightly turbulent flow in tubes only when the internal diameter of the tube is greater than about 2 cm and when there is a large difference in the temperature between the tube wall and the fluid. The often considerable increase in the heat transfer coefficient which accompanies this is expressed through the following equations which are based on measurements and in which the value of Gr often approaches that of Re. In all the dimensionless groups, including Gr, the internal diameter of the tube d is the distinguishing length.

Between $Re = 1600$ and $Re = 4600$, Watzinger and Johnson [W3] give the equation

$$Nu = 0.255Gr^{0.25}Re^{0.07}Pr^{0.37} \tag{2-43}$$

for heat transfer in *vertical pipes* where forced and free flow are equidirectional. Above $Re = 4600$ the influence of free convection is so slight that it can be disregarded.

Observations by Colburn and Hougen [C5] in *vertical tubes* and by Nusselt [N4] in *horizontal tubes* have been represented by Colburn utilizing an equation which McAdams [7] has altered to give

$$Nu = 1.62(Re\,Pr\,d/L)^{1/3}\left(\frac{\eta_{fl}}{\eta_w}\right)^{1/3}(1+0.015Gr^{1/3}) \tag{2-44}$$

For *horizontal tubes* the following equations have been formulated on the basis of further measurements:

The Eq. (2-45) of Kern and Othmer [K8] was developed for large differences in temperature between the tube wall and the fluid:

$$Nu = 1.86(Re\,Pr\,d/L)^{1/3}(\eta_{fl}/\eta_w)^{0.14}\left(\frac{2.25(1+0.010Gr^{1/3})}{\log Re}\right) \tag{2-45}$$

Métais [M4] represented his observations of the heating or cooling of water in a tube with an internal diameter of 50 mm for Re between 500 and 2500 by the equation:

$$Nu = 1.345Re^{0.34}Pr^{0.32}(Pr_{fl}/Pr_w)^{0.188}Gr^{0.06} \tag{2-46}$$

For *vertical tubes* Kirschbaum [K10] states the equation

$$Nu = 0.032[Re+f(Gr/Re)]^{0.8} \cdot Pr^{0.37}(L/d)^{-0.054} \tag{2-47}$$

The function $f(Gr/Re)$, which is different for upward and downward flow, can only be extracted indirectly from a diagram (see Gröber-Erk-Grigull [2], 3rd edition, 1961, reprint p. 214, Fig. 96).

2-9 HEAT TRANSFER IN ROUGH AND CURVED TUBES

Heat Transfer in Rough Tubes

All the heat transfer equations introduced so far apply to gases and liquids flowing in tubes which are perfectly smooth.

There exists a series of measurements dealing with the influence of the *roughness of the tube or duct walls on heat transfer*; these measurements principally relate to turbulent flow. According to the experimental work carried out by Böhm [B10] and the observations by Schefels [S3] of the heat transfer of gases in walled ducts, the heat transfer coefficient increases with the roughness of the walls. For naturally rough brickwork where the joins have been carefully smoothed, the heat transfer coefficient is about 10 per cent higher than for completely smooth duct walls. By artificially increasing the roughness by means of projecting edges and the like, Böhm noticed a further increase in the heat transfer coefficient of about 60 per cent for laminar flow, and about 100 per cent for turbulent flow.

On the other hand, when working with liquids flowing in various rough metal tubes, Pohl [P7] noticed that the heat transfer coefficients decreased with increasing roughness. However, in earlier experiments with liquids, for example those conducted by Stanton [S24] and Lorenz [L6], a slight increase in the heat transfer coefficient was observed for increasing tube wall roughness.

Numer [N3] found that the heat transfer coefficient for artificially rough tubes was up to three times its value for smooth tubes. Similarly, in numerous other research papers, an enhancement in heat transfer due to the presence of roughness has been reported; this is described in publications by Brauer [B16], Dipprey and Sabersky [D4], Kolar [K13], Sheriff and Gumley [S13, 14], Gowen and Smith [G5, 6], and others.

The important results from recent measurements are displayed in Fig. 2-13; these are taken from the experimental results reported by Dipprey and Sabersky [D4] although similar results are to be found in the publications of Sheriff and Gumley [S13, 14]. In this diagram $St = Nu/RePr$ is shown as a function of Re; the lower straight line represents heat transfer in smooth tubes. It is well known that, with small values of Re the dimensionless group St, and therefore also Nu, have the same value for rough tubes as they have for tubes with smooth walls. However, the larger the ratio of the average rise in roughness ε to the internal diameter of the tube d, the smaller the value of Re at which the curve begins to rise from the line corresponding to smooth tube behaviour. Beyond a maximum the curve changes into another, almost straight line. In these circumstances where St, and therefore Nu becomes greater than its value for the smooth tube, the heat transfer coefficient increases as a result of roughness.

This diagram is in remarkable agreement with the curve for the coefficient of resistance ψ, measured by Nicuradse [N153], in rough tubes as shown in Fig. 4-2, page 119. The coefficient ψ is displayed graphically as a function of Re; there are a number of curves, each of which corresponds to a specific value of ε/d. The overall form of these curves is S-shaped. The value of ψ decreases initially for increasing values of Re until a minimum value is reached when the curve becomes a horizontal line which corresponds to the "fully rough region". For a given value of ε/d, the S-shaped part of the curve lies in the same region of Re for both St and ψ in the two diagrams.

To the right of the maximum of a curve in Fig. 2-13, the results of Dipprey and

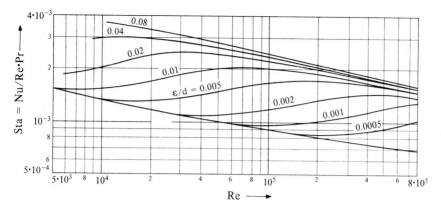

Figure 2-13 Heat transfer in rough tubes.

Sabersky can be represented by the following equation which has been developed by the author but which has not been published until now:

$$Nu_{\text{rough}} = Nu_{\text{smooth}}[1.87 + 0.54 \lg(1000\varepsilon/d)] \qquad (2\text{-}48)$$

Heat Transfer in Bent Tubes

It will be shown in more detail later in the section dealing with the pressure drop which occurs with flow in tubes, that double turbulence occurs in the cross-sections of curved tubes; see Fig. 4-5, page 125. As such turbulence reduces the thickness of the boundary layer, it causes an increase in the heat transfer compared with the equivalent process in straight tubes. It can be expected that this influence will have less and less effect relative to an increasing Reynolds number; this is because the increase in heat transfer will become due, more and more, to the rising turbulence.

The influence of the radius of the bend in a tube on heat transfer can easily be measured. The apparently large difference between experimental results can be explained, apart from the degree of roughness in the tubes, probably partly by the fact that when a tube is bent, the originally round cross-section does not always retain its shape and small folds may even occur in the side of the tube around the centre point of the curve. In spite of this a clear picture emerges from the measurements.

Using measurements taken in gases flowing in curved tubes Jeschke [J2] has developed an equation which is valid for when Re is greater than 10^5

$$\alpha_{kr} = (1 + 3.54d/D)\alpha_{ger} \qquad (2\text{-}49)$$

in which D is the diameter of the bend, α_{ger} is the average heat transfer coefficient in straight tubes and α_{kr} is the average heat transfer coefficient in curved tubes, these tubes having the same diameter d. For example, from Eq. (2-49), the heat transfer coefficient in curved tubes is 7 per cent higher than in straight tubes when $D = 50d$, and 18 per cent higher when $D = 20d$. This equation was confirmed in principle, using new measurements, by Rogers and Mayhew [R9].

On the other hand, the widely scattered experimental values obtained by Seban and McLoughlin [S11], with low Reynolds numbers, are reproduced more accurately if the constant in Eq. (2-49) is raised from 3.54 to 5 or 6. According to Mori and Nakyama [M6], whose work was more theoretical than experimental, an increase in this constant of even up to 8 or 10 would be appropriate. It is also known from the test values obtained by Woschni [W11], which are probably very exact, that the numerical value in Eq. (2-49) increases with decreasing Re, up to the critical Reynolds number.

The author has found, but has not yet had these findings published, that on average, the measured values in the completely turbulent region, can be expressed by the equation

$$\alpha_{kr} = (1 + 21/Re^{0.14} \cdot d/D) \cdot \alpha_{ger} \qquad (2\text{-}50)$$

This equation agrees very well with Jeschke's Eq. (2-49) when Re is greater than 10^5. Near to Re_{kr}, on the other hand, the expression $21/Re^{0.14}$ attains a value of about 6.

E. F. Schmidt [S8] together with Srinivasan, Naudapurkar and Holland [S22] arrived at values of the same order of magnitude. Essentially they assembled a large number of values which are found in the literature. The values given in both publications can be more successfully reproduced by means of an equation such as (2-49) or (2-50) if d/D is replaced by $(d/D)^n$, where n can lie between 0.5 and 0.75. In particular Schmidt attributes greater influence upon heat transfer to the bend in the tube than do Srinivasan et al.

In laminar flow, which will not be dealt with here, the influence of the bend in a tube on heat transfer is considerably greater; see, for example, [S11] and [W11].

It must be noted that with bent tubes, the critical Reynolds number is higher than with straight tubes, see page 20.

2-10 LITERATURE ON HEAT TRANSFER IN TUBES AND DUCTS

See Foreword, Sec. 1-5

A

1. Abbrecht, P. H.; Churchill, S. W.: The Thermal Entrance Region in Fully Developed Turbulent Flow. A. I. Ch. E. J. 6 (1960) 2, 268–273.
2. Abecassis, G.; Cornu, G.; Taccoen, L.: Contribution à l'étude des transferts thermiques unilatéraux en écoulement turbulent dans un espace annulaire lisse—Cas des métaux liquides. Rev. gén. Therm. 7 (1968) 83, 1219–1226.
3. Ackermann, G.: Wärmeübergang und molekulare Stoffübertragung im gleichen Feld bei großen Temperatur- und Partialdruckdifferenzen. VDI-Forschungsheft 382 Berlin (1937); see also Z. VDI., supplement "Verfahrenstechnik" (1937) No. 1, pp. 36–67.
4. Allen, R. W.; Eckert, E. R. G.: Friction and Heat-Transfer-Measurements to Turbulent Pipe Flow of Water (Pr = 7 and 8) at Uniform Wall Heat Flux. Trans. ASME, Ser. C 86 (1964) 3, 301–310 and J. Heat Transfer. (1964).

5. Altenkirch, E.: Eine allgemeine Gleichung für den Wärmeübergang im glatten Rohr. Kältetech. 5 (1953) 253–354.

6. Azer, N. Z.; Chao, B. T.: Turbulent Heat Transfer in Liquid Metals—Fully Developed Pipe Flow with Constant Wall Temperature. Int. J. Heat Mass Transfer 3 (1961) 2, 77–83.

B

1. Back, L. H.; Cuffel, R. F.; Massier, P. F.: Laminar, Transition and Turbulent Boundary-Layer Heat-Transfer Measurements with Wall Cooling in Turbulent Airflow through a Tube. Trans. ASME, Ser. C 91 (1969) 4, 477–487.

2. Baehr, H. D.: Zur Darstellung des Wärmeübergangs bei freier Konvektion durch Potenzgesetze. Chem. Ing. Tech. 26 (1954) 269.

3. Baehr, H. D.: Gleichungen für den Wärmeübergang bei hydrodynamisch nicht ausgebildeter Laminarströmung in Rohren. Chem. Ing. Tech. 32 (1960) 2, 89 and 90.

4. Baehr, H. D.; Hicken, E.: Neue Kennzahlen und Gleichungen für den Wärmeübergang in laminar durchströmten Kanälen. Kältetechn. Klimatisierung 21 (1969) 2, 34–38.

5. Bankston, C. A.; McEligot, D. M.: Turbulent and Laminar Heat Transfer to Gases with Varying Properties in the Entry Region of Circular Ducts. Int. J. Heat Mass Transfer 13 (1970) Feb., 319–344.

6. Barrow, H.; Humphreys, J. F.: The Effect of Velocity Distribution on Forced Convection Laminar Flow Heat Transfer in a Pipe at Constant Wall Temperature. Heat and mass transfer 3 (1970) 227–231.

7. Barthels, H.: Darstellung des Wärmeüberganges in konzentrischen Ringspalten unter Benutzung der Analogie zwischen Impuls- und Wärmeaustausch. Diss. TH Aachen (1967).

8. Beck, F.: Wärmeübergang und Druckverlust in senkrechten konzentrischen und exzentrischen Ringspalten bei erzwungener Strömung und freier Konvektion. Chem. Ing. Tech. 35 (1963) 837–844.

9. Blasius, H.: Das Ähnlichkeitsgesetz bei Reibungsvorgängen in Flüssigkeiten. Phys. Z. 12 (1911) 1175, or Forschungsarbeit Ing.-Wes., number 131 (1913).

10. Böhm, H.: Versuche zur Ermittlung der konvektiven Wärmeübergangszahlen an gemauerten engen Kanälen. Arch. Eisenhüttenwes. 6 (1932/33) 423–431.

11. Böhm, J.: Der Wärmeübergang bei zähen Flüssigkeiten. Wärme 65 (1942) 221–228 (Review of a range of works).

12. Böhm, J.: Messungen des Wärmeübergangs im laminaren Strömungsgebiet mit Rizinusöl. Wärme 66 (1943) 144–152.

13. Böhm, J.: Der Wärmeübergang im Übergangsgebiet von laminarer zu turbulenter Strömung (Measurements in ethylene glycol, ethylene glycol–water mixtures and calcium chloride–water mixtures). Wärme 67 (1944) 3–11.

14. Boussinesq, J.: Calcul de la moindre longueur que doit avoir un tube …, pour qu'un régime sensiblement uniforme s'y établisse and Comptes Rendus 113 (1891) 9 and 49. ·

15. Bradley, D.; Entwistle, A. G.; Developed Laminar Flow Heat Transfer from Air for Variable Properties. Int. J. Heat Mass Transfer 8 (1956) 621–638. (Detailed theory in particular the influence of temperature dependence upon the physical properties and of free convection upon the temperature profile).

16. Brauer, H.: Strömungswiderstand und Wärmeübergang bei Ringspalten mit rauhen Kernrohren. Atomkernenergie 6 (1961) 152–161 and 207–211.

17. Bridgman, P. W.; Holl, H.: Theorie der physikalischen Dimensionen. Leipzig and Berlin: B. G. Teubner 1932.

18. Brown, W. G.; Grassmann, P.: Der Einfluß des Auftriebs auf Wärmeübergang und Druckgefälle bei erzwungener Strömung in lotrechten Rohren. Forschg. Ing.-Wes. 25 (1959) 69–78.

19. Brown, M.: Heat and Mass Transfer in a Fluid in Laminar Flow in a Circular or Flat Tube. Am. Inst. Chem. Engrs. J. 6 (1960) 179–183.

20. Burck, E.: Der Einfluß der Prantdl-Zahl auf den Wärmeübergang und Druckverlust künstlich aufgerauhter Strömungskanäle. Wärme- und Stoffübertragung 2 (1969) 2, 87–98.

C

1. Cheesewright, R.: Turbulent Natural Convection from a Vertical Plane Surface. Trans. Am. Soc. Mech. Engrs., Ser. C. 90 (1968) 1–8.
2. Chen, Sh. D.; Lee, L. P.: Fully Developed Forced Convection Turbulent Flow Heat Transfer in Circular Pipes and Between Parallel Plates of Liquid Metals. Bull. College Engng. nat. Taiwan Univ. 12 (1968) 79–96.
3. Chmiel, H.; Rautenbach, R.; Schümmer, P.: Zur Theorie des Wärmeübergangs in der turbulenten Rohrströmung viscoelastischer Flüssigkeiten. Chem. Ing. Tech. 44 (1972) 543–545.
4. Colburn, A. P.; Drew, T. B.; Worthington, H.: Ind. Eng. Chem. 39 (1947) 958.
5. Colburn, A. P.; Hougan, O. A.: Studies in Heat Transmission III (Flow of Fluids at Low Velocities). Ind. Eng. Chem. 22 (1930) 534–539.
6. Colburn, A. P.: A Method of Correlating Forced Convection Heat Transfer Data and a Comparison with Fluid Friction. Trans. Am. Inst. Chem. Engrs. 29 (1933) 174–218.
7. Comings, E. W.; Clapp, J. T.; Taylor, J. F.: Air Turbulence and Transfer Processes. Ind. Eng. Chem. 40 (1948) 1076–1082.
8. Cope, W. F.: The Friction and Heat Transmission Coefficients of Rough Pipes: Proc. Inst. Mech. Engrs. 145 (1941) 99–105.

D

1. Davis, L. P.; Perona, J. J.: Development of Free Convection Flow of a Gas in a Heated Vertical Open Tube. Int. J. Heat Mass Transfer 14 (1971) 7, 889.
2. Deissler, R. G.; Eian, G. C.: Analytical and Experimental Investigation of Heat Transfer with Variable Fluid Properties. NACA TN 2629 (1952).
3. Depew, C. A.; August, S. E.: Heat Transfer due to Combined Free and Forced Convection in a Horizontal and Isothermal Tube. Trans. Am. Soc. Mech. Engrs., Ser. C 93 (1971) 4, 380–384.
4. Dipprey, D. F.; Sabersky, R. H.: Heat and Momentum Transfer in Smooth and Rough Tubes at Various Prandtl Numbers. Int. J. Heat Mass Transfer 6 (1963) 5, 329–353.
5. Dittus, F. W.; Boelter, L. M. K.: Heat Transfer in Automobile Radiators of the Tubular Type. Univ. Cal. Publ. Engrg. 2 (1930) 443–461.
6. Drew, T. B.; Hogan, J. J.; McAdams, W. H.: Heat Transfer in Streamline Flow. Ind. Eng. Chem. 23 (1931) 936–945.

E

1. Eagle, A.; Ferguson, R. M.: On the Coefficient of Heat Transfer from the Internal Surface of Tube Walls. Proc. Roy. Soc. (A) 127 (1930) 540–566.
2. Eckert, E.: Der Wärmeübergang bei Kühlen und Heizen. Z. VDI 80 (1936) 137 and 138.
3. Eckert, E.: Technische Strahlungsaustauschrechnungen. Berlin: VDI-Verlag 1937.
4. Eckert, E.: Wärmeübertragung bei turbulenter Strömung. Z. VDI 85 (1941) 581–583.
5. Eckert, E.: Die Berechnung des Wärmeübergangs in der laminaren Grenzschicht umströmter Körper. VDI-Forschungsheft 416, Berlin 1942.
6. Eckert, E. R. G.: Convective Heat Transfer for Mixed, Free and Forced Flow through Tubes. Trans. Am. Soc. Mech. Engrs. 46 (1954) 51.
7. Eckert, E. R. G.; Ervine, T. F. jun.: Simultaneous Turbulent and Laminar Flow in Ducts with Noncircular Cross Sections. J. Aeronautical Sc. 22 (1955) No. 1.
8. Eckman, V. W.: On the Change from Steady to Turbulent Motion of Liquids. Arkiv Mat., Astron. och Fysik, 6 (1911) 5.

F

1. Friend, W. L.; Metzner, A. B.: Turbulent Heat Transfer inside Tubes. J. Amer. Inst. Chem. Engrs. 4 (1958) 393–402.
2. Fujii, T.; Takeuchi, M.; Fujii, M.; Suzaki, K.; Uehara, H.: Experiments on Natural Convection Heat Transfer from the Outer Surface of a Vertical Cylinder to Liquids. Int. J. Heat Mass Transfer 13 (1970) 753–787.

G

1. Gersten, K.: Wärme- und Stoffübertragung bei großen Prandtl-bzw. Schmidt-Zahlen. Wärme- und Stoffübertragung 7 (1974) 65–70 (V-shaped flow).
2. Glaser, H.: Regeneratoren mit bewegter Speichermasse. Forschung Ingenieurwes. (1951) 1, 9–15.
3. Gnielinski, V.: Eine Weiterentwicklung der Gleichung von H. Hausen für den Wärmeübergang im turbulent durchströmten Rohr unter Berücksichtigung neuerer Meßwerte bei hohen Reynolds-Zahlen. Vortrag in Bad Mergentheim am 2. IV. 1973 bei der Sitzung des Ausschusses "Wärme- u. Stoffübertragung" der Verfahrenstechnischen Gesellschaft (VDI). Published under the title: Neue Gleichungen für den Wärme- und Stoffübergang in turbulent durchströmten Rohren und Kanälen. Forschung Ingenieurwesen 41 (1975) No. 1.
4. Gnielinski, V.: In the new edition of the VDI-Wärmeatlas. New equations for the heat and mass transfer in turbulent flow through tubes. Lecture at the annual meeting of industrial engineers in Berlin.
5. Gowen, R. A.; Smith, J. W.: The Effect of the Prandtl Number on Temperature Profiles for Heat Transfer in Turbulent Pipe Flow. Chem. Eng. Sci. 22 (1967) 12, 1701–1711.
6. Gowen, R. A.; Smith, J. W.: Turbulent Heat Mass Transfer from Smooth and Rough Surfaces. Int. J. Heat Mass Transfer 11 (1968) 1657–1674.
7. Grätz, L.: Über die Wärmeleitfähigkeit von Flüssigkeiten. Ann. Physik (N. F.), 18 (1883) 79–94 and 25 (1885) 337–357.
8. Grass, G.: Wärmeübergang an turbulent strömende Gase im Rohreinlauf. Allgemeine Wärmetechnik 7 (1956) 58–64.
9. Grass, G.: Der Einfluß von Störungen der Strömung im Einlauf auf den Wärmeübergang bei Flüssigkeiten in Rohren und Ringspalten. Atomenergie 3 (1958) Number 10.
10. Grass, G.; Herkenrath, H.; Hufschmidt, W.: Anwendung des Prandtlschen Grenzschichtmodells auf den Wärmeübergang an Flüssigkeiten mit stark temperaturabhängigen Stoffeigenschaften bei erzwungener Strömung. Wärme- und Stoffübertragung 4 (1973) 113–119.
11. Grassmann, P.: Stoff- und Wärmeaustausch bei unmittelbarer Berührung beider Fluide. Chem. Ing. Tech. 44 (1972) 291–295.
12. Gregorig, R.; Trommer, H.: Verminderung des Wärmedurchgangs bei ungleichmäßiger Geschwindigkeitsverteilung und ungenauer Rohrteilung. Schweiz. Bauz. 70 (1952) 151–155 and 174–176.
13. Gregorig, R.: Verallgemeinerter Ausdruck für den Einfluß temperaturabhängiger Stoffwerte auf den turbulenten Wärmeübergang. Wärme- und Stoffübertragung 3 (1970) 26–40.
14. Gregorig, R.: Ausgebildete turbulente Rohrströmung für sehr große Reynolds-Zahlen. Das Prinzip von Hamilton. Wärme- und Stoffübertragung 5 (1972) 73–80.
15. Gregorig, R.: Verallgemeinerter Ausdruck für den Einfluß temperaturabhängiger Stoffwerte auf den turbulenten Wärmeübergang. Wärme- und Stoffübertragung 3 (1970) 26–40.
16. Grigull, U.; Tratz, H.: Thermischer Einlauf in ausgebildeter laminarer Rohrströmung. Int. J. Heat Mass Transfer 8 (1965) 669–678.
17. Grigull, U.; Tratz, H.: Wärmeübergang an Quecksilber bei laminarer und turbulenter Rohrströmung. Wärme- und Stoffübertragung 1 (1968) 2, 61–71.

H

1. Hackl, A.; Gröll, W.: Zum Wärmeverhalten zähflüssiger Öle. Verfahrenstechnik 3 (1969) 141–145.
2. Hahnemann, H.: Neue experimentelle Untersuchungen über die Entstehung der turbulenten Rohrströmung. Forschung 8 (1937) 226–237.
3. Hahnemann, H. W.: Näherungstheorie für den Wärmeübergang in Spaltströmungen. Forsch. Ing.-Wes. 33 (1967) 1, 1–12.
4. Hahnemann, H. W.: Zur Theorie des Wärme- und Stoffübergangs in turbulenter Strömung. Forsch. Ing. Wes. 34 (1968) 3, 90.
5. Hanratty, T. J.; Rosen, E. M.; Kabel, R. L.: Effect of Heat Transfer on Flow Field at Low Reynolds numbers in Vertical Tubes. Ind. Eng. Chem. 50 (1958) 815–820.
6. Hartmann, H.: Wärmeübergang bei laminarer Strömung durch Ringspalte. Chem. Ing. Tech. 33 (1961) 1, 22–26.

7. Hausen, H.: Darstellung des Wärmeübergangs in Rohren durch verallgemeinerte Potenzbeziehungen. Z. VDI. Supplement "Verfahrenstechnik" 1943, Number 4, pp. 91–98.
8. Hausen, H.: Wärmeübertragung im Gegenstrom, Gleichstrom und Kreuzstrom. Berlin, Göttingen, Heidelberg, München: Springer and Bergmann 1st ed. 1950.
9. Hausen, H.: Zur Frage nach dem gleichwertigen Durchmesser bei der Wärmeübertragung in einseitig beheizten Spalten. Abhandlungen der Braunschweigischen Wissenschaftlichen Gesellschaft X (1958) 150–165.
10. Hausen, H.; Düwel, L.: Zur Frage nach dem gleichwertigen Durchmesser bei der Wärmeübertragung in einseitig beheizten Spalten. Kältetech. 11 (1959) 242–249.
11. Hausen, H.: Neue Gleichungen für die Wärmeübertragung bei freier und erzwungener Strömung. Allgem. Wärmetechnik 9 (1959) 75–79.
12. Hausen, H.: Bemerkungen zur Veröffentlichung von A. Hackl und W. Gröll: Zum Wärmeverhalten zähflüssiger Öle (see [H 1]). Verfahrenstechnik 3 (1969) 8, 355 and 11, 480.
13. Hausen, H.: Erweiterte Gleichung für den Wärmeübergang bei turbulenter Strömung. Wärme- und Stoffübertragung 7 (1974) 222–225.
14. Helmholz, H.: A theory in respect of geometrically similar movements of fluid bodies and also application to the problem of steering air balloons. Monthly report of the Capital Academy of Scientists at Berlin from 26 June 1873, pp. 501–514.
15. Hicken, E.: Wärmeübergang bei ausgebildeter laminarer Kanalströmung für am Umfang veränderllche Randbedingungen. Dissertation Braunschweig 1966.
16. Hicken, E.: Wärmeübergangsmessungen in rechteckigen Kanälen bei nur teilweise beheiztem Umfang. Wärme- und Stoffübertragung 1 (1968) 3, 185–189.
17. Hicken, E.: Der Wärmeübergang im thermischen Anlauf rechteckiger Kanäle bei nur teilweise beheiztem Umfang. Wärme- und Stoffübertragung 5 (1972) 213–219.
18. Hermann, R.; Burbach, Th.: Strömungswiderstand und Wärmeübergang in Rohren. Diss. Leipzig 1930.
19. Hobler, T.; Koziol, K.: Wärmeübergang in beidseitig abwechselnd eingebuchteten Rohren. Angew. Chemie, Ser. B (Polen) (1965) 1, 3–43.
20. Hofmann, E.: Wärmedurchgangszahlen von Steilrohrverdampfern. Z. ges. Kälte-Ind. 44 (1934) 190–194.
21. Hofmann, E.: Der Wärmeübergang bei der Strömung im Rohr. Z. ges. Kälte-Ind. 44 (1937) 99–107.
22. Hofmann, E.: Über die Gesetzmäßigkeiten der Wärme- und Stoffübertragung auf Grund des Strömungsvorganges im Rohr. Forsch. Ing.-Wes. 11 (1940) 159–169.
23. Hufschmidt, W.; Burck, E.; Riebold, W.: Die Bestimmung örtlicher und mittlerer Wärmeübergangszahlen in Rohren bei hohen Wärmestromdichten. Int. J. Heat Mass Transfer 9 (1966) 539.
24. Hufschmidt, W.; Burck, E.: Der Einfluß temperaturabhängiger Stoffwerte auf den Wärmeübergang bei turbulenter Strömung von Flüssigkeiten in Rohren bei hohen Wärmestromdichten und Prandtl-Zahlen. Int. J. Heat Mass Transfer 11 (1968) 6, 1041–1048.
25. Hull, T. A.; Tsao, P. H.: Proc. Roy. Soc. (London) A 191 (1947) 6.
26. Humble, L. V.; Lowdermilk, W. H.: Desmon, L. G. Measurements of Average Heat Transfer and Friction Coefficients for Subsonic Flow of Air in Smooth Tubes at High Surface and Fluid Temperatures. NACA Report 1020 (1951).

I

1. Ivanovski, M. N.: Rapid Method of Measuring the Average Heat Transfer Coefficient in a Tube. —TR— 4511, 90–103.

J

1. Jauernick, R.: Über den örtlichen Widerstandsbeiwert und die Wärmeübergangszahlen in Rohrbündeln bei hohen Reynoldschen Zahlen. Zeitschr. "Die Wärme" 61 (1938) 738–743 and 751–756.
2. Jeschke, H.: Wärmeübergang und Druckverlust in Rohrschlangen. Supplement "Technische Mechanik" in Z. VDI. 69 (1925) 24–28.

3. Jodlbauer, K.: Das Temperatur- und Geschwindigkeitsfeld um ein geheiztes Rohr bei freier Konvektion. Forsch. Geb. Ing. Wes. 4 (1933) 157–172.
4. Johannsen, K.; Bartsch, G.: Bilateraler Wärmeübergang bei erzwungener turbulenter Strömung flüssiger Metalle zwischen parallelen Platten. Wärme 76 (1970) 1/2, 32–38.
5. Jones, A. S.: Extensions to the Solution of the Graetz Problem. Int. J. Heat Mass Transfer 14 (1971) 4, 619.
6. Jung, J.: Wärmeübergang und Reibungswiderstand bei Gasströmung in Rohren bei hohen Geschwindigkeiten. VDI-Forschungsheft 380 (1936).

K

1. Kármán, Th. v.: Analogy between Fluid Friction and Heat Transfer. Trans. Amer. Soc. mech. Engrs. 61 (1939) 705 and 710, and Engineering 148 (1939) 210–213.
2. Kast, W.: Zur Frage der Analogie zwischen Wärme- und Stoffaustausch. Wärme- und Stoffübertragung 5 (1972) 15–21.
3. Kaul, V.; Kiss, M. v.: Forced Convection Heat Transfer and Pressure Drop in Artificially Roughened Flow Passages. Neue Technik 7 (1964) B 6, 297–309.
4. Kaye, W. A.; Furnas, C. C.: Heat Transfer Involving Turbulent Fluids. Ind. Eng. Chem. 26 (1934) 783–786.
5. Kays, W. M.; Nicoll, W. B.: Laminar Flow Heat Transfer to a Gas with Large Temperature Differences. Trans. Am. Soc. Mech. Engrs. Ser. C 85 (1963) 4, 329–338.
6. Kays, W. M.; Leuna, E. Y.: Heat Transfer in Annular Passages—Hydrodynamically Developed Turbulent Flow with Arbitrarily Prescribed Heat Flux. Int. J. Heat Mass Transfer 6 (1963) 7, 537–557.
7. Keil, R. H.; Baird, H. I.: Enhancement of Heat Transfer by Flow Pulsation. Ind. Engng. Chem. Process Des. Devel. 10 (1971) 4, 473–478.
8. Kern, D. Q.; Othmer, D. F.: Effect of Free Convection on Viscous Heat Transfer in Horizontal Tubes. Trans. Am. Inst. Chem. Engrs. 39 (1943) 517–535.
9. Kirillov, V. V.; Malyugin, Yu. S.: Local Heat Transfer During the Flow of a Gas in a Pipe at High Temperature Differences. High Temperature 1 (1963) 2, 227–231.
10. Kirschbaum, E.: Neues zum Wärmeübergang mit und ohne Änderung des Aggregatzustandes. Chem. Ing. Tech. 24 (1952) 393–400.
11. Koch, B.: Turbulenter Wärmeaustausch im Rohr. Arch. ges. Wärmetechn. 1 (1950) 2–8.
12. Kokorev, L. S.; Ryaposov, V. N.: Turbulent Heat Transfer During the Flow of a Heating Medium of Small Prandtl Number along a Tube. Int. chem. Eng. 2 (1962) 4, 514–519.
13. Kolar, V.: Heat Transfer in Turbulent Flow of Fluids through Smooth and Rough Tubes. Int. J. Heat Mass Transfer 8 (1965) 639–653.
14. Kraus, W.: Temperatur- und Geschwindigkeitsfeld bei freier Konvektion um eine waagerechte quadratische Platte. Phys. Z. 41 (1940) 126–150.
15. Kraußold, H.: Die Wärmeübertragung bei zähen Flüssigkeiten in Rohren. VDI-Forschungsheft 351 (1931).
16. Kraußold, H.: Neue amerikanische Untersuchungen über den Wärmeübergang an Flüssigkeiten bei laminarer Strömung. Forschung 3 (1932) 21–24.
17. Kraußold, H.: Die Wärmeübertragung an Flüssigkeiten in Rohren bei turbulenter Strömung. Forschung Ing.-Wes. 4 (1933) 39–44.
18. Kraußold, H.: Der konvektive Wärmeübergang. Die Technik 3 (1948) 205–213 and 257–261. (Most important equations.)
19. Krischer, O.: Wärmeaustausch in Ringspalten bei laminarer und turbulenter Strömung. Chem. Ing. Tech. 33 (1961) 1, 13–19.
20. Krischer, O.: Wärme- und Stoffaustausch bei umströmten oder durchströmten Körpern verschiedener Form. Chem. Ing. Tech. 33 (1961) 155–162. (Detailed description of the help of a characteristic length and other characteristic values.)
21. Kropholler, H. W.; Carr, A. D.: The Prediction of Heat and Mass Transfer Coefficients for Turbulent Flow in Pipes at all Values of the Prandtl or Schmidt Number. Int. J. Heat Mass Transfer 5 (1962) 1191–1205.

22. Kubair, V.; Kuloor, N. R.: Heat Transfer to Newtonian Fluids in Coiled Pipes in Laminar Flow. Int. J. Heat Mass Transfer 9 (1966) 63–75.
23. Kuiken, H. K.: Free Convection at Low Prandtl Numbers. J. Fluid Mech. 37 (1969) Pt. 4, 785–798.
24. Kuprianoff, J.: Neue Formen der Prandtlschen Gleichung für den Wärmeübergang. Z. tech. Phys. 16 (1935) 13.
25. Kutateladze, S. S.; Kirdyashkin, A. G.; Ivakin, V. P.: Turbulent Natural Convection on a Vertical Plate and in a Vertical Layer. Int. J. Heat Mass Transfer 15 (1972) I, 193–202.

L

1. Lawn, C. J.: Turbulent Heat Transfer at Low Reynolds Numbers. Trans. ASME, Ser. C91 (1969) 4, 532–536.
2. Lévêque, M. A.: Les lois de la transmission de la chaleur par convection. Ann. mines (12). 13 (1928) 201, 305 and 381.
3. Linke, W.: Hydraulische Durchmesser und Anlaufströmungen bei Wärmeaustauschern. Arch. ges. Wärmetech. 1 (1950) 161–169.
4. Linke, W.; Kunze, H.: Druckverlust und Wärmeübergang im Anlauf der turbulenten Rohrströmung. Allgemeine Wärmetechnik 4 (1953) 73–79.
5. Lloyd, J. R.; Sparrow, E. M.; Eckert, E. R. G.: Laminar Transition and Turbulent Natural Convection Adjacent to Inclined and Vertical Surfaces. Int. J. Heat Mass Transfer 15 (1972) 457–474.
6. Lorenz, H.: Wärmeabgabe und Widerstand von Kühlerelementen. Abhandl. Aerodyn. Inst. Aachen, Heft 13 (1933), 12; see also H. Lorenz, Beitrag zum Problem des Wärmeüberganges in turbulenter Strömung. Z. techn. Phys. 15 (1934) 155–162 and 201–206.
7. Lorenz, H. H.: Der Wärmeübergang von einer ebenen, senkrechten Platte an Öl bei natürlicher Konvektion. Z. tech. Phys. 15 (1934) 362–366.
8. Lyon, R. N.: Liquid Metal Heat-Transfer Coefficients. Chem. Eng. Progr. 47 (1951) 75–79.

M

1. Malina, I. A.; Sparrow, E. U.: Variable-Property, Constant Property and Entrance-Region Heat Transfer Results for Turbulent Flow of Water and Oil in a Circular Pipe. Chem. Engg. Sc. 19 (1964) 953–962.
2. Mattioli, G. D.: Theorie der Wärmeübertragung in glatten und rauhen Rohren. Forsch. Ing. Wes. 11 (1940) 149–158.
3. Meißner, W.; Schubert, G. U.: Kritische Reynoldssche Zahl und Entropieprinzip. Ann. Phys., 6. Following. 3 (1948), 163–182.
4. Métais, B.: Wärmeübergang bei strömenden Flüssigkeiten im waagerechten Rohr mit Eigenkonvektion. Chem. Ing. Tech. 32 (1960) 535–539.
5. Morgan, A. I. jr.; Carlson, R. A.: Wall Temperature and Heat Flux Measurement in a Round Tube. Trans. ASME, Heat Transfer, Ser. C 83 (1961) 2, 105–110.
6. Mori, Y.; Nakayama; W.: Study on Forced Convective Heat Transfer in Curved Pipes. Int. J. Heat Mass Transfer 10 (1967) 1, 37–59 and 5, 681–695.

N

1. Nesselmann, K.: Wärmeübergang und Druckverlust in konzentrischen und exzentrischen Ringspalten bei erzwungener Strömung und freier Konvektion. VDI-Z. 104 (1962) 18, 824.
2. Novotny, J. L.; McComas, S. T.; Sparrow, E. M.; Eckert, E. R. G.: Heat Transfer for Turbulent Flow in Rectangular Ducts with Two Heated and Two Unheated Walls. A. I. Ch. E. J. 10 (1964) 4, 466–470.
3. Nunner, W.: Wärmeübergang und Druckabfall in rauhen Rohren. VDI-Forschungsheft 455 (1956) 5–39.
4. Nußelt, W.: Der Wärmeübergang in Rohrleitungen. Forsch. Arb. Ing. Wes. No. 89, Berlin 1909, Habilitationsschrift, also Z. VDI. 63 (1909) 1750–1755 and 1808–1812.

5. Nußelt, W.: Abhängigkeit der Wärmeübergangszahl von der Rohrlänge. Z. VDI. 54 (1910) 1154–1158.
6. Nußelt, W.: Das Grundgesetz des Wärmeüberganges. Gesundh. Ing. 38 (1915) 477–482 and 490–496.
7. Nußelt, W.: Der Wärmeübergang im Rohr. Z. VDI. 61 (1917) 685–689.
8. Nußelt, W.: Die Wärmeübertragung an Wasser im Rohr. Publication honouring the hundredth anniversary of the Techn. Hochschule Karlsruhe 1925.
9. Nußelt, W.: Der Einfluß der Gastemperatur auf den Wärmeübergang im Rohr. Techn. Mech. und Thermodynamik 1 (1930) 227–290.
10. Nußelt, W.: Wärmeübergang, Diffusion und Verdunstung. Z. ang. Math. Mech. 10 (1930) 105–121.
11. Nußelt, W.: Der Wärmeaustausch zwischen Wand und Wasser im Rohr. Forsch. Ing. Wes. 2 (1931) 309–313.

O

1. Owen, P. R.; Thomson, W. R.: Heat Transfer across Rough Surfaces. J. Fluid Mechanics 15 (1963) Pt. 3, 321–334.

P

1. Parknin, W.; Jahn, M.; Reineke, H. H.: Forced Convection Heat Transfer in the Transition from Laminar to Turbulent Flow in Closely Spaced Circular Tube Bundles. Lecture to the 5th Int. Heat Transfer Conference in Tokyo 1974.
2. Peterka, J. A.; Richardson, P. D.: Natural Convection from a Horizontal Cylinder at Moderate Grashof Numbers. Int. J. Heat Mass Transfer 12 (1969) 749–752.
3. Petukhov, B. S.; Popov, V. N.: The Theoretical Calculation of Heat Flux and Frictional Resistance in Laminar Flow in Tubes of an Incompressible Liquid with Variable Physical Properties. High Temperature 1 (1963) 2, 205–212.
4. Petukhov, B. S.; Kirillov, V. V.; Chu Tzu-Hsiang; Maidamik, V. N.: An Experimental Investigation of the Effect of the Temperature Factor on the Heat Exchange Occurring during the Turbulent Flow of Gas in Pipes. Teplofizika Vysokikh Temperatur 3 (1965) 102–108, engl. translation. High Temperature 3 (1965) 1, 91–96.
5. Petukhov, B. S.: Heat Transfer and Friction in Turbulent Pipe Flow with Variable Properties. Advances in Heat Transfer. New York 1970.
6. Pitschmann, P.: Wärmeübergang von elektrisch beheizten horizontalen Drähten an Flüssigkeiten bei Sättigung und natürlicher Konvektion. Diss. TH München 1969.
7. Pohl, W.: Einfluß der Wandrauhigkeit auf den Wärmeübergang an Wasser. Forsch. Ing. Wes. 4 (1933) 230–237. With comments by L. Prandtl: Forsch. Ing. Wes. 5 (1934) 5.
8. Pohlhausen, E.: Der Wärmeaustausch zwischen festen Körpern und Flüssigkeiten mit kleiner Reibung und Wärmeleitung. Z. ang. Math. Mech. 1 (1921) 115–121.
9. Prandtl, L.: Eine Beziehung zwischen Wärmeaustausch und Strömungswiderstand der Flüssigkeiten. Phys. Zeitschrift, 11 (1910) 1072–1078.—See also Prandtl, L.: Führer durch die Strömungslehre, 3rd edition Braunschweig: Vieweg 1949, pp. 372f.
10. Prandtl, L.: Bemerkungen über den Wärmeübergang im Rohr. Phys. Zeitschrift 29 (1928) 487.
11. Presser, K. H.; Pietralla, G.; Harth, R.: Wärmeübergang und Druckverlust an innenbeheizten Ringspalten bei Hochdruckgaskühlung. Atomkern-Energie 12 (1967) 1/2, 43–54.
12. Presser, K. H.: Experimentelle Prüfung der Analogie zwischen konvektiver Wärme- und Stoffübertragung an einem Pfeilrippenrohr. Chem. Ing. Tech. 41 (1969) 21, 1176.

Q

1. Quack, H.: Natürliche Konvektion innerhalb eines horizontalen Zylinders bei kleinen Grashof-Zahlen. Wärme- und Stoffübertragung 3 (1970) 134–138.
2. Quarmby, A.; Anand, R. K.: Turbulent Heat Transfer in Concentric Annuli with Constant Wall Temperatures. Trans. ASME, Ser. C 92 (1970) 1, 33–45.

R

1. Rauber, A.: Wärmeübergang bei unterkühltem Verdampfen an im Ringspalt aufwärts strömendes Wasser für höhere Systemdrücke und Heizflächenbelastungen. Dissertation Kalrsruhe 196: VDI-Z. 107 (1965) 1228.
2. Reichardt, H.: Die Wärmeübertragung in turbulenten Reibungsschichten. Z. ang. Math. Mech. 20 (1940) 297–328 (discussed by Eckert, E. in Z. VDI. 85 (1941) 581–583).
3. Reichardt, H.: Der Einfluß der wandnahen Strömung auf den turbulenten Wärmeübergang. Mitt. Max-Planck-Inst. Strömungsforschung No. 3, Göttingen 1950.—Die Grundlagen des turbulenten Wärmeüberganges. Arch. ges. Wärmetech. 2 (1951) 129–142.
4. Renz, U.; Vollmert, H.: Der Wärme- und Stoffaustausch im Übergangsgebiet laminar/turbulent. Int. J. Heat Mass Transfer 18 (1975) 1009–1014. (Calculation and measurement are in close agreement.)
5. Reynolds, O.: An Experimental Investigation of the Circumstances which Determine whether the Motion of Water Shall be Direct or Sinuous, and the Law of Resistance in Parallel Channels. Phil. Trans. Roy. Soc. London 1883 or Scient. Papers II, p. 51. See also Proc. Manchester Lit. a. Phil. Soc. 8 (1874) 9.
6. Rieque, R.; Siboul, R.: Etude expérimentale de l'échange thermique à flux élevé avec l'eau en convection forcée à grande vitesse dans des tubes de petit diamètre avec et sans ébullition. Int. J. Heat Mass Transfer 15 (1972) I, 579–592.
7. Rieger, M.: Experimentelle Untersuchung des Wärmeübergangs in parallel durchströmten Rohrbündeln bei konstanter Wärmestromdichte im Bereich mittlerer Prandtl-Zahlen. Int. J. Heat Mass Transfer 12 (1969) 11, 1421–1447.
8. Rietschel, H.: Untersuchungen über Wärmeabgabe, Druckhöhenverlust und Oberflächentemperatur bei Heizkörpern unter Anwendung großer Luftgeschwindigkeiten. Mitt. Prüf.-Anst. f. Heizungs- u. Lüftungseinr. d. TH Berlin, 19, No. 3.
9. Rogers, G. F. C.; Mayhew, Y. R.: Heat Transfer and Pressure Loss in Helically Coiled Tubes with Turbulent Flow. Int. J. Heat Mass Transfer 7 (1964) 1207–1216.

S

1. Sauer, H. J. jr.; Burford, L. W.: Heat Transfer Coefficients and Friction Factors for Longitudinally Grooved Tubes. Trans. ASME, Ser. C 91 (1969) 3, 455–457.
2. Schack, A.: Der Wärmeübergang in Rohren und an Rohrbündeln. Arch. Eisenhüttenwes. 13 (1939/40) 4, 155–169.
3. Schefels, G.: Reibungsverluste in gemauerten engen Kanälen und ihre Bedeutung für die Zusammenhänge zwischen Wärmeübergang und Druckverlust in Winderhitzern. Arch. Eisenhüttenwes. 6 (1932/33) 477–486.
4. Schlünder, E. U.: Über eine zusammenfassende Darstellung der Grundgesetze des konvektiven Wärmeübergangs. Verfahrenstechnik 4 (1970) 11–16.
5. Schmidt, B. H.: Luftdurchlässigkeit von Wasserröhrchenkühlern im Kreuzstrom. Z. VDI. 76 (1932) 273–276.
6. Schmidt, E.; Beckmann, W.: Das Temperatur- und Geschwindigkeitsfeld vor einer wärmeabgebenden senkrechten Platte bei natürlicher Konvektion. Tech. Mech. Thermodynamik 1 (1930) 341–349 and 391–406.
7. Schmidt, E.; Wenner, K.: Wärmeabgabe über den Umfang eines angeblasenen geheizten Zylinders. Forschung 12 (1941) 65–73.
8. Schmidt, E. F.: Wärmeübergang und nicht-isothermer Druckverlust bei erzwungener Strömung in schraubenförmig gekrümmten Rohren. Diss. Braunschweig 1965.
9. Schumacher, K.: Großversuche an einer zu Studienzwecken gebauten Regenerativkammer. Arch. Eisenhüttenw. 4 (1930/31) 63–74.
10. Schwier, K.: Der Wärmeübergang im horizontalen Rohr bei laminarer Strömung und seine Beeinflussung durch freie Konvektion und durch die Temperaturabhängigkeit der Stoffwerte. Fortschr.-Ber. VDI-Z. R. 6, No. 6, 77 S.

11. Seban, R. A.; McLaughlin, E. F.: Heat Transfer in Tube Coils with Laminar and Turbulent Flow. Int. J. Heat Mass Transfer 7 (1963) 387–395.
12. Senftleben, H.: Wärmeabgabe von Körpern verschiedener Form in Flüssigkeiten und Gasen bei freier Strömung. Allgem. Wärmetech. 10 (1961) 10, 192–199.
13. Sheriff, N.; Gumley, P.; France, J.: Heat Transfer Characteristics of Roughened Surfaces. Chem. Process Engng. 45 (1964) 624–629.
14. Sheriff, N.; Gumley, P.: Heat Transfer and Friction Properties of Surfaces with Discrete Roughnesses. Int. J. Heat Mass Transfer 9 (1966) 1297–1320.
15. Sherwood, T. K.; Petrie, J. M.: Heat Transmission to Liquids Flowing in Pipes. Ind. Eng. Chem. 24 (1932) 736–745.
16. Sherwood, T. K.; Kiley, D. D.; Mangsen, G. E.: Heat Transmission to Oil Flowing in Pipes. Ind. Eng. Chem. 24 (1932) 273–277.
17. Sieder, E. N.; Tate, G. E.: Heat Transfer and Pressure Drop of Liquids in Tubes. Ind. Eng. Chem. 28 (1936) 1429–1435. (See the extract of W. Wachendorf in Verfahrenstechnik 1937, pp. 133 and 134.)
18. Smithberg, E.; Landis, F.: Friction and Forced Convection Heat Transfer Characteristics in Tubes with Twisted Taps Swirl Generators. Trans. ASME, Ser. C 86 (1964) 1, 39–49.
19. Soenecken, A.: Der Wärmeübergang von Rohrwänden an strömendes Wasser. VDI-Forschungsheft 109 (1911).
20. Sparrow, E. M.; Chen, T. S.: Mutual Dependent Heat and Mass Transfer in Laminar Duct Flow. Am. Inst. Chem. Engrs. J. 15 (1963) 3, 434–441.
21. Sparrow, E. M.; Lloyd, J. R.; Hixon, C. W.: Experiments on Turbulent Heat Transfer in an Asymmetrically Heated Rectangular Duct. Trans. ASME, Ser. C 88 (1966) 2, 170–174.
22. Srinivasan, P. S.; Nandapurkar, S. S.; Holland, F. A.: Pressure Drop and Heat Transfer in Coils. Trans. Instn. chem. Engrs. (1968) 218, CE 113–CE 119.
23. Staniszewski, B.: Übersicht der Arbeiten aus dem Gebiet der Wärmeübertragung bei freier Konvektion. Luft- und Kältetechn. 3 (1967) 209–213.
24. Stanton, T. E.: On the Passage of Heat between Metal Surfaces and Liquids in Contact with them. Phil. Trans. Roy. Soc. Lond. (A), 190 (1897) 67. See also: Stanton, T. E.: Friction. London: Longmans 1923.
25. Steimle, F.: Zusammenhang zwischen Wärmeübergang und Druckabfall turbulenter Strömungen. Kältetech.-Klimatisierung 23 (1971) 126–128; more fully in: Abhandlungen des Deutschen Kältetechnischen Vereins No. 20, Karlsruhe 1970.
26. Stender, W.: Der Wärmeübergang an strömendes Wasser in vertikalen Rohren. Berlin 1924.
27. Stephan, K.: Wärmeübergang und Druckabfall bei nicht ausgebildeter Laminarströmung in Rohren und in ebenen Spalten. Chem. Ing. Tech. 31 (1959) 773–778.
28. Stephan, K.: Wärmeübertragung laminar strömender Stoffe in einseitig beheizten oder gekühlten Kanälen. Chem. Ing. Tech. 32 (1960) 6, 401–404.
29. Stephan, K.: Gleichungen für den Wärmeübergang laminar strömender Stoffe in ringförmigen Querschnitten. Chem. Ing. Tech. 33 (1961) 338–343.
30. Stephan, K.: Wärmeübergang bei turbulenter und bei laminarer Strömung in Ringspalten. Chem. Ing. Tech. 34 (1962) 207–212.
31. Stone, J. P.; Ewing, C. T.; Miller, P. R.: Heat Transfer Studies on some Stable Organic Fluids in a Forced Convection Loop. J. Chem. Engng. Data 7 (1962) 4, 519–525.
32. Subbotin, V. I.; Ušakov, P. A.; Gabrianovič, B. N.; Talanov, V. D.; Sviridenko, I. P.: Wärmeaustausch bei durch runde Rohre strömenden flüssigen Metallen. J. Ing.-Phys. (russ.) (1963) 4, 16–21.
33. Sugawara, S.; Michiyoshi, I.: Heat Transfer by Natural Convection in Laminar Boundary Layer on Vertical Flat Wall. Mem. Fac. Eng. Kyoto Univ. 13 (1951) 149–161.

T

1. Touloukian, Y. S.; Hawkins, G. A.; Jakob, M.: Heat Transfer by Free Convection from Heated Vertical Surfaces to Liquids. Trans. Am. Soc. mech. Engrs. 70 (1948) 13–18.
2. Tritton, D. J.: Transition to Turbulence in Free Convection Boundary Layers on an Inclined Heated Plate. J. Fluid Mechanics 16 (1963) Pt. 3, 417–435.

3. Tseng, C. M.; Besant, R. W.: Transient Heat and Mass Transfer in Fully Developed Laminar Tube Flows. Int. J. Heat Mass Transfer 15 (1972) I, 203–216.

U

1. Ulrichson, D. L.; Schmitz, R. A.: Laminar Flow Heat Transfer in the Entrance Region of Circular Tubes. Int. J. Heat Mass Transfer 8 (1965) 2, 253–258.
2. Upmalis, A.: Wärmeübergang und Wärmedurchgang bei Glasrohren. Wärme 74 (1968) 3, 76–79.

V

1. Vliet, G. C.: Natural Convection Local Heat Transfer on Constant-Heat-Flux Inclined Surfaces. Trans. Am. Soc. Mech. Engrs. Ser. C 91 (1969) 511–516.

W

1. Walker, R. A.; Bott, T. R.: Effect of Roughness on Heat Transfer in Exchanger Tubes. Chem. Engr. (1973) 271, 151–156.
2. Warner, Ch. Y.; Arpach, V. S.: An Experimental Investigation of Turbulent Natural Convection in Air at Low Pressure along a Vertical Heated Flat Plate. Int. J. Heat Mass Transfer 11 (1968) 397–406.
3. Watzinger, A.; Johnson, D. G.: Wärmeübertragung von Wasser an Rohrwand bei senkrechter Strömung im Übergangsgebiet zwischen laminarer und turbulenter Strömung. Forschg. Ing.-Wes. 9 (1938) 182–196; 10 (1939) 182–196.
4. Weder, E.: Messung des gleichzeitigen Wärme- und Stoffübergangs am horizontalen Zylinder bei freier Konvektion. Wärme- und Stoffübertragung 1 (1968) 10–14.
5. Weinbaum, S.: Natural Convection in a Horizontal Circular Cylinder. J. Fluid Mechanics 18 (1964) Pt. 3, 409–437.
6. Wilcox, W. R.: Simultaneous Heat and Mass Transfer in Free Convection. Chem. Eng. Sci. 13 (1961) 3, 113–119.
7. Williamson jr., K. D.; Bartlit, J. R.; Thurston, R. S.: Studies of Forced Convection Heat Transfer to Cryogenic Fluids. Chem. Eng. Progr. Sympos. Ser. 64 (1968) 87, 103–110.
8. Wilson, N. W.; Medwell, J. O.: An Analysis of Heat Transfer for Fully Developed Turbulent Flow in Concentric Annuli. Trans. ASME, Ser. C 90 (1968) 1, 43–50.
9. Winkler, K.: Wärmeübergang in Rohren bei hohen Reynoldsschen Zahlen. Forsch.-Ing.-Wes. 6 (1935) 261–268.
10. Wohl, M. H.: Heat Transfer in Laminar Flow. Chem. Eng. 75 (1968) 14, 81–86.
11. Woschni, G.: Untersuchung des Wärmeüberganges und des Druckverlustes in gekrümmten Rohren. Diss. Dresden 1959.

2-11 HEAT TRANSFER BY CROSS FLOW

Heat exchangers which are operated in cross flow usually contain tube assemblies; the gas flowing on the outside impinges perpendicularly, or nearly perpendicularly on these tubes. With constant fluid velocity, this type of flow gives rise to an increase in turbulence compared with the case where the gas flows along the tubes and parallel to them. It can thus be deduced that the heat transfer coefficient at the outer surface of a tube is higher for cross flow than for parallel flow.

Heat Transfer from Gases in Cross Flow

The superiority of heat transfer by cross flow rather than parallel flow was first established by Rietschel [R44], Thoma [T43] and Reiher [R42, 43] from measure-

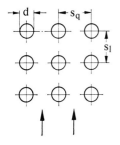

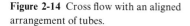

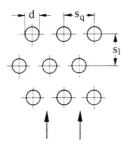

Figure 2-14 Cross flow with an aligned arrangement of tubes.

Figure 2-15 Cross flow with a staggered arrangement of tubes.

ments taken in air and other gases. From their results, which were obtained, for the most part, from tube assemblies operating with cross flow, it can be deduced that in processes which are otherwise similar, staggered arrangements of tubes will facilitate more heat transfer than will "in-line banks" of tubes; (see Figs. 2-14 and 2-15). This advantage, however, will be offset by an increased drop in pressure; this will be shown in Sec. 4-4.

Later, most importantly, Pierson [P41] and Huge [H50] carried out extensive tests on the heat transfer of air in cross flow. As well as varying the fluid velocity, they also varied the inter-tube distance over a wide range and observed their net influence on heat transfer. The results of their tests have been reproduced comprehensively by Grimison [G47] using an equation of the type

$$Nu = KRe^m \tag{2-51}$$

Table 2-3 Constants for equations for heat transfer in cross flow, from measurements by Pierson and Huge; assembled by Grimison. a and b **are spacing ratios at right angles to and parallel to the direction of flow**

$a =$	1.25		1.5		2		3	
	K	m	K	m	K	m	K	m
b	in-line arrangement of tubes							
1.25	0.348	0.592	0.275	0.608	0.100	0.704	0.0633	0.752
1.5	0.367	0.586	0.250	0.620	0.101	0.702	0.0678	0.744
2	0.418	0.570	0.299	0.602	0.229	0.632	0.198	0.648
3	0.290	0.601	0.357	0.584	0.374	0.581	0.286	0.608
b	staggered arrangement of tubes							
0.6							0.213	0.636
0.9					0.446	0.571	0.401	0.581
1.0			0.497	0.558				
1.25	0.518	0.556	0.505	0.554	0.519	0.556	0.522	0.562
1.5	0.451	0.568	0.460	0.562	0.452	0.568	0.488	0.568
2	0.404	0.572	0.416	0.568	0.482	0.556	0.449	0.570
3	0.310	0.592	0.356	0.580	0.440	0.562	0.421	0.574

Table 2-4 Ratio of the heat transfer at tube assemblies with a small number of tube rows to heat transfer with 10 rows of tubes (average)

| Number of tube rows | In-line arrangement of tubes | | | Staggered arrangement of tubes | | |
| | according to Reiher | according to Scholz | | according to Reiher | according to Scholz | |
		$Re < 10^5$	$Re > 10^5$		$Re < 10^5$	$Re > 10^5$
2	0.91	1.05	0.87	0.68	0.80	0.69
5	0.97	1.0	1.0	0.90	1.05	1.06

where the constants K and m are dependent on the inter-tube distances in a manner which will be discussed later. Here the outer diameter of the tube is inserted into both the Nusselt number and the Reynolds number as the diameter d; the highest average velocity in the main direction of flow is introduced into Re as w; it follows that the narrowest free flow inter-tube distance in a row of tubes, measured perpendicularly to the direction of flow, is a determining factor for this. Velocity, like density, is based on the average temperature of the gas flowing through the tubes; thermal conductivity and viscosity, on the other hand, are based on the average value of the temperature of the tube wall and the mean gas temperature. The influence of the inter-tube distances is expressed, according to Grimison, by the "spacing ratios" a and b, which are determined by the ratio of the distance s between the centres of two neighbouring tubes, to the outer diameter of the tube d. $a = s_q/d$ is the spacing ratio in a row of tubes at right angles to the main flow, $b = s_l/d$ is the spacing ratio in the direction of flow (see Figs. 2-14 and 2-15). Table 2-3 gives the values of constants K and m for in-line and staggered arrangements of tubes and for various spacing ratios a and b. These values apply to tube assemblies of 10 or more tubes lying behind one another in the direction of flow.

Table 2-4, derived from observations made by Reiher [R42] and Scholz [S54] gives an indication of what can be expected by way of alterations to the heat transfer coefficients when the number of rows of tubes is reduced but the process remains otherwise unaltered.

Grimison correlated the test measurements of Pierson and Huge by a second method and derived the expression (2-52).

$$Nu = 0.32 f_a \cdot Re^{0.61} \cdot Pr^{0.31} \tag{2-52}$$

The layout factor f_a, which is determined mainly by the arrangement of the tubes, was set out graphically by Grimison for the various values of Re dependent on a and b. The curves thus obtained, between which interpolation is very difficult, can be represented by Eq. (2-52) in slightly modified form [H45]; for an in-line arrangement:

$$Nu = 0.34 \cdot f_a \cdot Re^{0.60} Pr^{0.31} \tag{2-53}$$

can be used with

$$f_a = 1 + \left(a + \frac{7.17}{a} - 6.52\right)\left(\frac{0.266}{(b-0.8)^2} - 0.12\right)\sqrt{\frac{1000}{Re}} \qquad (2\text{-}54)$$

for a staggered arrangement:

$$Nu = 0.35f_a \cdot Re^{0.57} Pr^{0.31} \qquad (2\text{-}55)$$

can be employed with

$$f_a = 1 + 0.1a + 0.34/b \qquad (2\text{-}56)$$

The expressions for f_a in Eqs. (2-54) and (2-56) presuppose that the values for a and b lie somewhere between 1.25 and 3. With an in-line arrangement it should be noted that with $a = 1.25$ the same values for f_a are obtained from Eq. (2-54) as are calculated with $a = 1.5$. In place of Eqs. (2-53) and (2-54) it would be simpler, though less accurate, to write the following expression for the in-line arrangement:

$$Nu = \left[0.34Re^{0.60} + \left(a + \frac{7.17}{a} - 6.52\right)\left(\frac{5.75}{b-1.12} - 4.77\right)\right]Pr^{0.31} \qquad (2\text{-}57)$$

This equation deviates by a maximum of ± 3 per cent from Eqs. (2-53) and (2-54) and can be considered to be accurate enough for all practical purposes.

At about the same time as Pierson and Huge were doing this work, Glaser [G42] measured, in regenerators, heat transfer in cross flow in 7 mm thick iron rods which were helically coiled in several layers which were aligned above one another. He obtained results which were very consistent with those of Pierson and Huge. This is illustrated by Fig. 2-16 which also shows that in the case being considered the heat transfer coefficient is two to three times larger with cross flow

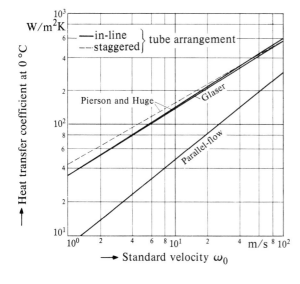

Figure 2-16 Heat transfer in tubes with cross flow and parallel flow.

than with parallel flow. See Sec. 18-2, page 463 for the procedure for measuring heat transfer in regenerators.

Experimental values by Bressler [B52] are up to 10 per cent lower; those by Hammeke, Heinecke and Scholz [H42] are correspondingly higher than the values obtained by Pierson and Huge and thus, also by Grimison. These differences can probably be attributed to the various levels of turbulence in the air flowing round the pipes, to the unequal roughness of the tube surfaces and to small geometrical variations. Comparison with experimental values of Benke [B43] shows further that Eqs. (2-53), (2-54) and (2-57) for in-line arrangements are valid for the region determined by Grimison, that is up to $a = 5$ and $b = 22$. They can also be used for staggered arrangements with spacing ratios above $a = 3$ and $b = 3$ as in this region, the difference in heat transfer between in-line and staggered arrangements disappears. Confined to the region of up to $a = 3$ and $b = 3$, Eqs. (2-53) and (2-54) can be extrapolated up to $Re = 200,000$ without any substantial increase in error.

According to the experimental results obtained by Hammeke, Heinecke and Scholz [H42] which have already been mentioned, the heat transfer coefficient rises more quickly above $Re = 200,000$, with increasing Reynolds number, than below it. With an in-line arrangement Nu is proportional to $Re^{0.82}$ and in a staggered arrangement is proportional to $Re^{0.96}$ for $Re > 200,000$, and finally, with a "crossed in-line" arrangement, where tubes are set in a row perpendicular to the preceding row, it is proportional to $Re^{0.85}$. Brauer [B51] deduced, both from his own and from other people's measurements, that the most advantageous ratio of heat transfer to pressure loss is to be found in this cross-latticed tube arrangement.

Furthermore Groehn and Scholz [G48] have measured heat transfer in a cross-counterflow heat exchanger constructed of coils of tubes (see Sec. 8-3) and in which, these tubes were alternately wound to the right and to the left in successive layers as is customary. They established in accord with Glaser's results (Fig. 2-16) that the values for heat transfer and pressure drop observed in assemblies of straight tubes, could be transferred unchanged to tube coils of this type.

A further paper by Groehn and Scholz [G49] showed that heat transfer could be increased considerably, up to a maximum of 50 per cent, by means of "trip fins", 0.5 to 1 mm high, arranged axially and attached to the outside of the pipe which is at right angles to the direction of flow.

Among the numerous other studies concerned with heat transfer in tube assemblies used in cross flow, mention must be made, above all, of the work of Hirschberg [H48]. He has developed a remarkable method of calculation from his own experimental results.

New research work into heat transfer and pressure drop in tubes and tube assemblies in cross flow has been carried out by Niggeschmidt [N42]. The heat transfer values measured by him are presented in the following empirical equations. In these he has not, as is usual, inserted the velocity of flow in the narrowest cross-section between two neighbouring pipes into Re but has introduced the velocity of flow from the initial direction w_0. He denotes the Reynolds number formed with w_0 by Re_0.

His measurements in a single tube agree with the equation given by F. Brandt

[B47]

$$Nu = (1.506 + 0.158Re_0{}^{0.42})^2 \text{ (single tube in cross flow)} \qquad (2\text{-}58)$$

According to Niggeschmidt the same equation can be used for rows of tubes if the expression on the right hand side is multiplied by an "arrangement factor" which is dependent only on the transverse spacing ratio a. Equation (2-59) is thus obtained for single tubes.

$$Nu = (1.506 + 0.158Re_0{}^{0.4})^2 \cdot \frac{a - \pi/8}{a - \pi/4} \text{ (single row of tubes)} \qquad (2\text{-}59)$$

On the other hand Niggeschmidt developed the following new equations for tube assemblies:

for banks of tubes in an in-line arrangement

$$Nu = (1.517 + 0.205Re_0{}^{0.38})^2 \cdot \frac{a}{a - \pi/4} \qquad (2\text{-}60)$$

for assemblies of tubes with a staggered arrangement

$$Nu = (1.878 + 0.256Re_0{}^{0.36})^2 \cdot \frac{a}{a - \pi/4} \qquad (2\text{-}61)$$

Niggeschmidt's experimental results are in close agreement with those of Bressler which have already been mentioned. Moreover Niggeschmidt [N42] has taken measurements for partly staggered tube assemblies. It is worthy of note that in these new measurements, which have just been discussed, more notice has not been taken of the strong dependence of the spacing ratios a and b, as established by Grimison [Eqs. (2-54) and (2-56)]. Rather the Eqs. (2-59), (2-60) and (2-61) are terminated by factors which embody slight modifications to a.

Heat Transfer in Liquids in Cross Flow

Ulsamer [U41] studied the heat transfer in liquids in cross flow for a single tube. The results of his tests with air, water, paraffin and transformer oil have been combined in Eq. (2-62)

$$Nu = 0.6 \cdot Re^{0.5} \cdot Pr^{0.31} \qquad (2\text{-}62)$$

Heat transfer in a tube belonging to a bank of tubes across which a liquid passes in cross flow, can be calculated, thanks to the similarity principle, using the empirical Eqs. (2-53) to (2-57) or even (2-58) to (2-61), as long as the dependence on Pr is allowed for by means of a suitable exponent of Pr.

2-12 HEAT TRANSFER IN FINNED TUBES

The increase in the heat transfer coefficient, which is caused by cross flow at the outer side of the tubes is, above all, of importance if with parallel flow there occurs

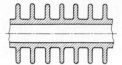

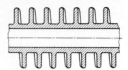

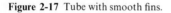

Figure 2-17 Tube with smooth fins.   **Figure 2-18** Tube with helically wound fins.

a high heat transfer coefficient on the inside but a low heat transfer coefficient on the outside. Such differences in the heat transfer coefficient can be due to the conditions of flow, differences in pressure and temperature and, most important, to the physical properties of the fluids. For example, with parallel flow the heat transfer coefficient internally can be two to three times greater if a gas at high pressure is flowing inside the tube and a gas at low pressure is flowing on the outside of the tube. The difference between inner and outer heat transfer coefficients can be almost eliminated by a cross flow arrangement, or even better a cross counterflow arrangement (Sec. 8-3).

The difference in the heat transfer coefficient is usually considerably larger if a liquid flows inside the tube and a gas flows on the outside. In these circumstances cross flow or cross counterflow alone are not sufficient to equalize the two heat transfer coefficients. A further improvement can be obtained if the operation of cross flow is augmented by fins, that is by an artificial increase in the outer tube surface.

The fins usually consist of smooth, circular sheet metal discs which are fixed to the tubes equally spaced apart (Fig. 2-17). The fins may be cast onto the tubes which are made of cast iron or cast steel. The fins are frequently produced from sheet metal strips which can be wound round the tube in the form of a helicoidal surface, so that they are always at right angles to the tube wall (Fig. 2-18).[16]

Occasionally, fins are arranged on the inside of tubes, in an axial direction. In addition fins are to be found on flat surfaces, particularly on the exterior of vertical walls of air cooled containers or domestic central heating radiators. However in the following paragraphs only tubes with externally arranged fins will be considered.

The heat which a fin gives up to the surrounding gas, or takes from it, must first flow from the tube wall through the fin root and then to the interior of the fin, or vice versa. This produces a radial temperature drop in the fin. This has the effect that on average, there is a smaller temperature difference between the surrounding gas and the surface of the fin than between the gas and the fin root and the tube surface. Thus the increase in outer surface effected through the use of fins is not quite matched by the increase in the heat transfer on the outside of the tube.

The ratio of the temperature difference available, on average, at the fin surface, to the temperature difference at the fin root, will be called the "fin efficiency". This fin efficiency η_R can be calculated theoretically if it is assumed that the gas temperature is constant and the heat transfer coefficient α_R has the same value

---

[16] Tubes with such fins are often falsely called "spiral finned tubes"; they should be called "helical finned tubes" or "wound finned tubes".

uniformly across the fin surface. Such calculations will be discussed in more detail in Part Three of this book in Sec. 8-4. The diagrams presented in this section enable the fin efficiency to be read off for fins with a right-angular or triangular cross-section, with various fin dimensions and with various values of the heat transfer coefficient.

If it is further assumed that the heat transfer coefficient is α_R at the surface F_{Tube} of the tube between the fins, then the quantity of heat which is transferred from the finned tube to the surrounding gas in unit time can be calculated from Eq. (2-63), where $\Theta_0 - \theta_a$ is the temperature difference between the outer surface of the tube wall and the gas. Θ_0 is also the surface temperature at the fin root.

$$\dot{Q} = \alpha_R(F_{Tube} + \eta_R F_{Fin})(\Theta_0 - \theta_a) \tag{2-63}$$

It will be shown subsequently that the heat transfer coefficient can be determined quite simply from experimental results. First it must be established that the heat transfer coefficient at the fin surface, is in no way constant even with the turbulent flow conditions that usually prevail. It increases, rather, from the value at the fin root up to twice or three times the amount at the fin edge. This is mainly explained by the fact that the boundary layer is thicker at the fin root than it is at the fin edge. An example of how the heat transfer coefficient is distributed over the fin surface is shown in Fig. 2-19 obtained from measurements taken by Krückels and Kottke [K52].

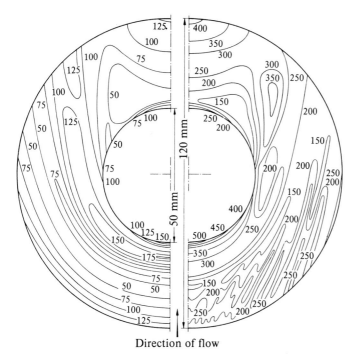

Direction of flow

Figure 2-19a and b Distribution of the heat transfer coefficient α at a round, single fin. Velocity of flow: left 2 m/s, right 10 m/s, α in Kcal/(m²hK); values in W/(m²K) about 16.3 per cent higher.

In spite of these differences, for practical calculations it is desirable to know an average value α_R which can then be directly inserted into Eq. (2-63). In order to determine such an average value Th. E. Schmidt [S52] developed the empirical Eq. (2-64) from experimental results obtained from tube assemblies by eight different workers.

$$Nu_d = C \cdot (Re_d)^{0.625}(F/F_0)^{-0.375} \cdot Pr^{1/3} \tag{2-64}$$

where C takes the value

> 0.30 with an in-line arrangement
> 0.45 with a staggered arrangement.

The suffix d indicates that the characteristic length of the outer tube diameter d is to be inserted into Nu and Re. The velocity of flow in the narrowest cross-section of a row of tubes is standard for Re. F is the complete outer surface of the finned tube, F_0 is the surface of a finless tube of the same length with the same outer diameter d. On average Eq. (2-64) successfully replicates the experimental results and has been confirmed as adequate upon application.[17]

2-13 HEAT TRANSFER IN PACKED BEDS

Packed beds are of technical significance in the form of solid beds and fluidized beds. In fluidized beds, chemical processes and drying processes take place at great speed thanks to the rapid movement of small solid particles. Although as a result, unusually favourable rates of heat transfer are to be found in fluidized beds, they do not play a significant part in processes involving heat transfer alone. On the other hand, packed beds are employed for heat transfer in numerous industrial processes, particularly as the packings and heat storing masses in regenerators. They are used in this way, for example to preheat the air which is to be blown into blast furnaces, Siemens-Martin Furnaces, and glass furnaces and they are used also for heat transfer in gas turbine installations and for cooling air and other gases in low temperature technology. In the blast furnace itself heat is transferred from the blast flowing upwards to the granular material which consists of ore, coke and fluxing materials and which is moving slowly downwards.

The mode of operation of regenerators and their calculation will be dealt with in detail in Part Four of this book. In the following paragraphs will be discussed the construction of the heat storing mass used in regenerators and the heat transfer coefficients which determine the rate of the heat transfer with the gas which flows through the heat-storing mass.

The Design of Regenerator Heat-Storing Masses

The heat-storing mass of a regenerator can be designed in very many different ways. Typical cross-sectional shapes are shown in Figs. 2-20 and 2-21. The most simple

---

[17] From a personal communication from Professor Dr. K. Stephan, Stuttgart.

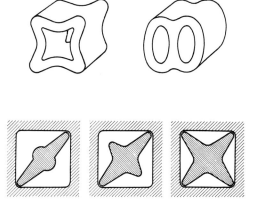

Figure 2-20 Packing bricks.

arrangement is formed by smooth walled long channels. The free flow cross-sectional area of such channels is often reduced by inserted packing bricks such as breakfast roll shaped bricks, spiral bricks and rod shaped bricks; in this way the velocity of flow is increased as also are the heating surface area and the thermal capacity of the heat-storing mass (see Fig. 2-20).

By gradually altering the size or shape of these packing bricks, a uniform channel cross-sectional area can be maintained while at the same time, the free flow area is increased in the direction of rising temperature. In so doing, advantage can be taken of the influence of radiative heat transfer which increases rapidly with temperature. Checkers can also be constructed (Fig. 2-21, top) from single, square bricks; these checkers can be arranged in-line ("free movement") or in staggered formation. The design of checkers can be varied in many ways through the use of curve shaped bricks (for example see Fig. 2-21, centre). Here again the choice of various thicknesses of brick make it possible to enlarge the free flow cross-section in the upwards direction.

Many packing materials can be used as heat-storing masses when working with low temperatures or temperatures which are not too high; these packing materials

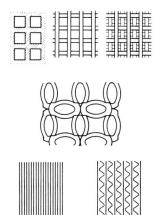

Figure 2-21 Geometries of the cross-sections of the heat-storing masses of regenerators.

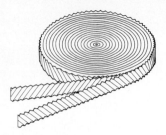

Figure 2-22 Heat-storing mass for low temperatures constructed from corrugated sheet metal according to Fränckl.

can be balls, Raschig rings, pebbles or other granular material. A very fine sectioning and thus particularly favourable heat transfer conditions with low pressure drops are obtained with metal heat-storing masses which consist of thin sheet metal (see Fig. 2-21, bottom). Such sheets can be arranged parallel to each other and only a slight distance apart. The sheets, or at least some of them, are usually corrugated in order to increase both rigidity and heat transfer as for example, in the Ljungström combustion air pre-heaters (Fig. 11-3) or in low temperature technology regenerators (see Figs. 11-1 and 2-22). Figure 2-22 shows the type of heat-storing mass suggested by M. Fränkl [F41] which has been used successfully for many years in low temperature technology. These heat-storing masses are constructed of sheet metal strips which are made of aluminium or steel coated with aluminium, and which are diagonally corrugated and are arranged in such a way that the corrugations on the alternate metal strips cross. About 200 checkers, 2 cm high, constructed from such metal strips, form the packing for a regenerator. Very finely subdivided heat storing masses have recently been produced from very strong glass ceramic materials as displayed in Fig. 2-23;[18] the figure is shown as an example of the heat-storing mass of a rotary regenerator for a gas turbine installation. Although the width of the small ducts amounts to 1 mm and less, manufacturers claim that such heat-storing masses can withstand temperatures of up to 1100°C. Regenerators for the Philips gas refrigerator, which was developed for the production of very low temperatures, have extremely small dimensions and their heat-storing masses consist of thin copper wires coiled together.

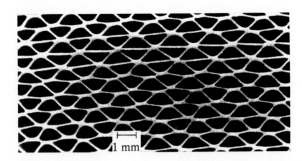

1 mm

Figure 2-23 Ceramic heat-storing mass for high temperatures.

[18] Made available by the Motor & Turbine Union GmbH of Munich.

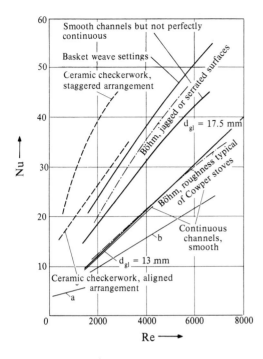

Figure 2-24 Heat transfer in various types of regenerator packing from measurements taken by Yazicizade compared with some of Böhm's (-·-··-) and Langhan's (- - - -) values. a, b heat transfer in tubes with laminar and turbulent flow.

Heat Transfer Measurements in Packed Beds

The heat transfer between the various types of heat storing masses and the gases flowing through them has been measured many times. If the heat storing mass contains only smooth channels with a uniform cross-section the heat transfer can be calculated according to the equations given in Sec. 2-9, for tubes and ducts. This also applies when the channel cross-section is made narrower by regular packing bricks (see Fig. 2-20) as long as the cross-section does not vary, or varies only gradually, along the length of the channel. Böhm [B10] has obtained measurements from walled, rectangular channels with surfaces of varying roughness; Yazicizade [Y41] has taken measurements from model channels. Moreover Yazicizade has examined a discontinuous channel arrangement as well as a basket work arrangement which differ from simple channels in that the walls usually have openings distributed along them. Yazicizade's results are represented by the solid lines in Fig. 2-24. By way of comparison, Böhm's experimental values are given by dash-and-dot lines; the dashed lines show a few values obtained by Langhans [L42] which are taken from Fig. 2-25 which will be discussed shortly.

Yazicizade does not use the true heat transfer coefficient α in the values of Nu but employs instead the heat transfer coefficient $\bar{\alpha}$ which is 10 per cent to 20 per cent smaller than α. The form of $\bar{\alpha}$, which relates to the average temperature of heat storing mass, is to be found in Eq. (13-3) which appears in Part Four of this book. However the values of Nu presented by other authors use the true heat transfer coefficient α. The equivalent diameter, determined by Eq. (2-65) as a generalization

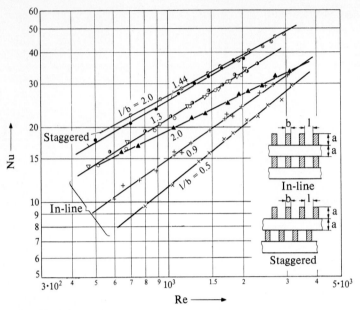

Figure 2-25 Heat transfer in heat-storing masses in the form of checkerwork according to Langhan's measurements.

of Eq. (2-14), is used as diameter in Nu and Re; it is also employed in the experimental measurements shown in the diagram that follows.

$$d_{gl} = 4V/F \tag{2-65}$$

Here, in a given volume of the packing, V is the free volume between the heat-storing elements and F is the surface of the heat-storing mass which is washed over by the gas.

Kistner [K44] measured the heat transfer in heat storing masses constructed in the form of checkerwork. Later Langhans [L42] carried out measurements in this type of heat storing mass in small model experimental regenerators. His results, which are in close agreement with those of Kistner, are reproduced in Fig. 2-25. According to Langhans, the fluid velocity w refers to that cross-section through which one can see, if one considers two layers of bricks laid on top of each other in the direction of flow. In this case also, Nu contains the true heat transfer coefficient, α.

Furnas [F42] provided the first measurements in irregularly shaped, generally broken up packing material. A larger number of further experimental results of this type have been compiled by Weishaupt [W45], shown in Fig. 2-26, where, in addition, Yazicizade's test data relating to pebbles [Y41] are given as curves a and b. In this diagram the quantity of heat, which is transferred in one second, in 1 m^3 of filling at a temperature difference of 1 K is displayed, in the units of $J/(m^3 \, s \, K)$, as dependent on $\dot{m}/d$, where $\dot{m}$ is the mass flowrate of the gas flowing through 1 m^2 of the complete cross-sectional area of the filling [units of $\dot{m}$ are thus $kg/(m^2 \, s)$], and d

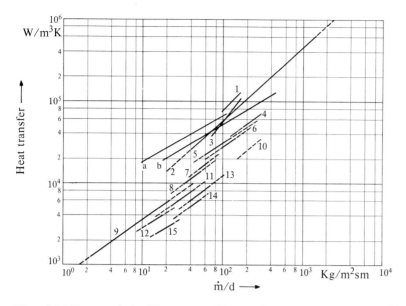

Figure 2-26 Heat transfer in different types of heat-storing mass as compiled by Weishaupt. 1. basalt 10 mm; 2. spherical filling 1.6 to 6.3 mm; 3. pebble grit 10 mm; 4. iron ore 8 mm; 5. iron ore 11 mm; 6. steel spheres 19 mm; 7. steel spheres 32 mm; 8. iron ore 38 mm; 9. pebbles 8–33 mm; 10. cinders 8 mm; 11. limestone 19 mm; 12. limestone 41 mm; 13. cinders 19 mm; 14. cinders 19 mm; 15. cinders 70 mm; (a) quartzite chippings 11.5 mm; (b) quartzite chippings 9.1 mm.

is the average particle diameter, measured in metres (m). Polthier [P42, 43] has measured the heat transfer in a heated sphere located in a packed bed consisting of non-heated spheres of a similar size.

Numerous other measurements have been taken in spherical fillings. The results of 244 tests carried out in fillings made of spheres and in other packing materials of various forms have been gathered together by Barker [B41]. Figures 2-27 and 2-28 are taken from his paper; in these graphs $Nu/RePr^{1/3}$ is shown as dependent on Re. In the dimensionless groups the average particle diameter is used for d, the "void fluid velocity" is used for w, that is the fluid velocity in ducts which are free from packing material. On the whole, these form a family of curves which, in some degree, lie close to each other. In Fig. 2-28 there are individual curves which lie substantially lower down the graph than the main body of curves. These individual curves represent results from not very precise experiments, carried out at high temperatures under non-equilibrium conditions. In these cases, only the chronologically varying exit temperature of a gas flowing through a previously heated mass of heat storing material is observed; see Sec. 13-4 and Fig. 13-7. On the other hand, the expected heat transfer coefficient can, in many cases, be evaluated from the upper curves. Moreover, the experimental results which have appeared subsequent to Barker's paper, fit in very well with the groups of curves in Figs. 2-27 and 2-28. By way of an example the experimental values obtained by Malling and Thodos [M42] are set out in Fig. 2-27 and are shown as a dot and dash line.

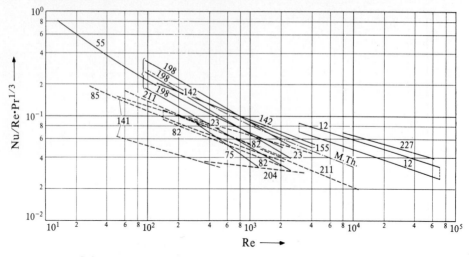

—— Spheres in a more or less regular arrangement;
‒ ‒ ‒ ‒ Other types of packing material;
—‧—‧— Values from Malling and Thodos.
The numbers relate to the table and the bibliography
provided by Barker [B41].

Numbers from Barker	Type of packing material	Measurements mm	Temperature range °C
12	spheres	49	116
55	spheres	16	12 to 25
75	spheres	17	21
142	spheres	38	18 to 60
155	spheres	10	25 to 42
198	spheres	16	near room temperature
227	spheres	100	—
23	cylinder	4 to 6 }	21
	spheres	5 to 9 }	
82	Raschig rings	5 to 17	27 to 100
85	spheres	5 to 8 }	21 to 50
	pellets	6 to 10 }	
	cylinder	6 to 10 }	
141	gravel	8 to 33	38 to 121
204	granules	2 to 6	38(?)
211	Raschig rings	13 to 30 }	17 to 42
	Prym rings	56 }	
	Berl saddles	6 to 13 }	

Figure 2-27 Heat transfer in heat-storing masses in regenerators.

Glaser [G42] measured the heat transfer and pressure drop in regenerator fillings developed for low temperature technology as illustrated in Fig. 2-22, using experimental regenerators through which air or nitrogen flowed. The test procedures employed by him, and later also by Langhans [L42] and Yazicizade [Y41]

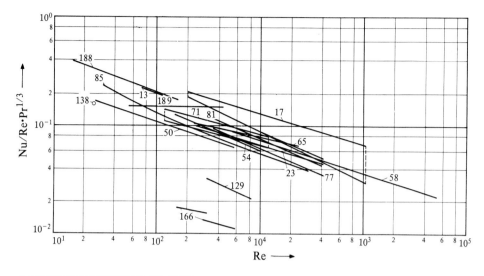

Number from Barker	Diameter mm	Temperature range °C	Number from Barker	Diameter mm	Temperature range °C
13	4	60 to 93	77	2.3 to 12	15 to 71
17	3, 4, 6	32 to 150	81	10 to 52	27 to 93(?)
23	5	21	85	5, 6, 8	21 to 65
50	2.1	21 to 32	129	11	1200
54	3 to 6	27 to 50	138	0.7	38
58	—	—	166	8,13	38 to 1150
65	9	21 to 93	188	5	38 to 480
71	19, 32, 49	38 to 710	189	1.6, 3.6	16 to 100

Figure 2-28 Heat transfer on irregularly packed spheres. The numbers relate to the table and the bibliography of Barker [B41].

in their experiments, which have already been mentioned, will be discussed at the end of Part Four of this book in Sec. 18-2. The results of Glaser's heat transfer measurements are shown in Fig. 2-29 which displays at the top the direct experimental values for sheet metal with fine and coarse corrugations, while at the bottom of the figure are shown the same but smoothed out curves for a metal sheet width b of 25 mm and different values for the equivalent diameter d_{gl}, introduced in Eq. (2-65). In this way the heat transfer coefficient α is plotted as dependent on the fluid velocity w_0 converted to 0°C and 1 bar. w_0 is thus determined when the gas flows parallel to the length of the regenerator through the free flow cross-sectional area, that is the area remaining after the cross-sectional area of the metal packing has been subtracted from the total area of cross-section of the regenerator. In spite of the low fluid velocity which exists in the laminar region, α has a relatively high value. In addition, it is noticeable that the larger the value of α, the smaller the value of b/d_{gl}; this means that for a metal sheet width b, which is almost constant in all experiments, the larger values of α are associated with larger values of the

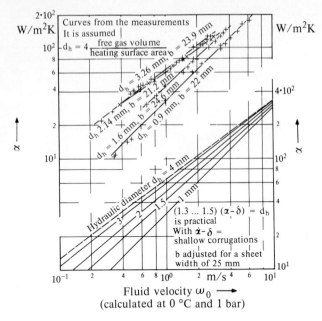

Figure 2-29 Heat transfer for corrugated sheets from Fig. 2-22; from Glaser's measurements.

equivalent diameter d_{gl} of the corrugations. These two phenomena are explained by the fact that because of the shortness of individual ducts, the gas always impinges on new metal sheet edges; the resultant turbulent initial disturbance is super-imposed on the mainly laminar flow and this considerably increases the heat transfer. In small, wide ducts these disturbances fade away more slowly than in narrow ducts and this results in an increase in α for increasing d_{gl}.

Equations for Heat Transfer in Packed Beds with Spheres

Jeschar [J42] established Eqs. (2-66), (2-67) and (2-68) for the replication of his own experimental results and a series of results obtained by other workers, for packed beds of spheres.

$$Nu' = 1.25 + Re'^{0.5} + 0.005 Re' \tag{2-66}$$

with

$$Re' = \frac{1}{1-\varepsilon} \frac{wd}{v} = \frac{1}{1-\varepsilon} Re \tag{2-67}$$

$$Nu' = \frac{\varepsilon}{1-\varepsilon} \frac{\alpha d}{\lambda} = \frac{\varepsilon}{1-\varepsilon} Nu \tag{2-68}$$

In these equations ε is the porosity, and again w is what the fluid velocity would be if the duct were empty, and d is the diameter of the spheres.[19] From these equations

[19] In a personal note dated 23 March, 1973, Jeschar brought to my attention that he prefers the constant 1.25 in Eq. (2-66) rather than the expression $2\varepsilon/(1-\varepsilon)$ which was used originally. Jeschar himself used the notation Re' and Nu' in the opposite sense to that used in Eqs. (2-66) to (2-68).

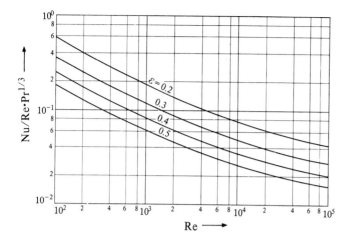

Figure 2-30 Heat transfer in a packed bed with spheres using Jeschar's equation with $Pr = 0.73$. The porosity is shown as ε.

Fig. 2-30 shows $Nu/RePr^{1/3}$ as a function of Re; this corresponds exactly to the representation used by Barker in Figs. 2-27 and 2-28. The considerable influence of the porosity is shown in Fig. 2-30. This figure also probably makes clear how wide is the distribution of the experimental results included in Barker's diagrams.

In order to clarify Re and Nu in Eqs. (2-67) and (2-68) it should be pointed out that the average true velocity in the voids amounts to w/ε and from Eq. (2-65), the equivalent diameter works out to be $2/3 \cdot \varepsilon/(1-\varepsilon) \cdot d$ with sphere diameter d. If this is inserted into the normal defining equations for Re and Nu the expressions (2-67) and (2-68) are generated if the factor $2/3$ is omitted.

On the basis of his own measurements Jeschar has extended the right-hand side of his Eq. (2-66), by generalizing the original expression, in such a way that it is still applicable for multi-granular packing. From these same measurements Jeschar has further concluded that for heat transfer in a blast furnace, this equation is equally applicable to the situation where the coke and iron are layered alternately one on top of the other or where they are mixed together.

2-14 LITERATURE ON HEAT TRANSFER IN CROSS FLOW, IN FINNED TUBES AND IN PACKED BEDS

See Foreword, Sec. 1-5.

A

41. Agnew, J. B.; Potter, O. E.: Heat Transfer Properties of Packed Tubes of Small Diameter. Trans. Instn. chem. Engrs. 48 (1970) 1, T 15–T 20.
42. Arthur, J. R.; Linnet, J. W.: The Interchange of Heat between a Gas Stream and Solid Granules. J. Chem. Soc. (1947) 416.
43. Austin, A. A.; Beckmann, R. B.; Rothfus, R. R.; Kermode, R. I.: Convective Heat Transfer in Flow Normal to Banks of Tubes. Ind. Eng. Chem., Process Design Development 4 (1965) 4, 379–387.

B

41. Barker, J. J.: Heat Transfer in Packed Beds. Ind. Engng. Chem. 57 (1965) 43–51.
42. Barnett, P. G.: The Influence of Wall Thickness, Thermal Conductivity and Method of Heat Input in the Heat Transfer Performance of some Ribbed Surfaces. Int. J. Heat Mass Transfer 15 (1972) I, 1159–1170.
43. Benke, R.: Der Wärmeübergang von Rohrelementen an Luft im Kreuzstrom bei größeren Abstandsverhältnissen. Arch. Wärmewirtschaft 19 (1938) 287–291.
44. Beveridge, G. S. G.; Haughey, D. P.: Axial Heat Transfer in Packed Beds (between 20 and 650°C). Int. J. Heat Mass Transfer 15 (1972) I, 953–968.
45. Bowers, Th. G.; Reintjes, H.: A Review of Fluid-to-Particle Heat Transfer in Packed and Moving Beds. Chem. Eng. Progr. Sympos. Ser. Heat Transfer 57 (1961) 32, 69–74.
46. Bradshaw, A. V.; Johnson, A.; McLachlan, N. H.; Chiu, Y. T.: Heat Transfer between Air and Nitrogen and Packed Beds of Non-Reacting Solids. Trans. Instn. chem. Engrs. 48 (1970) 3, T77–T84.
47. Brandt, F.: veröffentlicht in Schumacher, A.; Waldmann, H.: Wärme- und Strömungstechnik im Dampferzeugerbau. Grundlagen und Berechnungsverfahren. Essen: Vulkan-Verlag 1972; in particular pp. 32–39. See also [N 42] p. 98.
48. Brauer, H.: Spezialrippenrohre für Querstrom-Wärmeaustauscher. Kältetechnik 13 (1961) 8, 274–279.
49. Brauer, H.: Wärme- und strömungstechnische Untersuchungen an quer angeströmten Rippen-rohrbündeln. Parts 1 and 2. Chem. Ing. Tech. 33 (1961) 5, 327–335; 6, 431–438.
50. Brauer, H.: Wärmeübergang und Strömungswiderstand bei fluchtend und versetzt angeordneten Rippenrohren. Dechema-Monogr. 40 (1962) 41–76 and 616–671.
51. Brauer, H.: Strömungswiderstand und Wärmeübergang bei quer angeströmten Wärmeaustauschern mit kreuzgitterförmig angeordneten glatten und berippten Rohren. (Vortr. a. d. Jahrestreffen d. Verfahrensingenieure, 6.—8. 10. 1963, Hannover.) Chem. Ing. Tech. 36 (1964) 3, 247–260.
52. Bressler, R.: Wärmeübergang und Druckverlust bei Rohrbündel-Wärmeübertragern. Diss. Munich 1956. Summary in Forschung Ing. Wes. 24 (1958) 90–103.

C

41. Chukhanov, Z. F.: Heat and Mass Transfer between Gas and Granular Material. Int. J. Heat Mass Transfer 6 (1963) 8, 691–701.
42. Chukhanov, Z. F.: Heat and Mass Transfer between Gas and Granular Material—Part II. Int. J. Heat Mass Transfer 13 (1970) 1805–1818.
43. Chukhanov, Z. F.: Heat and Mass Transfer between Gas and Granular Material—Part III. Int. J. Heat Mass Transfer 14 (1971) 3, 337.
44. Cler, M.; Swetchine, D.; Viannay, S.; Pirovano, A.: Essais comparatifs de faisceaux de tubes à ailettes extérieures hélicoidales. Echange thermique, Perte de pression. Rev. gén. Therm. 12 (1973) 133, 23–39.

F

41. Fränkl, M.: DRP 490878 and 492431.
42. Furnas, C. C.: Heat Transfer from a Gas Stream to a Bed of Broken Solids. Ind. Engng. Chem. 22 (1930) 26.

G

41. Gillespie, B. M.; Crandall, E. D.; Carberry, J. J.: Local and Average Interphase Heat Transfer Coefficients in a Randomly Packed Bed of Spheres. Amer. Inst. Chem. Engrs. J. 14 (1968) 3, 483–490.
42. Glaser, H.: Wärmeübergang in Regeneratoren. Z. VDI., Supplement "Verfahrenstechnik" (1938) 112–125.
43. Glaser, H.: Instationäre Messung der Wärmeübertragung von Raschig-Ringschüttungen. Chem. Ing. Tech. 27 (1955) 637–643.

44. Glaser, H.: Messung des Wärmeüberganges an Kugelschüttungen nach zwei verschiedenen Verfahren. Dechema-Monogr. 40 (1962) 616–641, 77–96.
45. Glaser, H.: Wärmeübergang an Kugelschüttungen. Chem. Ing. Tech. 34 (1962) 7, 468–472.
46. Goppelsröder, H.: Wärmeübergang an quer angeströmten Glattrohrbündeln. Chemiker Ztg. 86 (1962) 10, 357–363.
47. Grimison, E. D.: Correlation and Utilization of New Data on Flow Resistance and Heat Transfer for Cross Flow of Gases over Tube Banks. Trans. Amer. Soc. mech. Eng. 59 (1937) 583–594.
48. Groehn, H. G.; Scholz, F.: Änderung von Wärmeübergang und Strömungswiderstand in querangeströmten Rohrbündeln unter dem Einfluß verschiedener Rauhigkeiten sowie Anmerkungen zur Wahl der Stoffwertbezugstemperaturen. Lecture to the 4th Int. Heat Transfer Conference in Versailles 1970.
49. Groehn, H. G.; Scholz, F.: Vorteile von Stolperrippenrohren in Wärmeaustauschern mit quer durchströmten Rohrbündeln. Brennstoff-Wärme-Kraft 23 (1971) 5 and 6. (Longitudinal fins of a height of 0.5–2 mm increase the heat transfer coefficient by up to 50 % with an aligned arrangement without increasing the pressure.)
50. Groehn, H. G.; Scholz, F.: Investigations on Steam Generator Models of In-Line Tube Arrangement and Multistart Helices Design in Pressurized Air and Helium. Mitteilung der Kernforschungsanlage Jülich GmbH, Institut für Reaktorbauelemente. Conference on Component Design of High Temperature Reactors Using Helium as a Coolant. London May 1972. The Inst. of Mech. Eng.
51. Gupta, A. S.; Thodos, G.: Mass and Heat Transfer through Fixed and Fluidized Beds. Chem. Eng. Progr. 58 (1962) 7, 58–62.
52. Gupta, A. S.; Thodos, G.: Direct Analogy between Mass and Heat Transfer to Beds of Spheres. A. I. Ch. E. J. 9 (1963) 6, 751–754.

H

41. Hahne, E.: Conductive Heat Transfer on Finned Tubes and Three-Tube Arrangements. Lecture to the meeting of commissions B1, B2, E1 of the Internationalen Kälteinstituts Sept. 1972 in Freudenstadt.
42. Hammeke, K.; Heinecke, E.; Scholz, F.: Wärmeübergangs- und Druckverlustmessungen an quer angeströmten Glattrohrbündeln, insbesondere bei hohen Reynolds-Zahlen. Int. J. Heat Mass Transfer 10 (1967) 427–446.
43. Handley, D.; Heggs, P. J.: Momentum and Heat Transfer Mechanisms in Regular Shaped Packings. Trans. Inst. chem. Engrs. 46 (1968) 9, T251–T264.
44. Hartmann, D.: Wärmeübergang und Druckverlust in Ringspalten mit berippten Innenrohren unter besonderer Berücksichtigung der Steigung schraubenlinienförmig verlaufender hoher Rippen. Diss. TH München (1965) 125 p.
45. Hausen, H.: Gleichungen zur Berechnung des Wärmeübergangs im Kreuzstrom an Rohrbündeln. Kältetechnik-Klimatisierung 23 (1971) 3, 86–89. Correction to this 6, 196.
46. Hicken, E.: Ein einfaches Näherungsverfahren zur Ermittlung der Wärmeübertragung von Rippen bei temperaturabhängigen Randbedingungen. Wärme- und Stoffübertragung 1 (1968) 4, 254–256.
47. Hilpert, R.: Wärmeabgabe von geheizten Drähten und Rohren im Luftstrom. Forschung 4 (1933) 215.
48. Hirschberg, H. G.: Wärmeübergang und Druckverlust an quer angeströmten Rohrbündeln. Abh. Deutsch. Kältetechn. Verein No. 16, Karlsruhe: F. C. Müller 1961.
49. Hofmann, E.: Wärmeübergang und Druckverlust bei Querströmung durch Rohrbündel. Z. VDI. 84 (1940) 97–101.
50. Huge, E. C.: Experimental Investigation of Heat Transfer and Flow Resistance in Cross Flow of Gases over Tube Banks. Trans. Amer. Soc. mech. Eng. 59 (1937) 573–582.

J

41. Jaeschke, L.: Über den Wärme- und Stoffaustausch und das Trocknungsverhalten ruhender luftdurchströmter Haufwerke aus Körpern verschiedener geometrischer Form in geordneter und ungeordneter Verteilung. Diss. d. TH Darmstadt 1960.

42. Jeschar, R.: Wärmeübergang in Mehrkornschüttungen aus Kugeln. Arch. Eisenhüttenw. 35 (1964) 6, 517–526.

K

41. Kast, W.: Wärmeübergang an Rippenrohrbündeln. Chem. Ing. Tech. 34 (1962) 8, 546–551.
42. Kast, W.: Messung des Wärmeübergangs in Haufwerken mit Hilfe einer temperaturmodulierten Strömung. Allgem. Wärmetech. 12 (1965) 6–7, 119–125.
43. Kiss, M. von: Wärmeübergang und Druckverlustmessungen an Längsrippen. Neue Technik 6 (1964) B 6, 310–317.
44. Kistner: Bestimmung der Wärmeübergangszahlen und Druckverluste bei doppelt versetzter und nicht versetzter Rostpackung. Arch. Eisenhüttenwes. 3 (1929/30) 751–768.
45. Klemm, B.: Beitrag zur Frage des Druckverlustes und des Wärmeübergangs in gasdurchströmten Ringspalten mit drahtbewickelten Kernrohren. Diss. TH München (1964) 86 p. Wärme 71 (1964) 1, 1–12.
46. Klier, R.: Wärmeübergang und Druckverlust bei quer angeströmten, gekeuzten Rohrgittern. Int. J. Heat Mass Transfer 7 (1964) 7, 783–799.
47. Kling, G.: Das Wärmeleitvermögen eines von Gas durchströmten Kugelhaufwerks. Forschung 9 (1938) 82–90.
48. Klinkenberg, A.; Harmens, A.; Unsteady State Heat Transfer in Stationary Packed Beds. Chem. Eng. Sci. 11 (1960) 4, 260–266.
49. Klinkenberg, A.; Harmens, A.: Unsteady State Heat Transfer in Stationary Packed Beds. Chem. Eng. Sci. 16 (1961) 2, 133–134.
50. Klinkenberg, A.: Equations for Transient Heat Transfer in Packed Beds. A. I. Ch. E. J. 8 (1962) 5, 703–704.
51. Kremer, H.; Scholand, E.: Wärmeübergang von strömenden heißen Gasen an Rippenrohre. Chem. Ing. Tech. MS 031/74; Synopse in Chem. Ing. Tech. 46 (1974) No. 5.
52. Krückels, W.; Kottke, V.: Untersuchung über die Verteilung des Wärmeübergangs an Rippen und Rippenrohr-Modellen. Chem. Ing. Tech. 42 (1970) 6, 355–362.
53. Krujilin, G.: Technical Physics of the USSR, 5 (1938) 289.
54. Kühl, H.: Probleme des Kreuzstrom-Wärmeaustauschers. Berlin, Göttingen, Heidelberg: Springer 1959.
55. Kühns, W.: Zur Bestimmung des Rippenwirkungsgrades von Kreisrippenrohren. CZ-Chemie-Techn. 2 (1973) 2, 63–66.

L

41. Landsberg, R.: Heat and Mass Transfer on the Outside of Cool Air Ducts. Bulletin Research Council of Israel 5 (1956) 147–150.
42. Langhans, W. U.: Wärmeübertragung und Druckverlust in Regeneratoren mit rostgitterartiger Speichermasse. Arch. Eisenhüttenwes. 33 (1962) 347–353 and 441–451.
43. Littlefield, J. M.; Cox, J. E.: Optimization of Annular Fins on a Horizontal Tube. Heat and mass transfer 7 (1974) 87–93 (with minimal material use the fins arranged next to one another must be narrower and thicker than a fin which stands alone).
44. Littman, H.; Barile, R. G.; Pulsifer, A. H.: Gas-Particle Heat Transfer Coefficients in Packed Beds at Low Reynolds Numbers. Ind. Eng. Chem., Fundamentals 7 (1968) 4, 554–561.
45. London, A. L.; Shah, R. K.: Offset Rectangular Plate-Fin Surfaces—Heat Transfer and Flow-Friction Characteristics. Trans. ASME, Ser. A 90 (1968) 3, 218–228.
46. London, A. L.; Young, M. B. O.; Stang, J. H.: Glass-Ceramic Surfaces, Straight Triangular Passages—Heat Transfer and Flow Friction Characteristics. Trans. ASME, Ser. A 92 (1970) 4, 381–389.
47. Lund, G.; Dodge; B. F.: Fränkl Regenerator Packings Heat Transfer and Pressure Drop Characteristics. Ind. Eng. Chem. 40 (1948) 1019–1032.

M

41. Maclaine-Cross, I. L.; Banks, P. J.: Coupled Heat and Mass Transfer in Regenerators-Prediction Using an Analogy with Heat Transfer. Int. J. Heat Mass Transfer 15 (1972) I, 1225–1242.
42. Malling, G. F.; Thodos, G.: Analogie between Mass and Heat Transfer in Beds of Spheres. Ind. J. Heat Mass Transfer 10 (1967) 489–498.
43. Mayer, E.: Verbesserung des Wärmeüberganges in Rohren durch Einbauten. Techn. Rdsch. Sulzer (1968) Forschungsh. 21–24.
44. Mayinger, F.; Schad, O.: Örtliche Wärmeübergangszahlen in querangeströmten Stabbündeln. Wärme- und Stoffübertragung 1 (1968) 1, 43–51.

N

41. Naegelin, R.; Cadelli, N.; Ciszewski, G.: Messung von Wärmeübergang und Druckabfall an Stäben mit Längsrippen. Neue Technik 7 (1964) B6, 321–331.
42. Niggeschmidt, W.: Druckverlust und Wärmeübergang bei fluchtenden, versetzten und teilversetzten querangeströmten Rohrbündeln. Dissertation Darmstadt 1975.
43. Nordon, P.; McMahon, G. B.: The Theory of Forced Convective Heat Transfer in Beds of Fine Fibres. Pt. 1 and 2. Int. J. Heat Mass Transfer 6 (1963) 6, 455–465 and 467–474.

P

41. Pierson, O. L.: Experimentelle Untersuchung über den Einfluß der Rohranordnung auf den Konvektionsübergang und den Strömungswiderstand bei Strömung von Gasen senkrecht zu Rührbündeln. (Babcock and Wilcox Co., New York), Trans. Amer. Soc. mech. Eng. 59 (1937) 563–572.
42. Polthier, K.: Druckverlust und Wärmeübergang in gleichmäßig durchströmten Schüttsäulen aus unregelmäßigen Teilchen. Arch. Eisenhüttenwes. 37 (1966) 5, 365–374.
43. Polthier, K.: Strömung und Wärmeübergang in Schüttungen mit rotationssymmetrischen Oberflächenprofilen und Kornverteilungen. Arch. Eisenhüttenwes. 37 (1966) 6, 453–462.
44. Presser, K. H.: Experimentelle Prüfung der Analogie zwischen konvektiver Wärme- und Stoffübertragung an einem Pfeilrippenrohr. Chem. Ing. Tech. 41 (1969) 21, 1176–1181.
45. Presser, K. H.: Stoffübergang und Druckverlust an parallel angeströmten Stabbündeln in einem großen Bereich von Reynolds-Zahlen und Teilungsverhältnissen. Int. J. Heat Mass Transfer 14 (1971) 9, 1235.
46. Pruschek, R.: Der Transport von Wärme und Stoff in der turbulenten Strömung durch Füllkörperrohre. Part 1: Theorie und Versuche, Versuchsergebnisse, Part 2: Die Auswirkung des turbulenten Wärmetransports in einem Füllkörperrohr mit wärmeproduzierenden Füllkörpern (Kugelhaufen reaktor). Forsch. Ing.-Wes. 29 (1963) 1, 11–19; 2, 47–49.
47. Pulsifer, A. H.: Gas-Particle Heat Transfer Coefficients in Packed Beds. Diss. Syracuse Univ. (1965).

R

41. Radestock, J.; Jeschar, R.: Theoretische Untersuchung der gegenseitigen Beeinflussung von Temperatur- und Strömungsfeldern in Schüttungen. Chem. Ing. Tech. 43 (1971) 24, 1304–1310.
42. Reiher, H.: Wärmeübergang von strömender Luft an Rohre und Rohrbündel im Kreuzstrom. Forschungsarb. Ing.-Wes. No. 269, Berlin: VDI-Verlag 1925.
43. Reiher, H.; Neidhardt, G.: Temperaturen und Wärmeaustausch an einem gußeisernen Wärmeaustauscher. Arch. Wärmewirtsch. B 7 (1926) 153–157.
44. Rietschel, H.: Untersuchungen über Wärmeabgabe, Druckhöhenverlust und Oberflächentemperatur bei Heizkörpern unter Anwendung großer Luftgeschwindigkeiten. Mitt. Prüf-Anst. f. Heizungs- u. Lüftungseinr. d. TH Berlin, 19, No. 3.
45. Roetzel, W.: Querangeströmte Rohrbündelwärmeaustauscher. Maschinenmarkt 76 (1970) 523–526.

S

41. Saunders, O. A.; Ford, H. J.: Heat Transfer in the Flow of Gas through a Bed of Solid Particles. J. Iron Steel Inst. London 141 (1940) 291.
42. Sawitzki, P.: Zweidimensionale Bestimmung des Wirkungsgrades gerader Rechteckrippen. Kältetech.-Klimatisierung 24 (1972) 315–317.
43. Sawitzki, P.: Zweidimensionale Temperaturfelder in geraden Rechteckrippen. Wärme- und Stoff-übertragung 5 (1972) 253–256.
44. Schack, K. R.: Experimentelle und theoretische Untersuchung des turbulenten Wärmeübergangs an Rippen. Diss. Aachen 1958.
45. Schad, O.: Zum Wärmeübergang an elliptischen Rohren. Diss. TH Stuttgart 1967.
46. Schlünder, E. U.: Wärme- und Stoffübertragung in mit Schüttungen gefüllten Rohren. Chem. Ing. Tech. 38 (1966) 11, 1161–1168.
47. Schlünder, E. U.; Hennecke, F. W.: Wärmeübertragung zwischen gasdurchströmten Füllkörper-schüttungen und darin eingebetteten Rohren. Chem. Ing. Tech. 40 (1968) 21–22, 1067–1071.
48. Schlünder, E. U.: Wärmeübergang an bewegte Kugelschüttungen bei kurzfristigem Kontakt. Chem. Ing. Tech. 43 (1971) 11, 651–654.
49. Schmidt, E.; Helweg, E.: Temperaturverteilung in den Blöcken im Stoßofen. Forsch. Ing. Wes. 4 (1933) 238–248.
50. Schmidt, E.; Wenner, K.: Wärmeabgabe über den Umfang eines angeblasenen geheizten Zylinders. Forschung Ing. Wes. 12 (1941) 65–73.
51. Schmidt, Th. E.: Die Wärmeleitung von berippten Oberflächen. Abh. Deutsch. Kältetechn. Vereins, No. 4, Karlsruhe 1950.
52. Schmidt, Th. E.: Der Wärmeübergang an Rippenrohre und die Berechnung von Rohrbündel-Wärmeaustauschern. Kältetechn. 15 (1963) 98–102 and 370–378.
53. Schmidt, Th. E.: Verbesserte Methoden zur Bestimmung des Wärmeaustausches an beripten Flächen. Kältetechnik 18 (1966) 4, 135–138.
54. Scholz, F.: Einfluß der Rohrreihenzahl auf den Druckverlust und Wärmeübergang von Rohrbündeln bei hohen Reynolds-Zahlen. Chem. Eng. Ing. Tech. 40 (1968) 20, 988–995.
55. Schumacher, R.: Wärmeübergang an Gase in Füllkörper- und Kontaktrohren. Chem. Ing. Tech. 32 (1960) 9, 594–597.
56. Stachiewicz, J. W.: Effect of Variation of Local Film Coefficients on Fin Performance. Trans. ASME, Ser. C 91 (1969) 1, 21–26.
57. Stephan, K.: Wärmeleistung von Rippenrohren bei unvollkommener Befestigung der Rippen. Kältetechnik 18 (1966) 2, 41–48.
58. Straub, D.; Schaber, A.; Giesen, H.: Temperaturverteilung und Rippenwirkungsgrad bei veränderlicher Wärmeübergangszahl. Kältetechnik 18 (1966) 2, 48–51.

T

41. Theoclitus, G.: Heat-Transfer and Flow-Friction Characteristics of Nine-Pin-Fin Surfaces. Trans. ASME, Ser. C 88 (1966) 4, 383–390.
42. Thomas, D.: Wärmeübergang, Druckverlust und wirtschaftliche Wärmeübertragung bei Drall-strömungen. Chem. Ing. Tech. 42 (1970) 9/10, 680–686.
43. Thoma, H.: Hochleistungskessel. Berlin: Springer 1921.
44. Türk, R.: Modellversuche zur Ermittlung der Wärmeübergangsverhältnisse im Gitterwerk von Regenerativkammern. Radex-Rdsch. (1961) 1, 506–525.

U

41. Ulsamer, J.: Die Wärmeabgabe eines Drahtes oder Rohres an einen senkrecht zur Achse strömenden Gas- oder Flüssigkeitsstrom. Forschung 3 (1932) 94–98.
42. Upmalis, A.: Wärmedurchgang bei Verbundrippenrohren. Brennstoff-Wärme-Kraft 12 (1960) 1, 21–23.

W

41. Waldmann, H.: Druckabfall und Wärmeübergang von Mehrkornschüttungen aus unregelmäßigen Partikeln. Chem. Ing. Tech. 43 (1971) 1 and 2, 37–41.
42. Walker, G.; Vasishta, V.: Heat-Transfer and Flow-Friction Characteristics of Dense-Mesh Wire-Screen Stirling-Cycle Regenerators. Advances Cryogenic Engng. 16 (1971) H-3, 324–332.
43. Walker, V.; White, L.; Burnett, P.: Forced Convection Heat Transfer for Parallel Flow through a Roughened Rod Cluster. Int. J. Heat Mass Transfer 15 (1972) I, 403–424.
44. Webb, R. L.; Eckert, E. R. G.; Goldstein, R. J.: Heat Transfer and Friction in Tubes with Repeated-Rib Surface. Int. J. Heat Mass Transfer 14 (1971) 4, 601.
45. Weishaupt, J.: Messungen an Regeneratoren von Groß-Sauerstoff-Anlagen. Kältetechnik 5 (1953) 99–103.
46. Weiss, S.: Wärmeübergang und Strömungswiderstand bei Rippenrohren im Längsstrom. Chem. Tech. 16 (1964) 1, 7–17.
47. Weyrauch, E.: Über die Verwendung niedrig berippter Rohre in der Kältetechnik. Kälte 15 (1962) 9, 471, 472, 474–476.
48. Weyrauch, E.: Der Einfluß der Rohranordnung auf den Wärmeübergang und Druckverlust bei Querstrom von Gasen durch Rippenrohrbündel. Kältetechnik 21 (1969) 3, 62–65.
49. Whitaker, St.: (Forced Convection Heat Transfer Correlations for Flow in Pipes, past Flat Plates, Single Cylinders, Single Spheres and for Flow in Packed Beds and Tube Bundles). A.I.Ch.E. J. 18 (1972) 2, 361–371.
50. Winding, C. C.; Chency, A. J.: Mass and Heat Transfer in Tube Banks. Ind. Eng. Chem. 40 (1948) 1087–1093.

Y

41. Yazicizade, A. Y.: Untersuchung der Wärmeübertragung und des Druckabfalls in Regeneratoren mit körniger oder schachtartig aufgebauter Speichermasse. Diss. Hannover 1965. Glastechnische Berichte 39 (1936) 203–217.

2-15 HEAT TRANSFER WITH SIMULTANEOUS CHANGES OF STATE, ESPECIALLY WITH CONDENSATION AND EVAPORATION

When one of the fluids taking part in the exchange of heat, condenses or evaporates, the heat transfer coefficients are, as a rule, considerably larger than when no change of state occurs.

In recent years an extraordinarily large number of research papers dealing with both these cases has appeared. Therefore in what follows only the most important of these papers will be considered.

Heat Transfer by Condensation

The heat transfer coefficients in the case of condensation can be calculated over a wide range using the equations developed theoretically by Nusselt [N65], at least for the case of "film condensation", that is when the condensate drips onto the wall in the form of a thin laminar film. The equation of Nusselt (2-69) for the average heat transfer coefficient is valid for the condensation of static saturated steam on a vertical plane wall of height h.

$$Nu = 0.943 \sqrt[4]{\frac{gr\rho^2\lambda^3}{\eta h(\theta_s - \Theta_0)}} \tag{2-69}$$

In Eq. (2-69) the condensation enthalpy is denoted by r, the density, thermal conductivity and viscosity of the liquid are denoted by ρ, λ and η. The temperature of the saturated steam is θ_s and Θ_0 is the temperature of the wall surface. Equation (2-69) is valid, with adequate accuracy, for the inner or outer wall surface of a not too narrow, vertical, stationary tube. For condensation in a horizontal tube with an outer diameter d, Nusselt developed the corresponding relationships in the form of Eq. (2-70).

$$Nu = 0.725 \sqrt[4]{\frac{gr\rho^2\lambda^3}{\eta d(\theta_s - \Theta_0)}} \tag{2-70}$$

Heat transfer coefficients of order of magnitude, 5000 to 10,000 watt/(m²K) are predicted by these equations for water vapour. On average the heat transfer coefficients measured for water vapour and fumes of organic material are higher than indicated by Eqs. (2-69) and (2-70). However the experimental values fluctuate quite a lot so that a deviation of up to $+20$ per cent and -10 per cent from the equations must be expected.

The liquid film can be turbulent with larger quantities of condensate, such as occur, for example, with large temperature differences and high walls. Here the heat transfer coefficient is increased by 25 to 40 per cent according to the calculations of Grigull [G70] and the observations of Shea and Krase [S75]. Turbulence can be expected when the dimensionless group, Eq. (2-71) exceeds 300 in value.

$$\frac{\alpha \cdot (\theta_s - \Theta_0) \cdot h}{\eta \cdot r} \tag{2-71}$$

The quantity α is obtained from Eq. (2-69).

Grigull [70, 71] has proposed Eq. (2-72) for the replication of theoretical and experimental results in the area of turbulent films of condensate.

$$\alpha = 0.003 \sqrt{\frac{g\rho^2\lambda^3(\theta_s - \Theta_0)h}{\eta^3 \cdot r}} \tag{2-72}$$

Even higher heat transfer coefficients of up to 80,000 watt/(m²K) and above have been obtained by condensation in the form of drops, in which the condensate is precipitated in the form of numerous small drops. The appearance of condensation in drop form depends primarily upon the condition of the surfaces and the material of which the wall is made. In spite of the many efforts that have been made to develop a satisfactory theory explaining condensation in drop form, and to promote this condensation by making additions to the condensing vapour, it has not proved possible to maintain condensation in drop form over a length of time, with any degree of certainty. Thus it is advisable to use instead the heat transfer coefficients which are valid for film condensation.

Usually film condensation only occurs in the condensation of vapours from fluids other than water. If the film is laminar, then, Eqs. (2-69) and (2-70) can also be

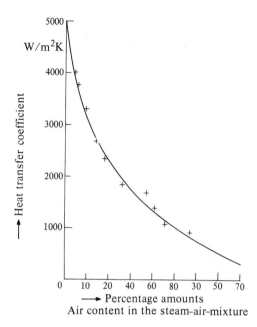

—▶ Percentage amounts
Air content in the steam–air-mixture

Figure 2-31 Heat transfer coefficient for the cooling of a saturated hydrogen–air mixture.

used here with the same limit on the degree of accuracy previously discussed. Gregorig, Kern and Turek [G67] have suggested a relationship of greater accuracy for heat transfer by condensation, on the basis of their own measurements.

As has already been mentioned in the introduction, it is worth noting that in practice, it is often important, that even in heat exchangers through which gases are flowing, certain components of a gas mixture often condense, or conversely liquids evaporate into gases. In such cases heat will be transferred not only for the heating and cooling of the gas but also for the condensation or evaporation itself. From earlier, more theoretical considerations [J71, N65] relating to the processes involved, Jaroschek [J69] has measured the heat transmission coefficient for condensating water vapour in the presence of air, and Lüder [L70] has measured the heat transfer coefficient for the condensation of water vapour, ammonia and methane in the presence of air as well as the heat transfer coefficient for water vapour in the presence of hydrogen. Figure 2-31 illustrates the heat transfer coefficients for hydrogen–air mixtures which can be calculated approximately using the measurements of Jaroschek [J69]. Lüder's [L70] measurements basically give the same heat transfer coefficient curve, depending on how it is synthesized.

The values of Fig. 2-31 are in close agreement with the measurements of Renker [R62] which he obtained for a gas mixture inside a tube, flowing at a velocity of 5 m/s. At higher fluid velocities α is larger; according to Renker it is about four times larger with an air content of 50 moles, for the flow velocity of 25 m/s. Values obtained by Kirschbaum and Wetjen [K70] as well as by Stewart and others [S94] from observations on an annular slot, are of the same order of magnitude on average as the data displayed in Fig. 2-23. Values determined by Schrader [S69] in a tube bundle condenser, are somewhat lower.

The extent to which the heat transfer coefficient decreases with increasing air content can be seen from Fig. 2-31. All measurements confirm the view of Nusselt [N65] and Ackermann [A62, 63] that the steam, before it condenses, must diffuse through an air or gas layer which lies upon the surface where condensation takes place.

Heat Transfer by Evaporation

Heat transfer, and therefore also the heat transfer coefficient in evaporation, are more dependent than in other processes, on the temperature difference $\Delta\theta = \Theta_0 - \theta_f$, which in this case exists between the surface of the wall and the evaporating liquid. In Fig. 2-32 the heat flux $\dot{q}$ is shown as a function of this temperature difference; this heat flux will be called technically the "heating surface area load" and is equal to the quantity of heat released from the unit solid surface area in unit time. Four regions can be differentiated in Fig. 2-32. Heat will be transferred between A and B, mainly by free or forced convection.

The region where bubbles rise in the boiling liquid which extends from B to C is of technical importance. Steam bubbles develop at particular positions on the heating surface area and these quickly increase and rise up through the liquid. The number of steam bubbles which develop rises with increasing temperature difference; consequently the liquid becomes more agitated and heat transfer increases considerably.

The regions where the film evaporates extend from C through D to E and it is in these regions that "cushions"* of steam develop between the heating surface area and the liquid. These gradually grow together to form a trapped film of steam. If the heating surface area temperature is not held constant forcibly, the conditions between C and D are unstable to the extent that, for example, when $\dot{q} = $ const. such a situation can arise at a point between D and E. The sharp rise in heating surface area temperature that thus takes place, particularly from point C onwards, often

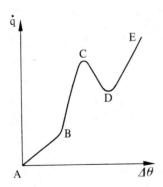

Figure 2-32 Heat flux $\dot{q}$ for the evaporation of liquids, shown dependent on the temperature difference between the wall surface and boiling liquid.

* Editor's note: Hausen uses the word "polster" which might be translated as pillow, cushion or bolster.

has the effect that the wall melts or is burnt through. For this reason, in American terminology, condition C is called "burn out". Point D is called the permitted freezing point.

Heat transfer in the region of bubbles which arise in the course of the process of boiling has been the subject of detailed investigation. W. Fritz [F68] conducted a survey of the most important works appearing up to 1963. This survey shows that the heat transfer coefficients measured at smooth plates, in vessels, and on the outside of horizontal and vertical tubes, all mutually agree within a 20 per cent distribution of the experimental results.

The endeavour to reproduce the heat transfer coefficients measured in various evaporating liquids using generally valid similarity relationships has led to some remarkable estimates if not also to a fully satisfactory result. Stephan [S86] has established the determining dimensionless groups and by adapting his own and other workers' measurements taken in various evaporating liquids has developed the similarity relationship (2-73) which is valid for low pressures.

$$\frac{\alpha \cdot d}{\lambda} = c \left[\frac{\dot{q} \cdot d}{\lambda T_S} \right]^{n_1} \cdot \left[\frac{d \cdot T_S \lambda}{v \sigma} \right]^{n_2} \cdot \left[\frac{R_p \rho_d r}{d \cdot \rho_f (f \cdot d)^2} \right]^{0.133} \tag{2-73}$$

In (2-73) $\dot{q}$ denotes the heat flux, T_S denotes the bubble temperature, λ, ρ_f and v denote the thermal conductivity, density and kinematic viscosity of the liquid; ρ_d is the density of the steam, r denotes the evaporation enthalpy, σ is the surface tension, R_p denotes the smoothing depth from *DIN* 4762 and finally d and f denote the diameter of the disruption of flow and the frequency of the bubbles respectively. Thus for evaporation at horizontal plane plates the values $c = 0.013$, $n_1 = 0.8$ and $n_2 = 0.4$ can be employed while for evaporation at horizontal tubes $c = 0.071$, $n_1 = 0.7$ and $n_2 = 0.5$ can be used. Taking $c = 10^{-5}$, $n_1 = 0.7$ and $n_2 = 0.8$ Fedders [F61] was also able to demonstrate, using Eq. (2-73) the pressure dependence of heat transfer in the evaporation of water in tubes almost up to the critical pressure. However in doing this the accuracy of the replication of experimental results for atmospheric pressure was reduced. This pressure dependence, calculated using the method of Fedders has been confirmed also for other liquids through measurements taken by Bier, Gorenflo and Wickenhäuser [B71]. Similarity relationships have been outlined by F. Müller [M74], Vaihinger and Kaufmann [V61], as well as by other authors.

The practical application of such equations in many cases may be rendered difficult by the fact that some of the values occurring in these equations, such as R_p, d and f, are not easy to determine, particularly in less well known liquids. Less complicated equations are thus needed to satisfy practical requirements, even if these simpler equations are not as accurate. The following simple equation, developed by Fritz [F68] in 1963 is suited to this purpose and on average reproduced as well as is possible the then known experimental values obtained for evaporating water at low pressures. Fritz's equation is:

$$\alpha = A \cdot \left(\frac{\dot{q}}{\dot{q}_0} \right)^{0.72} \cdot \left(\frac{p}{p_0} \right)^{0.24} \tag{2-74}$$

in which A has the value $1.95 \, W/(m^2 K)$ for water. $\dot{q}_0$ is a heat flux of $1 \, W/m^2$ and p_0 is a pressure of 1 bar. Equation (2-74) can be applied satisfactorily also to other liquids if the value of A is selected from the following table [H64].

Most of the values given in Table 2-5 are extracted from the diagrams provided by Fritz [F68]. Figure 2-33 consists of one of these diagrams which deals with the heat transfer coefficients for freezing mixtures in the case of bubble evaporation.

Table 2-5 Constant A in Eq. (2-74) for different evaporating liquids

Liquid	$\left(\dfrac{\alpha}{\alpha_{Water}}\right)\dot{q}$	A $W/(m^2 K)$
Water	1.0	1.95
Solutions in water		
24% NaCl	0.61	1.18
10% Na_2SO_4	0.94	1.83
20% Sugar	0.80^{20}	1.56^{20}
40% Sugar	0.53^{20}	1.03^{20}
60% Sugar	0.37 to 0.42^{20}	0.72 to 0.82^{20}
26% Glycerine	0.83	1.62
55% Glycerine	0.75	1.46
Inorganic liquids		
Ammonia	1.15	2.24
Sulphur Dioxide	0.78	1.52
Organic liquids *(including freezing mixtures)*		
Paraffin	0.5 to 0.7	0.9 to 1.4
Propane	0.55	1.07
n-Butane	1.07	2.10
Methyl Alcohol	0.72	1.40
Ethyl Alcohol	0.63	1.24
Isopropanol	0.70	1.36
Higher Alcohols	0.5 to 0.8	0.9 to 1.6
Acetone	0.69	1.35
Benzene	0.47	0.92
Toluene	0.41	0.80
Carbontetrachloride	0.43	0.85
Methyl Chloride	0.94	1.82
Methylene Chloride	0.62	1.19
R11 (Freon 11)	0.88	1.72
R12 (Freon 12)	0.58	1.13

[20] Values according to Averin and Krushilin. Similar values can be expected from F. Mayinger and E. Hollbon who have carried out measurements in 69 per cent sugar solution but these have not yet been published.

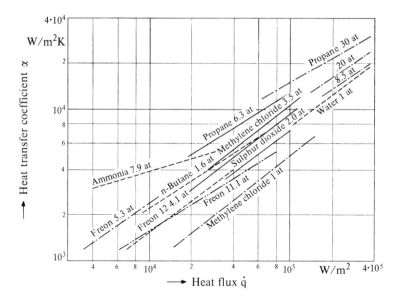

Figure 2-33 Heat transfer in the bubble evaporation of different liquids, in particular freezing mixtures.

Even in the bubbles of boiling liquid metals, which have become important through their use in cooling within nuclear reactors, Eq. (2-74) provides the correct order of magnitude of the heat transfer coefficients. However these heat transfer coefficients exhibit somewhat irregular behaviour, when examined in detail, as can be seen from Fig. 2-34 which originates in the paper of W. Fritz [F68].

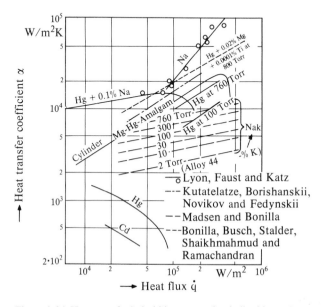

Figure 2-34 Heat transfer in bubble evaporation in liquid metals.

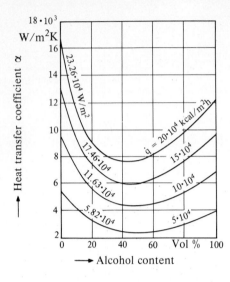

Figure 2-35 Heat transfer in the boiling of a liquid ethanol–water mixture.

In boiling mixtures the heat transfer coefficient is smaller than it is in pure liquids. The value of the heat transfer coefficient must be determined experimentally from case to case. By way of an example, Fig. 2-35 displays the heat transfer coefficient of a boiling ethanol–water mixture [F68].

In low temperature technology the heat transfer coefficient for evaporation can be considerably enhanced by attaching a very thin, porous layer to the heating surface area [W64].

The Pressure Dependence of Heat Transfer in Bubble Evaporation

Equation (2-73), using the constants of Fedder, shows a good agreement with experimental measurements in the case of evaporating water up to a pressure of 150 bar. On the other hand, Eq. (2-74), which is represented by the bottom, solid line in Fig. 2-36, is only valid up to a pressure which amounts to about a tenth of the critical pressure p_k. The following, extended Eq. (2-75) [H64] can be used for higher pressures up to about $0.9 p_k$.

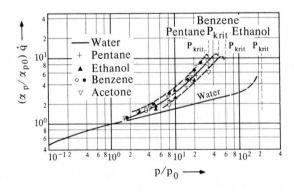

Figure 2-36 Pressure dependence of the heat transfer coefficient in bubble evaporation in water and organic liquids.

$$\alpha = A\left[\left(\frac{\dot{q}}{\dot{q}_0}\right)^{0.72}\cdot\left(\frac{p}{p_0}\right)^{0.24} + f\left(\frac{p}{p_k}\right)\right] \qquad (2\text{-}75)$$

with

$$f\left(\frac{p}{p_k}\right) = 33{,}200\frac{p}{p_k} + \frac{4260}{0.36 + p/p_k} - 11{,}850 \qquad (2\text{-}76)$$

With this equation, for example, the dashed curve for benzene, pentane and ethanol in Fig. 2-36 can be replicated very satisfactorily for a heat flux $\dot{q}$ of about 40,000 W/(m²K).

It is simpler, and in many cases sufficiently accurate to calculate $f(p/p_k)$ using Eq. (2-77)

$$f\left(\frac{p}{p_k}\right) = 25{,}500\left(\frac{p}{p_k} - 0.1\right) \text{ for } \frac{p}{p_k} > 0.1 \qquad (2\text{-}77)$$

$f(p/p_k) = 0$ between $p = 0$ and $p = 0.1$.

The equations above assume ideally smooth surface areas. Stephan [S86], as well as Danoliva and Belskij [D63], among others, have investigated the influence of increased roughness, which, in general, favours heat transfer. From Eq. (2-73) α is proportional to the function $R_p^{0.133}$ of the smoothing depth.

Even though boiling takes place at a constant temperature, if the heat necessary for evaporation is extracted from the fluid, the wall temperature and with it also the temperature difference $\Delta\theta$ between wall and evaporating liquid vary spatially. In these circumstances the heat flux $\dot{q}$, and thus the heat transfer coefficient, is dependent also upon wall position. The means whereby the effective average temperature difference can be calculated with spatially varying heat transfer coefficients, and hence also the total heat transfer, will be discussed in Part Three of this book; see Sec. 5-13. The differences relating to the logarithmic temperature difference are illustrated diagrammatically on sheet A25 in the second edition of the VDI-Wärmeatlas (VDI Heat Atlas).

In a survey carried out by Slipcevic [S79] on the evaporation of freons, it is shown that with a constant temperature of evaporation for a freon and constant heat flux, a higher heat transfer coefficient occurs in a horizontal bundle of tubes than in a single, horizontal tube. The transfer coefficients increase from the bottom-most row of tubes up to the top-most row of tubes and can even be considerably larger at the bottom row of tubes than they are for a single tube. This can be attributed to free convection in the boiling liquid. Thus, in the device overall, increases in the heat transfer coefficient of about 20 to 60 per cent can be expected, and indeed the lower the evaporation temperature the greater will be this increase.

Maximum Heat Flux

Because of its importance for the cooling of nuclear reactors the maximum heat flux $\dot{q}_{max}$ at point C in Fig. 2-32 has been measured for the evaporation of a large number of liquids. The following Table 2-6 gives an insight into the orders of magnitude; the table was compiled by Fritz [F68].

Table 2-6 Maximum heat flux $\dot{q}_{max}$ of boiling liquids at the "burn out" point

Liquids	Absolute pressure (atmospheres)	Heat flux $\dot{q}_{max}$ 10^4J/(m²s)
Water	1	70 to 140
	1	115
	10	185
	30	290
	50	395
	100	370
	200	185
Benzene	1	45 to 70
Methyl Alcohol	1	50
Ethyl Alcohol	1	60
1-Butyl Alcohol	1	45
1-Propyl Alcohol	1	45
Ethylene Glycol	1	70
Pentane	2	40
	12	60
1-Heptene	1	35
Acetone	1	45

Kutateladze [K84] has developed the following Eq. (2-78) in order to calculate the maximum heat flux $\dot{q}_{max}$.

$$\dot{q}_{max} = \sqrt{K} \cdot r \cdot \sqrt{\rho_d} \cdot \sqrt[4]{\sigma(\rho_f - \rho_d) \cdot g} \qquad (2\text{-}78)$$

where K is an empirically determined constant lying between 0.13 and 0.19 with the average value 0.14, r is the evaporation enthalpy, ρ_d and ρ_f are the densities of the vapour and the liquid, σ is the surface tension and g is the acceleration due to gravity. Equation (2-79) which was similarly developed theoretically, largely by Zuber [Z61] agrees, over quite a large range with Eq. (2-78)

$$\dot{q}_{max} = \frac{\pi}{24} \sqrt{\frac{\rho_f}{\rho_f + \rho_d}} \cdot r \cdot \sqrt{\rho_d} \sqrt[4]{\sigma(\rho_f - \rho_d) \cdot g} \qquad (2\text{-}79)$$

A short development of this equation is to be found in [T67].

Questions relating to heat transfer by condensation and evaporation will not be considered further as, as has already been pointed out in the introduction, evaporators and condensers represent an area of technical specialization which has been dealt with by many people. It only remains to point out, that fundamentally, the relationships developed in this book can be calculated without difficulty, even for evaporators and condensers, if the relevant heat transfer coefficients are used. The calculation is simple, that is as long as the heating or the cooling of the vapour or liquid does not take place. Indeed it is even simpler than for gas operated heat exchangers, because the temperature of the evaporating or condensing material does not change, or changes only very slightly. Of course in many cases, such as steam boilers for example, it is necessary to take into consideration the influence of radiative heat transfer in the gas (see Sec. 3-1 ff.).

2-16 LITERATURE ON HEAT TRANSFER WITH CHANGES IN STATE, PARTICULARLY WITH CONDENSATION AND EVAPORATION

See Foreword, Sec. 1-5.

A

61. Ackermann, G.: Theorie der Verdunstungskühlung. Ing. Arch. 5 (1934) 124.
62. Ackermann, D.: Beitrag zur Berechnung des Wärmeübergangs bei Kondensation in Anwesenheit von Inertgas. Wärme- und Stoffübertragung 1 (1968) 4, 246–250.
63. Ackermann, D.: Wärme- und Stoffübergang bei der Kondensation eines turbulent strömenden Dampfes in Anwesenheit von Inertgas. Diss. Stuttgart 1972.
64. Ackermann, D.: Wärme- und Stoffübergangskoeffizienten bei der Abkühlung von Dampf-Gas-Gemischen in Oberflächenkondensatoren. Chem. Ing. Tech. 44 (1972) 274–280.
65. Alty, T.; Macky, C. A.: The Accommodation Coefficient and the Evaporation Coefficient of Water. Proc. Roy. Soc. (A) 149 (1935) 104–116; also Proc. Roy. Soc. (A) 161 (1937) 68–79.
66. Anderès, G.: Einfluß der Oberflächenspannung auf den Stoffaustausch zwischen Dampfblasen und Flüssigkeit. Chem. Ing. Tech. 34 (1962) 597–602.
67. Auracher, H.: Wasserdampfdiffusion und Reifbildung in porösen Stoffen. VDI-Forschungsheft 566 (1974). Kurzer Auszug in Forschung Ing.-Wes. 40 (1974) 196.
68. Averin, E. K.; Krushilin, C. E.: Einfluß der Oberflächenspannung und der Viscosität auf die Wärmeübergangszahl beim Sieden von Wasser (russ.). Izvest. Acad. Nauk. SSSR Otdel. Tech. Nauk. (1950) No. 10, 131–137. Extract in Chem. Ing. Tech. 28 (1956) 431.

B

61. Bachner, D.; Koelzer, W.; Müller, D.: Untersuchungen über die Kondensation verschiedener Gase (CO$_2$, N$_2$, H$_2$, Xe). Forschungsber. Nordrhein-Westf. (1966) 1643, 37 p.
62. Bähr, A.: Bedeutung der von Dampfblasen erzeugten Mikrokonvektion für die Wärmeübertragung beim Sieden. Chem. Ing. Tech. 38 (1966) 9, 922–925.
63. Bandel, J.; Schlünder, E. U.: Druckverlust und Wärmeübergang bei der Verdampfung siedender Kältemittel im durchströmten Rohr. Chem. Ing. Tech. 45 (1973) 345–350.
64. Beer, H.; Durst, F.: Mechanismen der Wärmeübertragung bei Blasensieden und ihre Simulation. Chem. Ing. Tech. 40 (1968) 13, 632–638.
65. Beer, H.: Beitrag zur Wärmeübertragung beim Sieden. Progr. Heat and Mass Transfer. Vol. 2, Oxford 1969, pp. 311–370. (Heat transfer to steam bubbles.)
66. Berenson, P. J.: Experiments on Pool-Boiling Heat Transfer. Int. J. Heat Mass Transfer 5 (1962) Oct. 985–999.
67. Bergles, A. E.; Rohsenow, W. M.: The Determination of Forced-Convection Surface-Boiling Heat Transfer. Trans. ASME, Ser. C 86 (1964) 3, 365–372.
68. Best, R.; Burow, P.; Beer, H.: Bubble Boiling Heat Transfer as a Result of Hydrodynamic Effects. Lecture at the Conference of Commissions B1, B2, E1 of the Internationalen Kälteinstituts in Freudenstadt 1972.
69. Bewilogua, L.; Knöner, R.; Wolf, G.: Heat Transfer in Boiling Hydrogen, Neon, Nitrogen and Argon. Cryogenics 6 (1966) 1, 36–39.
70. Bewilogua, L.; Knöner, R.: Contribution to the Problem of Heat Transfer in Low-Boiling Liquids. J. Amer. chem. Soc. 90 (1968) 12, 3086–3087.
71. Bier, K.; Gorenflo, D.; Wickenhäuser, G.: Zum Wärmeübergang beim Blasensieden in einem weiteren Druckbereich. Chem. Ing. Tech. 45 (1973) 935–942.

72. Blatt, T. A.; Adt, R. R. jr.: An Experimental Investigation of Boiling Heat Transfer and Pressure-Drop Characteristics of Freon 11 and Freon 113 refrigerants. Am. Inst. Chem. Engrs. J 10 (1964) 369–373.
73. Bode, H.: Wärme- und Stoffübergang in der Umgebung wachsender Dampfblasen. Wärme- und Stoffübertragung 5 (1972) 134–140.
74. Bonilla, Ch. F.; Percy, Ch. W.: Wärmeübertragung auf kochende binäre Flüssigkeitsmischungen. Trans. Amer. Inst. chem. Engrs. 37 (1941) 685–706. (Phys. Ber. 1943, p. 1363.)
75. Borchmann, J.: Zur Kondensation schnell strömender Dämpfe in Ringspalten. Chem. Ing. Tech. 38 (1966) 8, 832–837.
76. Borchmann, J.: Zur Kondensation von R 11 und Wasserdampf bei hohen Dampfgeschwindigkeiten. Kältetechnik 19 (1967) 7, 208–213.
77. Bosnjaković, F.: Verdampfung und Flüssigkeitsüberhitzung. Tech. Mech. Thermodynam. 1 (1930) 358–362.
78. Brauer, H.: Berechnung des Wärmeüberganges bei ausgebildeter Blasenverdampfung. Chem. Ing. Tech. 35 (1963) 11, 764–774.

C

61. Cess, R. D.; Sparrow, E. M.: Film Boiling in a Forced-Convection Boundary-Layer Flow. J. Heat Transfer 83 (1961) 370–376.
62. Chawla, J. M.: Wärmeübergang und Druckabfall in waagerechten Rohren bei der Strömung von verdampfenden Kältemitteln. Forsch. Ing.-Wes. VDI-Forsch.-H. 523 (1967) 36 p. (Spatial heat transfer coefficient of evaporating freon 11.)
63. Chawla, J. M.: Wärmeübergang und Druckabfall in waagerechten Rohren bei der Strömung von verdampfenden Kältemitteln. Kältetechnik 19 (1967) 8, 246–252.
64. Chawla, J. M.: Wärmeübergang und Druckverlust bei der Kältemittelverdampfung im waagerechten Strömungsrohr. Chem. Ing. Tech. 40 (1968) 5, 229–234.
65. Chawla, J. M.: Wärmeübertragung in durchströmten Kondensatorrohren. Chem. Ing. Tech. 43 (1971) 14, 838.
66. Chawla, J. M.: Impuls- und Wärmeübertragung bei der Strömung von Flüssigkeits-Dampf-Gemischen. Chem. Ing. Tech. 44 (1972) 118–120.
67. Chawla, J. M.: Wärmeübergang in durchströmten Kondensatorrohren. Kältetech.-Klimatisierung 24 (1972) 233–240.
68. Claassen, H.: Verdampfen und Verdampfer mit senkrechten Heizrohren. Magdeburg: Schallehn & Vollbrück 1938.
69. Colburn, A. P.; Drew, T. B.: The Condensation of Mixed Vapors. Trans. Am. Inst. Chem. Engrs. 33 (1937) 197–215.
70. Cole, R.; Shulman, H. L.: Bubble Growth Rates at High Jakob Numbers. Int. J. Heat Mass Transfer 9 (1966) 12, 1377–1390.
71. Cole, R.; Rohsenow, W. M.: Correlation of Bubble Departure Diameters for Boiling of Saturated Liquids. Chem. Eng. Progr. Sympos. Ser. 65 (1969) 92, 211–213.
72. Cryder, D. S.; Finalborgo, A. C.: Heat Transmission from Metal Surfaces to Boiling Liquids. Trans. Am. Inst. Chem. Engrs. 33 (1937) 346–361.

D

61. Dallmeyer, H.: Über die gleichzeitige Wärme- und Stoffübertragung bei der Kondensation eines Dampfes aus einem Gemisch mit einem nichtkondensierenden Gas in laminarer und turbulenter Strömungsgrenzschicht. Diss. TH Aachen 1968, 97 p.
62. Dallmeyer, H.: Stoff- und Wärmeübertragung bei der Kondensation eines Dampfes aus einem Gemisch mit einem nichtkondensierenden Gas in laminarer und turbulenter Strömungsgrenzschicht. VDI-Forschungsheft No. 539 (1970) 5–24.
63. Danoliva, G. N.; Belskij, V. K.: Die Wärmeabgabe beim Sieden von Freon 113 and 12 an Rohren mit verschieden rauher Oberfläche. Cholod. Technika 42 (1965) 4, 24–28 (Russian).

64. Denny, V. E.; Jusionis, V. J.: Effects of Noncondensible Gas and Forced Flow on Laminar Film Condensation. Int. J. Heat Mass Transfer 15 (1972) I, 315–326.
65. Dornieden, M.: Wärmeübergang bei der Kondensation von Dämpfen an Rohrbündeln unter Berücksichtigung des Druckverlustes. Chem. Ing. Tech. 44 (1972) 269–273.
66. Dornieden, M.: Zur Berechnung ein- und mehrgängiger Rohrbündel-Kondensatoren. Chem. Ing. Tech. 44 (1972) 618–622.

E

61. Emmerson, G. S.: Heat Transmission with Boiling. Nuclear Eng. 5 (1960) 54, 493–499.
62. Estrin, J.; Hayes, T. W.; Drew, T. B.: The Condensation of Mixed Vapors. A. I. Ch. E. J. 11 (1965) 5, 800–803.

F

61. Fedders, H.: Messung des Wärmeüberganges beim Blasensieden von Wasser an metallischen Rohren. Diss. Tech. Un. Berlin. Bericht Kernforschungsanlage Jülich No. 740-RB (1971).
62. Forster, H. K.; Zuber, N.: Dynamics of Vapor Bubbles and Boiling Heat Transfer. Am. Inst. Chem. Engrs. Journ. 1 (1955) 531–535.
63. Frank, A.: Wärme- und Stoffaustausch zwischen Dampfblase und Flüssigkeit bei Stickstoff/Sauerstoff-Gemischen. Chem. Ing. Tech. 23 (1960) 5, 330–335.
64. Frederking, T.: Wärmeübergang bei der Verdampfung der verflüssigten Gase Helium und Stickstoff. Forsch.-Ing.-Wes. 27 (1961) 1, 17–30; 2, 58–62.
65. Fritz, W.; Homann, F.: Über die Temperaturverteilung im siedenden Wasser. Phys. Z. 37 (1936) 873–878.
66. Fritz, W.: Film- und Tropfenkondensation von Wasserdampf. Z. VDI Beihefte Verfahrenstechnik No. 4 (1937) 127–132.
67. Fritz, W.: Verdampfen und Kondensieren. Verfahrenstechnik (1943) 1, 1–14.
68. Fritz, W.: Grundlagen der Wärmeübertragung beim Verdampfen von Flüssigkeiten. Chem. Ing. Tech. 35 (1963) 11, 753–764; also numerous bibliographies.

G

61. Gorenflo, D.: Zum Wärmeübergang bei der Blasenverdampfung an Rippenrohren. Diss. TH Karlsruhe 1966.
62. Gorenflo, D.: Zur Druckabhängigkeit des Wärmeüberganges an siedende Kältemittel bei freier Konvektion. Chem. Ing. Tech. 40 (1968) 15, 757–762.
63. Grassmann, P.; Karagunis, A.; Kopp, J.; Frederking, F.: Wärmeübergang an flüssiges Helium bei Blasen- und stabiler Filmverdampfung. Kältetechnik 10 (1958) 206–208.
64. Gregorig, R.: Beitrag zur rechnerischen Erfassung der Analogie zwischen Tropfenkondensation und Verdampfung. Kältetechnik 6 (1954) 2–7.
65. Gregorig, R.: Einfluß der Heizwand-Eigenschaften auf den Mechanismus der Blasenverdampfung. Chem. Ing. Tech. 39 (1967) 1, 13–20.
66. Gregorig, R.: Zur Thermodynamik der existenzfähigen Dampfblase an einem aktiven Verdampfungskeim. Verfahrenstechnik 1 (1967) 9, 389–392.
67. Gregorig, R.; Kern, J.; Turek, K.: Improved Correlation of Film Condensation Data Based on a more Rigorous Application of Similarity Parameters. Heat and mass transfer 7 (1974) 1–13.
68. Greitzer, E. M.; Abernathy, F. H.: Film Boiling on Vertical Surfaces. Int. J. Heat Mass Transfer 15 (1972) I, 475–492.
69. Griffith, P.: Bubble Growth Rates in Boiling. Trans. Am. Soc. Mech. Engrs. 80 (1958) 721–727.
70. Grigull, U.: Wärmeübergang bei der Kondensation mit turbulenter Wasserhaut. Forsch. Ing.-Wes. 13 (1942) 49–57; Z. VDI 86 (1942) 444–445.
71. Grigull, U.: Wärmeübergang bei Filmkondensation. Forsch. Ing.-Wes. 18 (1952) 10–12.
72. Grigull, U.; Abadzic, E.: Blasen- und Filmsieden von Kohlendioxid im kritischen Gebiet. Forschung Ing.-Wes. 31 (1965) 27–30.

73. Grigull, U.; Abadzic, E.: Heat Transfer from a Wire in the Critical Region. Proc. Inst. Mech. Engrs. 182 (1967/68) Pt. 3, I, 52–57.

H

61. Haffner, H.: Zum Wärmeübergang an Kältemittel bei hohen Drücken. Chem. Ing. Tech. 44 (1972) 286–291.
62. Hamburger, L. G.: On the Growth and Rise of Individual Vapour Bubbles in Nucleate Pool Boiling. Int. J. Heat Mass Transfer 8 (1965) 11, 1369–1386.
63. Hartmann, H.: Wärmeübergang bei der Kondensation strömender Sattdämpfe in senkrechten Rohren. Chem. Ing. Tech. 33 (1961) 5, 343–348.
64. Hausen, H.: Näherungsgleichung zur Berechnung der Wärmeübertragung bei Blasenverdampfung bis in die Nähe des kritischen Punktes. Wärme- und Stoffübertragung 3 (1970) 41–43.
65. Heimbach, P.: Zur Frage des Wärmeüberganges bei strömenden Wasserdampf-Luft-Gemischen mit großen Dampfanteilen. Linde-Ber. aus Techn. u. Wiss. (1960) 10, 39–43.
66. Heimbach, P.: Wärmeübergangskoeffizienten für die Verdampfung von R 22 und R 13 an einem überfluteten Rippenrohr-Bündel. Linde-Ber. Techn. u. Wissensch. No. 29 (1971) 33–42.
67. Heimbach, P.: Wärmeübergangskoeffizienten für die Verdampfung von Kältemmitel-Öl-Gemischen an einem überfluteten Glattrohrbündel. Kältetech.-Klimatisierung 24 (1972) 287–295.
68. Heimbach, P., Sürth/Köln: Wärmeübergangskoeffizienten für die Verdampfung von Kältemittel-Öl-Gemischen an einem überfluteten Rippenrohr-Bündel. Linde-Ber. (1972) 32, 3–10.
69. Henrici, H.: Kondensation von Frigen 12 und Frigen 22 an glatten und berippten Rohren. Diss. TH Karlsruhe (1961) 1–110.
70. Henrici, H.: Kondensation von R 11, R 12 und R 22 an glatten und berippten Rohren. Kältetechnik 15 (1963) 8, 251–256.
71. Henrici, H.; Hesse, G.: Untersuchungen über den Wärmeübergang beim Verdampfen von R 114 und R 114-Öl-Gemischen an einem horizontalen Glattrohr. Kältetechn.-Klimatisierung 23 (1971) 54–58.
72. Hicken, E.; Garnjost, H.: Zur Kondensation an der Wand dampfdurchströmter Kanäle mit veränderlichem Querschnitt. Wärme- und Stoffübertragung 4 (1971) 60–68.
73. Hildebrandt, U.: Experimentelle Untersuchung des Wärmeüberganges an Helium I bei Blasenverdampfung in einem senkrechten Rohr. Wärme- und Stoffübertragung 4 (1973) 142–151.
74. Hofmann, E.: Wärmeübergangszahlen verdampfender Kältemittel. Kältetechnik 9 (1957) 7–12.
75. Hofmann, E.: Beitrag zur Berechnung von Flüssigkeitskühlern mit verdampfendem Kältemittel in Rohren. Kältetechnik-Klimatisierung 23 (1971) 90–96.
76. Hommann, G.: Der Wärmeübergang bei der Verdampfung von Kältemitteln in horizontalen glatten und berippten Rohren. Luft- und Kältetechn. 6 (1970) 2, 90–93.

I

61. Insinger, Th. H.; Bliss, H.: Transmission of Heat to Boiling Liquids. Trans. Am. Inst. Chem. Engrs. 36 (1940) 491–516. (Report relating to this in Chem. Fabr. 14 (1941) 407.)
62. Ivey, H. J.: Relationships between Bubble Frequency, Departure Diameter and Rise Velocity in Nucleate Boiling. Int. J. Heat Mass Transfer 10 (1967) 8, 1023–1040.

J

61. Jakob, M.; Fritz, W.: Versuche über den Verdampfungsvorgang. Forsch. Ing.-Wes. 2 (1931) 435–447.
62. Jakob, M.; Erk, S.; Eck, H.: Der Wärmeübergang beim Kondensieren strömenden Dampfes in einem vertikalen Rohr. Forsch. Ing. Wes. 3 (1932) 161–170.
63. Jakob, M.; Linke, W.: Der Wärmeübergang von einer waagerechten Platte an siedendes Wasser. Forsch. Ing.-Wes. 4 (1933) 75–81.
64. Jakob, M.; Linke, W.: Der Wärmeübergang beim Verdampfen von Flüssigkeiten an senkrechten und waagerechten Flächen. Phys. Z. 36 (1935) 267–280.

65. Jakob, M.; Erk, S.; Eck, H.: Verbesserte Messungen und Berechnungen des Wärmeüberganges beim Kondensieren strömenden Dampfes in einem vertikalen Rohr. Phys. Z. 36 (1935) 73–84.
66. Jakob, M.: Heat Transfer in Evaporation and Condensation. Mech. Eng. 58 (1936) 643–660 and 729–739.
67. Jakob, M.: The Influence of Pressure on Heat Transfer in Evaporation. Proc. 5. Intern. Congress Applied Mech. 1938 p. 561.
68. Jones, W. P.; Renz, U.: Condensation from a Turbulent Stream onto a Vertical Surface. Int. J. Heat Mass Transfer 17 (1974) 1019–1028.
69. Jaroschek, K.: Einfluß des Luftgehaltes im Heizdampf auf den Wärmeübergang in Wärmeaustauschern. Z. VDI supplement "Verfahrenstechnik" (1939) 5, 135–140.
70. Johnson, H. A.: Transient Boiling Heat Transfer to Water. Int. J. Heat Mass Transfer 14 (1971) 67–82.
71. Josse, E.: Versuche über Oberflächenkondensation, insbesondere für Dampfturbinen. Z. VDI 53 (1909) 322–330, 376–383 and 406–412.

K

61. Kast, W.: Wärmeübergang bei Tropfenkondensation. Chem. Ing. Tech. 35 (1963) 163–168.
62. Kast, W.: Bedeutung der Keimbildung und der instationären Wärmeübertragung für den Wärmeübergang bei Blasenverdampfung und Tropfenkondensation. Chem. Ing. Tech. 36 (1964) 933–940.
63. Kast, W.: Probleme des Wärmeübergangs bei Blasenverdampfung und Tropfenkondensation. Chem. Tech. 16 (1964) 10, 601–604.
64. Kast, W.: Theoretische und experimentelle Untersuchung der Wärmeübertragung bei Tropfenkondensation. Fortschr. Ber. VDI-Z. R 3, No. 6 and VDI-Z. 107 (1965) 480.
65. Kindler, H.: Messung des örtlichen Wärmeübergangs bei der Kondensation von gesättigtem Wasserdampf in einem waagerechten Rohr. Diss. TU Braunschweig 1970.
66. Kiper, A. M.: Minimum Bubble Departure Diameter in Nucleate Pool Boiling. Int. J. Heat Mass Transfer 14 (1971) 7, 931.
67. Kirschbaum, E.: Heizwirkung von kondensierendem Heiß- und Sattdampf. Arch. Wärmewirtsch. 12 (1931) 265–266.
68. Kirschbaum, E.; Kranz, B.; Starck, D.: Wärmeübergang am senkrechten Verdampferrohr. VDI-Forsch.-Heft No. 375, 1–8. Berlin 1935.
69. Kirschbaum, E.: Neues zum Wärmeübergang mit und ohne Änderung des Aggregatzustandes. Chem. Ing. Tech. 24 (1952) 393–400.
70. Kirschbaum, E.; Wetjen, K.: Wärmeübergang bei Filmkondensation strömenden lufthaltigen Wasserdampfes am senkrechten Rohr. Chem. Ing. Tech. 25 (1959) 565–568.
71. Kirschbaum, E.: Der Wärmeübergang im senkrechten Verdampferrohr in dimensionsloser Darstellung. Chem. Ing. Tech. 27 (1955) 248–257.
72. Kirschbaum, E.: Der Verdampfungsvorgang bei Selbst-Umlauf im senkrechten Rohr. Dechema-Monogr. 40 (1962) 616–641, 121–147.
73. Koh, J. C. Y.; Sparrow, E. M.; Hartnett, J. P.: The Two Phase Boundary Layer in Laminar Film Condensation. Int. J. Heat Mass Transfer 2 (1961) 1/2, 69–82.
74. König, A.: Der Einfluß der thermischen Heizwandeigenschaften auf den Wärmeübergang bei Blasenverdampfung. Wärme- und Stoffübertragung 6 (1973) 38–44.
75. König, A.; Gregorig, R.: Über das Abreißen von Dampfblasen beim Behältersieden. Wärme- u. Stoffübertragung 6 (1973) 165–174.
76. Körner, M.: Messung des Wärmeübergangs bei der Verdampfung binärer Gemische. Wärme- und Stoffübertragung 2 (1969) 178–191.
77. Kollera, M.; Grigull, U.: Untersuchung der Kondensation von Quecksilberdampf. Wärme- und Stoffübertragung 4 (1971) 244–258.
78. Kopp, J. H.: Wärme- und Stoffaustausch bei Mischkondensation. Brennstoff-Wärme-Kraft 18 (1966) 3, 128 and 129.
79. Kosky, P. G.; Lyon, D. N.: Pool Boiling Heat Transfer to Cryogenic Liquids. Am. Inst. Chem. Engrs. Journal 14 (1968) 372–387.

80. Kotake, S.: On the Mechanism of Nucleate Boiling. Int. J. Heat Mass Transfer 9 (1966) 8, 711–728.
81. Krischer, S.; Grigull, U.: Mikroskopische Untersuchung der Tropfenkondensation. Wärme- und Stoffübertragung 4 (1971) 48–59.
82. Kroger, D. G.; Rohsenow, W. M.: Condensation Heat Transfer in the Presence of a Non-Condensable Gas. Int. J. Heat Mass Transfer 11 (1968) 1, 15–26.
83. Kurihara, H. M.; Myers, J. E.: The Effects of Superheat and Surface Roughness on Boiling Coefficients. A. I. Ch. E. J. 6 (1960) 1, 83–91.
84. Kutateladze, S. S.: Heat Transfer in Condensation and Boiling. USAEC Report AEC-tr-3770 (1952).
85. Kutateladze, S. S.: Boiling Heat Transfer. Int. J. Heat Mass Transfer 4 (1961) Dec., 31–45.

L

61. Lee, J.: Turbulent Film Condensation. A. I. Ch. E. J. 10 (1964) 4, 540–544.
62. Le Fevre, E. J.; Rose, J. W.: Heat Transfer Measurements During Dropwise Condensation of Steam. Int. J. Heat Mass Transfer 7 (1964) 272 and 273.
63. Le Fevre, E. J.; Rose, J. W.: An Experimental Study of Heat Transfer by Dropwise Condensation. Int. J. Heat Mass Transfer 8 (1965) 1117–1133.
64. Leniger, H. A.; Veldstra, J.: Wärmedurchgang in einem senkrechten Verdampferrohr bei natürlichem Umlauf. Chem. Ing. Tech. 34 (1962) 1, 21–26.
65. Leniger, H. A.; Veldstra, I.: Wärmedurchgang im senkrechten Verdampferrohr bei Zwangsumlauf und Entspannungsverdampfung. Chem. Ing. Tech. 34 (1962) 6, 417–422.
66. Linke, W.: Zum Wärmeübergang bei der Verdampfung von Flüssigkeitsfilmen. Kältetechnik 5 (1953) 275–279.
67. Lippert, T. E.; Dougall, R. S.: A Study of the Temperature Profiles Measured in the Thermal Sublayer of Water, Freon-113, and Methyl Alcohol During Pool Boiling. Trans. ASME, Ser. C 90 (1968) 3, 347–352.
68. Lotz, H.: Wärme- und Stoffaustauschvorgänge in bereifenden Lamellenrippen-Luftkühlern im Zusammenhang mit deren Betriebsverhalten. Kältetech.-Klimatisierung 23 (1971) 208–217.
69. Lotz, H.: Programmierte Berechnung bereifender Rippenluftkühler. Kältetech. Klimatisierung 24 (1972) 275–285.
70. Lüder, H.: Wärmeübergang bei der Kondensation von Dämpfen aus Gasdampfgemischen. Vortrag auf der VDI-Hauptversammlung in Dresden 1939 (Vorbericht Z. VDI 83 (1939) 596).
71. Lyon, D. N.: Pool Boiling of Cryogenic Liquids. Chem. Eng. Progr. Sympos., Ser. 64 (1968) 87, 82–92.

M

61. Madejski, J.: Über die Wärmeübertragung bei der Kondensation von Dämpfen in Anwesenheit inerter Gase. Chem. Ing. Tech. 29 (1957) 801–813.
62. Marschall, E.: Wärmeübergang bei der Kondensation von Dämpfen aus Gemischen mit Gasen. Kältetechnik 19 (1967) 8, 241–245.
63. Marschall, E.: Wärmeübergang bei der Kondensation von Dämpfen aus Gemischen mit Gasen. Abh. dt. kältetechn. Ver. (1967) No. 19, 87 p.
64. Marschall, E.; Meyder, R.: Stoffübergang bei Kondensation in Anwenseheit nicht kondensierbarer Gase. Wärme- und Stoffübertragung 3 (1970) 191–196.
65. Marschall, E.; Hickmann, R. S.: Laminar Gravity-Flow Film Condensation of Binary Vapor Mixtures of Immiscible Liquids. Trans. ASME. J. Heat Transfer (1973) 1–5.
66. Marschall, E.; Meyder, R.: On the Condensation Mass Transfer in the Presence of Non-condensables. Wärme- und Stoffübertragung 3 (1973) 191–196.
67. Mayinger, F.; Lahrs, J.: Impuls- und Wärmetransport in konzentrischen und exzentrischen Ringspalten. Chem. Ing. Tech. 47 (1975) 197.
68. McFadden, P. W.; Grassmann, P.: The Relation between Bubble Frequency and Diameter during Nucleate Pool Boiling. Int. J. Heat Mass Transfer 5 (1962) 169–173.

69. Meisenburg, S. J.; Boarts, R. M.; Badger, W. L.: The Influence of Air in Steam on the Steam Film Coefficient of Heat Transfer. Trans. Am. Inst. Chem. Engrs. 31 (1934/35) 622–638.
70. Mikic, V. B.: On Mechanism of Dropwise Condensation. Int. J. Heat Mass Transfer 12 (1969) 1311–1323.
71. Mikic, B. B.; Rohsonow, W. M.: A New Correlation of Pool-Boiling Data Including the Effect of Heating Surface Characteristics. Trans. ASME, Ser. C 91 (1969) 2, 245–250.
72. Mills, A. F.; Seban, R. A.: The Condensation Coefficient of Water. Int. J. Heat Mass Transfer 10 (1967) 12, 1815–1827.
73. Müller, H.: Untersuchungen über den Wärmeübergang an siedendem Wasser in horizontalen Rohren. Wärme 68 (1961) 50–52.
74. Müller, F.: Wärmeübergang bei der Verdampfung unter hohen Drücken. VDI-Forschungsheft 522, Düsseldorf 1967.
75. Muthoo, H. K. K.: Boiling Heat Transfer. Chem. Process. Eng. 42 (1961) 8, 348–350.

N

61. Nagle, W. N.; Bays, G. S.; Blenderman, L. M.; Drew, T. B.: Heat Transfer Coefficients during Dropwise Condensation of Steam. Trans. Am. Inst. Chem. Engrs. 31 (1934/35) 593–604.
62. Nishikawa, K.; Yamagata, K.: On the Correlation of Nucleate Boiling Heat Transfer. Int. J. Heat Mass Transfer 1 (1960) 219–235.
63. Nishikawa, K.; Kusuda, H.; Yamasaki, K.: Growth and Collapse of Bubbles in Nucleate Boiling. Bull. JSME 8 (1965) 30, 205–210.
64. Nitschke, K.: Untersuchung der Intensität des Wärmeaustausches beim Verdampfen organischer Flüssigkeiten im Behälter und in Rohren. Chem. Tech. 18 (1966) 4, 223–229.
65. Nußelt, W.: Die Oberflächenkondensation des Wasserdampfes. Z. VDI 60 (1916) 541–546; 569–575.

O

61. Ouwerkerk, van H. J.: The Rapid Growth of a Vapour Bubble at a Liquid-Solid Interphase. Int. J. Heat Mass Transfer 14 (1971) 9, 1415.

P

61. Panitsidis, H.; Gresham, R. D.; Westwater, J. W.: Boiling of Liquids in a Compact Plate-Fin Heat Exchanger. Int. J. Heat Mass Transfer 18 (1975) 37–42.
62. Pitschmann, P.; Grigull, U.: Filmverdampfung an waagerechten Zylindern. Wärme- und Stoffübertragung 3 (1970) 75–84.
63. Ponter, A. B.; Diah, I. G.: Condensation of Vapors of Immiscible Binary Liquids on Horizontal Copper and Polytetrafluoroethylene-Coated Copper Tubes. Wärme- und Stoffübertragung 7 (1974) 94–106.

R

61. Rant, Z.: Verdampfen in Theorie und Praxis. Dresden und Leipzig: Steinkopf 1959.
62. Renker, W.: Der Wärmeübergang bei der Kondensation von Dämpfen in Anwesenheit nicht kondensierender Gase. Chem. Technik 7 (1955) 451–461.
63. Rettig, H.: Die Verdampfung von Tropfen an einer heißen Wand unter erhöhtem Druck. Diss. TH Stuttgart 1966.
64. Roetzel, W.: Heat Transfer in Laminar Film Condensation. Variable Viscosity and Subcooling. Wärme- und Stoffübertragung 6 (1973) 127–132.
65. Rohsenow, M. W.; Clark, J. A.: A Study of the Mechanism of Boiling Heat Transfer. Trans. Am. Soc. Mech. Engrs. 73 (1951) 609–620.
66. Rohsenow, H.: Nucleation at Heating Surfaces. Ind. Eng. Chem. 57 (1965) 5, 12–14.
67. Rose, J. W.: Condensation of a Vapour in the Presence of a Non-Condensing Gas. Int. J. Heat Mass Transfer 12 (1969) 2, 233–237.

S

61. Sallaly, M.: Die Wärmeübertragung bei der Blasenverdampfung von Flüssigkeiten an künstlichen Siedekeimen. Chem. Eng. Sci. 21 (1966) 4, 367–380.
62. Schlünder, E. U.; Chawla, J. M.: Örtlicher Wärmeübergang und Druckabfall bei der Strömung verdampfender Kältemittel in innenberippten, waagerechten Rohren. Kältetechnik 21 (1969) 5, 136–139.
63. Schlünder, E. U.: Über den Wärmeübergang beim Blasensieden. Verfahrenstechn. 4 (1970) 11, 493–497.
64. Schmidt, E.: Verdunstung und Wärmeübergang. Gesundheitsingenieur 52 (1929) 525.
65. Schmidt, E.; Schurig, W.; Sellschopp, W.: Versuche über die Kondensation von Wasserdampf in Film- und Tropfenform. Tech. Mech. Thermodyn. 1 (1930) 53–63.
66. Schmidt, E.; Behringer, Ph.; Schurig, W.: Wasserumlauf in Dampfkesseln. VDI-Forsch.-Heft No. 365. Berlin 1934.
67. Schmidt, Th. E.: Der Wärmeübergang bei der Kondensation in Behältern und Rohren. Kältetechnik 3 (1951) 282–288.
68. Schneider, H. W.; Chawla, J. M.: Wärmeübergang und Druckabfall beim unterkühlten Sieden in senkrechten Rohren. Chem. Ing. Tech. 47 (1975) 207.
69. Schrader, H.: Einfluß von Inertgasen auf den Wärmeübergang bei der Kondensation von Dämpfen. Chem. Ing. Tech. 38 (1966) 1091–1094.
70. Schulenberg, F.: Wärmeübergang und Druckverlust bei der Kondensation im senkrechten Rohr. Chem. Ing. Tech. 41 (1969) 7, 443.
71. Schulenberg, F.: Wärmeübergang und Druckverlust bei der Kondensation von Kältemitteldämpfen in luftgekühlten Kondensatoren. Kältetech. Klimatisierung 22 (1970) 75–81.
72. Schwartz, F. L.; Siler, L. G.: Correlation of Sound Generation and Heat Transfer in Boiling. Trans. ASME, Ser. C 87 (1965) 4, 436–438.
73. Selin, G.: Kondensation von Wasserdampf in Tropfenform. Dechema-Monogr. 40 (1962) 149–154, 616–641.
74. Sernas, V.; Hooper, F. C.: The Initial Vapor Bubble Growth on a Heated Wall During Nucleate Boiling. Int. J. Heat Mass Transfer 12 (1969) 12, 1627–1639.
75. Shea, F. L.; Krase, N. W.: Dropwise and Film Condensation of Steam. Trans. Amer. Inst. Chem. Engrs. 36 (1940) 463–490.
76. Slipcevic, B.: Verdampfung und Verflüssigung in waagerechten glatten Rohren. Kälte 18 (1965) 9, 481–483.
77. Slipcevic, B.: Über die Verdampfung von Frigenen. Kälte 21 (1968) 9, 468–471.
78. Slipcevic, B.: Bemessung von Verdampfern und Kondensatoren unter Berücksichtigung örtlich veränderlicher Wärmedurchgangskoeffizienten. Kältetech. Klimatisierung 22 (1970) 424–429.
79. Slipcevic, B.: Wärmeübertragung bei der Verdampfung von Frigenen. Verfahrenstechnik 5 (1971) 29–35.
80. Slipcevic, B.: Wärmeübergang beim Sieden von R-Kältemitteln in horizontalen Rohren. Kältetech.-Klimatisierung 24 (1972) 345–351.
81. Sparrow, E. M.; Eckert, E. R. G.: Effects of Superheated Vapor and Noncondensible Gases on Laminar Film Condensation. Am. Inst. Chem. Engrs. Journal 7 (1961) 473–477.
82. Sparrow, E. M.; Lin, S. H.: Condensation Heat Transfer in the Presence of a Noncondensable Gas. Trans. ASME, Ser. C 86 (1964) 3, 430–436.
83. Sparrow, E. M.; Marschall, E.: Binary, Gravity Flow Film Condensation. J. of Heat Transfer, May 1969, 205–211.
84. Stender, W.: Der Wärmeübergang bei kondensierendem Heißdampf. Z. VDI 69 (1925) 905–909.
85. Stepánek, J.; Heyberger, A.; Vesely, V.: Wärmeübergung am waagerechten Rohr bei Kondensation gesättigter und überhitzter Dämpfe. Int. J. Heat Mass Transfer 12 (1969) 2, 137–146.
86. Stephan, K.: Mechanismus und Modellgesetz des Wärmeübergangs bei der Blasenverdampfung. Chem. Ing. Tech. 35 (1963) 11, 775–784.
87. Stephan, K.: Beitrag zur Thermodynamik des Wärmeübergangs beim Sieden. Abh. deutsch. Kältetech. Ver. No. 18 Karlsruhe 1964. See also Chem. Ing. Tech. 35 (1963) 775–784.

88. Stephan, K.: Einfluß des Öls auf den Wärmeübergang von siedendem Frigen 12 und Frigen 22. Kältetechnik 16 (1964) 162–166.
89. Stephan, K.: Stabilität beim Sieden. Brennstoff-Wärme-Kraft 17 (1965) 12, 571–578.
90. Stephan, K.: Übertragung hoher Wärmestromdichten an siedende Flüssigkeiten. Chem. Ing. Tech. 38 (1966) 112–117.
91. Stephan, K.; Körner, M.: Berechnung des Wärmeübergangs verdampfender binärer Flüssigkeits-gemische. Chem. Ing. Tech. 41 (1969) 409–417.
92. Stephan, K.; Körner, M.: Blasenfrequenzen beim Verdampfen reiner Flüssigkeiten und binärer Flüssigkeitsgemische. Wärme- und Stoffübertragung 3 (1970) 185–190.
93. Stewart, J. K.; Cole, R.: Bubble Growth Rates during Nucleate Boiling at High Jakob Numbers. Int. J. Heat Mass Transfer 15 (1972) 655–664.
94. Stewart, P.; Clayton, J.; Loya, B.; Hurd, S.: Condensing Heat Transfer in Steam-Air Mixtures in Turbulent Flow. Ind. Engng. Chem., Proc. Design and Develop. 3 (1964) 48–54.
95. Struve, H.: Der Wärmeübergang an einen verdampfenden Rieselfilm. VDI-Forsch.-H. 534 (1969) 36 pages.
96. Struve, H.: Beitrag zur Bemessung von Rieselverdampfern. Kältetech.-Klimatisierung 24 (1972) 241–252.

T

61. Tamir, A.; Taitel, Y.; Schlünder, E. U.: Direct Contact Condensation of Binary Mixtures. Int. J. Heat Mass Transfer 17 (1974) 1253–1260.
62. Tanner, D. W.; Pope, D.; Potter, C. J.; West, D.: Heat Transfer in Dropwise Condensation at Low Steam Pressures in the Absence and Presence of Noncondensible Gas. Int. J. Heat Mass Transfer 11 (1968) 181–190.
63. Thomas, D.: Blasen- und Filmverdampfung bei Wasser unter atmosphärischem Druck und in der Nähe des kritischen Punktes. Brennstoff-Wärme-Kraft 19 (1967) 1, 14–19.
64. Thomas, D. G.: Enhancement of Film Condensation Heat Transfer Rates on Vertical Tubes by Vertical Wires. Ind. Eng. Chem. Fundamentals 6 (1967) 1, 97–103.
65. Thomas, D. G.: Enhancement of Film Condensation Rate on Vertical Tubes by Longitudinal Fins. A. I. Ch. E. J. 14 (1968) 4, 644–649.
66. Thomas, D. G.: Prospects for Further Improvement in Enhanced Heat Transfer Surfaces. (Evaporation and Condensation.) Desalin 12 (1973) 2, 189–215.
67. Tong, L. S.: Boiling Heat Transfer and Two-Phase Flow. New York: John Wiley 1967.
68. Turek, K.: Wärmeübergang und Druckverluste bei der Filmkondensation strömenden Sattdampfes an horizontalen Rohrbündeln. Chem. Ing. Tech. 44 (1972) 280–285.
69. Turner, R. H.; Millsand, A. F.; Denny, V. E.: The Effect of Non-Condensible Gas on Laminar Film Condensation of Liquid Metals. Trans. ASME, J. Heat Transfer (1973) 6–11.

U

61. Umur, A.; Griffith, P.: Mechanism of Dropwise Condensation. Trans. Amer. Soc. Mech. Engrs. Ser. C 87 (1965) 275–282.
62. Upmalis, A.: Wärmeübergang an siedendes Wasser in rauhen Stahlrohren. Wärme- und Stoffüber-tragung 6 (1973) 160–164.

V

61. Vaihinger, D.; Kaufmann, W. D.: Zum Druckeinfluß auf den Wärmeübergang bei ausgebildeter Blasenverdampfung. Chem. Ing. Tech. 44 (1972) 921–927.

W

61. Wenzel, H.: Neue Wärmeübergangsmessungen bei Tropfenkondensation im Vergleich mit der Theorie. Brennstoff-Wärme-Kraft 18 (1966) 440–445.

62. Wenzel, H.: Erweiterte Theorie des Wärmeübergangs bei Tropfenkondensation. Wärme- und Stoffübertragung 2 (1969) 6–18.
63. Wenzel, H.: On the Condensation Coefficient of Water Estimated from Heat-Transfer Measurements during Dropwise Condensation. Int. J. Heat Mass Transfer 12 (1969) 125.
64. Wett, T.: High-Flux Heat Exchange Surface Allows Area to be Cut by over 80 %. The Oil and Gas Journ. 27 (1971) Dec. 118–120.
65. Winkelsesser, G.: Die Wärmeabgabe von strömendem Heiß- und Sattdampf. Dechema-Monogr. 20 No. 244, Weinheim 1952.
66. Winterton, R. H. S.: Effect of Gas Bubbles on Liquid Metal Heat Transfer. Int. J. Heat Mass Transfer 16 (1973) 549–554.

Z

61. Zuber, N.: Hydrodynamic Aspects of Boiling Heat Transfer. Trans. ASME 80 (1958) 711–720.

THREE

THE INFLUENCE OF THERMAL RADIATION ON HEAT TRANSFER

3-1 THE ABSORPTION AND EMISSION OF RADIATION

The equations for heat transfer indicated so far only take into consideration the effect of thermal conduction and convection. With very high temperatures, such as occur, for example, in the iron and steel industry or in boiler plants, the thermal radiation of the gases and of the solid walls containing them, share quite considerably in the transfer of heat. The following works deal with the influence of gas radiation on heat transfer: first there are those by Nernst [N101], Nusselt [N102], Schack [S101], and other researchers, and then above all those by E. Schmidt [S107], Hottel and Mangelsdorf [H105] and Eckert [E102].

Heat transfer by gas radiation is dependent upon the fact that water vapour and carbon dioxide, which are almost always contained in the hot gas, are capable of both emitting and absorbing radiation. On the other hand, gases which contain one or two atoms, such as oxygen, nitrogen and hydrogen, etc., do not radiate and are almost completely permeable to thermal radiation; thus they do not take part in the exchange of heat by radiation. However other gases containing more atoms do radiate, for example the hydrocarbons; in combustion these gases are burnt at high temperatures to such an extent that they are no longer able to make any noticeable contribution to heat transfer by radiation.

Gas radiation can be increased still further in combustion gases, basically by the radiation of glowing, tiny soot particles which are emitted when combustion is incomplete. In this case, we speak of luminous flames. This phenomenon can be generated by the addition of hydrocarbons to the combustible gases (the so-called carburation). This type of radiation, which is also called flame radiation, is important in the firing of steam boilers and other industrial equipment. However it hardly ever applies to heat exchangers, because in general, combustion no longer takes place here. For this reason this type of radiation will not be dealt with any further. By way of exception, cases where consideration of this type of radiation appears to be important in heat exchangers, can be found in the attached bibliography.[1]

In heat exchangers, the radiation emitted by the gases will be absorbed by the solid walls of the ducts through which they are flowing. Conversely these walls also take part in the transfer of heat by means of their own radiation. In what follows, the laws of radiation of solid bodies will be discussed followed by the laws developed for gas radiation.

Radiation from the Surface of Solid Bodies

The quantity of heat radiating through unit contact area of a solid body or of a gas in unit time, is known as the emissivity.

That body which emits the largest possible amount of energy is also in a position to completely absorb all the radiation impinging upon it. Such a body is said to be "completely black". The energy emitted from such a body per unit surface area and in unit time is given by Eq. (3-1), obtained from the Stefan–Bolzmann Law of Radiation.

$$E_s = C_s(T/100)^4 \tag{3-1}$$

In this equation T denotes the absolute temperature and C_s denotes the radiation coefficient of the completely black body. C_s has the value 5.67 if the units of E_s are W/m².[2] The energy E_s, given by Eq. (3-1), which thus represents the emissivity of a completely black body, will radiate out in all directions into the complete semi-infinite space. The energy which radiates in a direction perpendicular to the surface, is related to the solid angle 1 and amounts to E_s/π.

The radiation of a black body consists of rays of varying wavelengths λ. This distribution of the energy of this radiation in the different wavelength regions, that is the so-called spectral energy distribution of black body radiation, is determined by Planck's Law of Radiation:

$$dE_s = E_{s\lambda}\, d\lambda = 0.374 \cdot 10^{-15} \frac{\lambda^{-5}}{\exp\dfrac{0.01438}{\lambda T} - 1} d\lambda \text{ in W/m}^2 \text{ with } \lambda \text{ in m,} \tag{3-2}$$

where dE_s denotes the radiation energy per unit area and per unit time falling in the small wavelength region between λ and $\lambda + d\lambda$ at temperature T of the black body. The quantity dE_s relates to the unpolarized radiation which is emitted into the total semi-infinite space.

---

[1] Above all, the corresponding sections in Schack's [5] and Eckert's [3] books must be mentioned; these also contain comprehensive bibliographies.

[2] The unit of C_s is $\dfrac{\text{W}}{\text{m}^2}\left(\dfrac{100}{\text{K}}\right)^4$. If this is multiplied by $(\text{K}/100)^4$, that is the unit of $(T/100)^4$, the correct unit of E_s will be obtained.

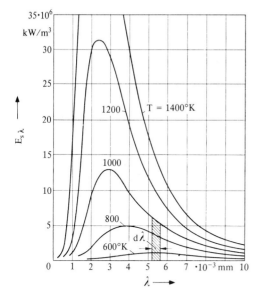

Figure 3-1 Energy distribution of black body radiation.

Figure 3-1, in which $E_{s\lambda}$ is shown as a function of λ, displays the energy distribution given by Eq. (3-2) for various temperatures T. It can be seen that not only the absolute values of the energy increase sharply with rising temperature, but also the distribution of energy for different wavelengths varies with temperature. The radiation energy increases, with rising temperature, much more quickly in the small wavelength regions than in the longer wavelength regions. The highest value of the energy is therefore displaced, with rising temperature, in the direction of decreasing wavelength. For the wavelength λ_m at which $E_{s\lambda}$ has a maximum, *Wien's Displacement Law* specifies that $\lambda_m T = 0.002896 \text{ m K}$.

An *arbitrary* non-black body absorbs only a certain portion of the radiation falling on it. Apart from cases in which part of the radiation is able to penetrate the body, the remainder of the radiation which has not been absorbed will be completely reflected. The ratio ε of the absorbed radiation energy to the incident radiation energy is called the *absorption coefficient* (also, less suitably, absorptivity*). The absorption coefficient is dependent on the type of radiating body and the character of its surface area. Only completely black bodies have an absorption coefficient of 1.

The less a body is able to absorb, the less it is able to radiate. This fact is expressed in *Kirchhoff's Law*, according to which, at a given temperature, the emissivity E of an arbitrary body is the same as the product of its absorption coefficient ε and the emissivity E_s of the completely black body. On this basis ε is also known as the emission coefficient*.

---

* Editor's note: The terms *emissivity* and *absorptivity* are common to English scientific texts but are not used by Hausen.

It follows from this, bearing in mind Eq. (3-1), that an arbitrary body radiates the quantity of heat E in unit time per unit surface area, given by Eq. (3-3) below.

$$E = \varepsilon E_s = \varepsilon C_s (T/100)^4 \tag{3-3}$$

Using $C = \varepsilon C_s$ as the radiation coefficient of a non-black body, (3-4) can be used in place of Eq. (3-3)

$$E = C(T/100)^4 \tag{3-4}$$

Table 3-1 shows the radiation coefficients C and the absorption coefficients ε of different solid materials.

In heat exchanger calculations it is usually sufficient to assume that the absorption coefficient ε is independent of the wavelength, that is, to consider the

Table 3-1 Absorption coefficients ε and radiation coefficients $C = \varepsilon C_s$ of solid materials

Material and its surface composition	t in °C	ε in per cent	C in $\dfrac{W}{m^2}\left(\dfrac{100}{K}\right)^4$
Completely black bodies		100	5.67
Aluminium, polished	37.8	4.5	0.26
	538	7.8	0.45
Aluminium, oxidised	37.8	8 to 20	0.46 to 1.15
	538	18 to 33	1.04 to 1.10
Copper, polished	37.8	4	0.23
	538	4	0.23
Brass, polished	37.8	7	0.40
	260	10	0.58
Brass, oxidised	37.8	46	2.65
	538	67	3.87
Iron, polished smooth	20	24	1.38
Iron, rolling skin	20	77	4.44
Iron, skin (of a casting)	37.8	81	4.67
Iron, red with rust	37.8	69	3.98
Iron, very rusty	20	85	4.90
Nickel, polished	37.8	5	0.28
	538	10	0.58
Nickel, oxidised	538	46 to 66	2.65 to 3.81
Zinc	37.8	2	0.12
	538	4	0.23
Sandstone	37.8	83	4.79
	538	90	5.19
Ceramic material	260	59	3.40
	538	36	2.08
Firebricks	538	63 to 84	3.64 to 4.85
	1093	77 to 91	4.44 to 5.25
Wood	70	91	5.25
Paper	37.8	93	5.37
	538	76	4.39

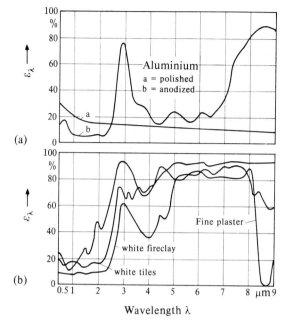

(a)

(b)

Wavelength λ

Figure 3-2a and b Absorption coefficient ε_λ of different solid surfaces dependent on wavelength λ.
(a) Aluminium, polished and anodized.
(b) Non-metallic surfaces.

absorbing or radiating body as "*grey*". On the other hand, if ε is dependent on wavelength, as almost always happens in reality, its surface area is said to be "*coloured*". In certain circumstances ε can deviate from a constant average value and Figs. 3-2a and 3-2b indicate the extent of this deviation for aluminium, fire clay, white tiles and plaster.

The radiation energy calculated from Eqs. (3-3) or (3-4) is emitted from the surface in question into the total semi-infinite space situated in front of the surface. The distribution of this energy in the various beam directions is determined, approximately, by the *Lambert Cosine-Law*. If a surface emits the energy $E_n\, d\Omega$ in unit time, in the perpendicular direction, into the solid angle $d\Omega$, according to this law, its radiation at an angle ϕ will be given by (3-5).

$$E_n \cos \phi \cdot d\Omega \qquad (3\text{-}5)$$

Deviations occur, even from this law, in particular for rays whose direction takes them close to the surface area, i.e., for the so-called oblique rays. Metals radiate in

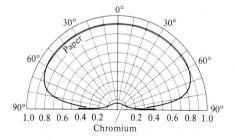

Figure 3-3 Direction distribution of thermal radiation from a non-conducting material (paper) and a metal (chromium).

this direction significantly more than indicated by Lambert's Law while non-metals such as wood, glass and paper radiate less than indicated. Two examples of this are shown in Fig. 3-3. In this diagram the ratio to the radiation from the completely black body, i.e., the absorption or emission coefficient ε_ϕ, is illustrated graphically for every direction, specified by the angle ϕ to the normal of the surface. If the Lambert Cosine-Law is totally valid a full semi-circle will result corresponding to a constant absorption ratio ε_ϕ. In technical calculations it is usually assumed that Lambert's Law is satisfied.

Radiation from Water Vapour and Carbon Dioxide

An important difference in the behaviour of gases compared to solid bodies lies in the fact that the radiation from a gas which falls upon such gases and penetrates them, is only able to be absorbed within a restricted wavelength region. In all other wavelength regions the gases are almost totally permeable to radiation. By way of example, radiation within a very small wavelength region between λ and $\lambda + d\lambda$ will be considered as it penetrates water vapour or carbon dioxide. Part of this radiation will be absorbed by the gas while the rest permeates through the gas and then leaves it again. The part of the incident radiant energy which is absorbed by the gas is known as the absorption coefficient (or also "*absorptivity*") ε_λ of the gas at wavelength λ. ε_λ has a finite value only within the wavelength region in which the gas absorbs radiation; outside this region ε_λ is so small as to disappear. Figures 3-4 and 3-5, which show ε_λ as a function of λ, illustrate the absorption regions, also called "*absorption bands*", of carbon dioxide and water vapour. Finer details are suppressed for the sake of clarity. From these diagrams it will be seen that carbon dioxide has three main absorption bands which stretch from 2.65 to 2.8 µm, from 4.15 to 4.45 µm and from 13 to 17 µm. The infra-red spectrum of water vapour exhibits four principal bands in the wavelength regions of about 1.7 to 2 µm, from 2.3 to 3.4 µm, from 4.4 to 8.5 µm and from 12 to 30 µm. Smaller absorption bands are shown as dotted lines.

Nevertheless, the absorption coefficient ε_λ shown in Figs 3-4 and 3-5 only applies to a certain gas pressure and to a layer of gas of certain thickness. ε_λ

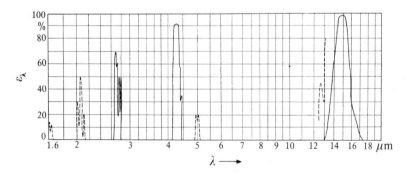

Figure 3-4 Absorption spectrum of carbon dioxide.

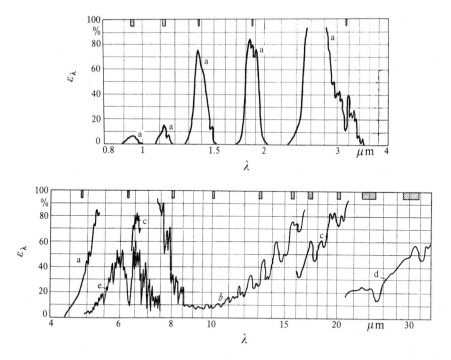

Figure 3-5 Absorption spectrum of water vapour. a, b, c, d, e relate to different thicknesses of the layers of water vapour or to different temperatures.

becomes larger with increasing pressure, or partial pressure, p of the absorbing gas and with increasing layer thickness s of the gas. According to Beer's Law, given T and λ, ε_λ should be dependent only upon the product $p \cdot s$ of the pressure and the layer thickness. If, for example, the partial pressure of carbon dioxide is reduced by a half, according to this law if the radiation path is doubled, then as much radiation will be absorbed as at the original partial pressure. It thus follows from this law that ε_λ is only dependent upon the quantity of gas which absorbs radiation. With constant total pressure *Beer's Law* is almost completely valid for carbon dioxide but not for water vapour, that is at least not for a mixture of water vapour and nitrogen. If the total pressure is increased, deviations from Beer's Law also occur for carbon dioxide; see, for example, [M101].

Where Beer's Law holds true, radiation of wavelength λ and initial energy $J_{\lambda 0}$ is weakened by absorption to J_λ when passing through a gas with a layer thickness s; J_λ is related to $J_{\lambda 0}$ by Eq. (3-6).

$$J_\lambda = J_{\lambda 0} \cdot \exp\left(-a_\lambda p s\right) \tag{3-6}$$

In (3-6) a_λ is known as the "*extinction coefficient*". According to the hypothesis embodied in this law, a_λ is a function of λ but not of p and s. However Eq. (3-6) can be applied more generally, for example to mixtures of water vapour and nitrogen, if a_λ is considered to be dependent also on p (or s). It will be seen from

Eq. (3-6) that $J_{\lambda 0} - J_\lambda$ is the quantity of energy absorbed; in Eq. (3-7) the absorption coefficient ε_λ is developed.

$$\varepsilon_\lambda = \frac{J_{\lambda 0} - J_\lambda}{J_{\lambda 0}} = 1 - \exp\left(-a_\lambda ps\right) \tag{3-7}$$

In order to determine the total effect of gas radiation the *absorption coefficient ε of the gas for the total radiation over all wavelengths penetrating the gas* is needed. This absorption coefficient, which is also known as the *"degree of blackness"*, can be calculated on the whole from the values of ε_λ, because the radiation energy at the various wavelengths can be integrated and the value of the integral for all the incident radiation energy can be inserted into the coefficient. Such calculations have been carried out by Schack [S101]; these enabled approximate values of ε to be determined by theoretical means for the first time.

3-2 CALCULATION OF THE TOTAL RADIATION OF CARBON DIOXIDE AND WATER VAPOUR

Prior to the calculations of Schack mentioned above, W. Nusselt [N102, 103] measured the total radiation of combustion gases, which consisted in the main of equal quantities of CO_2 and CO, up to a temperature of 2250°. In order to be able to determine the heat transfer by radiation of other arbitrarily different gas mixtures, E. Schmidt [S107] and E. Eckert [E102, S109] in Germany and Hottel and Mangelsdorf [H105] and M. McCaig [M101] in America, have measured the total radiation of water vapour and carbon dioxide, as well as of their mixtures with nitrogen or air, over a wide range of temperatures of up to 1300° and with varying layer thicknesses and partial pressures. Hottel and Mangelsdorf found, to some extent, different values for emissivity but basically arrived at the same results as Schmidt and Eckert [S109].[3]

The results of the German and American measurements can be uniformly represented by Figs. 3-6 and 3-7. In these diagrams the product $p \cdot s$ of the total or partial pressure p of the radiating gas and the beam length s is shown in cm · bar as the abscissa and the emission coefficient ε at various temperatures is shown in °C as the ordinate.

Since carbon dioxide complies with Beer's Law, Fig. 3-6 is applicable also to mixtures of carbon dioxide with other non-radiating gases. For this the partial pressure of carbon dioxide has, quite simply, to be inserted for p. The curves in the upper part of Fig. 3-7, on the other hand, only relate to the radiation of *pure* water vapour. In mixtures containing nitrogen, the radiation of water vapour is less than

---

[3] Hottel and Mangelsdorf took their measurements in mixtures containing air while for Eckert's measurements, pure nitrogen was added to the mixture. Perhaps the differences in their results can be explained by the fact that the oxygen in air influences the radiation of carbon dioxide and water vapour in a different manner than does nitrogen. The deviations between the different measurements are also dealt with by H. C. Hottel and R. B. Egbert [H106].

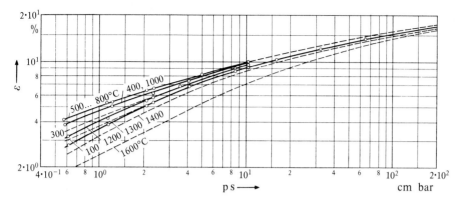

Figure 3-6 Emission coefficient ε of carbon dioxide; from Eckert.

that which would be expected from Beer's Law. Thus for gas mixtures the values of ε read off from the upper part of Fig. 3-7 must be multiplied by a correction factor f. This factor can be derived from the curves in the bottom right-hand corner of the figure; these show f as a function of $p \cdot s$ for different ratios of the water vapour partial pressure p to the total pressure P. These curves, determined from measurements by Hottel and Mangelsdorf [H105] and by Eckert, disregard the small dependence upon temperature found by Eckert [E102].

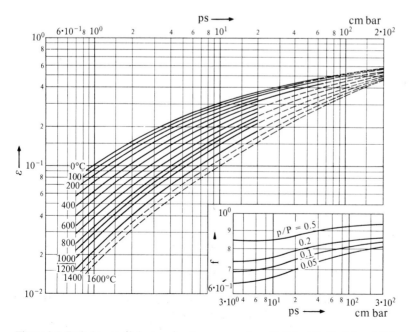

Figure 3-7 Emission coefficient ε of water vapour; from E. Schmidt, Hottel and Eckert: p = partial pressure of the water vapour; P = total pressure; s = layer thickness; f = correction factor.

In other respects Figs. 3-6 and 3-7 are valid for a total pressure of 1 bar abs. and for other pressures which are fairly close to this. At higher pressures an increase in radiation is to be expected.

Radiation from Carbon Dioxide–Water Vapour Mixtures

When carbon dioxide and water vapour are both present in a gas mixture the emission coefficient ε_{CO_2} and ε_{H_2O} cannot simply be added together. Rather, the combined emission or absorption coefficient $\varepsilon_{CO_2+H_2O}$ is somewhat smaller than the sum of ε_{CO_2} and ε_{H_2O} because the bands partially overlap and each of the two radiating gases weakens the radiation of the other by absorption. However the weakening effect is so slight that Eckert does not discuss the matter further in the second edition of his book [3] while Schack [5] considers that a correction for this effect is not needed.

The Influence of the Geometry of the Mass of Radiating Gas

In order to be able to read off ε from Figs. 3-6 and 3-7 it is necessary to know what is the beam length s. The beam length can easily be determined if the gas layer is uniformly thick in all directions. However this is only possible in the special case where the mass of gas is hemispherical in form and where a very small irradiated surface dF is considered at the centre of the base of this hemisphere; see Fig. 3-8a. With other geometries of the mass of gas, as, for example in Fig. 3-8b, the actual shape can be replaced by a hemisphere, the gas contents of which emit the same radiation energy in unit time to the surface area under consideration. Thus the radius of this hemisphere represents the effective beam length of the gas. On this basis, Nusselt [N103], Jakob [J101], E. Schmidt [S108], Hottel [H107] and Eckert [3, 3rd edition, p. 239] calculated s for various cases by integration over all directions of radiation. Their most important results are compiled in the following table.

The effective beam length s for gas radiation.

(a) Ducts:

Circular cross-section of diameter d		$s = 0.95d$
Slot/slit cross-section of width a		$s = 1.8a$
Bundle of tubes with an outer tube diameter d and inter-tube distances a:		
triangular arrangement	$a = d$	$s = 3.0a$
	$a = 2d$	$s = 3.8a$
square arrangement	$a = d$	$s = 3.5a$

(b) Gas chambers of which the dimensions are of the same magnitude in all directions:

Sphere of diameter d	$s = 0.65d$
Cube with sides of length a	$s = 0.66a$
Closed circular cylinder, height $h = $ diameter d	$s = 0.77d$

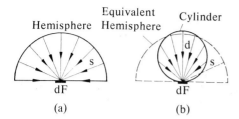

Figure 3-8a and b Irradiation of a surface element dF. (a) By hemispherical gas body with radius s. (b) By a cylindrical gas body with diameter d.

In most cases, according to Port [P105] and Hausen [H103], s can be determined approximately as follows from the equivalent diameter d_{gl}. The following formulae are applicable to the geometries specified:

(a) for ducts with cross section F_q and circumference u_q,

$$d_{gl} = \frac{4F_q}{u_q}, s = 0.9 \, d_{gl} \text{ on average.}$$

(b) for gas chambers of volume V and surface area F

$$d_{gl} = \frac{4V}{F}, s = d_{gl} \text{ on average.}$$

Having determined the effective beam length s in this way, the emission and absorption coefficients of carbon dioxide and water vapour for arbitrary partial pressures and temperatures, can be extracted from Figs. 3-6 and 3-7. Note should be taken of the following considerations when fixing the effective temperatures.

Temperature Dependence of the Absorption and Emission Coefficients of Gases

In the considerations above relating to the total radiation of carbon dioxide and water vapour, it was tacitly assumed that the absorption coefficient of a gas is equal to its emission coefficient and that Kirchhoff's Law also applies to gases. However this only occurs when the solid body which receives radiation from the gas and which then emits radiation to the gas, has the same temperature as the gas. The fact that, in this particular case, Kirchhoff's Law must be fulfilled, results from the consideration that the gas and the solid body can only be in temperature equilibrium in the absence of outside influences, if the gas receives as much radiation as energy from the solid body as it emits to it. In this case, if $p \cdot s$ is disregarded, the emission coefficient of the gas is determined by the gas temperature T_g and can thus be denoted by ε_g. It follows that ε_g can then be read off directly from Figs. 3-6 and 3-7 for the gas temperature T_g.

On the other hand, it is remarkable that when T_g and the wall temperature T_w are not the same, the absorption coefficient of the gas depends only relatively slightly on T_g but strongly on T_w. The slight dependence on T_g corresponds to the slight temperature dependence of the absorption bands in Figs. 3-4 and 3-5. The strong dependence on T_w is explained by the fact that the spectral energy distribution of the radiation sent into the gas by the walls varies considerably with the wall temperature T_w, as has already been shown by Fig. 3-1, for black body radiation. The fragment of energy radiating from the wall, which just falls onto the wavelength regions in which the gas absorbs radiation, thus varies a great deal with the wall temperature. The corresponding amount of the energy absorbed in the various wavelength regions also varies with T_w.

This can be made clear by an example. In Fig. 3-1, in the wavelength region represented by the shaded portion, about 10 per cent of the energy of black body radiation occurs at 600 K while 5 per cent of the energy occurs at 1000 K. If, for example, the absorption coefficient of the gas in this wavelength region amounted to 60 per cent, in the first case, the absorbed portion of the total radiation energy would amount to 6 per cent while in the second case it would amount to 3 per cent. Similar reasoning can be applied to the other absorption regions. In order to emphasize these ideas, namely that the absorption coefficient of the gas is strongly dependent on T_w and to a lesser extent on T_g, this coefficient should be denoted by ε_{gw}.

The influence of the two temperatures T_g and T_w on the absorption coefficient ε_{gw} has been measured by Hottel and Mangelsdorf [H105] and Hottel and Egbert [H106]. According to their observations, the influence of T_g can be disregarded when $T_w < T_g$ or when T_w is only slightly higher than T_g. Thus the absorption coefficient of the gas is only dependent on the wall temperature T_w, and ε_w can be used instead of ε_{gw}. It follows that ε_w can be read off directly from Fig. 3-6 or 3-7 for the wall temperature T_w.

For the case where T_w is considerably higher than T_g Hottel et al. [H105, 106] have represented their results using the equations given below. ε_w^* is the emission coefficient of the gas at wall temperature T_w but with a density equal to that which the gas would possess at temperature T_g, and at an effective partial pressure p_g. The density would attain this value at T_w if the partial pressure was given by Eq. (3-8)

$$p_w = p_g \cdot \frac{T_w}{T_g} \tag{3-8}$$

Thus if ε_w^* is fixed for this partial pressure and for T_w from Figs. 3-6 and 3-7 Hottel et al. calculate the absorption coefficient ε_{gw} of carbon dioxide using Eq. (3-9), and of water vapour employing Eq. (3-10)

$$\varepsilon_{gw} = \varepsilon_w^* \cdot \left(\frac{T_g}{T_w}\right)^{0.65} \qquad \text{carbon dioxide} \tag{3-9}$$

$$\varepsilon_{gw} = \varepsilon_w^* \left(\frac{T_g}{T_w}\right)^{0.45} \qquad \text{water vapour} \tag{3-10}$$

As has already been pointed out, it is only necessary to use these equations in the few cases where T_w is considerably higher than T_g.

3-3 RADIATION BETWEEN A GAS AND A SOLID WALL

Radiative Heat Transfer with a Black Wall

A gas is considered at temperature T_g which is completely surrounded by a black wall at temperature T_w. From knowledge of the emission coefficient ε_g and the absorption coefficient ε_{gw} the exchange of radiation between the gas and the wall, with spatially constant temperatures T_g and T_w, can be calculated as follows. From Eq. (3-3) the gas radiates the energy E_g towards a unit surface area of the wall in unit time:

$$E_g = \varepsilon_g \cdot C_s \cdot \left(\frac{T_g}{100}\right)^4 \tag{3-11}$$

On the other hand the gas absorbs an amount of energy E_w, given by (3-12), from the black body radiation emitted by a unit surface area of the wall.

$$E_w = \varepsilon_{gw} \cdot C_s \left(\frac{T_w}{100}\right)^4 \tag{3-12}$$

The radiation energy which will be transferred in unit time between gas and the surface F of the wall completely surrounding the gas, is given by (3-13).

$$\dot{Q}_s = F(E_g - E_w) = F \cdot C_s\left[\varepsilon_g\left(\frac{T_g}{100}\right)^4 - \varepsilon_{gw}\left(\frac{T_w}{100}\right)^4\right] \tag{3-13}$$

In this equation, as mentioned above, ε_{gw} can usually be replaced by ε_w.

Heat Transfer by Radiation Between a Gas and a Grey Wall with an Arbitrary Absorption Coefficient

If the absorption coefficient ε_{wall} of the wall varies only slightly, that is 20 per cent at the most, from the absorption coefficient of a black wall, the radiative heat transfer can be calculated simply in the following way. The wall only absorbs a portion ε_{wall} of the radiation falling upon it from the gas; the wall also emits a smaller proportion ε_{wall} of energy to the gas. From Eqs. (3-11) and (3-12) it follows that the two amounts thus decrease in the ratio $\varepsilon_{wall} : 1$. The exchange of radiation between the gas and a grey wall with the absorption coefficient $\varepsilon_{wall} > 0.8$ can thus be calculated approximately using Eq. (3-14)

$$\dot{Q}_s = F \cdot \varepsilon_{wall} \cdot C_s\left[\varepsilon_g\left(\frac{T_g}{100}\right)^4 - \varepsilon_{gw}\left(\frac{T_w}{100}\right)^4\right] \qquad \text{where } \varepsilon_{wall} > 0.8 \tag{3-14}$$

On the other hand, if a greater accuracy is required or if $\varepsilon_{wall} < 0.8$ the radiation reflected from the wall, and the remaining wall radiation which is left after the absorption by the gas, play an important part which should not be overlooked. This energy also takes part in the exchange of radiation in which part of the reflected radiation is absorbed by the gas and in which the residual radiation which is not taken up by the gas, is partly absorbed and partly reflected by other sections of the wall. For the remaining non-absorbed radiation, this process is repeated for successively smaller quantities of energy so that, in the end, the total energy transferred by radiation is represented by an infinite series.

The summation of the terms of this series has been carried out exactly by Elgeti [E106]. The two following approximation methods have been derived from his results; with these the exchange of radiation can be calculated fairly quickly, and with sufficient accuracy. As Eckert [E102] before him, Elgeti replaced Eq. (3-13) with (3-15) for the exchange of radiation with a grey wall.

$$\dot{Q}_s = FC_s \left[\bar{\varepsilon}_g \left(\frac{T_g}{100} \right)^4 - \bar{\varepsilon}_{gw} \left(\frac{T_w}{100} \right)^4 \right] \tag{3-15}$$

In Eq. (3-15) $\bar{\varepsilon}_g$ and $\bar{\varepsilon}_{gw}$ take the place of ε_g and ε_{gw} or ε_w in Eq. (3-13). The use of $\bar{\varepsilon}_g$ and $\bar{\varepsilon}_{gw}$ instead of ε_g and ε_{gw} or ε_w expresses the influence of the secondary reflections and absorptions. As the transformations of ε_{gw} or ε_w to $\bar{\varepsilon}_{gw}$ or $\bar{\varepsilon}_w$ and of ε_g to $\bar{\varepsilon}_g$ are to be effected in a similar manner, the indices g and w will be dropped in order to simplify matters. For mixtures of CO_2 and H_2O $\varepsilon_{CO_2+H_2O}$ is calculated by the simple addition of ε_{CO_2} and ε_{H_2O}.

From the first set of transformations of Elgeti together with the values of ε obtained from either or both of Figs. 3-6 or 3-7, the ratio $\varepsilon/\varepsilon_\infty$ can be obtained, in which ε_∞ denotes the absorption coefficient of an infinitely thick gas layer. From Elgeti [E106] the following can be inserted for ε_∞:

$$\text{with } CO_2: \qquad \varepsilon_\infty = 0.23$$
$$\text{with } H_2O: \qquad \varepsilon_\infty = 0.90$$
$$\text{with } CO_2 + H_2O: \quad \varepsilon_\infty = 0.98$$

With the known absorption coefficient of the wall ε_{wall} the absorption ratio $\bar{\varepsilon}/\varepsilon_\infty$ can then be read off as ordinate value from Fig. 3-9. Finally multiplication by ε_∞ gives the required values of $\bar{\varepsilon}$, i.e., of $\bar{\varepsilon}_g$, $\bar{\varepsilon}_{gw}$ or $\bar{\varepsilon}_w$.

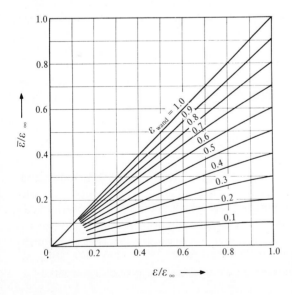

Figure 3-9 Diagram for the determination of the values of $\bar{\varepsilon}$ in Eq. (3-15) from ε and ε_∞ with known emission coefficient ε_{wall}.

It is even simpler, although not quite so accurate, to express the exchange of radiation which has been accumulated by the reflections and the additional absorptions, by means of an apparently increased layer thickness s^*. s^* can be calculated from the real layer thickness s and the absorption ratio ε_{wall} of the wall using Eq. (3-16).

$$s^* = s/\varepsilon_{wall}^{0.85} \tag{3-16}$$

With $p \cdot s^*$ an emission or absorption coefficient ε^* can be read off from Figs. 3-6 and 3-7 for the required temperature T_g or T_w and thus Eq. (3-17) can finally be obtained.

$$\bar{\varepsilon} = \varepsilon_{wall} \cdot \varepsilon^* \tag{3-17}$$

With simultaneous radiation of carbon dioxide and water vapour ε^* is the sum of $\varepsilon_{CO_2}^*$ and $\varepsilon_{H_2O}^*$. Elgeti [E106] has shown that this process is also almost always sufficiently accurate, principally because the experimentally determined curves in Figs. 3-6 and 3-7 are subject to great uncertainty.

Spatially Variable Gas and Wall Temperatures

In a heat exchanger the gas temperature T_g, and usually the wall temperature T_w, vary from one position to another, for example in the longitudinal direction of a tube. For this reason the intensity of the radiation also varies strongly from one position to another. In such cases the total radiation energy which is transferred can be obtained, in principle, by integration or by summation over many small quantities of radiation. Nevertheless, by using the Gaussian Method of Quadrature [G101], Hausen and Binder [H104] have shown that for most cases it is sufficient to calculate the quantity of heat transferred by radiation at only two appropriately selected positions. If the gas in the heat exchanger cools down from T_{g1} to T_{g2} these positions are determined to be those at which the gas temperature has the following values:

$$T_{g1}^* = T_{g1} - 0.2(T_{g1} - T_{g2}) \tag{3-18}$$

and

$$T_{g2}^* = T_{g2} + 0.2(T_{g1} - T_{g2}) \tag{3-19}$$

At these positions the radiation $\dot{q}_{s1}^*$ and $\dot{q}_{s2}^*$ which is transferred per unit surface area can be calculated using Eq. (3-15). From this the average value $\dot{q}_{sm}$ which applies to the entire surface area can be formed from Eq. (3-20).

$$\frac{1}{\dot{q}_{sm}} = \frac{1}{\dot{q}_{s1}^*} + \frac{1}{\dot{q}_{s2}^*} \tag{3-20}$$

Thus, Eq. (3-21) finally yields the total radiation being transferred in unit time.

$$\dot{Q}_s = F \cdot \dot{q}_{sm} \tag{3-21}$$

By this method $\dot{Q}_s$ can be obtained to an accuracy of usually about 2 per cent.

In Eqs. (3-18) and (3-19) the numerical factor has been set intentionally at 0.2 instead of at the value 0.2113, as in the Gaussian Quadrature Method, because, compared with precise calculations, the energy of thermal radiation can be computed with sufficient accuracy using the value 0.2. This is due to the fact that the thermal radiation is more or less proportional to the fourth power of absolute temperature, while the Gaussian process computes the integral exactly over a second order function using only two data points.

Roetzel [R101] has suggested that, for the Gaussian quadrature, the temperatures T_{g1}^* and T_{g2}^* can be determined logarithmically by the following equations:

$$\lg (T_{g1}^* - T_w) = \lg (T_{g1} - T_w) - 0.211[\lg (T_{g1} - T_w) - \lg (T_{g2} - T_w)] \quad (3\text{-}22)$$

$$\lg (T_{g2}^* - T_w) = \lg (T_{g2} - T_w) + 0.211[\lg (T_{g1} - T_w) - \lg (T_{g2} - T_w)] \quad (3\text{-}23)$$

in which T_w denotes the wall temperature which is, in the first instance, taken to be constant. With a spatially variable wall temperature the values at the positions in question are inserted in place of T_w. If the quantities of heat being transferred by radiation are calculated at temperatures which have been determined in this way, the total amount of heat transferred by radiation is obtained from Eqs. (3-20) and (3-21). In the extreme case of heat transfer by radiation alone, using the Stefan–Boltzmann Law, with $T_{g1} = 1300\,\text{K}$, $T_{g2} = 400\,\text{K}$ and the wall temperature, which is considered here to be constant, $T_w = 300\,\text{K}$, Roetzel was able to obtain a remarkably higher degree of accuracy with his own equations than with Eqs. (3-18) and (3-19).

Equations (3-18), (3-19) and (3-20) or (3-21) and (3-22) can also be used for calculations if, in addition to radiation, convection also takes part in heat transfer. In this case $\dot{q}_{s1}^*$ and $\dot{q}_{s2}^*$ can be replaced by the total quantities of heat transferred ($\dot{q}_1^*$ and $\dot{q}_2^*$) at the positions specified. The total transmission of heat can thus be calculated accordingly.

Determination of the Radiative Component α_s of Heat Transfer Coefficients

The influence of thermal radiation upon heat transfer, which may be very considerable at high temperatures, can be investigated mathematically by introducing a heat transfer coefficient α_s which represents this radiation, the value of which can be added to the value of the convective heat transfer coefficient. Using α_s, the quantity of heat, which is transferred per unit surface area and per unit time between a gas and a wall by means of radiation, can be expressed by

$$\dot{q}_s = \Delta \dot{Q}_s / \Delta T = \alpha_s (T_g - T_w) \quad (3\text{-}24)$$

If this is compared with (3-15), then Eq. (3-25) is obtained.

$$\alpha_s = \frac{C_s}{T_g - T_w} \left[\bar{\varepsilon}_g \left(\frac{T_g}{100} \right)^4 - \bar{\varepsilon}_w \left(\frac{T_w}{100} \right)^4 \right] \quad (3\text{-}25)$$

For small temperature differences $T_g - T_w$, $\bar{\varepsilon}_w$ can be set equal to $\bar{\varepsilon}_g$. Defining

$T_m = (T_g + T_w)/2$, we can develop the approximation:

$$\frac{T_g^4 - T_w^4}{T_g - T_w} = \left(\frac{dT^4}{dT}\right)_{T_m} = 4T_m^3 \qquad \text{(see Footnote 4)}$$

Thus Eq. (3-25) becomes

$$\alpha_s = \frac{4C_s}{100} \cdot \bar{\varepsilon}_g \left(\frac{T_m}{100}\right)^3 \qquad (3\text{-}26)$$

If α_k is the component in the heat transfer coefficient, which represents convection and thermal conduction, the total heat transfer coefficient α becomes the sum of α_k and α_s:

$$\alpha = \alpha_k + \alpha_s \qquad (3\text{-}27)$$

In order to estimate the order of magnitude of the influence of thermal radiation on the transfer of heat overall, the radiation of a gas is examined which contains 70 per cent nitrogen, 15 per cent carbon dioxide and 15 per cent water vapour. This gas flows at atmospheric pressure through a tube with a diameter of 100 mm and with a wall absorption coefficient $\varepsilon_{wall} = 0.9$. The velocity of flow calculated at 0°C, amounts to $w_0 = 2.5$ m/s. If the temperature difference between gas and wall is not too large, then following from the relationships above, the radiative component of the heat transfer amounts to 6 per cent at 0°C, 39 per cent at 400°C and 81 per cent at 1200°C.

3-4 LITERATURE ON RADIATIVE HEAT TRANSFER

See Foreword, Sec. 1–5.

B

101. Birkebak, R. C.; Eckert, E. R. G.: Effects of Roughness of Metal Surfaces on Angular Distribution of Monochromatic Reflected Radiation. Journ. Heat. Transfer 87 (1965) 85–94.
102. Beér, J. M.; Siddall, R. G.: Verfahren zur Voraussage der Wärmeübertragung durch Strahlung in Flammen. Chem. Ing. Tech. 46 (1974) 47–55.

C

101. Codegone, C.: Die Wärmestrahlung der Flammen in nicht isothermen Hohlräumen. Int. J. Heat Mass Transfer 5 (1962) 121–127.

E

101. Eckert, E.: Messung der Reflexion von Wärmestrahlen an technischen Oberflächen. Forsch. Ing. Wes. 7 (1936) 265–270.
102. Eckert, E.: Messung der Gesamtstrahlung von Wasserdampf und Kohlensäure in Mischung mit nichtstrahlenden Gasen bei Temperaturen bis zu 1300°C. VDI-Forschungsheft 387, Berlin 1937.

---

[4] The margin of error in this equation is less than 1 per cent for T_g/T_w between 0.9 and 1.1. The same margin of error applies to Eq. (3-26) in the case where the difference between ε_g and ε_w does not produce a greater margin of error.

103. Eckert, E.: Technische Strahlungsaustauschmessungen. Berlin: VDI-Verlag (1937.

104. Edwards, D. K.: Journ. Opt. Soc. Am. 50 (1960) 617–666. Journ. Heat Transfer 84 (1962) 1–11. (Radiation properties of the gas from monochromatic measurements.)

105. Edwards, D. K.: Radiative Transfer Characteristics of Materials. J. Heat Transfer 91 (1969) 1.

106. Elgeti, K.: Ein neues Verfahren zur Berechnung des Strahlungsaustausches zwischen einem Gas und einer grauen Wand. Brennstoff-Wärme-Kraft 14 (1962) 1–6.

G

101. Gaußsches Integrationsverfahren, siehe z.B. Hütte, Des Ingenieurs Taschenbuch, vol. I, 28th ed. 1955, S. 502–504.

102. Günther, R.: Das Entwerfen von Flammen und Feuerräumen. Chem. Ing. Tech. 46 (1974) 56–62.

H

101. Habib, I. S.; Greif, R.: Heat Transfer to a Flowing Non-Gray Radiating Gas. Int. J. Heat Mass Transfer 13 (1970) 1571–1582.

102. Hahne, E.; Tratz, H.: Temperaturverlauf und Wärmeabgabe für einen Stab bei gleichzeitiger Wärmeleitung und Wärmestrahlung. Wärme- und Stoffübertragung 1 (1968) 52.

103. Hausen, H.: Briefliche Mitteilung an Professor Eckert. Vgl. Eckert, E.: Einführung in den Wärme- und Stoffaustausch, 3rd ed. Berlin, Göttingen, Heidelberg: Springer 1966 p. 239.

104. Hausen, H.; Binder, J. A.: Vereinfachte Berechnung der Wärmeübertragung durch Strahlung von einem Gas an eine Wand. Int. J. Heat Mass Transfer 5 (1962) 317–327.

105. Hottel, H. C.; Mangelsdorf, G.: Heat Transmission by Radiation from Nonluminous Gases II. Experimental Study of Carbon Dioxide and Water. Trans. Amer. Inst. Chem. Engrs. 31 (1935) 517–549.

106. Hottel, H. C.; Egbert, R. B.: Trans. Amer. Soc. mech. Engrs. 63 (1941) 293–307.

107. Hottel, H. C.; Egbert, R. B.: Radiant Heat Transmission from Water Vapor. Trans. Am. Inst. Chem. Engrs. 38 (1942) 531–568. (Effective layer thickness for various arrangements.)

108. Howell, J. R.; Perlmutter, M.: Monte Carlo Solution of Thermal Transfer through Radiant Media between Gray Walls. J. Heat Transfer, C 86 (1964) 116–122.

J

101. Jakob, M.: In Eucken, A., und Jakob, M.: Der Chemie-Ingenieur: vol. I, part 1, Leipzig. 1933, pp. 300–303. (Even layer.)

K

101. Kast, W.: Die Erhöhung der Wärmeabgabe durch Strahlung bei mehrfachen Reflexionen zwischen strahlenden Flächen. Fortschr. Ber. VDI Z. Series 6, No. 5 (1965).

102. Koritnig, O. Th.: Die Wärmeübertragung in technischen Feuerungen durch Karburierung. Wärme 65 (1942) 281 and 282.

103. Krinninger, H.: Messung des Emissionsvermögens austhenitischer Stähle für den schnellen natriumgekühlten Brutreaktor. Wärme- and Stoffübertragung 3 (1970) 139–145.

L

101. Landfermann, C. A.: Über ein Verfahren zur Bestimmung der Gesamtstrahlung von Kohlensäure und Wasserdampf in technischen Feuerungen. Diss. TH Karlsruhe 1948 (theoretical).

102. Leckner, B.: Radiation from Flames and Gases in a Cold Wall Combustion Chamber. Int. J. Heat Mass Transfer 13 (1970) 185–197.

M

101. McCaig, M.: Ultraabsorption von Wasserdampf und Kohlensäure. London, Edinburgh, Dublin, Phil. Mag. J. Sci. 34 (7) (1943) 321–342.

N

101. Nernst, W.: Beitrag zur Strahlung der Gase. Phys. Zeitschr. (1904) 777.
102. Nußelt, W.: Der Wärmeübergang in der Verbrennungskraftmaschine. VDI-Forschungsheft 264 (1923) and Z. VDI 67 (1923) 692–708.
103. Nußelt, W.: Die Gasstrahlung bei der Strömung im Rohr. Z. VDI. 70 (1926) 763–765.

P

101. Papula, L.: Verfahren zur Erzeugung leuchtender Flammen bei methanhaltigen Gasen durch Selbstkarburierung des Methans im Ofenraum. Z. VDI 91 (1949) 208.
102. Patat, F.: Wärmeübergang bei leuchtenden und nichtleuchtenden Flammen. Verfahrenstechnik 1942, No. 3, 90 and 91.
103. Penner, S. S.: Quantitative Molecular Spectroscopy and Gas Emissivities. Reading, Mass, Addison-Wesley Publ. 1959.
104. Pich, R.: Der Wärmeaustausch durch Strahlung zwischen zwei Flächen, von denen die größere die kleinere umschließt. Wärme 69 (1962) 28–31.
105. Port, F. J.: Heat Transmission by Radiation from Gases. Sci. D. Thesis, Massachusetts Inst. of Techn. 1939. (Equivalent layer thickness.)

R

101. Roetzel, W.: Berücksichtigung veränderlicher Wärmeübergangskoeffizienten und Wärme-kapazitäten bei der Bemessung der Wärmeaustauscheı. Wärme- und Stoffübertragung 2 (1969) 163–170.

S

101. Schack, A.: Über die Strahlung der Feuergase und ihre praktische Berechnung. Z. techn. Phys. 5 (1924) 267–278, see also Schack, A.: Der industrielle Wärmeübergang. Düsseldorf: Verlag Stahleisen 1929, pp. 206–225.
102. Schack, A.: Zur Extrapolation der Messungen der ultraroten Strahlung von Kohlensäure und Wasserdampf. Z. techn. Physik 22 (1941) 50–56.
103. Schack, A.: Die Strahlung der Feuergase. Arch. Eisenhüttenwesen vol. 13 (1939/40) No. 6, 241/48. (Report in Z. VDI 85 (1941) 197).
104. Schack, K.: Berechnung der Strahlung von Wasserdampf und Kohlendioxid. Chem. Ing. Tech. 42 (1970) 53–58.
105. Schimmel, W. P.; Novotny, J. L.; Kast, S.: Effect of Surface Emittance and Approximate Kernels in Radiation-Conduction Interaction. Wärme- und Stoffübertragung 3 (1970) 1–6.
106. Schmidt, E.: Wärmestrahlung technischer Oberflächen bei gewöhnlicher Temperatur. Beiheft z. Gesundh.-Ing. Series 1 Issue 20, München 1927.
107. Schmidt, E.: Messung der Gesamtstrahlung des Wasserdampfes bei Temperaturen bis 1000°C. Forschung 3 (1932) 57–70.
108. Schmidt, E.: Die Berechnung der Strahlung von Gasräumen. Z. VDI 77 (1933) 1162–1164.
109. Schmidt, E.; Eckert, E.: Die Wärmestrahlung von Wasserdampf in Mischung mit nichtstrahlenden Gasen. Forsch. Ing.-Wes. 8 (1937) 87–90.
110. Schwiedessen, H.: Anteil von Konvektion, Wand- und Gasstrahlung bei der Wärmeübertragung in Industrieöfen. Z. VDI 82 (1938) 404 and 405.
111. Schwiedessen, H.: Die Strahlung von Kohlensäure und Wasserdampf mit besonderer Berück-sichtigung hoher Temperaturen. Arch. Eisenhüttenw. 14 (1940/41) 9–14, 145–153 and 207–210. (Diagrams of total radiation up to 2100°C).
112. Sorofim, A. F.; Hottel, H. C.: Radiation Exchange among Non-Lambert Surfaces. Trans. Am. Soc. Mech. Engrs. Ser. C 88 (1966) 37–44.
113. Sparrow, E. M.: Heat Radiation between Simply Arranged Surfaces. Am. Inst. Chem. Engrs. J. 8 (1962) 12–18.

114. Sparrow, E. M.; Eckert, E. R. G.; Jonsson, V. K.: An Enclosure Theory for Radiative Exchange between Specularly and Diffusely Reflecting Surfaces. Trans. Am. Soc. Mech. Engrs. Ser. C 84 (1962) 294–300.
115. Splett, S.: Berechnung der Wärmestrahlung eines Gaskörpers an eine graue Wand. Brennstoff-Wärme-Kraft 17 (1965) 70 and 71.

T

101. Tingwaldt, C.: Die Absorption von Kohlensäure zwischen 300 und 1100°C. Phys. Z. 35 (1934) 715–720 and 39 (1938) 1–6.

V

101. Vortmeyer, D.; Börner, C. J.: Die Strahlungsdurchlaßzahl in Schüttungen. Chem. Ing. Tech. 38 (1966) 1077–1079.
102. Vossebrecker, H.: Zur Berechnung des Wärmetransports durch Strahlung mit der Monte-Carlo-Methode. Wärme- und Stoffübertragung 3 (1970) 146–152.

Z

101. Ziegler, A.: Der Einfluß der Karburierung und des Wasserdampfgehaltes von Heizgasen auf den Wärmeübergang im Siemens-Martin-Ofen. Bėr. Stahlwerks-Aussch. d. Ver. d. Eisenhüttenl. No. 96 (1925).

FOUR

THE PRESSURE DROP ACCOMPANYING FLUID FLOW THROUGH TUBES AND DUCTS

4-1 BASIC PROCESSES RELATING TO FLUID FLOW IN TUBES AND DUCTS

The fluid velocity of a gas or liquid flowing in a heat exchanger cannot rise above a certain limit. This is because the pressure drop, which accompanies this rise in fluid velocity, represents an energy loss for which economical or operational considerations impose a certain top limit. Thus it is important to be able to calculate the pressure drop in heat exchangers as well as the quantities of heat which are transferred.

However only the most important mathematical interrelationships in pressure drop will be discussed. For further details reference should be made to text books on fluid dynamics and other appropriate publications, for example the VDI-Wärmeatlas (VDI-Heat Atlas) [1], 2nd edition, 1974, pp. La1 to Lm2.

As with the laws of heat transfer, the laws of pressure drop in tubes and ducts vary depending on whether the flow is laminar or turbulent, see Sec. 2-2.

The Transition from Laminar to Turbulent Flow

As has already been mentioned in Sec. 2-2 the Reynolds number, which covers both laminar and turbulent flow, is known as the critical Reynolds number Re_{kr}.[1] When the entry of the fluid or gas into the tube is rough and greatly disturbed the critical Reynolds number is 2320, provided the internal diameter of the tube d is inserted into the Reynolds number as in Eq. (2-8). When the gas entry is placid, particularly when this is due to a rounded inlet piece, Re_{kr} can be considerably greater and

---

[1] From O. Reynolds [R153].

indeed, when all disturbances have been painstakingly eliminated, it can reach to over 50,000. For industrial applications, as a rule the calculation can be based on a critical Reynolds number of between 3000 and 4000.

Much theoretical and experimental work has been concerned with the question of the physical reasons for the sudden switch to turbulent flow. Prandtl [P151] and Tietjens [T152] were able to show that an increase in small disturbances is to be expected when the development of the velocity profile exhibits turning points. Such turning points can, for example, be generated by vortices which are superimposed on the laminar movement. Further contributions to the development of the theory in this area have been made by Heisenberg [H152], Tollmien [T153], Küchemann [K153], and others; see, for example, [P151, 152, 153].

Meissner and Schubert [M152] were able to calculate the entropy of laminar and turbulently flowing gases by combining an analysis of the hydrodynamics with considerations of the thermodynamics; the highest possible value of the entropy occurred when $Re < 1900$ for laminar flow and when $Re > 1900$ for turbulent flow. Thus, it follows from the second law of thermodynamics, that a sudden change over to turbulent flow can be expected when $Re > 1900$.

An experimental contribution of fundamental importance to the question of the development of turbulence was made by Schiller and his co-workers [S157]. These researchers observed the behaviour of flowing water immediately after it entered a tube. It is known that with sharp inlet edges the liquid suffers a constriction (contraction) shortly after entering the tube. Schiller and his co-workers were able to detect the formation of ring vortices at the constriction position which then move along the wall with the flow. With comparatively low fluid velocities such ring vortices occur a long way below the critical Reynolds number. However, because of the internal friction, the energy of the ring vortices fades away without any further disturbance to the laminar flow, so long as the dimensionless circulation of these vortices, which will be explained later, does not exceed the value 2340.

By the circulation of a vortex[2] is understood simply the product of the circumference and the circumferential velocity of the vortex, formed for an arbitrary cross-section of the vortex. The dimensionless circulation Z/v is the sum Z of the chronological mean circulations of all vortices within a tube section whose length is equal to the tube diameter d, divided by the kinematic viscosity v.

When, because of an increase in the fluid velocity, the dimensionless circulation reaches the critical value 2340, the flow becomes unstable and a turbulent mixing motion is introduced. Moreover, as Schiller and his co-workers were able to show, this critical value of the circulation in all types of tube entry examined by them, was reached just as the critical Reynolds number was attained, but was independent of the value of the critical Reynolds number itself. Indeed a value of Z/v considerably greater than 2340 was needed for the change to turbulence when Re_{kr} was only

---

[2] Usually the circulation is defined by the integral $\oint w \, ds$ which is formed along a closed line (w = velocity, ds = distance element, both considered as vectors).

slightly greater than 2320 which indicates the large degree of stability in the laminar flow before the switch to turbulence occurs. The fact that the critical value of Z/v is always 2340, brings light upon the processes which occur in rounded entry pieces and where the inflow of the liquid is very calm. Here vortices with a perceptible circulation only occur at considerably higher fluid velocities than with sharp entry edges, so that the value $Z/v = 2340$ is only reached with large Reynolds numbers.

The transfer from laminar to turbulent flow with increasing velocity in general does not happen suddenly. Rather there is a small intermediary transition region within which laminar and turbulent flow alternate, approximately periodically. Thus the change over from one type of flow to another can often take place once per second.

Starting Process and Fully-developed Flow

It is necessary in laminar flow, as in turbulent flow, to differentiate between the starting process and the fully-developed flow, which first appears at a specific distance from the inlet. The fluid usually flows into the tube or duct at a fluid velocity which is constant over the entire cross-section. The fluid then passes through the "starting distance", or hydrodynamic entry length. Here the parts flowing close to the wall become more and more delayed due to friction. On the other hand, in the centre portion of the tube cross-section, the fluid velocity increases since the average velocity w must indeed remain unaltered when variations in density or variations in the cross-section are disregarded. The velocity profile, which is obtained by plotting the velocity over a tube diameter, finally tends towards a specific final form which corresponds to the fully-developed condition of flow. In laminar flow the curve which reproduces the ultimate velocity distribution is a parabola (Fig. 4-1, left). On the other hand, in turbulent flow, the curve is flatter because of the equalizing effect of the mixing motion in the centre part of the tube, while close to the wall, in a region that is essentially within the boundary layer, the velocity decreases very quickly (see Fig. 4-1, right).

The transition taking place in this ultimate velocity distribution is, strictly speaking, asymptotic. In practice, one refers to fully-developed or turbulent flow as soon as the actual velocity in the centre of the tube differs by only 1 per cent from the corresponding velocity in the ultimate distribution. The distance between this

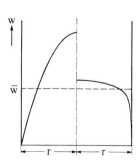

Figure 4-1 Velocity distribution over a tube cross-section. Left: fully developed laminar flow; Right: turbulent flow; $\bar{w}$ is the average fluid velocity.

position and the inlet can be considered as the length of the starting distance or the hydrodynamic entry length.* The pressure drop in the hydrodynamic entry length will be considered at the end of Sec. 4-2.

4-2 PRESSURE DROP IN SMOOTH TUBES AND DUCTS

The pressure drop in tubes can only be calculated from a strictly theoretical viewpoint in the case of fully-developed laminar flow. Approximate solutions are known for laminar and turbulent initial flow in the hydrodynamic entry length. However, in all other respects, our knowledge of pressure drop, as that of the heat transfer coefficients, comes, almost without exception, from measurements. In the case of pressure drop, as in that of heat transfer, the similarity principle largely determines the empirical equations which replicate the experimental results.

Considerations in Terms of the Similarity Principle, of the Pressure Drop in Fully-developed Laminar or Turbulent Flow

As the average fluid velocity w, the density ρ of the fluid, its viscosity η and the tube diameter d have a determining influence on pressure drop, the Reynolds number (see Sec. 2-4, Eq. (2-8)),

$$Re = \frac{\rho w d}{\eta} = \frac{w d}{v} \quad \text{with} \quad v = \frac{\eta}{\rho}$$

plays a fundamental role in pressure drop, as it does in heat transfer. A further dimensionless group exists, in which the pressure drop Δp is divided by the kinetic energy $\rho w^2/2$, related to the unit volume of the fluid, which is the same as atmospheric pressure. The "half-value" of this dimensionless group is known as the Euler number Eu. In addition, the ratio of the tube length L to the tube diameter d must occur as another dimensionless group. It follows from Sec. 2-4 that a similarity relationship of the form

$$\frac{2\Delta p}{\rho w^2} = \Phi\left(Re, \frac{L}{d}\right) = 2Eu \tag{4-1}$$

exists for the pressure drop in smooth tubes, in which Φ denotes an arbitrary function. If it is required to represent the fact, known empirically, that Δp is proportional to the tube length, Eq. (4-2) is developed which is only dependent on Re.

$$\Delta p = \psi(Re) \cdot \frac{\rho w^2}{2} \cdot \frac{L}{d} \tag{4-2}$$

---

* Editor's note: Hausen uses the term "Anlaufstrecke" which can be translated as "starting distance". In the English literature, the term "hydrodynamic entry length" is often used instead.

In engineering $\psi = \psi(Re)$ is usually called the resistance coefficient. The main aim of all pressure drop measurements is to determine the value of ψ as a function of Re.

If the cross-section of flow is not circular, it is advantageous and customary to introduce the equivalent (hydraulic) diameter d_{gl} into the Reynolds number, and into the final factor on the right hand side of Eq. (4-2), in place of the inner diameter of the tube, d, as in Eq. (2-14).

The Resistance Coefficient ψ—Reynolds Number Re Curve

In Fig. 4-2, $\psi(Re)$ is shown for fully-developed laminar and turbulent flow in straight tubes of circular cross-section, on the basis of known measurements [e.g. N102, H153, B151, P153]. The left leg of the curve, which extends as far as $Re = 2320$ or somewhat above, applies to laminar flow, while the remaining curves, further to the right, relate to turbulent flow. Equation 4-3 is applicable in general to smooth tubes and ducts, in the region of fully-developed laminar flow, as can be demonstrated theoretically.

$$\psi(Re) = \text{const}/Re \qquad (4\text{-}3)$$

The constant in this equation is dependent on the shape of the cross-section of flow. In circular tubes it has a value of 64. Using this value, it is possible to calculate the decline in the value of ψ for increasing Re which is shown graphically at the extreme left hand side of Fig. 4-2. With a square cross-section whose side has length a, the constant takes a value of 56.9 and for flow between flat plates at a distance a apart, it has a value of 96. Thus, from Eq. (2-14), the following can be inserted into the Reynolds number and into Eq. (4-2): with a square cross-section $d = d_{gl} = a$; for

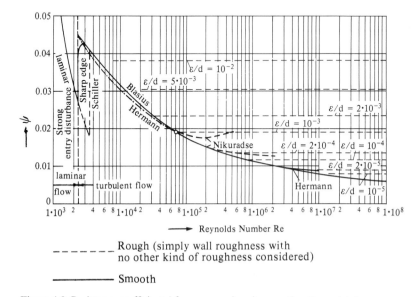

Figure 4-2 Resistance coefficient ψ for pressure drop in smooth and rough tubes.

flow between flat plates $d = d_{gl} = 2a$; see for example, [H151]. The relationship:

$$\Delta p = \frac{\text{const}}{2} \cdot \eta \cdot w \cdot \frac{L}{d^2}$$

follows from Eq. (4-2), (4-3) and Eq. (2-8) by regarding $v\rho = \eta$. Thus with fully-developed laminar flow, viscosity η and velocity w are proportional to the pressure drop.

If the Reynolds number exceeds the critical value which, in practical cases, is usually in the range between 3000 and 4000, the function $\psi(Re)$ increases with the appearance of turbulence (the vertical line in Fig. 4-2). If the Reynolds number then increases still further, $\psi(Re)$ decreases also in the *turbulent region*, but the decrease is slower than it is in the laminar region. The lower solid line in Fig. 4-2 illustrates this behaviour for *smooth tubes*. The slope of ψ, between the critical Reynolds number and $Re = 100,000$ (approximately), can be represented very closely for smooth tubes by the relationship (4-4) which was established by Blasius [B153].

$$\psi(Re) = \frac{0.3164}{\sqrt[4]{Re}} \tag{4-4}$$

With a larger Reynolds number $\psi(Re)$ decreases more slowly than predicted by Eq. (4-4). The following very precise equation developed by Hermann [H153] is valid for smooth tubes up to $Re = 2,000,000$:

$$\psi = 0.00540 + 0.3964 Re^{-0.300} \tag{4-5}$$

In the totally turbulent region, the pressure drop in smooth tubes can also be illustrated by the following relationship developed by Prandtl [P153]:

$$\frac{1}{\sqrt{\psi}} = 2\lg(Re\sqrt{\psi}) - 0.8 = 2\lg\frac{Re\sqrt{\psi}}{2.51} \tag{4-6}$$

It follows from this discussion that in order to calculate the pressure drop in smooth tubes, for a given Reynolds number, it is only necessary to extract the value of ψ from the solid curve in Fig. 4-2 or to calculate its value from one of the equations given, namely (4-3) to (4-6), and then to insert this value of ψ into Eq. (4-2).

Pressure Drop in Straight Tubes Taking the Starting Process into Consideration

From the theoretical considerations of Boussinesq [B156], the hydrodynamic starting length l_a, in laminar flow, can be calculated using:

$$l_a = 0.065 \cdot d \cdot Re \tag{4-7}$$

and this is in close agreement with experimental results. With turbulent flow, the starting length is usually shorter and is not, or only slightly, dependent upon the Reynolds number. According to Nikuradse's [N152] measurements, in turbulent flow, l_a can be set as equal to 50 to 80 times the inner tube diameter d.

It can be deduced from both calculations and measurements that the pressure drop is always greater in the starting length than it is in the fully-developed flow; see, for example, [L4]. If the entry process itself is taken into consideration, then this can be proved theoretically, in the following way. As the velocity directly after entry into the tube is usually greater than the velocity in the area just beyond the tube entry, the fluid must be accelerated by the entrance process. The work expended to achieve this acceleration will be $\rho w^2/2$ per unit volume if the energy of the fluid prior to entry, which is usually very small, is ignored. Thus, particularly in the laminar case, the kinetic energy of the fully-developed flow is greater than the energy at the entrance where the velocity is the same, at all positions on the tube cross-section. The work to achieve the necessary acceleration must, for this reason, be expended in the starting length. In addition, in the starting length, the deceleration of the velocity in the area close to the wall necessitates that frictional work be done. Finally, with an angular entry, there is usually, also, to be added a contraction loss, which is caused by the constriction of flow shortly after entry, as has been discussed previously in Sec. 4-1. According to calculations of Schiller [S155], with laminar flow and with a rounded inlet

$$2.16 \cdot w^2/2$$

and with turbulent flow and an angular inlet

$$1.4 \cdot w^2/2$$

are obtained for the total additional pressure loss which is caused by the processes described above, which occur at entry and in the starting length. With laminar flow, according to Hagen's measurements which were published in 1839 but are very reliable, the additional pressure loss increases to $2.7w^2/2$ if here also, an angular inlet is selected rather than a rounded one. The difference between an angular and a rounded inlet can be attributed to the contraction loss, mentioned above, which occurs at an angular inlet.

If, on the basis of these considerations, it is necessary to consider the additional pressure drop arising at the entrance and in the starting length, Eq. (4-2) should be expanded in the following way for pressure drop with an angular inlet:

$$\Delta p = \left[\psi(Re) \cdot \frac{L}{d} + 2.7 \right] \frac{\rho w^2}{2} \qquad (4\text{-}8)$$

With turbulent flow:

$$\Delta p = \left[\psi(Re) \cdot \frac{L}{d} + 1.4 \right] \frac{\rho w^2}{2} \qquad (4\text{-}9)$$

The numerical values 2.7 and 1.4 have also been confirmed, with a close degree of approximation, by recent measurements. In the second edition of his book, which appeared in 1973, Gregorig [G212] gives the values of 2.66 and 1.56 respectively for these. With a well rounded inlet, each of these is lowered by about 0.44. Nevertheless these values only play an important part in comparatively short tubes or in the special case where, in laminar flow, the Reynolds number considerably exceeds the value of 2320 through the use of a carefully rounded inlet and the avoidance (as much as possible) of all other inlet disturbances, thereby avoiding a change to turbulent motion. On the other hand, Eqs. (4-8) and (4-9) both assume that the tubes approximately contain the whole starting length and thus they are not too short. In heat exchangers it is often sufficient to calculate the pressure drop by means of the simpler Eq. (4-2), using Figs. 4-2 or 4-4 or even by employing one of the Eqs. (4-3) to (4-6).

If the value of the pressure drop, calculated from Eq. (4-2), (4-8) or (4-9), is compared with the pressure drop which has actually been observed in heat exchangers, the measured values are often considerably higher. In part this is explained by the fact that the calculation only takes into account the tubes of a heat exchanger, whereas practical measurements also include the pressure drop in the head as well as in the inlet and outlet columns of the heat exchanger. Linke [L152] has investigated the pressure drop in heat exchanger cowls and Roetzel [R154] has investigated the influence of the temperature dependence of the physical properties.

A further reason for the higher pressure drop simply lies in the fact that the tubes are rough rather than smooth. The following subsection deals with pressure drop in rough and curved tubes.

4-3 PRESSURE DROP IN ROUGH AND CURVED TUBES

Pressure Drop in Rough Tubes

It is not unreasonable to suggest that roughness on a tube wall will favour the development of turbulence. However, even in rough tubes, no critical Reynolds numbers appear that are lower than 2320. The change over to turbulence usually occurs at about the same Reynolds number as in smooth tubes, insofar as the entry conditions are the same [S153]. However in rough tubes the pressure drop in turbulent flow is always higher than in smooth tubes and, in general, the higher the Reynolds number, the greater the difference.

According to observations made by Hopf [H156], Schiller [S156] and Fromm [F153] it is necessary to distinguish between two types of roughness:

1. "shortwave" roughness peaks, called for short, "wall roughness", as shown in Fig. 4-3, top and middle, in the angular and the more rounded forms;
2. gently rounded "longwave" roughness, as shown in Fig. 4-3, bottom, called for short, "wall ripples".

In using this distinction, individual roughness elements lie close to each other in "wall roughness" but are further apart, relative to the size of the amplitude of the undulations, in "wall ripples".

In general, the resistance coefficient is dependent in addition to the Reynolds number, upon the relative roughness ε/d, where ε is the average height of the roughness ripple and d is the inner tube diameter up to that point.

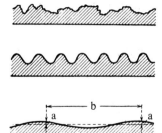

Figure 4-3 Two types of roughness in flow ducts. Top and middle: "shortwave" roughness; bottom: "longwave" roughness.

With wall roughness and with turbulent flow, the ψ-curve for a specific value of ε/d, passes only slightly above the curve for smooth tubes for small Reynolds numbers. Beyond a specific Reynolds number whose value is larger, the smaller the value of ε/d, the ψ-curve for wall roughness, having in some cases reached a minimum,[3] very quickly becomes a completely horizontal line. Such horizontal lines are plotted in Fig. 4-2 for different values of d/ε and thus for different relative roughnesses ε/d using Nikuradse's equation [N153]:

$$\psi = \frac{1}{(1.74 + 2\lg d/2\varepsilon)^2} \quad \text{or} \quad \frac{1}{\sqrt{\psi}} = 2\lg\frac{3.72d}{\varepsilon} \qquad (4\text{-}10)$$

Thus in this region ψ is only a function of ε/d. As a consequence, from Eq. (4-2), a completely square law of resistance exists for a precise value of ε/d, i.e., the pressure drop is exactly proportional to the square of the fluid velocity. It follows that in this case, because the influence of the Reynolds number, and thus that of the viscosity, disappears, the effect of the frictional forces becomes completely negligible compared with the forces of inertia of the turbulent mixing motion.

The horizontal lines in Fig. 4-2 can be applied, approximately, to commercial, copper drawn tubes, if it is noted that in such tubes ε usually has an order of magnitude of 0.01 mm. However, the values obtained in this manner are uncertain as copper pipes have other sources of roughness as well as roughness in the surface of wall itself.

On the other hand, with "wall ripples" of specific dimensions ε/d, the ψ-curve for the completely turbulent region has a slope similar to that for smooth tubes except that the greater the value of ε/d, the higher on the graph is the corresponding ψ-curve. All the possible intermediate types of roughness that can be thought of fall in between these two limiting cases. Colebrook and White [C153] have examined the basic shapes of the ψ-curves for various artificially produced, well defined types of roughness. Their results can be replicated satisfactorily employing the relationship (4-11) which is a combination of Eqs. (4-6) and (4-10) developed by Prandtl and Nikuradse.

---

[3] The thin, dashed line in Fig. 4-2 shows this type of variation, for example from the observations of Nikuradse in experiments involving sand roughness: this dashed line leaves the curve for smooth tubes just below $\psi = 0.02$.

$$\frac{1}{\sqrt{\psi}} = -2\lg\left[\frac{2.51}{Re\sqrt{\psi}} + \frac{\varepsilon}{3.72d}\right] \tag{4-11}$$

Graphical illustrations of ψ-values derived from this equation, and from the Prandtl–Colebrook equation, are to be found, for example, in [R151], pp. 162 and 190 to 198. An equation by Galavics [G151, R151] has more general applicability.

Most of the tubes used in engineering embody a mixture of roughness. The extent to which the form of the inner surface area approaches the limiting case of "wall roughness" or "wall ripples" depends, principally, upon the building material and composition of the wall. Tubes made of rusty iron or cast iron, for example, are close to the limiting case of wall roughness, while duct walls of planed wood or asphalted sheet iron belong, rather, to the other limiting case.

It would be of considerable practical importance at this point to discuss the behaviour of drawn steel tubes whose surface area composition is also mixed but

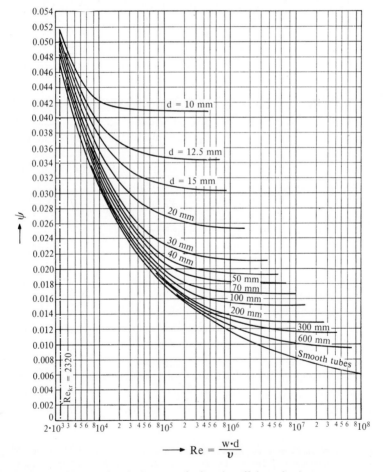

Figure 4-4 Pressure drop in drawn steel tubes. ψ coefficient of resistance.

tends more towards wall rippling. The values of ψ for straight steel tubes extracted from measurements taken by Fritzsche [F151], Zimmermann [Z152], Vuščović [V151] and Kamner[4] are illustrated in Fig. 4-4. These results compensate one another in such a way that by taking an average they permit a good repetition of the experimental results to be achieved. Indeed, Fritzsche's test values, which were determined specifically for rough tubes, were 10 per cent higher than the values displayed on the ψ-curve illustrated, for tubes of diameter 26 mm and 3 per cent higher for tubes of 39 mm diameter. Kamner's experimental values for tubes with a diameter of 10 mm on the other hand, lie about 7 per cent below the corresponding curve in Fig. 4-4.

Figure 4-4 shows that, in straight steel tubes, the curves for ψ, from the critical Reynolds number onwards, mostly lie somewhat higher than the curves for smooth tubes and that the curves for these tubes, as with wall rippling, change only gradually. The slope of the curves for steel tubes for very high values of Re cannot be predicted with any certainty, using previous experimental data. In Fig. 4-4 it is assumed that the curves gradually change to horizontal lines, and as such correspond to pure wall roughness. On the other hand, according to Zimmermann, after running through a flat minimum the curves climb again. Presumably, after this rise in the curve, it becomes horizontal again and the curve is thus flat but s-shaped.

Pressure Drop in Curved Tubes

In curved tubes the critical Reynolds number at which the transfer from laminar to turbulent flow takes place, is higher than in straight tubes. If D is the curve diameter and, as before, d is the internal diameter of the tube, according to White [W151] the critical Reynolds number is about 6000 when $D/d = 50$ and about 9000 when $D/d = 15$, compared with a value of from 2320 to about 4000 in straight tubes. Values of Re_{kr} measured by Taylor [T151], Adler [A152] and Woschni [W11] are in agreement with the values of White mentioned above. According to E. F. Schmidt [S161] this dependence of Re_{kr} upon d/D can be represented very precisely by means of Eq. (4-12)

$$Re_{kr} = 2300[1 + 8.6(d/D)^{0.45}]\qquad(4-12)$$

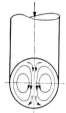

Figure 4-5 Double vortex in a curved tube.

---

[4] Prior to 1950, in the Linde A.G. Laboratory at Höllriegelskreuth near Munich, W. Kamner measured the pressure drop of nitrogen in steel tubes with an inside diameter of 10 mm in the range $Re = 2300$ to $Re = 20,000$. His results were never published.

This increase in the critical Reynolds number, which appears to be surprising at first, can perhaps be explained by the fact that in curved tubes lateral movements occur in the form of a double vortex in the region of laminar flow (see Fig. 4-5). These double vortices are initiated by that fact that the liquid particles are subject to the fastest flow in the middle of the tubes and are often subject to centrifugal forces in curved tubes; the liquid thus tends to move towards the sides of the tube which are opposite to the centre of curvature. Such lateral movements are presumably able to absorb part of the energy which disturbs the flow and thus delays the changeover to turbulence.

These double vortices could also be the basic reason why the pressure drop is greater in curved tubes than in straight tubes.

White [W151], and others, have measured the coefficient of resistance in laminar flow, which is larger in curved tubes, and have represented these measurements by means of an empirical equation.

Numerous measurements are available which were taken with turbulent flow and turning angle of 90°; other measurements are also available, but there are fewer of them, for turning angles larger and smaller than 90°, see for example, [W151, F152, H154, Z151]. As the pressure drop in straight tubes is usually fairly small, the precise measurement of this drop, which is, moreover, strongly dependent upon the technical construction and the roughness of the tube, is very difficult. The experimental values fluctuate widely and give different dependences upon the curve ratio D/d. On average, the results of the measurements can be combined in the following way. If Eq. (4-13) is developed in such a way that it takes into account the additional pressure loss caused by rerouting the flow, then ζ_u takes the value of 0.1, on average, for a turning angle of 90° in smooth tubes.

$$\Delta p_u = \zeta_u \frac{\rho w^2}{2} \left(= \psi_u \cdot \frac{\rho w^2}{2} \frac{L}{d} \right) \qquad (4\text{-}13)$$

With rough tubes and the same angle of curvature, ζ_u lies between 0.1 and 0.4, with an average value of 0.25. With a turning angle of 45° ζ_u is about six times bigger than the values given, and with a reversal angle of 180° it is about 1.4 to 1.5 times larger. From this it can be recognized that, as determined, principally by Fritzsche and Richter [F152], the first parts of the curve generate more additional pressure loss than do the later parts. Presumably this can also be explained by the fact that more energy is needed to initiate the transverse movements illustrated in Fig. 4-5 than to maintain them further along the curve. The value of ζ_u takes into account that an increased pressure loss is also present in part of the straight length of tube which is joined onto the curve, as it is here that the transverse movement first begins to fade away. It is evident that this loss has a proportionately greater influence on small angles of curvature than on large ones.

It is to be expected that, in coils of tubes with many helical coils, the pressure loss occurring over an average turn of 90° is less than when only one turn of 90° is present. This is confirmed by Jeschke's experiments [J2]; he has combined the results of his measurements taken in coiled, drawn wrought-iron tubes, with Reynolds numbers above 110,000, in Eq. (4-14).

$$\psi_{kr} = 0.0238 + 0.0891 d/D = 0.0238(1 + 3.74 d/D) \qquad (4\text{-}14)$$

In this equation ψ_{kr} denotes the total coefficient of resistance in curved tubes and D is, again, the curve diameter.

It is not easy to deduce from Eq. (4-14) that ψ_{kr}, if the assertion above is true, is smaller when the measurements correspond to curvatures of 90°. Indeed, it can be assumed that the constant term 0.0238 represents the value of ψ for a straight tube. It follows then from Eq. (4-14), and taking into account Eq. (4-13), that the relationship below can be derived for a number of coils n.

$$\zeta_u = \psi_u \frac{L}{d} = (\psi_{kr} - \psi)\frac{L}{d} = 0.0891 \frac{d}{D} \cdot \frac{\pi D}{d} \cdot n = 0.0891 \pi n.$$

From this, for a curve of 90°, i.e., for $n = 1/4$, it follows that $\zeta_u = 0.070$. However this is less than one third of the value measured, on average, for 90° curves with rough tube walls.

On the basis of later measurements, Woschni [W11] has confirmed that Eq. (4-14) provides a good approximation.

According to tests carried out by Jeschke with $Re = 110,000$ and less, ψ_{kr} climbs, through the curve of the tube, with decreasing Reynolds number, much more steeply than would be expected from Eq. (4-14). Thus, according to Jeschke, $\psi_{kr} - \psi$ is about seven times greater with $Re = 20,000$; with $Re = 40,000$ it is about four times greater and with $Re = 80,000$ about 1.5 times greater than the value predicted by Eq. (4-14). A similar, though somewhat smaller gradient for $\psi_{kr} - \psi$ has been found by E. F. Schmidt [S161] and also by Srinivasan, Nandapurkar and Holland [S22]. According to Schmidt the influence of tube curvature is more satisfactorily represented by an equation such as (4-14) if d/D is replaced by $(d/D)^{0.62}$ with a correspondingly altered constant. It is possible to achieve a similar result on the basis of the published results of the other authors mentioned [S22].

Woschni has gone on to measure the influence which the heating of the fluid has on pressure drop and heat transfer, and has found that this influence is better expressed by the factor $(\eta_{fl}/\eta_w)^{0.27}$ or $(\eta_{fl}/\eta_w)^{0.3}$ rather than the usual $(\eta_{gl}/\eta_w)^{0.14}$.

According to numerous authors, the influence of curvature on pressure drop is considerably greater for laminar flow than it is for turbulent flow, but this will not be discussed further.

4-4 PRESSURE DROP IN CROSS FLOW AND IN PACKED BEDS

Pressure Drop in Tubes Belonging to an Assembly of Tubes Arranged Across the Direction of Flow

Most research workers investigating heat transfer by cross flow have also measured the pressure drop in cross flow (see Sec. 2-11). Their results have been reproduced in the form:

$$\Delta p = n\psi \frac{\rho w^2}{2} \tag{4-15}$$

in which n is the number of the rows of tubes, lying one behind the other, in the direction of flow. ψ again is a dimensionless quantity which is primarily dependent on the Reynolds number and partly on the "ratios of division", i.e., on the ratio of the distance between the tube centres and the external tube diameter. As in Sec. 2-11, a is the transverse ratio of division, b is the longitudinal ratio of division. w denotes velocity in Eq. (4-15) and in the Reynolds number, since it also denotes the maximum fluid velocity between the tubes in the corresponding heat transfer equations; d is the external tube diameter.

The pressure drop measurements of Pierson [P41] and Huge [H50] can be replicated using the following equations by Jakob [J151];

with an aligned arrangement:

$$\psi = Re^{-0.15}\left[0.176 + \frac{0.32b}{(a-1)^{0.43+1.13/b}}\right] \qquad (4\text{-}16)$$

with a staggered arrangement:

$$\psi = Re^{-0.16}\left[1 + \frac{0.47}{(a-1)^{1.08}}\right] \qquad (4\text{-}17)$$

Thus, the physical properties occurring in the Reynolds dimensionless group and in Eq. (4-15) relate to a reference temperature θ_*, in the case where larger temperature differences exist between the fluid and the tube wall. θ_* can be calculated from the average temperature θ_m of the fluid and the surface area temperature Θ_0 of the wall. In an aligned arrangement θ_* can be found using:

$$\theta_* = \theta_m - 0.1(\theta_m - \Theta_0),$$

and in staggered arrangements:

$$\theta_* = \theta_m - 0.2(\theta_m - \Theta_0).$$

Equations (4-16) and (4-17) are valid for 10 and more rows of tubes lying one behind the other in the direction of flow. With only four rows of tubes ψ is about 8 per cent higher.[5]

In general, the measurements taken by Jauernick [J152] in water are in agreement with the tests carried out by Pierson and Huge and the formulae (4-16) and (4-17) which relate to them. However, according to Jauernick, they have higher values.

Reiher's [R42] measurements in aligned arrangements are in close agreement with Eqs. (4-16) and (4-17) but his values for staggered arrangements are higher by 50 per cent or more. Glaser's [G42] experimental results for aligned arrangements are 40 per cent to 50 per cent higher. On the other hand, Ter Linden [L151] and Brandt [B154] have measured lower values for flow resistance than have Pierson and Huge. Ter Linden's values are only slightly lower but the values obtained by Brandt are up to 50 per cent lower.

---

[5] Further equations of a similar type have been given by Hofmann [H49].

Deviations of up to ± 30 per cent from the respective average values are to be found in later experimental results, such as those of Bergelin [B152], Kays [K151], Dwyer [D151], Bressler [B52], Hammeke [H42], Scholz [S54], Groehn [G152], Achenbach [A151], London [L153], Brauer [B51, Klier [K46], Samoshka [S151] and others. The sometimes apparently complex dependence of the resistance coefficient on Re and the ratios of division a and b has been represented by W. Kast in the second edition of the VDI-Wärmeatlas (VDI Heat Atlas) [1], pp. Ld2 and Ld3 in two diagrams for aligned and staggered arrangements.

In rough tubes, calculations must be made with higher pressure drops than in smooth tubes. Thus for example, Wiener [W152] has measured a pressure drop which is 1.22 times higher in very rusty iron tubes than in smooth tubes.

Pressure Drop in Packed Beds

The shape of the element in a packed bed and the roughness of its surface area generally varies in such a way that the pressure drop in a gas flowing through a packed bed can be very variable. It is thus to be expected that a specific pressure-drop equation can only be valid for packing materials of the same or similar type.

In continuing an investigation begun by Ergun [E151], Brauer [B155] devised Eq. (4-18) for the pressure drop in a packing formed of spheres or of spherical packing materials:

$$\psi = \frac{160}{Re'} + \frac{3.1}{(Re')^{0.1}} \qquad (4\text{-}18)$$

where Δp, the pressure drop in a packing of height H, can be calculated using Eq. (4-19),

$$\Delta p = \psi \cdot \frac{1-\varepsilon}{\varepsilon^3} \cdot \rho w^2 \cdot \frac{H}{d} \qquad (4\text{-}19)$$

In this equation ε is the relative porosity, w is the fluid velocity in the duct when empty, d is the diameter of the sphere or average particle and

$$Re' = \frac{1}{1-\varepsilon} \cdot \frac{wd}{v}.$$

4-5 LITERATURE ON PRESSURE DROP

See Foreword, Sec. 1–5.

A

151. Achenbach, E.: Influence of Surface Roughness on the Flow through a Staggered Tube Bank. Wärme- und Stoffübertragung 4 (1971) 120–126. See also 2 (1969) 47–52.
152. Adler, M.: Strömung in gekrümmten Rohren. Z. angew. Math. Mech. 14 (1934) 257–275.

B

151. Bauer, B.; Galavics, F.: Experimentelle und theoretische Untersuchungen über die Rohrreibung. Zürich 1936.

152. Bergelin, P. O.; Brown, G. A.; Doberstein, S. C.: Trans. ASME 74 (1952) 953–960.
153. Blasius, H.: Das Ähnlichkeitsgesetz bei Reibungsvorgängen in Flüssigkeiten. Phys. Z. 12 (1911) 1175, oder Forschungsarbeit Ing. Wes. No. 131 (1913).
154. Brandt, H.; Dingler, J.: Wärme 59 (1936) 1–8; Brandt, H.: Über Druckverlust und Wärmeübertragung in Wärmeaustauschern. Diss. Techn. Hochschule Hannover 1934.
155. Brauer, H.: Druckverlust in Füllkörpersäulen bei Einphasenströmung. Chem. Ing. Tech. 29 (1957) 785–790.
156. Boussinesq, J.: Comptes Rendus 113 (1891) 9 and 49.

C

151. Chawla, J. M.: Reibungsdruckabfall bei der Strömung von Flüssigkeits-Gas-Gemischen in waagerechten Rohren. Forsch. Ing. Wes. 34 (1968) 47–54.
152. Chawla, J. M.: Reibungsdruckabfall bei der Strömung von Flüssigkeits-Gas-Gemischen in waagerechten Rohren (Review of the state of the art). Chem. Ing. Tech. 44 (1973) 58–63.
153. Colebrook, C. F.; White, C. M.: Experiments with Fluid Friction in Roughened Pipes. Proc. Roy. Soc. A 161 (1937) p. 367; see the extract of Homann, Fr. in Z. VDI 82 (1938) 405 and 406.

D

151. Dwyer, O. E.; Shehan, T. V.; Weismann, J.; Horn, F. L.; Schomer, R. T.: Ind. Eng. Chem. 48 (1956) 1836–1846.

E

151. Ergun, S.: Fluid Flow through Packed Columns. Chem. Engng. Progr. 48 (1952) 89–94.

F

151. Fritzsche, O.: Untersuchungen über den Strömungswiderstand der Gase in geraden zylindrischen Rohrleitungen. VDI-Forschungsheft 60 (1908).
152. Fritzsche, O.; Richter, H.: Beitrag zur Kenntnis des Strömungswiderstandes gekrümmter rauher Rohrleitungen. Forschung 4 (1933) 307–314.
153. Fromm, K.: Strömungswiderstand in rauhen Rohren. Z. ang. Math. Mech. 3 (1923) 339.

G

151. Galavics, F.: Die … Rauhigkeitscharakteristik zur Ermittlung der Rohrreibung … Schweiz. Arch. angew. Wiss. Tech. 55 (1939) 337–354. See Feuerungstech. 28 (1940) No. 6.
152. Groehn, H. G.; Heinecke, E.; Scholz, F.: Atomwirtschaft 14 (1969) 581–583.

H

151. Hahnemann, H.; Ehret, L.: Der Druckverlust der laminaren Strömung in der Anlaufstrecke von geraden ebenen Spalten. Jahrbuch der deutschen Luftfahrtforschung 1 (1941) 21–32 and (1942) 186–207.
152. Heisenberg, W.: Über Stabilität und Turbulenz von Flüssigkeitsströmen. Ann. d. Phys. (IV), 74 (1924) 597.
153. Hermann, R.; Burbach, Th.: Strömungswiderstand und Wärmeübergang in Rohren. Leipzig 1930.
154. Hofmann, A.: Der Druckverlust in 90°-Krümmern bei gleichbleibendem Kreisquerschnitt. Mitt. d. Hydr. Instituts der Techn. Hochschule München, No. 3 (1929) 45–67.
155. Hofmann, A.: Druckverlust und Wärmeübergang in durchströmten Rohren mit innenliegenden Drahtwendeln. Linde-Ber. from Techn. u. Wiss. (1962) 13, 25–31.
156. Hopf, L.: Die Messung der hydraulischen Rauhigkeit. Z. ang. Math. Mech. 3 (1923) 329.

J

151. Jakob, M.: Flow Resistance in Cross Flow of Gases over Tube Banks. Trans. Amer. Soc. Mech. Eng. 60 (1938) 384–386.

152. Jauernick, R.: Über den örtlichen Widerstandsbeiwert und die Wärmeübergangszahlen in Rohr-
    bündeln bei hohen Reynoldsschen Zahlen. Die Wärme 61 (1938) 738–743 and 751–756.

K

151. Kays, W. M.; London, A. L.; Lo, R. K.: Trans. ASME 76 (1954) 387–396.
152. Kirchbach, H.: Der Energieverlust in Kniestücken. Mitt. d. hydr. Inst. d. Techn. Hochsch.
    München, No. 3 (1929) 68–97.
153. Küchemann, D.: Störungsbewegungen in einer Gasströmung mit Grenzschicht. Z. angew. Math.
    Mech. 18 (1938) 207–222.

L

151. Ter Linden, A. J.: Der Strömungswiderstand eines Rohrbündels. Wärme 62 (1939) 319–323 (see
    also Schack, A.: Arch. Eisenhüttenwes. 13 (1939) 169.)
152. Linke, W.; Dia, T.: Die Bemessung der Zu- und Abflußhauben von Wärmeaustauschern. Kälte-
    technik 15 (1963) 85–91.
153. London, A. L.; Mitchell, J. W.; Sutherland, W. A.: Trans. ASME Ser. C 82, No. 3 (1960) 199–213.
154. Lorenz, H.: Strömung und Turbulenz. Handbuch der phys. und techn. Mechanik von Auerbach
    und Hort, Vol. VII, 1931.
155. Lorenz, H.: Die Energieverluste in Rohrerweiterungen und Krümmern. Z. ges. Kälteind. (1929)
    129.

M

151. Maurer, G. W.; Le Tourneau, B. W.: Friction Factors for Fully Developed Turbulent Flow in
    Ducts with and without Heat Transfer. Trans. Am. Soc. Mech. Engrs. Ser. D 86 (1964) 3, 627–636.
152. Meissner, W.; Schubert, G. U.: Kritische Reynoldssche Zahl und Entroepieprinzip. Ann. Phys. 6.
    Series 3 (1948) 163–182.
153. Möbius, H.: Experimentelle Untersuchung des Widerstandes und der Geschwindigkeitsverteilung
    in Rohren mit regelmäßig angeordneten Rauhigkeiten bei turbulenter Strömung. Phys. Zeitschr. 41
    (1940) 202–225.

N

151. Naegele, R.: Einfluß von Umlenkkammern auf Druckverlust und Wärmetransport von Rohrbündel-
    Wärmeübertragern. Verfahrenstechnik 5 (1971) 10, 401–405.
152. Nikuradse, J.: Gesetzmäßigkeiten der turbulenten Strömung in glatten Rohren. VDI-Forschungsheft
    356 (1932).
153. Nikuradse, J.: Strömungsgesetze in rauhen Rohren. VDI-Forschungsheft 361 (1933).
154. Noether, F.: Das Turbulenzproblem. Z. ang. Math. Mech. 1 (1921) 125–138 and 218, 219.

P

151. Prandtl, L.: Bemerkungen über die Entstehung der Turbulenz. Z. ang. Math. Mech. 1 (1921)
    431–436.
152. Prandtl, L.; Tietjens, O.: Hydro- u. Aeromechanik, vol. II. Berlin: Springer 1931, pp. 15–60.
153. Prandtl, L.: Neuere Ergebnisse der Turbulenzforschung. Z. VDI 77 (1933) 104–114.
154. Prandtl, L.: Führer durch die Strömungslehre. 7th ed. Braunschweig: Vieweg 1969, pp. 168–176.
    (Flow in tubes and ducts.)

R

151. Richter, H.: Rohrdynamik, 5th ed. Berlin, Heidelberg, New York: Springer-Verlag 1971.
152. Reynolds, O.: Proc. Manchester Lit. a. Phil. Soc. 8 (1874) 9.
153. Reynolds, O.: An Experimental Investigation of the Circumstances which Determine whether the
    Motion of Water Shall be Direct or Sinuous, and of the Law of Resistance in Parallel Channels.
    Phil. Trans. Roy. Soc. London 1883 or Scient. Papers vol. II p. 51.

154. Roetzel, W.: Calculation of Single Phase Pressure Drop in Heat Exchangers Considering the Change of Fluid Properties along the Flow Path. Heat and mass transfer 6 (1973) 3–13.

S

151. Samoshka, P. S.; Makaryavichyus, V. I.; Schlanchyauskas, A. A.; Zhyugzda, I. I.; Zhukanskas, A. A.: Int. Chem. Engng. 8 (1968) 388–392.
152. Schefels, G.: Reibungsverluste in gemauerten engen Kanälen. Arch. Eisenhüttenw. 6 (1932/33) 477–486.
153. Schiller, L.: Rauhigkeit und kritische Zahl. Ein experimenteller Beitrag zum Turbulenzproblem. Z. Phys. 3 (1920) 412.
154. Schiller, L.: Experimentelle Untersuchungen zum Turbulenzproblem. Z. ang. Math. Mech. 1 (1921) 436–444.
155. Schiller, L.: Die Entwicklung der laminaren Geschwindigkeit und ihre Bedeutung für die Zähigkeitsmessungen. Z. ang. Math. Mech. 2 (1922) 96–106 (particularly p. 102 and 106).
156. Schiller, L.: Über den Strömungswiderstand von Rohren verschiedener Querschnitte und Rauhigkeitsgrades. Z. ang. Math. Mech. 3 (1923) 2.
157. Schiller, L.: Strömungsbilder zur Entstehung der turbulenten Rohrströmung. Verh. 3. intern. Kongr. f. techn. Mech., Part 1, Stockholm 1931, p. 226.
158. Schiller, L.: Strömung in Rohren. Handbuch der Experimentalphysik, herausgegeben von W. Wien und F. Harms. Leipzig: Akad. Verl.-Ges. 1932, pp. 1–206, in particular pp. 175–201.
159. Schiller, L.: Wirbelphotographien als Mittel zur quantitativen Untersuchung der Turbulenzentstehung und des Widerstandes. Proc. 5th. Intern. Congr. Appl. Mech., New York 1938, p. 315.
160. Schiller, L.; Naumann, A.: Untersuchungen über den Mechanismus der turbulenten Rohrströmung. Ingenieurarchiv 11 (1940) 335–343.
161. Schmidt, E. F.: Wärmeübergang und nicht isothermer Druckverlust bei erzwungener Strömung in schraubenförmig gekrümmten Rohren. Diss. TH Braunschweig 1966, 100 p.
162. Sockel, H.: Turbulente Strömung in glatten Rohren mit Kreisquerschnitt. Österr. Ing. Z. 11 (1968) 12, 407–412.
163. Srinavasan; Nandapurkar; Holland: see [S 22].
164. Stanton, T. E.: Friction. London: Longmans 1923.

T

151. Taylor, G. J.: The Criterion for Turbulence in Curved Pipes. Proc. Roy. Soc. (A) 124 (1929) 243–249.
152. Tietjens, O.: Beiträge zur Entstehung der Turbulenz. Z. ang. Math. Mech. 5 (1925) 200–217. O. Tietjens: Ein allgemeines Kriterium der Instabilität laminarer Geschwindigkeitsverteilungen. Göttinger Nachrichten, Math.-Phys. Kl. I, 1 (1935) 79.
153. Tollmien, W.: Über die Entstehung der Turbulenzen. 1. Mitt. Göttinger Nachrichten. Math. phys. Klasse, 1929, p. 21, also Z. ang. Math. Mech. 25/27 (1947) 33 and 70.

V

151. Vuščović, J.: Der Strömungswiderstand von geraden Gasrohren. Mitt. d. Hydr. Instituts d. Techn. Hochschule München, No. 9 (1939) S. 35–50.

W

151. White, C. M.: Streamline Flow through Curved Pipes. Proc. Roy. Soc. (A) 123 (1929) 645.
152. Wiener, P.: Untersuchungen über den Zugwiderstand von Wasserrohrkesseln. Diss. Techn. Hochschule Aachen 1937.

Z

151. Zimmermann, E.: Der Druckabfall in 90°-Stahlrohrbögen. Arch. Wärmewirtschaft 19 (1938) 265.
152. Zimmermann, E.: Neue Ergebnisse der Druckabfallberechnungen für gerade Stahlrohrleitungen. Arch. Wärmewirtschaft 21 (1940) 6, 133–138.

THREE

RECUPERATORS

FIVE

TEMPERATURE DISTRIBUTIONS AND HEAT TRANSFER IN PARALLEL FLOW AND COUNTERFLOW

5-1 PRELIMINARY REMARKS

Theories relating to recuperators and regenerators will be described fully in Part Three and Part Four of this book. It has already been mentioned in the introduction, that by recuperators are to be understood heat exchangers in which the fluids between which heat is transferred flow through continuously and uniformly; in recuperators the heat is usually transferred through partition walls. Occasionally, however, immiscible materials, as for example, different liquids which are mutually insoluble, can be used in place of the partition walls. Regenerators, on the other hand, have heat storing masses and are reversed at specific intervals of time, if we disregard those heat storing masses which are rotated.

In order to achieve the aims of this book, which are emphasized in the Foreword, theories are developed here which are of particular value in that they clarify physically and determine numerically the processes which take place in heat exchangers. At the same time the main intention is to develop processes which enable heat exchanger calculations to be carried out simply and quickly in practical cases.

This third Part is devoted to recuperators. First the large, and very important, group of recuperators will be considered in which the materials between which heat is transferred, flow parallel to each other. In *co-current* or *parallel flow* the two materials move in the *same* direction, *counterflow* in the *opposite* direction. Counterflow is used in the majority of cases because it is fundamentally superior to parallel flow in the mechanisms of heat transfer, as will be established below.

To begin with, some typical forms of construction of recuperators, operated with the fluids flowing parallel to each other, will be described in order to facilitate in as clear a manner as possible the derivation of the equations which are relevant to parallel flow and counterflow.

5-2 ARRANGEMENTS WHICH PERMIT PARALLEL FLOW AND COUNTERFLOW[1]

The same types of recuperator construction are basically suitable for both parallel flow and counterflow. This is because the differences between the two types of flow only become evident in the size of the heat transferring surface (see Sec. 6-2 ff.) or in numerically different results in the heat transfer calculations. If, for the most part, only counterflow recuperators are discussed, the examples of operation under discussion can also be directly applied to parallel flow.

It is likely that the development of special heat exchanger equipment arose out of the experience of operating industrial furnaces. The concept of counterflow was first realized in kilns used for firing ceramic materials such as bricks, pottery and the like. Here the purpose of heat transfer in counterflow is to preheat and predry the material waiting to be fired using the hot waste gas escaping from the combustion zone of the kiln. For example, a well established process, which is still in use today, is that in which the ceramic materials are packed into a truck, and led through a very long, duct shaped furnace in counterflow to the waste gases which slowly sweep over them from the combustion zone. After firing, the materials usually move in counterflow to the fresh inflowing combustion air which not only cools the materials but is itself preheated by them. In both cases the heat is transferred through the surface areas of the solid materials. However this does not apply to equipment which is used exclusively for the transfer of heat between two fluids.

The first heat exchanger as such, arose out of the idea of passing through *brick walls* the heat contained in waste combustion gases to fresh air or other gases. An arrangement of historical interest consists of one in which such waste gas and air are led through smooth shafts which lie adjacent to each other and which can be easily constructed from single bricks (see Fig. 6-1, top). Recuperators of this type

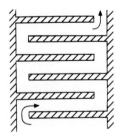

Figure 5-1 Brick recuperator with zig-zag path for the flow of gas.

---

[1] For heat exchanger arrangements, see also the descriptions by H. Kraussold in the periodical *Verfahrenstechnik* [K215] and in Lueger, *Lexikon der Verfahrenstechnik* [K216, L208].

are often used today in furnaces for industrial purposes, for example, for the annealing or preheating of metal blocks (pusher furnaces, soaking pits) and the like. Frequently the horizontal or vertical shafts are so arranged that, as in Fig. 5-1, they force the gas into a zig-zag motion.

Further, similarly arranged rows of shafts are situated both in front of and behind the rows of ducts shown and the different gases flow in alternating shafts. If the shafts for the various gases lie in the same direction, parallel-flow can be realized as can counterflow. The shafts which lie one behind another can also cross however, so that cross flow or even a complex so-called mixed circuit (flowpath) arrangement can be realized (see Sec. 8-3). Several other possibilities for the arrangement of the flow cross-section in brick recuperators will be discussed in Sec. 6-5.

Metallic heat exchangers, usually made of steel, are widely used for both gases and liquids provided that the temperatures are not too high as metals in general have only a limited resistance to such high temperatures. Most metal recuperators are constructed of tubes. They sometimes take the form of double-tube heat exchangers but for the most part metallic heat exchangers are made up of bundles of tubes.

In the double-tube heat exchanger which is shown in Fig. 5-2, each straight piece of the winding inner tube is surrounded by another, outer concentric tube; this is joined to the neighbouring outer tube by means of a short piece of pipe. In place of a single inner tube, several inner tubes can be arranged within a common outer tube. Such heat exchangers are suitable for quantities of liquids which are not too great, and most of all, for gases under high pressure, where, because of the usually high heat transfer coefficient, only a relatively small heating surface area is needed. On the other hand, several tubes arranged in parallel are required where large quantities of liquid are involved or particularly where large quantities of gas at low pressure must be handled. Such a tube bundle arrangement is shown in Fig. 5-3. The tubes are fixed at the top and the bottom to tube plates and along their lengths, they are surrounded by a common shell which serves to contain the gases

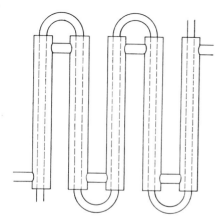

Figure 5-2 Double-tube heat exchanger.

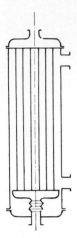

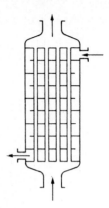

Figure 5-3 Tube bundle heat exchanger. The bottom tube plate is moveable to accommodate thermal expansion.

Figure 5-4 A counterflow heat exchanger in which cross flow is induced in the outer space by means of baffles.

which flow on the outside of the tubes. In order to avoid damage occurring through differential thermal expansion of the tube bundle and the shell, the lower tube plate is arranged so that it can move within the shell. The collecting tube connected to these plates must either be corrugated so that it can absorb any stresses or must be led outwards through a gland. Baffles are often fixed to the tube bundles (see Fig. 5-4) in order to approximate cross-counterflow.

The surfaces for heat transfer can also be constructed from plane or corrugated sheets instead of tubes (see Fig. 5-5). Recuperators made up from such sheet metal are often operated in cross flow. A special type of construction which is worthy of note is that where these thin sheets of metal are so bent and joined together that they form a multitude of hexagonal cross-sections which are arranged as in a honeycomb. Through such a recuperator, more than two gases can flow simultaneously partly in the same and partly in the opposite direction. Another variation of the plate recuperator is the spiral heat exchanger in which the heat transferring walls are built of spirally curved sheets which are only a small distance apart (Fig.

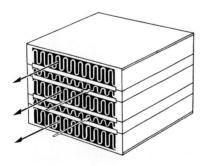

Figure 5-5 Recuperator constructed from smooth and corrugated sheets, often operated as a "reversing exchanger" in low temperature technology.

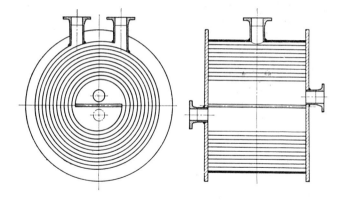

Figure 5-6 Spiral heat exchanger. (Manufactured in Germany by Alfa-Laval-Roca of Dühren.)

5-6). This spiral heat exchanger can be operated in parallel flow or counterflow. They have proved to be very useful in heat exchange between liquids and have the advantage that they can be easily cleaned once the front and rear walls have been removed.

Norton [N205] and others have described the heat exchanger for high temperatures in which pebbles moving downwards are heated by upward flowing waste combustion gases, or other very hot gases. The pebbles then pass into a second zone where they give up their heat in counterflow to the gas or liquid to be heated. Finally they are hauled up again and pass back into the first area thereby forming a closed circuit. The relationship between this arrangement and the furnaces described previously for firing ceramic materials is obvious.[2]

Worthy of note also are heat exchangers which have been developed for high pressure-synthesis technology in chemical engineering for temperatures of 200° to 500°C and higher and also for very high pressures of up to 1000 atm. Older types of construction, which to some extent seem complicated, have been described by W. Klempt [K213] for example. Nowadays the most simple types of construction are preferred with tube bundle arrangements for chemical reactions, such as those described, for example, in Lueger, Lexicon der Verfahrenstechnik [L208], p. 390, Figs. 7 and 8. The contact materials, which bring about the chemical reaction by catalysis, are usually to be found in the casing of the heat exchanger.

A particularly efficient counterflow heat exchanger has been produced for *low temperature technology*; this works at temperatures down to −200°C, but not

---

[2] Here it is a question of two counterflow heat exchangers connected one behind the other in which one of the fluids is solid, and thus it is possible to dispense with the partition walls. In these "pebble heaters" essentially the same processes take place as in recuperators and can be calculated in a similar manner. See for example H. Glaser [G204 or 303]. Nevertheless, occasionally they are not always calculated in quite the same way as recuperators. Moreover there are complex arrangements which come closer to the regenerator principle.

usually very much below this. These heat exchangers* serve to transfer the "cold" contained in cold gases, for example in oxygen and nitrogen, to compressed gases, for example to air.

Previously these heat exchangers were constructed in tube-bundle form* and thus corresponded, with certain differences to Fig. 5-3.

To a great extent, cross-counterflow heat exchangers are used where the gas flowing outwards impinges almost vertically on the tubes which are helically coiled in several layers. The advantage of increased heat transfer through cross flow is exploited in these heat exchangers and this will be considered in Sec. 8-3.

In the last thirty years moreover, the so-called reversing exchanger, which is reversed at regular time intervals, has gained considerably in importance; this exchanger operates at not too high pressures. At a reversal, air and nitrogen, for example, interchange their cross-sections of flow. As a rule these heat exchangers are manufactured from smooth and corrugated sheets as in Fig. 5-5. They have the advantage, in common with the regenerators dealt with in Part Four of this book, that the gases to be cooled, usually air, do not need to have water vapour and carbon dioxide extracted from them. Rather the solidified gases deposited in the reversing exchanger are automatically removed from the heat exchanger, as will be illustrated in Secs. 19-2 and 19-4 in the context of regenerators.

5-3 HEAT TRANSMISSION THROUGH PLANE AND CURVED WALLS

Relationships will be derived in the following material, which allow the temperature curve and the heat transfer to be determined in recuperators with a parallel flow of the gases or liquids. Thus the foundations will be laid for complete thermal calculations for recuperators.

In order to simplify the presentation, gases, rather than fluids, will be discussed, as heat exchangers have the largest dimensions for gases and are the more usual, that is if evaporators and condensers, which will not be dealt with here, are ignored. The equations which will be obtained can also be directly applied to heat transfer between two liquids or between a liquid and a gas.

In addition it can be assumed that the heat transfer coefficients α and α' for both gases have already been determined using the methods discussed in Part Two. If thermal radiation is ignored, it can generally be assumed, with sufficient accuracy, that these heat transfer coefficients have the same value at all positions in the recuperator. Moreover the thermal conductivity λ_s of the construction material and the thickness of the heat transferring partition wall are specified.

As has already been indicated in Sec. 2-1 of Part Two, with this data the heat transfer at an individual point in a heat exchanger can be obtained.

The most simple process of this type which has already been considered

---

* Editor's note: Hausen mentions here the abbreviation "Gegenströmer" which literally translated means "Counter-flow-er". This is not used in English.

previously is the heat transmission, through a plane wall, from a warm gas to a colder gas. Here the quantity of heat

$$\dot{Q} = kF(\theta - \theta'), \tag{5-1}$$

flows through the surface F of the plane wall in unit time where k is the heat transmission coefficient* and $\theta - \theta'$ is the temperature difference between the two gases (see Sec. 2-1, Eq. (2-5)). k is calculated from Eq. (2-6) in Sec. 2-1:

$$\frac{1}{k} = \frac{1}{\alpha} + \frac{\delta}{\lambda_s} + \frac{1}{\alpha'} \tag{5-2}$$

In most recuperators heat transfer takes place through curved walls, in particular through tube walls. Thus an expression for heat transmission through tube walls with a circular ring cross-section should be developed in place of Eq. (5-2) which only applies to plane walls.

A bundle of z equal tubes will be considered. α denotes the heat transfer coefficient on the inside, and α' denotes the heat transfer coefficient on the outside of the tube. d_i and d_a are the inner and outer diameters of the tube, r_i and r_a are the corresponding radii and L is the tube length. Thus, the inner surface area of the tubes available for heat transfer is

$$F_i = zd_i\pi L \tag{5-3}$$

while the outer surface area of the tubes available for heat transfer is

$$F_a = zd_a\pi L \tag{5-4}$$

Furthermore, the gas flowing inside the tubes has temperature θ and the gas flowing outside has temperature θ'. If, to begin with, we consider these temperatures to be constant, Eq. (5-5) represents the quantity of heat which is transferred between the first gas and the internal surface area in unit time (see Eqs. (2-1) and (2-2)):

$$\dot{Q} = \alpha F_i(\theta - \Theta_0) \tag{5-5}$$

and, correspondingly, Eq. (5-6) relates to the heat transfer on the outside:

$$\dot{Q} = \alpha' F_a(\Theta'_0 - \theta') \tag{5-6}$$

where in these two equations, as in Sec. 2-1, Θ_0 and Θ'_0 denote the surface temperatures of the tube walls.[3]

In order to investigate the flow of heat through the tube wall, a cylindrical surface of radius r should be imagined situated within the wall of one of the tubes, coaxial with its two surfaces (see Fig. 5-7). The area of this surface over the whole length of the tube, amounts to $2r\pi L$. If, in dealing with the flow of heat and the temperature distribution in the tube wall, it is assumed that the cylinder is symmetrical, the same temperature will occur at all corresponding points on the

---

* Editor's note: The expression "heat transmission coefficient" is discussed in a footnote in Sec. 2-1.
[3] With regard to the chosen notation, see Footnotes 2 and 3, pp. 11 and 12, and the list of symbols at the beginning of this book.

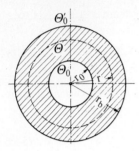

Figure 5-7 Temperature drop in a tube wall.

surface under consideration and in the radial direction the same temperature gradient $-d\Theta/dr$ will occur. This temperature gradient is a determining factor for the radial flow of heat. The quantity of heat $\dot{Q}$ flows through all z tubes in unit time; part of this flows through the tube area $2r\pi L$ under consideration and is represented by:

$$\frac{\dot{Q}}{z} = -\lambda_s \frac{d\Theta}{dr} 2r\pi L \tag{5-7}$$

in which λ_s denotes the thermal conductivity of the material of which the surface is constructed.

However this quantity of heat also flows in equilibrium through all other coaxial surfaces lying within the same tube wall but which have different radii r. Thus, reducing Eq. (5-7) to dr/r form and integrating from r_i to r_a with $\dot{Q}/z = $ const., Eq. (5-8) is obtained.

$$\ln\frac{r_a}{r_i} = \frac{\lambda_s z}{\dot{Q}} 2\pi L(\Theta_0 - \Theta'_0) \tag{5-8}$$

Further, if Eq. (5-8) is reduced to an expression equal to $\Theta_0 - \Theta'_0$, Eq. (5-5) to $\theta - \Theta_0$ and Eq. (5-6) to $\Theta'_0 - \theta'$, then by adding these together, Eq. (5-9) is obtained.

$$\theta - \theta' = \dot{Q}\left(\frac{1}{\alpha F_i} + \frac{1}{2\pi z L\lambda_s}\ln\frac{r_a}{r_i} + \frac{1}{\alpha' F_a}\right) \tag{5-9}$$

Furthermore, it is necessary to introduce a reference heating surface area F, the value of which can be arbitrarily selected from between F_i and F_a using Eqs. (5-3) and (5-4). This surface can be imagined to be like the surface illustrated in Fig. 5-7 and can be given a reference diameter d lying between d_i and d_a. Employing such a reference area, Eq. (5-10) can be developed from Eq. (5-9) for the plane wall, where k again denotes the heat transmission coefficient.

$$\dot{Q} = kF(\theta - \theta') \tag{5-10}$$

By inserting Eq. (5-9) into the last equation, Eq. (5-11) is obtained since $r_a/r_i = F_a/F_i$.

$$\frac{1}{kF} = \frac{1}{\alpha F_i} + \frac{1}{2\pi z L\lambda_s}\ln\frac{F_a}{F_i} + \frac{1}{\alpha' F_a} \tag{5-11}$$

This equation provides the required *relationship for the heat transmission coefficient k for tube walls with a circular ring cross-section.* The value of kF, which is required in Eq. (5-10), can be directly obtained from this relationship. However no specific value for k can be estimated unless the surface F is prescribed to which k should refer. F is often equated to the value of F_i or F_a since, as a rule, the surface with the greatest resistance to heat transfer is preferred. If, for example, $F = F_i$ is chosen for Eqs. (5-3) and (5-4), Eq. (5-11) is converted into the *Fourier equation for heat transfer through a tube wall*:

$$\frac{1}{k} = \frac{1}{\alpha} + \frac{d_i}{2\lambda_s}\ln\frac{d_a}{d_i} + \frac{1}{\alpha'}\frac{d_i}{d_a} \tag{5-12}$$

For many purposes this equation is more suitable than Eq. (5-11). However Eq. (5-11) can also be used with advantage, particularly with metal tubes, as the term λ_s can often be ignored because the thermal conductivity of metal is high. In this case Eq. (5-11) takes on the simple form:

$$kF = \frac{\alpha F_i \cdot \alpha' F_a}{\alpha F_i + \alpha' F_a} \tag{5-13}$$

Since the surface areas F, F_i and F_a are proportional to the diameters d, d_i and d_a respectively, Eq. (5-13) can also be written in the form

$$\frac{1}{kd} = \frac{1}{\alpha d_i} + \frac{1}{2\lambda_s}\ln\frac{d_a}{d_i} + \frac{1}{\alpha' d_a} \tag{5-14}$$

If, in addition, following the suggestion of Eckert [E202], the logarithmic average value of the diameters (5-15) is introduced, (δ is the wall thickness),

$$d_m = \frac{d_a - d_i}{\ln d_a/d_i} = \frac{2\delta}{\ln d_a/d_i} \tag{5-15}$$

Equation (5-16) may be developed which is similar formally to Eq. (2-6)

$$\frac{1}{kd} = \frac{1}{\alpha d_i} + \frac{\delta}{\lambda_s d_m} + \frac{1}{\alpha' d_a} \tag{5-16}$$

With the corresponding logarithmic average value for the surfaces

$$F_m = \frac{F_a - F_i}{\ln F_a/F_i} = z \cdot \pi \cdot L \cdot d_m \quad \text{(cylindrical walls)} \tag{5-17}$$

Equation (5-11) can be brought to its general, very usable form [H210]:

$$\frac{1}{kF} = \frac{1}{\alpha F_i} + \frac{\delta}{\lambda_s F_m} + \frac{1}{\alpha' F_a} \tag{5-18}$$

This equation is also valid for a spherical wall if the geometric mean (5-19) is used for F_m:

$$F_m = \sqrt{F_i \cdot F_a} = \pi \cdot d_i \cdot d_a \quad \text{(spherical walls)} \tag{5-19}$$

Equation (5-18) can also be applied to walls of arbitrary curvature since in all cases a suitable average value F_m can be estimated on the basis of Eqs. (5-17) and (5-19).

Application to Heat Exchangers

Having found the expressions for k or kF, Eq. (5-10) can generally be considered to be the *basic equation for heat transmission*. In its application to a heat exchanger it must, however, be remembered that the temperatures θ and θ' of the gases and thus the temperature difference $\theta - \theta'$ generally vary in the longitudinal direction within the heat exchanger. As has already been indicated, the use of Eq. (5-10) is valid primarily only for very small sections of the recuperator, over which sections integration must then take place. However the form of Eq. (5-10) can be retained for whole recuperators by introducing an average temperature difference $\Delta\theta_M = (\theta - \theta')_M$ and thus using Eq. (5-20) in place of Eq. (5-10):

$$\dot{Q} = kF\Delta\theta_M \tag{5-20}$$

The average temperature difference $\Delta\theta_M$ which is embodied within Eq. (5-20) plays a fundamental role in the considerations which follow. In determining the form of $\Delta\theta_M$ calculation procedures and equations will be derived from considerations of the temperature curve in the longitudinal direction along the length of the recuperator.

5-4 HEAT TRANSFER AND THE TEMPERATURE CURVE IN A RECUPERATOR WITH A FLUID OF CONSTANT TEMPERATURE

The calculation of the temperature curve and the average temperature difference is particularly simple if the temperature of one of the two materials taking part in heat transfer is constant. Evaporators and condensers belong to this type of heat exchanger; here fluid evaporates or steam condenses on one side of the heat transferring wall. During these changes in state the temperature remains constant, that is if the changes in pressure which are normally very slight are ignored. This situation arises approximately when warm air flows through a spiral tube which is cooled externally by water which is in a container. If the water, which is flowing to and fro uniformly, is thoroughly mixed, for example by means of tubes, an almost constant water temperature appears at equilibrium.

In order to be able to carry out more easily further considerations of the example mentioned above, we need to imagine that the spiral tube has been replaced by a straight, vertical tube which is surrounded by water at a constant temperature θ' between points a and b in Fig. 5-8. The air enters at the top with an entry temperature of θ_1. If $\theta_1 > \theta'$ the air is cooled on its way down by giving up heat to the water. If, however, the heat transmission coefficient k as well as θ' is assumed to be constant, with decreasing air temperature θ, not only the temperature difference $\theta - \theta'$ but also, according to Eq. (5-10), the quantity of heat transferred per unit surface of the duct walls will become smaller towards the bottom of the tube. Thus the air is cooled more and more slowly so that the curve of the temperature θ along the duct can be represented by a decreasingly curved line as in Fig. 5-9.

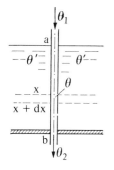

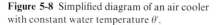

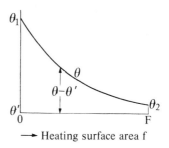

Figure 5-8 Simplified diagram of an air cooler with constant water temperature θ'.

Figure 5-9 Temperature curve in an air cooler as shown in Fig. 5-8.

The change in temperature of the air and the average temperature difference $\Delta\theta_M$ can be calculated in the following way. To begin with, a cross-section of the duct at a point in Fig. 5-8 will be considered which lies along line x and which is below the surface of the water. Here the air has temperature θ. Between a and x, the tube has the heat transferring surface area f, between x and $x+dx$ the surface area is df; these can be treated as parts of the reference heating surface area F mentioned in Sec. 5-3. If k is the heat transmission coefficient applicable to these surfaces, then from Eq. (5-10) the quantity of heat transferred in unit time through df will be

$$d\dot{q} = kdf(\theta - \theta') \tag{5-21}$$

This heat will be extracted from the air. If C is the heat capacity of the quantity of air* flowing through the duct in unit time and $-d\theta$ is the temperature drop of the air between x and $x+dx$, then it follows that:

$$d\dot{q} = -Cd\theta \tag{5-22}$$

Through Eqs. (5-21) and (5-22) it can be seen that the processes in heat exchangers will always be governed by a heat transmission equation of the same type as Eq. (5-21) and by a heat balance equation of the same type as Eq. (5-22). This will become more obvious in the general cases which will be treated later. If the two expressions obtained for $d\dot{q}$ in Eqs. (5-21) and (5-22) are set equal to each other, the integration from a to x yields (5-23) or (5-24) if k and C are assumed to be constant.

$$\int_a^x \frac{d\theta}{\theta - \theta'} = \ln \frac{\theta - \theta'}{\theta_1 - \theta'} = \frac{-kf}{C} \tag{5-23}$$

or

$$\theta = \theta' + (\theta_1 - \theta')\exp\left(-\frac{kf}{C}\right) \tag{5-24}$$

---

* It would be basically meaningful to mark with a dot the heat capacities C and C' of the gases relative to unit time corresponding to $\dot{Q}$, $\dot{q}$ and $\dot{m}$. However this idea has been abandoned because in complicated equations it would lead to an unsuitable notation. See Footnote 3, p. 141.

The temperature curve determined by this last equation is shown to be a function of f and is displayed in Fig. 5-9.

The average temperature difference $\Delta\theta_M$ is obtained by integrating from a to b in Eq. (5-23) instead of from a to x, that is over the entire length of the cooled tube. Equation (5-24) yields, first of all,

$$\int_a^b \frac{d\theta}{\theta - \theta'} = \ln\frac{\theta_2 - \theta'}{\theta_1 - \theta'} = -\frac{kF}{C} \tag{5-25}$$

where F is the total reference heating surface area between a and b and θ_2 is the temperature of the air emerging at b. Further the total quantity of heat transferred in unit time is given by Eq. (5-26) which corresponds to Eq. (5-22).

$$\dot{Q} = C(\theta_1 - \theta_2) \tag{5-26}$$

It is necessary to determine $\Delta\theta_M$ in such a way that it can be used within Eq. (5-20).

The equation

$$kF = C\frac{\theta_1 - \theta_2}{\Delta\theta_M}$$

follows by equating (5-20) to (5-26), and then by inserting Eq. (5-27) where

$$\Delta\theta_M = \frac{\theta_1 - \theta_2}{\ln\dfrac{\theta_1 - \theta'}{\theta_2 - \theta'}} \tag{5-27}$$

This equation for the average temperature difference is only a particular form of a more general relationship which has yet to be developed. However, from Eq. (5-27) it can be recognized that, in simple cases, $\Delta\theta_M$ can be determined solely from the temperature of the fluid, without reference to k or F.

5-5 TEMPERATURE CURVE FOR PARALLEL FLOW AND COUNTERFLOW USING THE HEAT BALANCE EQUATION

(Constant heat capacities C and C' of the gases.)

If, as is usually the case, the temperatures of the *two fluids* vary, a considerable difference in the temperature curve is displayed depending on whether operation is in parallel flow or in counterflow. These differences can be considerably clarified for the following simple reasons.

In parallel flow, the gases flow through the heat exchanger in the same direction. At the point of entry into the heat exchanger there is usually a large difference in the entrance temperatures between the two gases. As they pass in the same direction through the heat exchanger, the hotter gas cools down and the cooler gas becomes warmer. Thus the temperatures of the two gases approach a common average value. The steady reduction of the temperature difference between the two gases results in less and less heat, as is apparent from Eq. (5-21), being transferred by each unit surface area of the wall. As a later calculation will show, this results in an ever slower alteration in the temperature of the gas and thus, in a non-linear temperature curve in the longitudinal direction of the heat exchanger.

On the other hand, when the gases flow in opposite directions through the heat exchanger, the gas being warmed takes up as much heat as the cooling gas loses. Thus it is assumed in the following discussion that the heat exchanger does not lose any heat to its surroundings, nor does it receive any from them, and that the originally hot gas has at that point reached its lowest temperature. Thus, given suitable conditions, the temperature differences at both ends of the heat exchanger can be very small. Correspondingly at all other points in the heat exchanger only slight temperature differences can be expected. If, for example, the temperature differences at all points are the same, and if the same conditions apply at all points in the heat exchanger, the same quantity of heat will be transferred in unit time at all such points. It follows therefore that if their capacities are constant, the gases must rapidly and equally suffer overall alterations in temperature. Thus there results a straight line temperature curve in the longitudinal direction in the counterflow recuperator. In this particular case, given corresponding conservative dimensions of the heat exchanger, it is possible in principle, to arbitrarily approximate the exit temperature of each of the gases by the entry temperature of the opposite gas.

The differences in the temperature curve between parallel flow and counterflow which have been described, arise numerically, for a large part, from a heat balance equation which asserts that in equilibrium, at an arbitrary position in the heat exchanger, the gas being warmed takes up as much heat as the cooling gas loses. Thus it is assumed in the following discussion that the heat exchanger does not lose any heat to its surroundings, nor does it receive any from them, and that the quantity of heat conducted by the walls in the longitudinal direction is so small that it can be disregarded.[4]

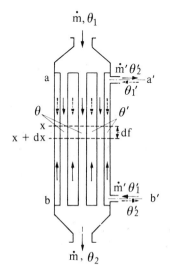

Figure **5-10** Tube bundle heat exchanger.

[4] The influence of the losses created by this will be handled separately in Secs. 7-2 and 7-3.

For the derivation of the heat balance equation, the heat exchanger illustrated in Fig. 5-10 will be examined. It consists of a bundle of straight parallel tubes which are surrounded by an outer tube. The quantity $\dot{m}$ of a gas with varying temperature θ flows in unit time through the inner tubes from top to bottom and in the space surrounding these tubes flows the quantity $\dot{m}'$ of another gas with a similarly varying temperature θ'. If the second gas flows upwards (solid arrows) counterflow occurs; if it flows downwards (dotted arrows) parallel flow takes place.

Between two infinitely close neighbouring cross-sections at position x one gas is cooled by $d\theta$ and the other gas is heated by $d\theta'$. Let C and C' be the heat capacities of the quantities of gas flowing through the heat exchanger in unit time. In what follows, these heat capacities will primarily be assumed to be constant. If, in addition, it is considered that $d\theta$ is negative, Eq. (5-28) yields the quantity of heat exchanged between these cross-sections in unit time:

$$d\dot{q} = -C\,d\theta = C'\,d\theta' \tag{5-28}$$

From this, Eq. (5-29) gives the total quantity of heat transferred between a and x in unit time obtained by integrating between the top end cross-section a and the point x under consideration, in the longitudinal direction.

$$\left.\begin{aligned}\dot{q} = C(\theta_1 - \theta) &= C'(\theta' - \theta'_1) \quad \text{with parallel flow} \\ &= C'(\theta'_2 - \theta') \quad \text{with counterflow}\end{aligned}\right\} \tag{5-29}$$

where θ and θ' again denote the temperatures in the cross-section x, θ_1 and θ'_1, denote the entry temperatures and θ_2 and θ'_2 denote the exit temperatures of the gases.

If we imagine the position x, and thus $\dot{q}$ to be variable, it follows from Eq. (5-29) that θ and θ' can be regarded as functions of $\dot{q}$. If then, for the cases where $C = C'$ the temperatures θ and θ' are displayed as dependent upon $\dot{q}$, for example, Figs. 5-11 and 5-12 are obtained. It can be recognized from this, that with parallel flow, the temperature difference $\theta - \theta'$ is large initially and then diminishes while, with counterflow, (assuming $C = C'$) a relatively small, constant temperature difference prevails throughout; this is in agreement with the discussion earlier in this chapter. If the heat exchanger is conservatively designed, a quantity of heat is transferred in counterflow which is about twice as large as that transferred in parallel flow for the same entry temperatures. Correspondingly, an almost doubled temperature difference between the gases obtains.

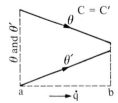

Figure 5-11 Temperatures θ and θ' in the cross-section x dependent on the quantity of heat $\dot{q}$ being transferred with parallel flow ($C = C'$).

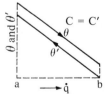

Figure 5-12 θ and θ' dependent on $\dot{q}$ with counterflow ($C = C'$).

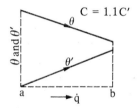

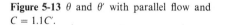

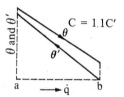

Figure 5-13 θ and θ' with parallel flow and $C = 1.1C'$.

Figure 5-14 θ and θ' with counterflow and $C = 1.1C'$.

Figures 5-13 and 5-14 show the corresponding temperature curve with $C = 1.1C'$. In this case, the gas with the higher heat capacity cools down less than the other gas warms up. In these cases therefore, the temperature difference cannot remain constant, even in counterflow, rather it must increase in the direction of the cold end of the exchanger.

Other remarkable conclusions can be extracted from Eq. (5-29). If $\dot{q}$ is eliminated, according to this equation, a unique *relationship exists between* θ *and* θ', based solely upon the *heat balance equation* provided that the heat capacities of both gases are given as are the gas temperatures in cross-section a or in any other cross-section of the heat exchanger. This relationship differs for parallel flow and counterflow.

Further, if $\pm C'(\theta_1 - \theta)$ is added to the centre and right-hand expressions in Eq. (5-29), Eq. (5-30) is obtained; this is expanded so that the temperature difference $\theta - \theta'$ remains alone on the left-hand side.

$$\left. \begin{array}{l} \theta - \theta' = \theta_1 - \theta_1' - \dfrac{C' + C}{C'}(\theta_1 - \theta) \quad \text{for parallel flow} \\[3mm] \theta - \theta' = \theta_1 - \theta_2' - \dfrac{C' - C}{C'}(\theta_1 - \theta) \quad \text{for counterflow} \end{array} \right\} \tag{5-30}$$

It follows from this that from the heat balance equation the variation of the *temperature difference between both gases is uniquely determined by the temperature* θ *of one of the gases.* Similarly $\theta - \theta'$ can also be shown to be dependent on θ' alone.

If Eq. (5-29) is applied to the temperatures in the end cross-section of the recuperator (bottommost cross-section in Fig. 5-10) Eq. (5-31) is obtained for the quantity of heat $\dot{Q}$ transferred in unit time in the complete heat exchanger.

$$\dot{Q} = C(\theta_1 - \theta_2) = C'(\theta_2' - \theta_1') \tag{5-31}$$

5-6 TEMPERATURE CURVE IN PARALLEL-FLOW AND COUNTERFLOW ALONG THE LENGTH OF THE HEAT EXCHANGER

(Heat capacities C and C' of the gases and the heat transmission coefficient k are constant.)

In the following text will be set out the means whereby the temperature curve in the longitudinal direction of the heat exchanger, that is in the direction of the longitudinal coordinate x, can be determined. In doing this it is much more convenient to introduce the heating surface area f in place of x since f extends from the initial cross-section a, to the point under consideration (see Figs. 5-8 and 5-10). As a rule f is proportional to the distance between these two cross-sections. This is the case, for example, in a tube assembly of z tubes with an overall uniform cross-section (Fig. 5-10),

$$f = z \cdot d \cdot \pi \cdot x,$$

when d denotes the arbitrarily selected diameter between d_i and d_a to which the heating surface area is relative (see Sec. 5-3). Thus the dependence on x can always be easily determined from the dependence of the temperature upon f.

From the heat transmission Eq. (5-21), the quantity of heat transferred in unit time at the point x through the heating surface area df is

$$d\dot{q} = kdf(\theta - \theta')\tag{5-32}$$

If Eq. (5-32) is divided on both sides by $\theta - \theta'$ and is then integrated from cross-section a to cross-section x, the very general relationship (5-33) is obtained.

$$\int_a^x \frac{d\dot{q}}{\theta - \theta'} = kf\tag{5-33}$$

From this equation the temperature curve as a function of f can be determined for parallel flow and cross flow under very different conditions. This will be set out in the following paragraphs for all cases in which C, C' and k are constant.

Special Case: $C = C'$ in Counterflow

The simplest case, and perhaps the most important, is that of *equal gas heat capacities* $\dot{C} = \dot{C}'$ with the gases in *counterflow*. With $C = C'$ Eq. (5-34) follows on from the second equation in (5-30).

$$\theta - \theta' = \theta_1 - \theta_2'\tag{5-34}$$

that is the temperature difference $\theta - \theta'$ has the same value $\theta_1 - \theta_2'$ at all points in the recuperator where $\theta_1 - \theta_2'$ is the temperature difference at the cross-section a in Fig. 5-10. Taking Eq. (5-34) and the heat balance equation $d\dot{q} = -Cd\theta$ (see Eq. (5-28)),

$$\int_a^x \frac{d\dot{q}}{\theta - \theta'} = -C \int_a^x \frac{d\theta}{\theta_1 - \theta_2'} = C\frac{\theta_1 - \theta}{\theta_1 - \theta_2'}$$

is obtained for the left-hand side of Eq. (5-33). Inserting this into (5-33)

$$\theta_1 - \theta = (\theta_1 - \theta_2')\frac{kf}{C}\tag{5-35}$$

is the equation for the temperature curve.

The temperature θ of the originally warmer gas thus *decreases linearly with f* as has already been discussed. Correspondingly,

$$\theta_2' - \theta' = (\theta_1 - \theta_2')\frac{kf}{C} \tag{5-36}$$

is obtained for the temperature of the cold gas from Eq. (5-35) using the second equation of (5-29) and setting $C = C'$. This also yields a linear temperature curve, as is illustrated by the solid lines in Figs. 5-12 and 5-15.

Equations (5-35) and (5-36) assume that the entry temperature θ_2' of the originally cooler gas is already known. Often, however, only the entry temperatures θ_1 and θ_1' are given. In order to find θ_2' in these cases, Eqs. (5-35) and (5-36) are applied to the total length of the recuperator and converting f into the total heating surface area F. Thus

$$\theta_1 - \theta_2 = (\theta_1 - \theta_2')\frac{kF}{C} \tag{5-37}$$

and

$$\theta_2' - \theta_1' = (\theta_1 - \theta_2')\frac{kF}{C} \tag{5-38}$$

are obtained. Equation (5-38) takes the form

$$\theta_1 - \theta_2' = \frac{\theta_1 - \theta_1'}{1 + \dfrac{kF}{C}} \tag{5-39}$$

by adding and subtracting θ_1 on the left-hand side. If $\theta_1 - \theta_2'$, and thus θ_2', has been calculated using this equation, then from Eqs. (5-35) and (5-36) the temperature curve in the recuperator can be determined.

Furthermore, the relationship

$$\theta_1 - \theta_2 = (\theta_1 - \theta_1')\frac{\dfrac{kF}{C}}{1 + \dfrac{kF}{C}} \tag{5-40}$$

is obtained from Eq. (5-37) by using Eq. (5-39) for the immediate determination of the exit temperature θ_2 of the originally warmer gas.

Parallel Flow and General Cases of Counterflow

In *parallel flow*, as in the general case $C \neq C'$ for *counterflow*, the temperature curve can be determined on the basis of the same considerations which have been employed in the simple case, $C = C'$, set out above. From this it would seem to be reasonable to handle parallel flow and counterflow together as in both cases the resultant relationships differ only in a few plus or minus signs and in the transposition of the entry/exit temperature of one of the gases.

Equation (5-41) is yielded by differentiating Eq. (5-30) and then comparing with the first expression for $d\dot{q}$ in Eq. (5-28):

$$
\left.
\begin{aligned}
d\dot{q} = -C\,d\theta &= -\frac{CC'}{C+C'}d(\theta-\theta') \quad &&\text{in parallel flow} \\[2mm]
&= +\frac{CC'}{C-C'}d(\theta-\theta') \quad &&\text{in counterflow}
\end{aligned}
\right\}
\quad (5\text{-}41)
$$

With these expressions for $d\dot{q}$ Eq. (5-42) below is obtained from Eq. (5-33) by integration:

$$
\left.
\begin{aligned}
\theta-\theta' &= (\theta_1-\theta_1')\exp\left[-\left(\frac{1}{C'}+\frac{1}{C}\right)kf\right] \quad &&\text{in parallel flow} \\[4mm]
\theta-\theta' &= (\theta_1-\theta_2')\exp\left[\left(\frac{1}{C'}-\frac{1}{C}\right)kf\right] \quad &&\text{in counterflow}
\end{aligned}
\right\}
\quad (5\text{-}42)
$$

It will be seen that the curve of the temperature difference $\theta-\theta'$ is primarily a function of f and is thus dependent upon the longitudinal coordinate x. Finally, if these relationships are inserted into Eq. (5-30)

$$
\left.
\begin{aligned}
\theta_1-\theta &= (\theta_1-\theta_1')\frac{C'}{C'+C}\left\{1-\exp\left[-\left(\frac{1}{C'}+\frac{1}{C}\right)kf\right]\right\} \quad &&\text{in parallel flow} \\[4mm]
\theta_1-\theta &= (\theta_1-\theta_2')\frac{C'}{C'-C}\left\{1-\exp\left[\left(\frac{1}{C'}-\frac{1}{C}\right)kf\right]\right\} \quad &&\text{in counterflow}
\end{aligned}
\right\}
\quad (5\text{-}43)
$$

are obtained. Correspondingly,

$$
\left.
\begin{aligned}
\theta'-\theta_1' &= (\theta_1-\theta_1')\frac{C}{C'+C}\left\{1-\exp\left[-\left(\frac{1}{C'}+\frac{1}{C}\right)kf\right]\right\} \quad &&\text{in parallel flow} \\[4mm]
\theta_2'-\theta' &= (\theta_1-\theta_2')\frac{C}{C'-C}\left\{1-\exp\left[\left(\frac{1}{C'}-\frac{1}{C}\right)kf\right]\right\} \quad &&\text{in counterflow}
\end{aligned}
\right\}
\quad (5\text{-}44)
$$

follow on from this for the second gas. Thus, *the curves for the temperatures θ and θ' of both gases are dependent on the heating surface area f and thus also upon the longitudinal coordinate of the heat exchanger*; they can be calculated from Eqs. (5-43) and (5-44) if the temperatures θ_1 and θ_1' or θ_1 and θ_2' are known for the cross-section a in Fig. 5-10.

However if only the entry temperatures of the gases are given, in counterflow at least one of the two *exit temperatures* θ_2 and θ_2' must be determined before the equations which have been obtained can be applied. In order to calculate the exit temperatures, relationships (5-45) and (5-46) are developed from Eqs. (5-43) and (5-44) with f set equal to the total heating surface area F:

$$\theta_1 - \theta_2 = (\theta_1 - \theta_1')\frac{C'}{C'+C}\left\{1 - \exp\left[-\left(\frac{1}{C'} + \frac{1}{C}\right)kF\right]\right\} \quad \text{in parallel flow}$$

$$\theta_1 - \theta_2 = (\theta_1 - \theta_2')\frac{C'}{C'-C}\left\{1 - \exp\left[\left(\frac{1}{C'} - \frac{1}{C}\right)kF\right]\right\} \quad \text{in counterflow}$$

$$(5\text{-}45)$$

$$\theta_2' - \theta_1' = (\theta_1 - \theta_1')\frac{C}{C'+C}\left\{1 - \exp\left[-\left(\frac{1}{C'} + \frac{1}{C}\right)kF\right]\right\} \quad \text{in parallel flow}$$

$$\theta_2' - \theta_1' = (\theta_1 - \theta_2')\frac{C}{C'-C}\left\{1 - \exp\left[\left(\frac{1}{C'} - \frac{1}{C}\right)kF\right]\right\} \quad \text{in counterflow}$$

$$(5\text{-}46)$$

In counterflow the exit temperature θ_2' can be determined from Eq. (5-46) by rearranging this equation in the following way:

$$\theta_1 - \theta_2' = (\theta_1 - \theta_1')\frac{1 - \dfrac{C}{C'}}{1 - \dfrac{C}{C'}\exp\left[\left(\dfrac{1}{C'} - \dfrac{1}{C}\right)kF\right]} \quad \text{in counterflow} \quad (5\text{-}47)$$

Having calculated θ_2', the temperature curve for the heat exchanger can be obtained employing Eqs. (5-43) and (5-44). It follows from Eq. (5-45) or (5-31) that the end temperature θ_2 of the originally warm gas is determined from θ_2'. However θ_2 can be calculated directly from the following relationship (5-48) which results from inserting Eq. (5-47) into Eq. (5-45):

$$\theta_1 - \theta_2 = (\theta_1 - \theta_1')\frac{1 - \exp\left[\left(\dfrac{1}{C'} - \dfrac{1}{C}\right)kF\right]}{1 - \dfrac{C}{C'}\exp\left[\left(\dfrac{1}{C'} - \dfrac{1}{C}\right)kF\right]} \quad \text{in counterflow} \quad (5\text{-}48)$$

The necessity, in counterflow, of determining θ_2' or θ_2 prior to the calculation of the temperature curve, lies in the fact that with given values of θ_1, θ_1', C and C' the temperature curve is only dependent on k in parallel flow whereas it is also

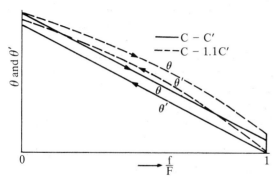

Figure 5-15 Temperature distribution in a recuperator with counterflow.

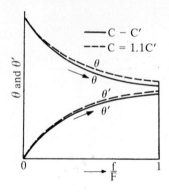

Figure 5-16 Temperature distribution in a recuperator with parallel flow.

dependent on the size of the total heating surface area F in counterflow. This dependence is shown in Eq. (5-47) in that this equation contains F in the denominator.

Figures 5-15 and 5-16 show the temperature curve, calculated using Eqs. (5-43) to (5-48) and (5-35) to (5-40), as a function of f for parallel flow and counterflow. As in Figs. 5-11 to 5-14, both $C = C'$ and $C = 1.1C'$ are again illustrated. In Fig. 5-15 it is conspicuous how strongly, in counterflow, the temperature curve deviates even for slight variations in C/C' from the value $C/C' = 1$, if the temperature differences $\theta - \theta'$ are small. This deviation in the curve is caused by the fact that, $\theta - \theta'$ quickly rises in proportion with increasing f, and, according to Eq. (5-32), the heat transfer and the temperature variation of the gases also increase in the same proportion.

Close Consideration of the Special Case where $C = C'$ in Counterflow

The relationships (5-35) to (5-40) derived above for the special case where $C = C'$ in counterflow can also be derived from the general equations for counterflow, namely Eqs. (5-43) to (5-48), by setting C equal to C' in these equations. However, in doing this there arises initially an indeterminate expression of the form $0/0$ because the exponent in Eqs. (5-43) to (5-48) becomes infinitely small. However, if the exponential function occurring in these equations is expanded into series form and this is then truncated after the second term, the equation:

$$1 - \exp\left[\left(\frac{1}{C'} - \frac{1}{C}\right)kf\right] = -\left(\frac{1}{C'} - \frac{1}{C}\right)k \cdot f = \frac{C' - C}{CC'}k \cdot f$$

is obtained for $\lim C \to C'$. Equations (5-43) to (5-48) can thus be converted, without difficulty, to Eqs. (5-35) to (5-40).

5-7 TEMPERATURE CURVE FOR THE WALLS THROUGH WHICH THE HEAT IS TRANSFERRED

If the curve of the gas temperatures θ and θ' is known along the length of the heat exchanger [Eqs. (5-43) and (5-44)] the temperature curve in the heat transferring walls can also be determined in a straightforward way. As before, heat conduction by the walls in the longitudinal direction of the heat exchanger is disregarded. If Eqs. (5-10), (5-5) and (5-6) are applied at a specific point in the exchanger, and if $df_i = \dfrac{F_i}{F} df$ and $df_a = \dfrac{F_a}{F} df$ is used for a very small part of the inner and outer tube surface respectively, the relationship (5-49) is obtained for the quantity of heat $d\dot{q}$ flowing through the surface element df in unit time:

$$
\left.
\begin{aligned}
d\dot{q} &= k(\theta - \theta') df, \\[2mm]
d\dot{q} &= \alpha \frac{F_i}{F} (\theta - \Theta_0) df, \\[2mm]
d\dot{q} &= \alpha' \frac{F_a}{F} (\Theta'_0 - \theta') df.
\end{aligned}
\right\}
\tag{5-49}
$$

Equations (5-50) and (5-51) for the surface temperature Θ_0 and Θ'_0 of the wall are derived from the first and second and then the first and last of these Eqs. (5-49).

$$
\Theta_0 = \theta - \frac{kF}{\alpha F_i} (\theta - \theta')
\tag{5-50}
$$

$$
\Theta'_0 = \theta' + \frac{kF}{\alpha' F_a} (\theta - \theta')
\tag{5-51}
$$

From this the average wall temperature $\Theta_M = \dfrac{\Theta_0 + \Theta'_0}{2}$ becomes:

$$
\Theta_M = \tfrac{1}{2}\left(1 - \frac{kF}{\alpha F_i} + \frac{kF}{\alpha' F_a}\right)\theta + \tfrac{1}{2}\left(1 - \frac{kF}{\alpha' F_a} + \frac{kF}{\alpha F_i}\right)\theta'
\tag{5-52}
$$

The relationships obtained for the wall temperatures Θ_0, Θ'_0 and Θ_M are equally valid for parallel flow and counterflow.

According to Eq. (5-52) the average wall temperature is closer to θ if $\alpha F_i > \alpha' F_a$ and closer to θ' if $\alpha F_i < \alpha' F_a$. With $\alpha F_i = \alpha' F_a$, Θ_M lies exactly midway between θ and θ'. In this case, the curve of Θ_M can easily be understood by imagining, for example, the mid lines between θ and θ' as shown in Figs. 5-15 and 5-16. It thus follows that in parallel flow the wall temperature Θ_M varies only slightly, or not at all. On the other hand in counterflow the wall temperature varies by approximately as much as the temperature of the gas. This has the result that, in parallel flow, no heat, or only a very small quantity of heat is conducted in the longitudinal direction of the wall. However in counterflow, under certain circumstances, the quantity of heat conducted in the longitudinal direction can be considerable so that losses can occur; this will be discussed in greater detail in Sec. 7-3, as has been already mentioned.

The almost constant wall temperature Θ_M in parallel flow could be one of the reasons why parallel flow might be preferred to counterflow which is otherwise superior. Thus it is possible with parallel flow, for example, to ensure that the material used for constructing the heat exchanger can withstand certain maximum temperatures above which it would become too frail and the danger of corrosion would be very great (see Sec. 7-1). On the other hand, when moist gases are being cooled, it could be undesirable for the wall temperature to drop below 0°C since the depositing of ice on the walls must be avoided.

5-8 AVERAGE TEMPERATURE DIFFERENCE $\Delta\theta_M$ IN PARALLEL FLOW AND COUNTERFLOW

(C, C' and k are constant.)

In order to find a general expression for $\Delta\theta_M$ Eq. (5-32) is first integrated from the entrance cross-section a to the exit cross-section b in Fig. 5-10. Equation (5-53) results from this which corresponds to Eq. (5-33); note that $k = \text{const.}$

$$\int_a^b \frac{d\dot{q}}{\theta - \theta'} = kF \tag{5-53}$$

If this is compared with Eq. (5-20) in which $\dot{Q}$ represents the total quantity of heat transferred in the heat exchanger in unit time, the generally valid *defining Eq.* (5-54) is obtained for $\Delta\theta_M$:

$$\frac{\dot{Q}}{\Delta\theta_M} = \int_a^b \frac{d\dot{q}}{\theta - \theta'} \tag{5-54}$$

In order to calculate $\Delta\theta_M$ from this equation it is necessary, to express $d\dot{q}$, under the integral sign, as a function of temperature; this is made possible with the help of Eq. (5-28) or with one of the relationships developed from it.

With *constant heat capacities C and C'* of the volumes of gas flowing through the heat exchanger in unit time, Eq. (5-54) can first be developed into the form

$$\frac{\theta_1 - \theta_2}{\Delta\theta_M} = \int_b^a \frac{d\theta}{\theta - \theta'} \tag{5-55}$$

by considering the heat balance Eqs. (5-28) and (5-31). Furthermore $\theta - \theta'$ is introduced into the integral of this equation from Eq. (5-30) by substituting the following expression from Eq. (5-31) for the ratio C/C' which occurs in Eq. (5-30). Upon effecting the integration,

$$\Delta\theta_M = \frac{\Delta\theta_a - \Delta\theta_b}{\ln\dfrac{\Delta\theta_a}{\Delta\theta_b}} \tag{5-56}$$

is finally obtained where $\Delta\theta_a$ and $\Delta\theta_b$ denote the temperature differences $\theta - \theta'$ in the entry and exit cross-sections a and b as in Fig. 5-10. Thus, from this,

in parallel flow $\Delta\theta_a = \theta_1 - \theta'_1$ and $\Delta\theta_b = \theta_2 - \theta'_2$,
in counterflow $\Delta\theta_a = \theta_1 - \theta'_2$ and $\Delta\theta_b = \theta_2 - \theta'_1$.

The Eq. (5-56) for the average temperature difference is valid both for parallel flow and counterflow for constant values of C and C'. The earlier Eq. (5-27) derived only for simple cases, is in agreement with this Eq. (5-56). If, besides the entry temperatures of the gases, the exit temperatures are also prescribed, the temperature differences $\Delta\theta_a$ and $\Delta\theta_b$ at the end of the heat exchanger are also known, and $\Delta\theta_M$ can be calculated very easily from Eq. (5-56).

If $\Delta\theta_a$ and $\Delta\theta_b$ deviate from one another only slightly, the arithmetical mean

$$\Delta\theta_M = \frac{\Delta\theta_a + \Delta\theta_b}{2} \tag{5-57}$$

can also be used for $\Delta\theta_M$ as a good approximation of the expression (5-56). This simple equation, which can be used almost immediately, is obtained from Eq. (5-56) by developing the limit for $\Delta\theta_a \to \Delta\theta_b$ by using the expression

$$\lim_{x \to 0} \ln(1+x) = x - \frac{x^2}{2}.$$

The accuracy of Eq. (5-57) is greater, the smaller the difference between $\Delta\theta_a$ and $\Delta\theta_b$. If Eq. (5-57) is used $\Delta\theta_M$ turns out to be 4 per cent too large when $\Delta\theta_a = 2\Delta\theta_b$ or $\Delta\theta_b = 2\Delta\theta_a$.

From Eqs. (5-56) and (5-57) can be drawn the important conclusion that counterflow is always superior to parallel flow if the required exit temperatures can be achieved with parallel flow. This is because, given equal entry and exit temperatures, in counterflow the average temperature difference $\Delta\theta_M$ is always greater than in parallel flow, as the following example shows. Let the entry temperatures of both gases be $\theta_1 = 100°$, $\theta'_1 = 0°$ and the exit temperatures be $\theta_2 = 52°$, $\theta'_2 = 48°$. Then, in parallel flow $\Delta\theta_a = 100°$, $\Delta\theta_b = 4°$ and thus from Eq. (5-56) $\Delta\theta_M = 29.8°$; on the other hand, in counterflow, $\Delta\theta_a = \Delta\theta_b = \Delta\theta_M = 52°$. It follows that in counterflow, using Eq. (5-20) the required capacity of the heat exchanger can be achieved with a considerably smaller heating surface area F or with a considerably smaller heat transmission coefficient k than would be needed for parallel flow.

Diagrams for the Determination of the Average Temperature Difference $\Delta\theta_M$

The calculation of the average temperature difference $\Delta\theta_M$ from Eq. (5-56) is so simple that generally, in the age of the electronic computer no other help is required. However, as in earlier times various diagrams were used to enable $\Delta\theta_M$ to be found, at least two will now be mentioned briefly.

Figure 5-17 shows a simple diagram for the determination of $\Delta\theta_M$ using the Eq. (5-56). The temperature differences $\Delta\theta_a$ and $\Delta\theta_b$ at the ends of the recuperator are plotted as coordinates; the larger of these temperature differences is indicated by $\Delta\theta_g$ and the smaller by $\Delta\theta_k$. The desired value of $\Delta\theta_M$ can be interpolated from the curves which are provided for constant values of $\Delta\theta_M$. A modification of this representation consists of plotting $\Delta\theta_a$ and $\Delta\theta_b$ logarithmically. This has the advantage that a very large range of values of $\Delta\theta_a$, $\Delta\theta_b$ and $\Delta\theta_M$ can be compiled together in a single diagram and each value can be read off equally accurately.

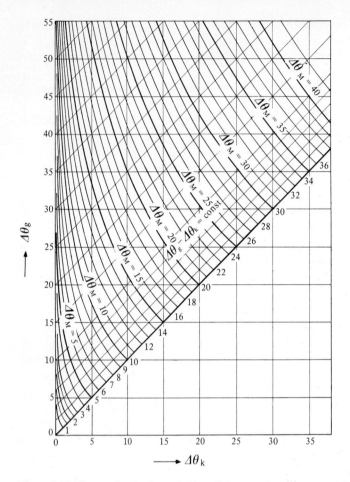

Figure 5-17 Diagram for the determination of the average temperature difference $\Delta\theta_M$ in parallel flow and counterflow from the larger and smaller temperature differences $\Delta\theta_g$ and $\Delta\theta_k$ at the ends of the heat exchanger.

Finally, Eq. (5-56) can also be represented by using a single line by plotting $\Delta\theta_M/\Delta\theta_a$ or $\Delta\theta_M/\Delta\theta_b$ as a function of $\Delta\theta_a/\Delta\theta_b$. Such a line, which is curved, is displayed in Fig. 5-18 where $\Delta\theta_k/\Delta\theta_g$ is set as abscissa and $\Delta\theta_M/\Delta\theta_g$ as ordinate; again $\Delta\theta_k$ denotes the smaller and $\Delta\theta_g$ the larger of the two temperature differences $\Delta\theta_a$ and $\Delta\theta_b$. Because on the whole, interpolation should be avoided, the required value can be read off from this diagram more accurately than it can from Fig. 5-17, but because of the indirect use of Eq. (5-56), the decrease in computational effort required will be only marginal.

Significance of the Integral in Eq. (5-55)

The integral in Eq. (5-55) corresponds to the integral $\int dy/(y-y^*)$ which is known in rectifier technology and in which y denotes the true composition of the steam rising

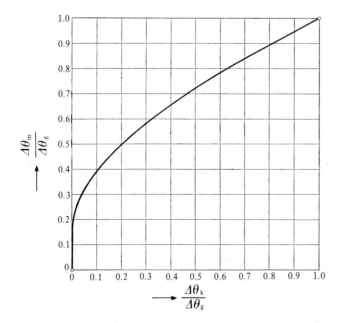

Figure 5-18 Curve for the determination of $\Delta\theta_M$ from the ratio of the temperature differences $\Delta\theta_k$ and $\Delta\theta_g$.

in the cross-section of the rectifying column under consideration; y^*, on the other hand, denotes the composition of a steam which is in equilibrium with the liquid which is trickling down through the same cross-section. The disturbance from equilibrium $y - y^*$ thus corresponds to the temperature difference $\theta - \theta'$ in a heat exchanger. In rectification this integral, which approximately represents a fundamental theoretical factor, is called the "*transfer unit coefficient*".* Similarly the integral in Eq. (5-55) can be understood to be the "transfer unit coefficient" in heat transfer. Accordingly a transfer unit is suggested when the integral formed for only one part of the heat exchanger, has the value 1, that is, when the temperature variation of the gas under consideration is equal to the average value of the temperature difference $\theta - \theta'$ in the part of the heat exchanger in question.

If C is constant it follows, from Eqs. (5-20) and (5-31), that the integral in Eq. (5-55) can also be set as equal to kF/C. kF/C can then also be considered as the "transfer unit coefficient". However, as a different coefficient kF/C' is usually obtained for the second gas and, moreover, as this analysis is only valid for constant values of C and C', it is doubtful whether such an unnecessary terminology (for kF/C and kF/C') should be introduced.

The application of these ideas as they apply to rectification is discussed in [H205] and [K210].

---

* Editor's note: Hausen uses the words, "Zahl der Übergangseinheiten" which we have translated as shown.

5-9 THE TWO PRINCIPAL METHODS OF CALCULATION FOR A HEAT EXCHANGER

It follows from previous considerations that the average temperature difference $\Delta\theta_M$ can be determined from the entry and exit temperatures of the fluid alone. It will be shown in Sec. 8-3 that this is also valid for cross flow. It is demonstrated in Sec. 5-12 that $\Delta\theta_M$ can also be calculated solely from the fluid heat capacities and the known temperatures, even when C and C' are dependent on the temperatures θ and θ' respectively. Finally, as the heat $\dot{Q}$ transferred in unit time is also dependent on these quantities the expression

$$\dot{Q}/\Delta\theta_M$$

is determined solely by these measurable items. From Eq. (5-20)

$$\underset{\text{(demand)}}{\dot{Q}/\Delta\theta_M} = \underset{\text{(output)}}{kF} \tag{5-58}$$

is obtained for this expression. The left-hand side of this equation can be considered as a combination of the demands made on the heat exchanger, which as a rule consist of the fact that with given entry temperatures, specific exit temperatures should be reached and a prescribed quantity of heat $\dot{Q}$ should be transferred. On the other hand the right-hand side can be considered as a measurement of the output of a given heat exchanger as this output is solely dependent on the heat transfer coefficient k and the heating surface area F. Following on from previous considerations, and as will be shown later in more detail, kF itself is principally determined by the dimensions of the heat exchanger and by the fluid velocity of the gases.

Equation (5-58) characterizes the two principal methods of calculation together, for heat exchangers when a *specific capacity is required*. $\dot{Q}$ and $\Delta\theta_M$ and thus $\dot{Q}/\Delta\theta_M$ are basically determined from the prescribed temperatures and the heat capacities of the gases. From Eq. (5-58) the required value of kF is then known immediately. The second main method of approach (see Sec. 6-2 and the sections which follow) is to examine which types of construction or which methods of operation must be used for this value of kF to be achieved.

The sequence in which the calculation is performed, is reversed if the question asked, *what quantity of heat* can be transferred in unit time in a *specific heat exchanger* and what end temperatures can the gases achieve. First kF must be calculated from the dimensions of the heat exchanger and the fluid velocity of the gases. The second calculation step is to determine the quantity of heat $\dot{Q}$ which can be transferred together with the exit temperatures of the gases from the value of $\dot{Q}/\Delta\theta_M$ obtained from Eq. (5-58).

As the quantity kF represents a measurement of the capacity of a heat exchanger, it is generally suitable for the classification of and the comparison of different heat exchangers. For this reason Kühne [K219, 220] has also suggested kF as the standard unit of comparison both for guarantee trials and the establishment of performance criteria for heat exchangers.

5-10 CALCULATION OF TWO OF THE FOUR ENTRY AND EXIT TEMPERATURES

It has been mentioned already on several occasions that two of the four entry and exit temperatures are often sought, given values of kF, C and C'. The fact that *two* of the temperatures can be determined is based, mathematically, on the fact that all the equations discussed up to now have been derived from *two* fundamental equations, namely a heat balance equation and a heat transmission equation.

In cases where k and F can be considered constant, as have C and C', the required end temperatures θ_2 and θ'_2 can be obtained from Eqs. (5-45) and (5-46) for parallel flow and from Eqs. (5-47) and (5-48) for counterflow. If Eqs. (5-45) and (5-46) are also used for counterflow from the four entry and exit temperatures any arbitrary pair of unknown values can then be obtained. This results in total in six different cases.

Another, more simple way in which the two arbitrary unknown temperatures can be determined is that in which the ratio

$$\tau = \frac{\Delta\theta_a}{\Delta\theta_b}$$

of the temperature differences at both ends of the recuperator can be calculated, basically, from k, F, C and C'.[5] By definition this ratio is specified by:

$$\text{for parallel flow} \quad \tau = \frac{\theta_1 - \theta'_1}{\theta_2 - \theta'_2} \tag{5-59}$$

$$\text{for counterflow} \quad \tau = \frac{\theta_1 - \theta'_2}{\theta_2 - \theta'_1} \tag{5-60}$$

Employing Eq. (5-42)

$$\ln \tau = \left(\frac{1}{C} \pm \frac{1}{C'}\right) kF \tag{5-61}$$

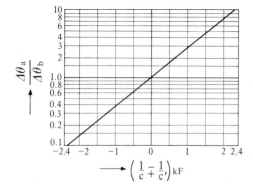

Figure 5-19 Ratio $\tau = \Delta\theta_a/\Delta\theta_b$ of the temperature differences at the ends of the heat exchanger as a function of $\left(\dfrac{1}{C} - \dfrac{1}{C'}\right) kF$ for counterflow and of $\left(\dfrac{1}{C} + \dfrac{1}{C'}\right) kF$ for parallel flow.

[5] Procedure first published by the Author in the first edition of this book.

or

$$^{10}\lg \tau = 0.4343 \left(\frac{1}{C} \pm \frac{1}{C'} \right) kF \tag{5-62}$$

can be obtained for τ, where the plus sign relates to parallel flow and the minus sign to counterflow. τ can easily be determined simply by calculation using one of these equations. However τ can also be read off from Fig. 5-19; this is somewhat easier but not as accurate.

The required temperatures are finally determined using the value of τ and the thermal balance (5-31). In counterflow, solutions for the six possibilities can be found by solving Eqs. (5-60) and (5-31):

(1) Case: θ_2 and θ_2' required.

$$\text{Solution:} \quad \theta_2 = \theta_1' + \frac{\frac{C'}{C} - 1}{\tau \frac{C'}{C} - 1} (\theta_1 - \theta_1'); \quad \theta_2' = \theta_1' + \frac{C}{C'} (\theta_1 - \theta_2).$$

(2) Case: θ_1 and θ_1' required.

$$\text{Solution:} \quad \theta_1 = \theta_2' + \tau \frac{\frac{C'}{C} - 1}{\frac{C'}{C} - \tau} (\theta_2 - \theta_2'); \quad \theta_1' = \theta_2' - \frac{C}{C'} (\theta_1 - \theta_2).$$

(3) Case: θ_1 and θ_2' required.

$$\text{Solution:} \quad \theta_1 = \theta_1' + \frac{\tau \frac{C'}{C} - 1}{\frac{C'}{C} - 1} (\theta_2 - \theta_1'); \quad \theta_2' = \theta_1' + \frac{C}{C'} (\theta_1 - \theta_2).$$

(4) Case: θ_2 and θ_1' required.

$$\text{Solution:} \quad \theta_2 = \theta_2' + \frac{1}{\tau} \frac{\frac{C'}{C} - \tau}{\frac{C'}{C} - 1} (\theta_1 - \theta_2'); \quad \theta_1' = \theta_2' - \frac{C}{C'} (\theta_1 - \theta_2).$$

(5) Case: θ_1 and θ_2 required.

$$\text{Solution:} \quad \theta_1 = \theta_1' + \frac{\tau \frac{C'}{C} - 1}{\tau - 1} (\theta_2' - \theta_1'); \quad \theta_2 = \theta_1 - \frac{C'}{C} (\theta_2' - \theta_1').$$

(6) Case: θ_1' and θ_2' required.

$$\text{Solution: } \theta_1' = \theta_1 - \frac{\tau - \dfrac{C}{C'}}{\tau - 1}(\theta_1 - \theta_2); \quad \theta_2' = \theta_1 + \frac{C}{C'}(\theta_1 - \theta_2).$$

By a different development of the relationships mentioned, progress can be made by finding the first of the required temperatures through the elimination of the second temperature in (5-31) and (5-60); in this case the other temperature required can be extracted from the heat balance (5-31). However the second temperature can also be obtained by elimination of the first temperature, so that for the first case:

$$\theta_2' = \theta_1' + \frac{\tau - 1}{\tau\dfrac{C'}{C} - 1}(\theta_1 - \theta_1')$$

is obtained. Such a solution is preferable when only this one temperature is required.

Other equations for parallel-flow can be obtained in a similar manner. As this is needed only infrequently it will not be discussed further.

It will be shown in the following paragraphs that the unknown temperatures, in addition to being obtained from τ, can also be determined from the efficiency of the heat exchanger.

5-11 EFFICIENCY OF HEAT EXCHANGERS IN PARALLEL FLOW AND COUNTERFLOW

(C, C' and k are constant.)

The efficiency of a heat exchanger can not only be defined differently but different views can also be taken as to whether such concepts are really necessary. The appropriateness of the several points of view in this matter will become self-explanatory later when it becomes clear what the concept of efficiency can and cannot accomplish.

That definition of efficiency which seems most appropriate for the calculation of heat exchangers on the basis of the equations developed so far will first be thoroughly discussed. This definition emerges when efficiency is so determined that it agrees as closely as possible with the usual concept of the efficiency of a boiler. As is known, the efficiency of a boiler is given by the ratio of the quantity of heat which is actually transferred through the boiler heating surface area to the total quantity of heat which is yielded directly from the boiler gases after a loss-free combustion. In this case the latter quantity of heat must be completely transferred in a loss-free boiler. Following from this, while conforming with fundamental principles, the efficiency of a heat exchanger can be defined as the ratio of the quantity of heat actually transferred to that quantity of heat which would be transferred in a perfect heat exchanger. Just what is such a perfect heat exchanger will be discussed in more detail. If, as was considered previously, the heat lost by a heat exchanger to its surroundings is ignored, it follows from Eq. (5-31) that the quantity of heat actually exchanged in unit time is given by:

$$\dot{Q} = C(\theta_1 - \theta_2) = C'(\theta'_2 - \theta'_1) \tag{5-63}$$

A perfect heat exchanger is one which is operated in counterflow and where $kF = \infty$. Thus, either k or F or both can be introduced as being infinitely large. In such a heat exchanger, only a finite quantity of heat can be transferred in unit time as long as finite quantities of gas with finite entry temperatures flow through it. It can be concluded from Eq. (5-20), that since $kF = \infty$, the average temperature difference $\Delta\theta_M$ must be infinitely small. However this is only possible when the temperature difference between the two gases at one point is infinitely small. With constant values of C and C' and with counterflow, it will seem from Eq. (5-56) that $\Delta\theta_M = 0$ if either $\Delta\theta_a = \theta_1 - \theta'_2 = 0$ or $\Delta\theta_b = \theta_2 - \theta'_1 = 0$, that is if the temperature difference at one end or the other of the heat exchanger disappears.

This means, that in a *perfect heat exchanger the gas with the smaller thermal capacity can be completely heated or cooled to the entry temperature of the other gas.* The quantity of heat transferred in a perfect heat exchanger in unit time is thus given by

$$\dot{Q}_{max} = C(\theta_1 - \theta'_1) \text{ with } C \leqslant C' \tag{5-64}$$

or

$$\dot{Q}_{max} = C'(\theta_1 - \theta'_1) \text{ with } C \geqslant C' \tag{5-65}$$

Given these relationships and then using Eq. (5-31) the following definition of the *efficiency of a heat exchanger* is yielded:

$$\eta_w = \frac{\theta_1 - \theta_2}{\theta_1 - \theta'_1} \text{ with } C \leqslant C' \tag{5-66}$$

or

$$\eta'_w = \frac{C}{C'} \frac{\theta_1 - \theta_2}{\theta_1 - \theta'_1} \text{ with } C \geqslant C' \tag{5-67}$$

The remarkable thing is that, following from this, the efficiency η_w of a heat exchanger is represented by different relationships depending upon whether C is smaller or larger than C'.

However, the efficiency η_w can be reduced to the single function η^* where

$$\eta^* = \frac{\theta_1 - \theta_2}{\theta_1 - \theta'_1} \tag{5-68}$$

This will be called the "efficiency function" which is related to via Eqs. (5-66) and (5-67) in the following way.

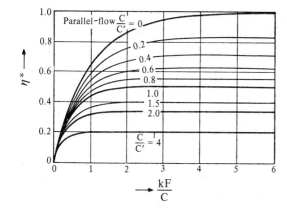

Figure 5-20 The efficiency function η^* for parallel flow.

$$\eta_w = \eta^* \text{ with } C \leqslant C', \tag{5-69}$$

$$\eta_w = \frac{C}{C'}\eta^* \text{ with } C \geqslant C'. \tag{5-70}$$

Possible means of evaluating the function η^* have been determined previously utilizing Eq. (5-45) or (5-40) or (5-48).

Figures 5-20 and 5-21 show the efficiency function η^* for parallel flow and counterflow as a function of kF/C for different ratios C/C'. Given Eqs. (5-66), (5-68) and (5-69) it will be seen that the efficiency can always be obtained from these diagrams directly when the temperature difference $\theta_1 - \theta_2$ is used to refer to the gas with the smallest thermal capacity so that C is always less than C'. It can be observed from Fig. 5-20 that in parallel flow, with increasing kF/C, η^* tends towards a limiting value which is less than unity and which can be computed using the limit obtained from Eq. (5-31)

$$\lim_{\theta_2' \to \theta_2} \eta^* = \frac{C'}{C + C'}$$

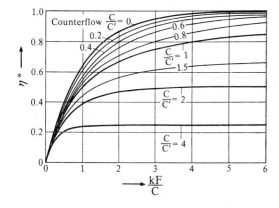

Figure 5-21 Efficiency function η^* for counterflow.

When $C = C'$ this limiting value is 0.5 so that in this case, an efficiency of, at most, 50 per cent can be obtained. In counterflow however, it will be noted from Fig. 5-21 that without specifying the value of C/C', η^* approaches the value 1 when $C \leqslant C'$; on the other hand η^* approaches the value C'/C when $C > C'$. The efficiency η_w itself, however, for $C > C'$, tends towards the limiting value $\eta'_w = 1$ as may be concluded from Eq. (5-70). The superiority of counterflow as opposed to parallel flow, as has been indicated several times already, is thus demonstrated by the considerably greater value of the efficiency. The differences between parallel flow and counterflow are slight only when the values of kF/C are very small, that is, in very small heat exchangers.

Figures 5-20 and 5-21 can be used, in the following manner, for the direct *calculation of the value of kF* which is critical for heat transfer calculations. If, for example, in addition to C and C', the entry temperatures θ_1 and θ'_1, as also the exit temperatures θ_2 and θ'_2, are specified, η^* can be calculated, basically using Eq. (5-68). Given the ratio C/C' and having obtained the value of η^* the value kF/C on the abscissa can then be extracted from the appropriate η^*-diagram. Thus kF can be determined immediately.

If, on the other hand, in addition to C, C' and kF only two of the temperatures are specified while the other two are unknown, the value of η^* must first be read off from the curve for the specified ratio C/C' for the appropriate value of kF/C. The two unknown temperatures can then be found from Eqs. (5-31) and (5-68).

This method of recuperator calculation has been systematically pursued further by Bošnjaković [B209] who has introduced the nomenclature "operation characteristic"* instead of efficiency. In his diagrams, which are very similar to Figs. 5-20 and 5-21, he has plotted a second set of curves for kF/C' next to the curves for C'/C and this considerably facilitates calculation. Moreover, he has developed appropriate representations which correspond to various complicated heat exchanger switching arrangements including cross flow, which will be dealt with later, and for recuperators with many passages. His method of heat exchanger calculation was included in the VDI-Wärmeatlas (VDI-Heat Atlas) [1, 2nd edition, pages N1–N12].

The efficiency, which is rigorously defined by Eqs. (5-66) and (5-68), has a unique value for each recuperator with given operational conditions as only one of the cases $C \geqslant C'$ or $C \leqslant C'$ will apply. However, it is more usual to understand Eqs. (5-66) and (5-67) simply as ratios of temperature difference. If Eq. (5-67) is rearranged, with repeated reference to Eq. (5-31) the following relationships are obtained for these temperature ratios.

$$\eta_w = \frac{\theta_1 - \theta_2}{\theta_1 - \theta'_1} \qquad (5\text{-}71)$$

$$\eta'_w = \frac{\theta'_2 - \theta'_1}{\theta_1 - \theta'_1} \qquad (5\text{-}72)$$

---

* Editor's note: Hausen uses the term "Betriebscharakteristik".

The expressions "temperature rise relationship" or even "heating relationship" have been suggested for these quantities.* As this method of analysis refers only to temperatures and no longer considers the transfer of thermal energy, it is really more correct to avoid the term "efficiency".

In considering the significance or meaning of efficiency, or of the thermal ratio, it must be emphasized that η_w or η'_w, can serve at most as a standard of reference for the efficiency of a heat exchanger if the value of this quantity is compared with the ratio C/C'. In order to determine an efficiency which is generally acceptable as a reference standard, it could be specified that any particular value of η_w or η'_w should be related to the ratio $C/C' = 1$. In so doing it is useful if the sum of the thermal capacity rates of the two gases (related to unit time), remains unchanged. This proves to be so in the reference case, where each of the two gas streams has a thermal capacity $0.5(C + C')$. However, this quantity varies so little that k and thus kF do not change noticeably; this assumption can always be shown to be true if C and C', and also α and α', differ only slightly. If kF is known, the efficiency $\eta_w = \eta^*$ can be obtained from Figs. 5-20 or 5-21 by choosing $2kF/(C+C')$ as abscissa and reading off the corresponding value of η^* from the curve for $C/C' = 1$. However, in counterflow, this value can also be obtained purely by calculation using the following Eq. (5-73) which, is obtained from the consideration of Eq. (5-68) in relation to Eq. (5-40).

$$\eta^* = \frac{2kF/(C+C')}{1+2kF/(C+C')} = \frac{1}{1 + \dfrac{C+C'}{2kF}} \tag{5-73}$$

These considerations can also be extended logically to the case where kF is not known but where, for example, the entry and exit temperatures of the two gases have been measured in a heat exchanger for given values of C and C'. Then, employing these temperatures, the required value of kF can be obtained from Eq. (5-46). The additional calculation necessary for the determination of the efficiency can then be carried out in the manner described above.

If the deduction of kF in this way proves not to be sufficiently accurate, new values of α and α' and thus also a new value for kF can be calculated, using a fluid velocity modified by multiplication by the ratio of $0.5(C+C')$ to C or C'. In this way the required efficiency is again obtained from Fig. 5-20 or 5-21 or using equation (5-73).[6]

Other Definitions of Efficiency

Other possible definitions of efficiency exist and these will now be indicated. Efficiency can be determined, for example, in such a way that it can be used to measure the rate of heat loss by a heat exchanger to its surroundings; this will be

---

* Editor's note: The expression "*thermal ratio*" is commonly used in English so that (5-71) might be called the "hot side thermal ratio" and (5-72) the "cold side thermal ratio".

[6] Considerations of a similar and related type have been employed also by H. Kühne [K220].

discussed later (see Sec. 7-2). These losses have the effect that less heat is taken up by the initially colder gas than is given up by the warmer gas. Thus the ratio of the quantity of heat absorbed by the cold gas to the quantity of heat given up by the hot gas can be termed as efficiency. This ratio is sometimes called the "chemical efficiency"* or the "commercial efficiency";* see [P205].

The definition which perhaps conforms most closely to normal linguistic usage is obtained, if the quantity of heat transferred in unit time is set against the energy simultaneously expended, usually in the form of work of compression which is necessary in order to overcome any pressure drop. Nevertheless, the evaluation of such an efficiency is not easy because ratios on both sides of the partition walls of the heat exchanger must be considered. However, in order to make a step in this direction, Grassmann [G205] considered initially only the processes taking place on one side of the partition walls and derived a relationship between the heat transfer coefficient and the energy in overcoming pressure drop. He thus proceeded from the usual power ratios for heat transfer and pressure loss; see Eq. (2-19) or (2-21) in Sec. 2-5 and Eq. (4-2) and (4-4) in Sec. 4-2. Thus he was able to show that with turbulent flow the heat transfer coefficient α is proportional to the power $(A/F)^{0.291}$, in which A is the energy expended in the manner discussed and F is the heating surface area. Here the physical properties and the tube diameter are assumed to be constant.

Following on from these considerations Grassmann [G207] compared the heat transfer coefficient, which had been calculated for smooth tubes, using his equation, with the heat transfer coefficients measured in an arbitrary heat exchanger with equal A/F. The latter heat transfer coefficient is usually smaller because in rough tubes the pressure loss permits only a low fluid velocity which can be used and a decrease in this velocity also reduces the value of α. A possible lining on the tube walls can further reduce the heat transfer coefficient. Grassmann called the ratio of the heat transfer coefficient measured in this way to the heat transfer coefficient calculated for smooth tubes, the "efficiency factor". He thus indicated that a corresponding efficiency factor can also be formed with the measured and calculated heat transfer coefficient k.

Moreover, Grassmann [G207] gives a thermodynamic definition of efficiency, with the help of the concept of exergy $E = H - H_0 - T_0(S - S_0)$, where the enthalpy H and the entropy S refer to the actual condition of the material and H_0 and S_0 refer to a fixed final condition at the ambient temperature T_0. H and S, and thus E, vary as heat is transferred to another material; exergy increases in material which is becoming warmer and decreases with material which is becoming cooler. According to Grassmann the ratio of exergy increase to exergy decrease can be regarded as the efficiency, although he did find this definition not to be suitable for practical application.

Glaser [G203] has suggested a "performance factor" which is obtained by

---

* Editor's note: Hausen uses the expressions "Unsetzungsgrad" and "Ausnutzungsfaktor" to denote what has been translated as "chemical efficiency" and "commercial efficiency" respectively.

dividing the specific heat transfer $\dot{Q}/\Delta\theta_M$ (i.e. the quantity of heat transferred per unit average difference in temperature) by the energy expended A to overcome the drop in pressure. He calculated this performance factor for various tube arrangements in parallel flow and cross flow as well as for tubes with a vortex configuration. He thus showed that this factor can serve as a standard for the estimation and comparison of heat exchangers. On the other hand Th. E. Schmidt [S212] considered it more useful to choose the "loss factor"* $\omega = A/\dot{Q}$ as a means of comparison. By dividing the ideal loss factor ω_{id} for laminar flow by the loss factor for the heat exchanger under consideration, he obtained an "efficiency factor" which is related to Grassmann's efficiency factor but which yields different numerical values. All making these many suggestions agree that heat exchangers are used for so many different purposes and that according to their area of application the heat transfer requirements, the necessary minimization of pressure loss, any economy in price and operating costs, the safe operating requirements and the like, vary greatly from one heat exchanger to another. Thus one definition may first seem suitable, then another appears to be more suitable although no clear preference can be given for either. Upon recognizing this fact Glaser [G204] came to the conclusion that efficiency or an efficiency factor cannot be suggested without at the same time first considering the entire operating context within which the heat exchanger is set. He thus suggests that, for example in the case of a gas turbine, the efficiency of the complete process should first be determined with a loss-free heat exchanger; subsequently taking into account all the losses in this exchanger, the efficiency is reevaluated. The ratio of the two efficiencies are considered then to be the efficiency factor for the exchanger.

The problems described which make the practical definition of efficiency to a certain degree difficult, can be avoided by considering the desired heat transfer and the admissible pressure loss as two separate requirements, both of which must be satisfied simultaneously by a heat exchanger. If these requirements are fulfilled precisely, as will be shown more clearly in Sec. 6-2 ff., various constructional possibilities can be evaluated and compared without reference to the idea of efficiency or efficiency factor. In this way, a design decision for a heat exchanger can be reached clearly and simply.

5-12 TEMPERATURE DISTRIBUTION, AVERAGE TEMPERATURE DIFFERENCE AND EFFICIENCY WITH TEMPERATURE DEPENDENT HEAT CAPACITIES C AND C'

Frequently the heat capacities of the gases are so strongly dependent upon temperature that the process described so far for calculating the temperature distribution, the average temperature difference $\Delta\theta_M$ and the efficiency using Eq. (5-66) is not sufficiently accurate. Sharp variations in the specific heat capacity are

---

* Editor's note: Hausen mentions the term "Verlustziffer".

shown, for example, by compressed air at low temperatures. In order to account for these variations in the integrals in Eqs. (5-33) and (5-54) the specific enthalpies h and h' of the gases, defined on the basis of the relationships $C d\theta = \dot{m} \, dh$ and $C' d\theta' = \dot{m}' \, dh'$, are introduced where $\dot{m}$ and $\dot{m}'$ denote the quantities of gas flowing through the heat exchanger in unit time. First, with h and h' Eqs. (5-28), (5-29) and (5-31) can be written in the form

$$d\dot{q} = -\dot{m} \, dh = \dot{m}' \, dh' \qquad (5\text{-}74)$$

$$\dot{q} = \dot{m}(h_1 - h) = \dot{m}'(h' - h'_1) \quad \text{in parallel flow} \qquad (5\text{-}75)$$

$$= (h'_2 - h') \quad \text{in counterflow} \qquad (5\text{-}76)$$

$$\dot{Q} = \dot{m}(h_1 - h_2) = \dot{m}'(h'_2 - h'_1) \qquad (5\text{-}77)$$

where the specific enthalpies at temperatures θ_1 and θ_2 or θ'_1 and θ'_2 are designated by h_1 and h_2 or h'_1 and h'_2. Using Eq. (5-74), Eq. (5-33) takes the form:

$$\dot{m} \int_x^a \frac{dh}{\theta - \theta'} = kf \qquad (5\text{-}78)$$

Here, as previously, k is considered to be constant.

Furthermore, using Eqs. (5-74) and (5-77), Eq. (5-54) for the average temperature difference, is converted into

$$\frac{h_1 - h_2}{\Delta\theta_M} = \int_b^a \frac{dh}{\theta - \theta'} \qquad (5\text{-}79)$$

see [H204]. Thus, using Eq. (5-78) for the temperature distribution in the heat exchanger, and employing Eq. (5-79) for the average temperature difference $\Delta\theta_M$, these items can be calculated if the integral of $dh/\theta - \theta'$ can be evaluated. This is usually only possible if approximate methods are used.

Step by Step Process

A generally practical *step by step process* results from the following consideration. The differential dh is converted to the finite difference Δh yielding the approximate equation below.

$$\int_x^a \frac{dh}{\theta - \theta'} = \sum_x^a \frac{\Delta h}{(\theta - \theta')} \quad \text{and} \quad \int_b^a \frac{dh}{\theta - \theta'} = \sum_b^a \frac{\Delta h}{(\theta - \theta')} \qquad (5\text{-}80)$$

It is then assumed that h is a function of θ and h' is a function of θ' where these relationships are set out in the status diagrams of the respective gases. In addition to the entry temperatures θ_1 and θ'_1, the exit temperatures θ_2 and θ'_2 together with the associated values of h_2 and h'_2 are also assumed. Then the summation in Eq. (5-80) can be evaluated numerically; this is shown diagrammatically in Fig. 5-22 for the case of counterflow.

For the first gas h is a function of θ (solid line). For the second gas, a

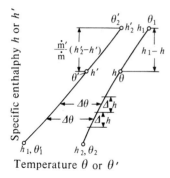

Figure 5-22 Determination of the average temperature difference in counterflow in general cases using approximate calculation of the integrals $\int \dfrac{dh}{\Delta\theta} = \Sigma \dfrac{\Delta h}{\Delta\theta}$.

transformed h', θ' curve is drawn as a dotted line, where the same abscissa scale is used for θ' and θ.* First h'_2 is fixed as principal datum point and set at θ'_2 but at the same level as h_1. Additional points for arbitrary positions (θ', h') of the second gas are obtained by plotting the points (θ', h'') where h'' is at height $\dot{m}'/\dot{m}(h'_2 - h')$ down from h'_2 as indicated in the diagram. From Eq. (5-77) the final point (θ'_1, h'_1) for the second gas then comes to rest automatically at the same height as point (θ_2, h_2) of the first gas. By plotting the curves in this way the general result is that the positions of both gases in an arbitrary cross-section of the heat exchanger always appear at the same heights on both curves. Thus, as can easily be seen from Fig. 5-22, the heat balance Eq. (5-66) is satisfied for these points. The horizontal distance between the points represents the temperature difference $\theta - \theta'$ in a cross-section of the heat exchanger. Drawing the curve for the second gas is simplified when its specific heat capacity c'_p is constant. Then only the final points of the dotted line need to be determined and then connected by a straight line, as has been indicated by M. Rabes [R201].

In order to evaluate the summation in Eq. (5-80) the enthalpy curve of the first gas is split up as in Fig. 5-22, into arbitrary sections Δh and the value of $\theta - \theta'$ is determined at the mid point of each section by measuring the horizontal distance from the other curve. In this manner the value of $\Delta h/\theta - \theta'$ can be found for each section. If the sections are so selected that $\theta - \theta'$ varies percentage-wise only very slightly from one section to the next, the required integral in Eq. (5-80) can be obtained very accurately by adding together all the values of $\Delta h/\theta - \theta'$. Thus, in this way, using Eqs. (5-78) and (5-79), a general process for determining the temperature distribution and the average temperature difference $\Delta\theta_M$ is provided.

The steps can be considerably larger, and the method be as accurate, if the

---

* Editor's note: This method was included as such in the 1950 edition of this book and appears almost unaltered in this edition. It therefore takes no account of the development of numerical methods for computers. Nowadays, the integral

$$\dot{m} \int_x^a \frac{dh}{\theta - \theta'}$$

would be evaluated employing Gaussian quadrature methods on the basis of given explicit algebraic relationships between gas specific heats and gas temperature. These quadrature methods are commonly available in standard computer program libraries.

temperature differences at both ends of a step are extracted from Fig. 5-22 and from these, using Eq. (5-56), the logarithmic average value of the temperature difference for this step is calculated.[7] This logarithmic average value can then be introduced into the summation of Eq. (5-80) in place of $\theta - \theta'$.

Finally, instead of using a step by step method, the integrals in Eq. (5-78) and (5-79) can also be evaluated in such a way that, having read off the values of $\theta - \theta'$ from Fig. 5-22, the reciprocals $1/\theta - \theta'$ are plotted on a graph as a function of h and the integrals measured employing a planimeter.[*][8]

Calculation of kF

Having determined $\Delta\theta_M$ from Eq. (5-79) and $\dot{Q}$ from Eq. (5-77) kF can be immediately obtained from Eq. (5-20) or (5-58).

For many calculations however it is useful to know that in order to find kF it is not necessary to completely work out the value $\Delta\theta_M$. Rather it is sufficient to determine the integral on the right-hand side of Eq. (5-79), for example by means of the step by step process illustrated in Fig. 5-22. Then, by using Eq. (5-78) for the total heating surface area F of the recuperator there directly follows

$$kF = \dot{m} \int_b^a \frac{dh}{\theta - \theta'} \tag{5-81}$$

Diagram for the Determination of $\Delta\theta_M$ With a Parabolic Curve for the Temperature Difference $\theta - \theta'$

A complete and exact equation for $\Delta\theta_M$ can be derived if the temperature difference $\theta - \theta'$ is known to be a quadratic function of the specific enthalpy in the form $\theta - \theta' = a + bh + ch^2$. This means that, as is shown in Fig. 5-23, the curve of $\theta - \theta'$ can be displayed as a parabolic curve dependent on h or h'. However, even in other

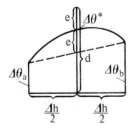

Figure 5-23 Diagram for determining the average temperature difference $\Delta\theta_M$ with a parabolic curve of $\Delta\theta = \theta - \theta'$.

---

[7] From a proposal by Johannes Wucherer of Höllriegelskreuth b. München.

[8] This last type of calculation could have been described using the basic ideas first accorded to Altenkirch [A201].

[*] Editor's note: Truly a pre-computer method!

cases, the temperature dependence of C and C' can often be accurately represented by assuming a $\theta - \theta'$ parabolic curve.

In Fig. 5-23, $\Delta\theta_a$ and $\Delta\theta_b$ are as previously, the values of $\theta - \theta'$ in the end cross-sections of the heat exchanger. At these positions the enthalpy of one of the gases is h_a or h_b. At the mid position $\frac{1}{2}(h_a + h_b)$, $\theta - \theta'$ has the value $\Delta\theta^*$. The length of the straight line shown in the diagram is given by

$$d = 2\Delta\theta^* - \tfrac{1}{2}(\Delta\theta_a + \Delta\theta_b) \tag{5-82}$$

This can most easily be found graphically by drawing the straight line e between the curved line and the straight connecting line joined to the upper corner, and then upwards once more beyond the curve as shown in Fig. 5-23. Given d

$$\theta - \theta' = \Delta\theta_a + 2(d - \Delta\theta_a)\frac{h - h_a}{h_b - h_a} + (\Delta\theta_a + \Delta\theta_b - 2d)\left(\frac{h - h_a}{h_b - h_a}\right)^2$$

results as the equation for the parabolic curve in Fig. 5-23. If this is inserted into Eq. (5-79) relationships (5-83) and (5-84) are obtained for $\Delta\theta_M$.[9]

$$\frac{\Delta\theta_M}{d} = \frac{2\sqrt{1 - \dfrac{\Delta\theta_a \cdot \Delta\theta_b}{d^2}}}{\ln \dfrac{1 + \sqrt{1 - \dfrac{\Delta\theta_a \cdot \Delta\theta_b}{d^2}}}{1 - \sqrt{1 - \dfrac{\Delta\theta_a \cdot \Delta\theta_b}{d^2}}}}, \quad \text{when } d^2 > \Delta\theta_a\Delta\theta_b \tag{5-83}$$

and[10]

$$\frac{\Delta\theta_M}{d} = \frac{\sqrt{\dfrac{\Delta\theta_a \cdot \Delta\theta_b}{d^2} - 1}}{\arctan \sqrt{\dfrac{\Delta\theta_a \cdot \Delta\theta_b}{d^2} - 1}}, \quad \text{when } d^2 < \Delta\theta_a\Delta\theta_b \tag{5-84}$$

[9] The first derivation can be found in [H207].

[10] The denominator on the right-hand side of Eq. (5-84) is first obtained in the form $\arctan a - \arctan b$. However, if the substitutions $a = \tan\alpha$ and $b = \tan\beta$ are inserted, it follows from the relationship

$$\tan(\alpha - \beta) = \frac{\tan\alpha - \tan\beta}{1 + \tan\alpha\tan\beta}$$

hat

$$\arctan a - \arctan b = \arctan\frac{a - b}{1 + ab};$$

this leads, by a suitable transformation, to Eq. (5-84).

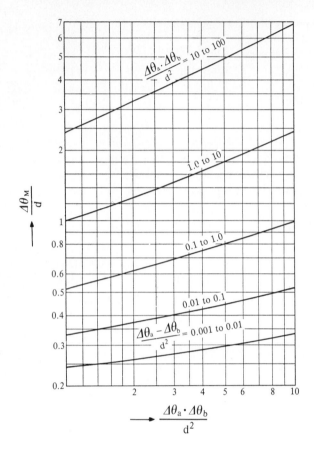

Figure 5-24 Diagram for determining the average temperature difference $\Delta\theta_M$ with a parabolic curve for $\Delta\theta = \theta - \theta'$.

From these equations it will be seen that $\Delta\theta_M/d$ is dependent only upon $\Delta\theta_a\Delta\theta_b/d^2$. It follows that $\Delta\theta_M$ can thus be determined very quickly using Fig. 5-24 where, using Eqs. (5-83) and (5-84), $\Delta\theta_M/d$ is displayed as a function of $\Delta\theta_a\Delta\theta_b/d^2$. Accordingly it is only necessary to calculate d and thus $\Delta\theta_a\Delta\theta_b/d^2$ in the manner prescribed and $\Delta\theta_M/d$ can be read off directly from the diagram. Having done this $\Delta\theta_M$ is then determined by multiplying by d. By breaking the horizontal scale at every power of ten and starting each corresponding curve from the left, greater accuracy can be obtained.

When, in a heat exchanger, the temperature difference $\theta - \theta'$ cannot be reproduced with sufficient accuracy by means of a parabola over the entire area, it is possible to think of the length of the heat exchanger as divided into two or more sections in each of which the process described above can be applied. Independent of whether such division into sections is required or not, it is often an advantage in this process if the temperature differences at both ends of the heat exchanger appear. However, if not too much importance is attached to this, it is advisable to use the process described for thermal radiation, which is based on Gaussian integration and which provides a higher degree of accuracy with the same number of divisions; see Sec. 3-3.

The Determination of Two Required Temperatures

In the case of temperature dependent gas thermal capacities, it has been assumed hitherto that both the entry temperatures and the exit temperatures of the gases are known. If, on the other hand, given kF, two of the temperatures must be calculated, then in general these temperatures can only be determined iteratively. To do this, the required temperatures are first fixed arbitrarily using Eq. (5-77). Employing these temperatures, the integral in Eq. (5-81) is computed in the manner described previously. The new temperatures obtained thus are used in the re-evaluation of the integral and the process continues iteratively until Eq. (5-81) is satisfied to a sufficient degree of accuracy.

This result can be obtained more speedily by using Fig. 5-24 or the approximate Gaussian integration procedure. In this case $\Delta\theta_M$ is determined with the estimated temperatures and tests are carried to find out how well Eqs. (5-20) or (5-58) are satisfied.

Efficiency

The heat exchanger efficiency $\eta_\omega = \dot{Q}/\dot{Q}_{max}$ can be calculated for various thermal capacities of gases. $\dot{Q}$, and thus $\dot{Q}_{max}$ can be determined by Eq. (5-77) in such a way that in completely counterflow operation the temperature difference $\theta - \theta'$ in at least one part of the heat exchanger becomes infinitely small. If $C < C'$ or $C > C'$ the value $\theta - \theta' = 0$ normally occurs at one or other end of the heat exchanger. $\dot{Q}_{max}$ is then determined using either $\dot{Q}_{max} = \dot{m}(h_1 - h_{min})$ or $\dot{Q}_{max} = \dot{m}(h'_{max} - h'_1)$, where h_{min} denotes the value of h when $\theta = \theta'_1$ and h'_{max} denotes the value of h' when $\theta' = \theta_1$. Note that θ_1 and θ'_1 represent again the entry temperatures of the two gases.

From these considerations are developed Eqs. (5-85) and (5-86) for the efficiency of a heat exchanger.

$$\eta_w = \frac{h_1 - h_2}{h_1 - h_{min}} \quad \text{when } C < C' \tag{5-85}$$

$$\eta'_w = \frac{h'_2 - h'_1}{h'_{max} - h'_1} \quad \text{when } C > C' \tag{5-86}$$

However, if for example $C < C'$ in the warmer part of a counterflow recuperator while on the other hand, $C > C'$ in the cooler part, the temperature difference $\theta - \theta' = 0$ can occur in the middle of the total heat exchanger. In this case $\dot{Q}_{max}$ can, in general, only be determined by numerical experiment.

With temperature dependent gas thermal capacities the determination of efficiency can thus sometimes be rendered quite difficult. Given temperature ratios developed in this way, hardly any advantage is secured by the introduction of efficiency in place of the average temperature difference $\Delta\theta_M$ or the integral

$$\int_b^a \frac{dh}{\theta - \theta'}$$

in heat exchanger calculations.

5-13 THE TEMPERATURE DISTRIBUTION AND THE AVERAGE TEMPERATURE DIFFERENCE WITH VARIABLE HEAT TRANSMISSION COEFFICIENT k

All the processes for calculating the temperature distribution, the average temperature difference, etc., which have been developed so far can also be used, with a few obvious generalizations, when the heat transmission coefficient k is variable in the longitudinal direction of the recuperator.

The method of calculation can be developed most simply when k is given as *a function of the longitudinal coordinates x or f.* This is the case for example, when the flow cross-section, and therefore the fluid velocity, vary in the longitudinal direction of the recuperator. Then employing the heat transfer equations in Sec. 2-5, corresponding variations in the heat transfer coefficients α and α' dependent on x or f also occur. If a possible temperature dependence of α and α' is ignored, it follows from Eq. (5-2), or Eqs. (5-11) to (5-18) inclusive, k is only dependent on x or f. In this case Eq. (5-32) can be integrated in the following way:

$$\int_a^x \frac{d\dot{q}}{\theta - \theta'} = \int_0^f k \cdot df \tag{5-87}$$

and

$$\int_a^b \frac{d\dot{q}}{\theta - \theta'} = \int_0^F k \cdot df \tag{5-88}$$

From these equations which determine (in the place of Eqs. (5-33) and (5-53)), the temperature distribution and the rate of heat transfer it can be recognized that all the methods of calculation discussed up to now remain relevant if, throughout, kf is replaced by $\int_0^f k\,df$ and kF is replaced by $\int_0^F k\,df$. These integrals can also be introduced into Eqs. (5-42) to (5-48) inclusive, which give the temperature distribution for constant thermal capacities C and C', and in the more general Eqs. (5-78) to (5-81). The calculation of the integral $\int k\,df$ thus does not present any major difficulties. Rather it is sufficient, as a first approximation, to assume k is linearly dependent on f.

More frequently k is a *function only of temperature.* This mainly occurs when large temperature variations are obtained in the gases in a heat exchanger with a uniform flow cross-section. In this case the heat transfer coefficients α and α' are, from Fig. 2-4 in Sec. 2-5, dependent on the temperatures θ and θ' of the gases, that is if these heat transfer coefficients are based only on thermal conduction and convection. This temperature dependence would be even greater if thermal radiation was known to have considerable influence on heat transfer in the case under examination; see for example, Eq. (3-25) or (3-26).

In such cases k is primarily a function of *both* the gas temperatures θ and θ'. However, if it is assumed that the entry temperatures of the gases are known, as well as their exit temperatures, θ' would clearly be a function of θ; see, for example, Eq. (5-29). Thus k can be seen as a function of θ alone. Since this also holds true for $\theta - \theta'$ as can be deduced from Eq. (5-30), Eq. (5-32) can be integrated in the form:

$$\int_a^x \frac{d\dot{q}}{k(\theta - \theta')} = f \tag{5-89}$$

and

$$\int_a^b \frac{d\dot{q}}{k(\theta - \theta')} = F \tag{5-90}$$

Correspondingly, Eqs. (5-78) and (5-81) also take the general form

$$\dot{m} \int_x^a \frac{dh}{k(\theta - \theta')} = f \tag{5-91}$$

and

$$\dot{m} \int_b^a \frac{dh}{k(\theta - \theta')} = F \tag{5-92}$$

In this way Eq. (5-89) or (5-91) determines the temperature distribution and Eq. (5-90) or (5-92) determines the rate of heat transfer in the complete recuperator. The integrals in these equations can usually only be worked out approximately.* The step by step process described in Fig. 5-22 is particularly suitable for this; the sum

$$\sum \frac{\Delta i}{\theta - \theta'}$$

would be formed and not

$$\sum \frac{\Delta i}{k(\theta - \theta')} \quad [11]$$

The average temperature difference $\Delta\theta_M$ also retains its significance in this general case and, because it is basically independent of k, can be determined by the process mentioned previously. For the calculation of a heat exchanger with a temperature dependent k, however, it is more advantageous to introduce the average value $(k(\theta - \theta'))_M$ of the product $k(\theta - \theta')$ in place of $\Delta\theta_M$. This average value is defined by the following equation:

$$\frac{\dot{Q}}{(k(\theta - \theta'))_M} = \int_a^b \frac{d\dot{q}}{k(\theta - \theta')} \tag{5-93}$$

or even by

$$\frac{h_1 - h_2}{(k(\theta - \theta'))_M} = \int_b^a \frac{dh}{k(\theta - \theta')} \tag{5-94}$$

A comparison of Eq. (5-90) with (5-93) shows that the quantity of heat transferred in the recuperator in unit time is given by

$$\dot{Q} = F(k(\theta - \theta'))_M \tag{5-95}$$

---

* Editor's note: Again, methods of numerical quadrature.
[11] The process outlined by Altenkirch [A201] in 1914 is based on the same considerations.

The calculation is considerably simplified if $k(\theta-\theta')$ can be represented sufficiently accurately by a quadratic function in h. Then, as is shown by comparing Eq. (5-94) with Eq. (5-79), the average value $(k(\theta-\theta'))_M$ can be determined by the process illustrated in Figs. 5-23 and 5-24, if in this process the value of $k(\theta-\theta')$ is used in place of $\theta-\theta'$ throughout.

However, accuracy can be increased here by using the Gaussian integration process described in Sec. 3-3. If $1/k(\theta-\theta')$ corresponds closely enough to a second degree function of h it is sufficient to calculate, accurately the "thermal flow densities"

$$(k(\theta-\theta_1'))_1^*$$

and

$$(k(\theta-\theta'))_2^*$$

at positions in the heat exchanger fixed by

$$h_1^* = h_1 - 0.2(h_1 - h_2)$$

and

$$h_2^* = h_2 + 0.2(h_1 - h_2)$$

Thus the relationship (5-96) is obtained for the average thermal flow density $(k(\theta-\theta'))_M$

$$\frac{2}{k(\theta-\theta')_M} = \frac{1}{(k(\theta-\theta'))_1^*} + \frac{1}{(k(\theta-\theta'))_2^*} \tag{5-96}$$

The total quantity of heat transferred in unit time is finally determined by Eq. (5-95). However, if the quadratic function is not sufficiently accurate, the required accuracy can be obtained by applying the process to two or more sections of the heat exchanger. However in place of such a subdivision it is possible to employ an expanded form of the Gaussian integration process, in which three or more data points are used; see Hütte I [25], 28th edition, p. 211.

The most general case where k is dependent on f as well as on θ and θ' must be applied when for example, not only the flow cross-section varies but also when large temperature variations occur. Often k is less dependent on f than on θ and θ'. As a rule it is sufficiently accurate to.consider k as the product of two factors k_T and k_f, of which k_T is only dependent on the temperatures and k_f is only dependent on the longitudinal direction of the recuperator. Equation (5-32) can then be integrated as follows:

$$\int_a^x \frac{d\dot{q}}{k_T(\theta-\theta')} = \int_0^f k_f \cdot df \tag{5-97}$$

$$\int_a^b \frac{d\dot{q}}{k_T(\theta-\theta')} = \int_0^F k_f \cdot df \tag{5-98}$$

If in Eq. (5-74), the specific enthalpy h is introduced, equations are obtained which correspond to Eqs. (5-91) and (5-92); these are used to determine the temperature distribution and the total quantity of heat transferred. The integrals on the left-hand side can be computed in a similar manner to that described already for the integrals in Eqs. (5-89) and (5-90).

If the heating surface area F, which is necessary for heat transfer, is required this must be determined iteratively; a value of F is first assumed, either arbitrarily or by assessment, as the upper limit of the integrals in Eq. (5-98) and this value of F is then varied systematically until the integrals on the right become equal to the integrals on the left-hand side.

Other methods whereby various heat transfer coefficients and physical properties can be considered, have recently been developed by Peters [P203] and Roetzel [R206, 210].

SIX

DETERMINATION OF THE SIZE AND CONSTRUCTION OF RECUPERATORS OPERATING IN PARALLEL FLOW AND COUNTERFLOW

6-1 PRELIMINARY REMARKS

The complete calculations for a recuperator operated in parallel flow or counterflow will be discussed below in their most important details. In Sec. 5-9 [Eq. (5-68)] it was shown that the heat transfer performance required of a heat exchanger can be simply and briefly expressed by the quantity

$$\dot{Q}/\Delta\theta_M = kF \tag{6-1}$$

or, in the general case, by the corresponding integral $\int k \, df$. Here $\dot{Q}$ denotes the quantity of heat being transferred in unit time, $\Delta\theta_M$ denotes the average temperature difference between two gases, k denotes the heat transmission coefficient and F denotes the total heat transferring surface; f is the heating surface area extending to the cross-section under consideration.

Once the required performance is known the question then arises what are to be the *size and construction of a heat exchanger* so that this *performance can be attained*. This chapter will be primarily concerned with this problem, which has been treated in Sec. 5-9 as only the second most important problem in the calculation of a heat exchanger. The problem can always be solved in numerous ways since basically this same performance can be attained with heat exchangers of varying constructions and different sizes. Thus, for example, a recuperator consisting of straight parallel tubes, as shown in Fig. 5-10, could either be constructed of a few very long tubes or alternatively of many short tubes. However, in place of these, many other arrangements are possible, for example, with cross flow or with combined cross and counterflow, or even with finned tubes (see Secs. 8-3 and 8-4). Furthermore, the question can be asked whether, in a particular case, a reversing regenerator (see Part Four of this book) might not be preferred to the uniformly operating recuperator.

In the face of this large number of possible solutions, the general process for fixing the dimensions consists in the first place, of taking approximate dimensions for a particular type of heat exchanger construction, these dimensions being based, if possible, on practical experience and then determining, by calculation, whether this yields the required value of kF.

This calculation can be avoided if, in choosing the dimensions, one of them, for example the length L of the tubes, is left free and the calculation is first carried out for the unit length; in this way kF/L is obtained. The necessary length of the heat exchanger can then easily be determined by dividing the required quantity for kF by the value for kF/L which has been obtained.

Accordingly, the method for the determination of kF must be specified on the supposition that the heat exchanger dimensions, given or assumed, and the fluid velocities of the gases are available.

There exists, moreover, a series of practical requirements which must be met in determining the dimensions and selecting the materials of construction; these requirements are given, in part, by the area of application but above all, by the temperature and pressure range in which the heat exchanger is likely to be operating. This is because temperature and pressure greatly influence the choice of construction materials and thus also determine, within certain limits, the shape and the order of magnitude of the size of the flow cross-section. In addition, it is necessary to take note of the physical properties of the construction materials, the danger of corrosion and other such factors.

6-2 DETERMINATION OF THE PERFORMANCE OR THE DIMENSIONS OF A PARALLEL FLOW OR COUNTERFLOW RECUPERATOR

The dimensions of the heat exchanger together with the quantities of gas $\dot{m}$ and $\dot{m}'$ flowing through the heat exchanger in unit time are both known. From these can be calculated the quantities kF or $\int k\,df$, which characterize the heat transfer performance of the heat exchanger. The most important data for this calculation can be obtained from considerations presented previously.

Heat Transferring Surfaces

An example will be considered of a recuperator constructed of straight parallel tubes (see Fig. 5-10), which is operated in counterflow or parallel flow. If, as previously (see Sec. 5-3), z is the number of tubes, d_i and d_a are the inner and outer diameters and L is the length of tubes taking part in the heat transfer process, it follows from Eqs. (5-3) and (5-4) that an inner heating surface area of

$$F_i = z \, d_i \pi L \tag{6-2}$$

and an outer heating surface area of

$$F_a = z \, d_a \pi L \tag{6-3}$$

are available for heat transfer.

There is a certain amount of doubt as to which section L should be taken as the actual length. Near the end cross-sections a and b, where the tubes are fastened together in tube assemblies or in other similar parts of the apparatus, the outer tube surface areas available to the gas flowing on the outside are presumably incomplete. Because of this, in practical calculations, often only the distance between the centre of the cross-section at the entrance and the centre of the cross-section at the exit of the outside gas is regarded as the operative length L (the length between a' and b' in Fig. 5-10). However, it is possible to take the view that L can be assumed to be equal to the entire distance between a and b. In this case the tubes would be partly subjected to cross flow at the entry and the exit positions of the gas; further it follows from Sec. 2-11, that higher heat transfer figures could be expected than for parallel flow. It is not altogether impossible that these end effects could result in a partial reduction in the undesirable influence of the possibly incomplete "wetting" of the solid surface by the gas, which has been mentioned previously. Moreover the tubes will be wetted along their entire length by the gas flowing on the inside of the tubes. However, the first hypothesis has the advantage of greater reliability or safety.

Fluid Velocities of the Gases

As well as F_i and F_a, according to Eqs. (5-11) and (5-13), the heat transfer coefficients α and α' of the inside and outside gases are the most important dimensions which determine kF. However, the determination of α and α' presupposes a knowledge of the fluid velocities w and w', which can easily be determined as follows. If the tube assembly consisting of z tubes is surrounded by a further tube which has an internal diameter D_i, Eq. (6-4) yields the inner flow cross-sectional area

$$F_q = z \, d_i^2 \frac{\pi}{4} \tag{6-4}$$

and

$$F'_q = (D_i^2 - z \, d_a^2) \frac{\pi}{4} \tag{6-5}$$

provides the outer flow cross-sectional area. Now ρ and ρ' are the densities of both the gases at temperatures θ and θ' at the pressures which are prevailing at the time.

The fluid velocities can be calculated from the gas mass flowrates $\dot{m}$ and $\dot{m}'$ using:

$$w = \frac{\dot{m}}{\rho \cdot F_q} \tag{6-6}$$

$$w' = \frac{\dot{m}'}{\rho' F_q'} \tag{6-7}$$

in which F_q and F_q' are determined by Eqs. (6-4) and (6-5). If for example, the dimensions of $\dot{m}$ and $\dot{m}'$ are given as kg/s, the dimension of F is m^2 and the dimensions of ρ and ρ' are kg/m^3, then w and w' would be obtained from Eq. (6-6) and (6-7) in m/s.

Determination of the Heat Transfer Coefficients α and α'

In addiion to the fluid velocities and the densities ρ and ρ', the calculation of the heat transfer coefficients α and α' also requires the values of the thermal conductivity λ, the viscosity η and the specific thermal capacity c_p of the fluid (see Sec. 2-4, and Sec. 2-5ff.). These physical properties, which also determine the kinematic viscosity $\upsilon = \eta/\rho$ and the thermal diffusivity $a = \lambda/\rho c_p$, can be obtained from collections of physical data such as the well known tables by Landolt-Börnstein [20] or the VDI Heat Atlas [1].

The diameter d which occurs in the dimensionless groups and also in the equations for the heat transfer coefficients, is, for the inside gas, equal to the internal diameter d_i of the heat transferring tube. However, contrariwise for the outside gas, the equivalent diameter d_{gl} from Eq. (2-14) should be introduced

$$d_{gl} = 4F_q'/U', \tag{6-8}$$

where F_q' denotes the cross-sectional area of flow from Eq. (6-5) and U' denotes the complete circumference of the cross-section of flow. In the tubular arrangement for recuperators (Fig. 5-10) considered there,

$$U' = (z d_a + D_i)\pi \tag{6-9}$$

results for U' for the diameter D_i of the outer tube. However, in certain cases, the thermal diameter d_{th} from Eq. (2-15) is used in place of d_{gl} when only the section U of the circumference which takes part in heat transfer is considered to be a determining factor. It is even better to choose an intermediate value which is dependent on Pr, which is mentioned in more detail on p. 20 and which is approximately represented by Eq. (2-16).

Normally the whole length of the tube in question through which the gas flows, is employed as the tube length L. Here the exact determination of L is not required because the influence of the dimensionless group L/d on the heat transfer coefficients is usually only very slight.

After these preliminaries, all the dimensions are known with which, using the equations in Sec. 2-5ff., the heat transfer coefficients α and α' in these inner and outer areas can be calculated.

Here care must be taken to see that the physical properties relative to the correct reference temperature are inserted into the dimensionless groups. This reference temperature is given in most of the equations mentioned in Sec. 2-5ff. In addition, as had already been mentioned in more detail in Sec. 2-5, the difference between heating and cooling must be taken into consideration. The usual dimensionless heat transfer equation first gives the value of Nu. The heat transfer coefficients are then obtained by solving for α the defining Eq. (2-8) for Nu.

The heat transfer coefficients α and α' determined in this way only take into account heat transfer by conduction and convection. If, however, thermal radiation has a marked influence, then using Eqs. (3-24) to (3-27) from Sec. 3-3 the corresponding values of α_s and α'_s can be added.

Calculation of kF

If F_i, F_a, α and α' have been determined and the thermal conductivity λ_s of the construction material of the tube or duct walls has been obtained from an accurate table, kF can be directly obtained for a recuperator constructed of tubes by employing Eq. (5-18) or (5-11) or one of the other equations given in Sec. 5-3.

If the recuperator is constructed of plane or curved plates or sheets of uniform thickness δ arranged parallel to each other, Eq. (5-2), which is valid for plane walls, can be used for determining k.

Methods for dealing with circumstance where α and α' together with C and C' are dependent on temperature or, for other reasons, vary from one recuperator position to another, were discussed in detail in Sec. 5-13. Here, because of the variation of k, only an average value of kF can be given but this can be avoided in the calculation process described in Sec. 5-13. However, if, in spite of this, the determination of the average value of k is required, for example in estimating the thermal performance of the heat exchanger, this can be obtained by dividing the values of $(k(\theta - \theta'))_M$ which has been calculated using Eq. (5-93) or (5-94) or another corresponding approximation method, by $\Delta\theta_M$. $\Delta\theta_M$ is normally determined employing Eq. (5-79) or, again, by an approximation method.

Calculation of Two Unknown Gas Temperatures

From the discussions above, the following problems can also be resolved. Given the dimensions of a recuperator and the quantities of gas flowing through it in unit time, and in addition given two of the four gas entry and exit temperatures, what are the values of the other two temperatures and what is the rate of heat transfer in such a recuperator? First kF is determined from the dimensions of the recuperator and the fluid velocities provided in the manner already described. In addition if the thermal capacities C and C' of the gas flow rates are constant the unknown temperatures can be determined by utilizing one of the Eqs. (5-45) to (5-48) and Eq. (5-31). Thus the two exit temperatures for example in counterflow, are obtained by calculating θ_2 from Eq. (5-48) and, employing this value, by computing θ'_2 from Eq. (5-31). However the problem generally can also be solved with the help of the ratio τ of the temperature differences at both ends of the heat exchanger which is determined using Eq. (5-60) or if the efficiency η^* is to be used, by employing Eq. (5-68). The calculation process necessary is discussed in Sec. 5-10 and Sec. 5-11.

With a knowledge of the four entry and exit temperatures, the rate of heat transfer $\dot{Q}$ in the recuperator is directly determined using Eq. (5-31).

6-3 INFLUENCE OF PRESSURE DROP IN THE CHOICE OF DIMENSIONS AND FLUID VELOCITIES

Once the thermal calculations for a recuperator have been completed, it is often necessary to find out whether the pressure drops of the gases passing through the recuperator are not too high. This is because, as has already been mentioned in Sec. 4-1f., each pressure drop involves a loss of energy. Thus, given the choice of various arrangements which all give the same value of kF, preference will generally be given to those with the smallest pressure drop, that is if they are not too expensive.

Equations for the calculation of pressure drop have been presented in Secs. 4-2 to 4-4. As the calculation is similar to that for determining the heat transfer coefficient, and moreover is in a simpler form, it is not necessary to go into it any further.

In what follows, several basic features relating to the significance and influence of pressure drop on the construction of the heat exchanger will be discussed. It will be sufficient here to limit this to pressure drop in tubes, thus disregarding the increases in pressure drop brought about by the additional losses which occur when the gas first enters the tubes and when the gas flows away again.

Above all, it is worth noting that the pressure drop in gases at higher pressures generally plays a much less significant role than it does in gases at lower pressures. This results from the following consideration. If turbulent flow is assumed, the term $\rho w^2 / 2$ in Eq. (4-2) from Sec. 4-2 is, above all, the determining factor for pressure drop. Since the density ρ is proportional to the pressure, and the velocity w, with the equivalent mass or mole flowrates of the gases for equal flow cross-sections, is inversely proportional to the pressure, ρw^2 and thus the pressure drop will be smaller the higher the pressure. However, higher values for pressure drop are usually allowable with higher pressures because an equivalent pressure drop at high pressures means a proportionally smaller energy loss than for lower pressures. It follows that, for example with the reversible isothermal compression of an ideal gas, the energy expenditure is proportional to the logarithm of the ratio of the initial and final pressure. From this the work which must be expended for compression of each 1 bar, but which is nullified by a pressure drop of one bar in the tube, becomes smaller with increasing absolute value of the pressure. Thus, often, relatively smaller flow cross-sections are associated with high pressures than with lower pressures.

With equal gas flowrates and equal flow cross-sections, the pressure drop is strongly dependent on temperature, because the fluid velocity and thus (together with other equivalent ratios) the pressure drop increase with rising temperature.

However, since, with high temperatures, the pressures which occur are generally lower and the quantities being processed are usually considerably larger than with low temperatures, particularly small pressure drops are sought in the case of high temperatures. This among other observations which will be discussed in Sec. 6-5 is one of the reasons why, as a rule, the cross-section of flow is chosen to be larger with higher temperatures.

Reciprocal Influence of Heat Transfer and Pressure Drop

In calculating pressure drop, it is often noticed that various measures which improve the exchange of heat also increase pressure drop. If, for example, the fluid velocities w and w' are increased while the tube diameters d_i and d_a remain the same, so that the number of tubes and the diameter of the outer casing can be decreased, it follows from the equations in Sec. 2-5 and Sec. 4-2 that an increase not only occurs in the heat transfer coefficients α and α' but also in the pressure drop Δp. The following conclusion can be drawn from this. The increase in α and α' means, because of the increase in k that goes with it, that the required value of kF can be obtained with a smaller heating surface area. This means a decrease in the cost of construction materials and thus a decrease in manufacturing costs. Thus a heat exchanger of a particular type of construction and with a pre-specified performance can be manufactured, as a rule, more cheaply, the higher the pressure loss which can be allowed.

Notice must be taken of this connection between heat transfer and pressure drop, above all, when a maximum value for the pressure drop is specified for one of the two gases or even for both of them. A process of heat exchanger calculation, which, in addition to starting from the desired temperature variations, also proceeds from the highest acceptable values of pressure drop, has been developed by Kühne [K223]. He uses a diagram in this, which connects the heat transfer resistance and the heat transferring surface and, as a parameter for the curves, obtains an expression in which the specified values of the fluid flowrates, the entry temperatures and the pressure drops are included. E. Schmidt has further shown [S211] how the dimensions of a heat exchanger can be calculated from the required rate of heat transfer and the energy loss caused by pressure drop, in such a way that either the lowest weight of heat transferring tube or the smallest possible volume for the heat exchanger is obtained. Such calculations are important if it is required to produce heat exchangers with the lowest possible expenditure on raw materials and thus the lowest possible cost or when a heat exchanger is required to take up the smallest amount of space possible, for example in aircraft hangars.

Because these types of process are based on apparently complex derived quantities which have first to be computed, the consideration of heat transfer and pressure drop separately, as described above, is often preferred. Of course the satisfaction of a simultaneous requirement of a specific heat transfer rate and a specific pressure drop can only be calculated iteratively. Here the process can again be used in which certain dimensions are assumed and kF and the expected values of Δp are calculated. Then the magnitude of the assumed dimensions must be altered and the calculation repeated until both requirements are satisfactorily and simultaneously met. In certain circumstances, in order to facilitate this extremely time consuming irksome iterative process* it is useful to examine what influence have any variations in the dimensions upon the heat transfer coefficient and the pressure drop, so long as radiation is ignored. Reference should be made to [S201].

By way of example the tube assembly arrangement will again be considered. To simplify matters, the ratio d_a/d_i of the inner and outer tube diameters, which are generally determined by the resistance to stress of the tube wall, will be taken as constant. In addition, only those cases will be considered in which the outer flow cross-section F'_q varies proportionally to the inner flow cross-section F_q, so that the fluid velocities inside and outside the tubes are always in the same unchanging ratio. Furthermore, if a large number of tubes z is assumed, such as occur in high performance recuperators, then, without the introduction of too much error, the equivalent diameter d_{gl} of the outer area can be assumed to be proportional to the tube diameters d_a and d_i. In so doing, the very slight influence of D_i on the flow circumference U', see Eq. (6-9), is ignored. The remaining measurements d_i, d_a and d_{gl}, which following from this are always in the same ratio, will simply be denoted by d.

With constant gas flowrates, w and w' must vary inversely with the quantity $z d^2$, while the heating surface areas F_i and F_a, and thus also F, are always proportional to $zd.L$. In addition, with fully developed turbulent gas flow, it follows from Eq. (2-18) that α and α' themselves are proportional to the value of $(L^{0.054} d^{1.732} z^{0.786})^{-1}$. Furthermore, if the term with λ_s is ignored in Eqs. (5-2) or (5-11) for k, this proportionality relationship which has been found, is also valid for the heat transmission coefficient k. Thus

$$kF = \text{const} \frac{L^{0.946} z^{0.214}}{d^{0.732}} \tag{6-10}$$

is obtained.

Working on the same assumptions it will be shown how Δp is dependent on L, d and z. According to Fig. 4-2, the pressure loss Δp is not, in general, shown over a larger range of the Reynolds number, as a product or a quotient of pure powers of w and d with constant exponents. However, in order to evaluate the overall influence, the approximate equation

$$\Delta p = \text{const} \frac{w^{1.8} L}{d^{1.2}} \tag{6-11}$$

is used; this can be applied successfully to smooth tubes for average values of the Reynolds number; it also satisfies the similarity relationship set out in Eq. (4-2). Considering again that the velocity w is inversely proportional to $d^2 z$, then according to Eq. (6-11)

$$\Delta p = \text{const} \frac{L}{d^{4.8} z^{1.8}} \tag{6-12}$$

is obtained.

---

* Editor's note: Such iterative processes can be refined using numerical analysis and can be readily programmed for a computer.

From Eq. (6-10) the following averages are suggested for increasing kF which cause at the same time a variation in Δp, specified by Eq. (6-12):

1. *Increase in tube length L, with unchanging values of z and d.* kF increases almost proportionally with L; at the same time pressure drop also increases proportionally to L.
2. *Increase in the number of tubes z with unchanging values of L and d.* In this case kF increases only very slowly while Δp decreases very quickly. Increasing the number of tubes also provides a means by which pressure drop can be lowered considerably without the value of kF being markedly altered. Doubling z lowers the pressure drop 0.287 times, while kF only increases 1.16 times.
3. *Decrease in tube diameter d with unchanging values of L and z.* If d is decreased by half, kF is increased by 1.66 times; in contrast Δp is increased by 27.8 times because of the quadrupling of the fluid velocity. This type of increase in kF is paid for by a quite disproportionately large increase in the pressure drop. Thus decreasing the tube diameter must only be considered when the initial pressure drop is very small or when the pressure drop plays a very minor role such as with high pressures for example.

Often, however, it is necessary to increase or decrease kF without at the same time *altering pressure drop*. This requirement can be met by assuming Δp to be constant in Eq. (6-12). At the same time, an appropriate one of the three measurements L, d or z is then eliminated from Eqs. (6-10) and (6-12). In this way it is possible to show, for example, that a doubling of kF with $\Delta p = $ const can be obtained in the following three, different ways:

(a) with $d = $ const L must be increased 1.9 times, z must be increased 1.4 times;
(b) with $L = $ const d must be lowered by 0.6 times and z must be increased by 4.1 times;
(c) with $z = $ const L must be increased by 2.4 times and d must be increased by 1.2 times.

Which of these three cases is the most suitable can depend on various circumstances, such as, for example, whether the available space or the possible tube lengths permit a lengthening of the heat exchanger. Often costs are a determining factor in the choice of building material. If, as assumed previously, the ratio d_i/d_a is unchanged, the weight of the material for the construction of the tube is proportional to zLd^2. The construction material needed then increases in the cases above

(a) by 2.7 times (2.7),
(b) by 1.5 times (2.4),
(c) by 3.4 times (2.9).

From this, with an unaltered tube length L, case (b) appears to be the most favourable. The relationships are not very different when, as often happens with thin tubes, the strength of the wall, and not the ratio of the diameters, remains unchanged. Then the figures in brackets in the table above apply.

Influence of an Alteration in the Outer Flow Cross-section

In the considerations so far it has been assumed that not only the ratio of the diameters d_i, d_a and d_{gl}, but also the ratio of the inner and outer flow cross-sections, remain unchanged. In order to examine the influence of a deviation from the last of these assumptions the changes will be examined in the heat exchange and pressure drop *if only the outer flow cross-section F_q' and thus d_{gl} are altered.* The values of z, d_i, d_a and L will thus be considered to be constant. If it is assumed, as previously, that the number of tubes z is large, it can also be assumed that, from Eq. (6-9) $U' = z\,d_a\pi$, i.e., U' can also be considered to be constant, so that, from Eq. (6-8) $d_{gl} = $ const F_q'. Thus, $w' = $ const$/d_{gl}$ is obtained from Eq. (6-7). Consequently, arising out of Eq. (2-18) in Sec. 2-5, α' will be inversely proportional to $d_{gl}^{0.946}$. In addition, as F_a remains unchanged,

$$\alpha' F_a = \text{const}/d_{gl}^{0.946} = \text{const}/F_q'^{0.946} \qquad (6\text{-}13)$$

results. Correspondingly, from Eq. (6-11)

$$\Delta p = \text{const}/d_{gl}^3 = \text{const}/F_q'^3 \qquad (6\text{-}14)$$

is obtained for Δp. If, for example, the outer flow cross-section F_q', and thus also the equivalent diameter d_{gl}, is doubled, it follows from Eqs. (6-13) and (6-14) that $\alpha' F_a$ is decreased 0.52 times and Δp is decreased by 1/8. The effect of this on the value of kF is mainly dependent on the ratio of αF_i to $\alpha' F_a$. According to Eq. (5-11), with $\alpha F_i = $ const, kF undergoes a percentage change which is only half as large as $\alpha' F_a$ when αF_i and $\alpha' F_a$ are approximately equal to each other. In other cases also, however, the influence on kF of a variation in F_q' is always less than its influence on $\alpha' F_a$. It follows from these considerations that *the pressure drop in the outer area can be greatly lowered by a proportionally small increase in the outer cross-section,* while kF will decrease only slightly. This possibility of reducing pressure drop is of particular importance when, for example, a higher gas pressure exists inside the tubes and a lower gas pressure exists in the outer area. This is because in this case, as has already been mentioned, with normal fluid velocities there exists a considerably higher pressure drop in the outer area than inside the tubes; it then often becomes a matter of reducing the pressure drop in the outer area. A similar situation occurs when a liquid is flowing inside the tubes and a gas is flowing on the outside.

6-4 THE IMPEDING OF HEAT TRANSFER AND THE INCREASE IN PRESSURE DROP DUE TO FLUID OR SOLID DEPOSITS

In certain circumstances heat transfer and, most of all, pressure drop, undergo marked changes when material contained in the gas is deposited on the tube walls as a liquid or in solid form. Such deposits are often intentional, particularly when heat exchangers are expressly used for the separation of the specific components of the gas by cooling. Drying out a gas by "cold" is an example of this. However such deposits are usually an unwanted side effect and nuisance as they unfavourably influence heat transfer and, in particular, pressure drop. Undesirable side effects occur, above all, in low temperature technology because air and other gases, which are cooled to very low temperatures of up to $-200°C$, almost always contain smaller or larger quantities of water vapour and carbon dioxide. Disregarding an eventual thawing of liquid water above $0°C$, not only water vapour but also carbon dioxide is deposited in solid form on the walls of the low temperature counterflow heat exchanger; see, for example, [H212, L204, D203, R204]. When continuously operating recuperators are used, as a rule equipment is provided for the prior removal of water vapour and carbon dioxide from the gases. However, with long operating periods, the danger of clogging cannot always be completely obviated as, in spite of these precautions, certain amounts of water vapour and carbon dioxide always remain in the gases.

Solid deposits occur particularly frequently during the evaporation of liquids, such as is exhibited, for example, in the encrustation of the evaporator installations in the sugar industry and the like; see, for example, [K208].

Basically, solid materials can be regarded as being contained in the gases as a matter of course; in particular at high temperatures this takes the form of flue ash or dust. However substantial quantities of dust are rarely found today in the gases arriving at a heat exchanger because electrical dust extractors are usually installed in large industrial plants.

The most disadvantageous effect on heat transfer of the deposits described, can be recognized in the following considerations. Often a loose deposit of low thermal conductivity is formed on the walls and this impedes the transmission of heat. If, for example, a plane wall of thickness δ had, on one side, a deposit of thickness δ_0 and thermal conductivity λ_0, the heat transmission coefficient k would be lowered in relation to Eq. (5-2), which applies to deposit-free walls, according to the following relationship:

$$\frac{1}{k} = \frac{1}{\alpha} + \frac{\delta_0}{\lambda_0} + \frac{\delta}{\lambda_s} + \frac{1}{\alpha'}$$

(6-15)

Correspondingly

$$\frac{1}{kF} = \frac{1}{\alpha F_i} + \frac{\delta_0}{\lambda_0 F_i} + \frac{\delta}{\lambda_s F_m} + \frac{1}{\alpha' F_a}$$

(6-16)

is obtained in place of Eq. (5-18) for tube walls with an internal deposit of thickness δ_0. It is assumed here that F_i is not greatly altered by the deposit. $\delta_0/\lambda_0 F_a$ would appear in Eq. (6-16) in place of $\delta_0/\lambda_s F_i$ for a deposit on the outside of the tubes. It follows that the influence of a solid or liquid deposit on the value of kF can be calculated using Eq. (6-15) or (6-16) and thus account can be taken of such deposits in the selection of the recuperator dimensions. In certain circumstances it must also be noted that the fluid velocity, and thus also the corresponding heat transfer coefficient which is included in the equation, is increased by the reduction in the inside cross-section.

Such deposits have a much stronger effect on pressure drop than they have on heat transfer. Indeed, if the thickness of the layer is increased by permanent deposits of solid material or by layers of loose snow-like* particles which accumulate at one position, the danger exists that the cross-section of flow will become more and more restricted. In this case not only does the pressure drop increase but also, after a certain length of time, blockages occur at one or more positions. The heat exchanger must then be taken out of operation and freed from these deposits.

6-5 CHOICE OF BUILDING MATERIALS AND THEIR INFLUENCE ON HEAT EXCHANGER CONSTRUCTION

It will now be discussed how the dimensions of a heat exchanger cannot be determined solely by calculation. Many different possibilities exist for the construction and the dimensions of a heat exchanger; these offer the same heat transfer efficiency and are in the area of permissible pressure drop and decisions must be made between these possibilities, usually on the grounds of practical considerations and experience. The most important determining principle in this will be discussed below.

First the choice of building material is dependent on the temperature and pressure range in which the heat exchanger operates and this choice influences the construction.

Three main temperature ranges can be distinguished. A middle range stretches from about -50 to $+500°C$. Thus it includes the heating and cooling of liquids and gases at moderately high or moderately low temperatures. For example, the utilization of the heat in the exhaust gases from a boiler plant for preheating feed water or for the heating of furnace combustion air or the cooling of air in cold stores belongs to this range. On the one hand the very high temperature range includes for example, that in blast furnaces and steel making plants and extends beyond $1500°C$. On the other hand in the low temperature technology range, industrial standard temperatures of far below $-200°C$ are produced.

In general the pressure load is considerably less at high temperatures than it is at low temperatures. This is shown in heat exchangers in the iron and steel industry where pressures above 1 or 2 bar occur only rarely, while in low temperature technology pressures of about 50 bar and more are customary. However chemical reactions at high pressures and high temperatures are often carried out in the chemical industry on a large scale. Thus heat exchangers intended for the chemical industry must be able to withstand both high temperatures and high pressures. If this aspect is ignored, it can be said that, with very high

---

* Editor's note: The German word *"Schneeschichten"* has been translated here by *"layers of snow-like particles"*.

temperatures, it is the temperature stability of the construction materials, which have to undergo only slight stress, that matters while with very low temperatures the resistance of the construction material to high pressures is important.

These differences in the temperature and pressure load affect the choice of materials of construction in the following way. In the average temperature region of between $-50°$ and $+500°C$ which was first mentioned, all the requirements regarding the strength and stability of the building material can usually be satisfied without difficulty. Thus here there is considerable freedom in the choice of building materials. In addition to bricks, metals are largely used, in particular cast iron and steel. Below 0°C it is generally only a question of metals being used as building materials.

Today, fireproof bricks are still increasingly being used as first choice for high temperatures, for example in the iron and steel industry, of up to 1500°C. Here bricks and ceramic materials are used predominantly at above 500°C although attempts have been made to introduce alloy steel as building material which has a resistance to temperature of up to about 1200°C, see, for example, [J203]. The properties of fireproof building materials will be further discussed shortly.

The practically exclusive use of metal as building material in low temperature technology is above all due to high pressures of the compressed gases; these pressures usually lie between 5 and 50 bar and in general only metals are able to give adequate resistance to these pressures. Moreover, a smaller flow cross-section can be produced with metals and, along with their high thermal conductivity, this favours heat transfer. With bricks and ceramic materials there is, in addition, the danger that, at low temperatures, moisture will penetrate into the pores or cracks and then freezing will cause them to shatter. Thus low and high alloyed steels must be considered. However, the copper, bronze and brass, which were used to a great extent previously, have now been almost entirely replaced by aluminium.

The effect that the choice of building material has on the flow cross-section is shown in Fig. 6-1 where several characteristic cross-section shapes for recuperators are illustrated. The use of the normal square bricks suggests the construction of

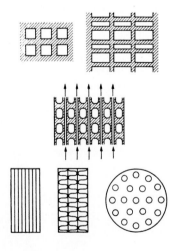

Figure 6-1 Various shapes of flow cross-sections for recuperators.

square or rectangular ducts (Fig. 6-1, top). Octagonal, round, elliptical or other shapes of cross-section can be produced by using shaped bricks; in this way the cross-sections of recuperators can easily be arranged so that the directions of flow of both gases cross one another (Fig. 6-1, centre).

Heat exchangers which are constructed in metal have quite different cross-section shapes, see Fig. 6-1, bottom; this results, in part, from the discussion in Sec. 5-2. *Recuperators* can, for example, be built from thin, plane or curved sheets arranged close together so that narrow, long cross-sections are formed. Here the first, third and fifth cross-sections would, for example, have hot gas flowing through them while the other cross-sections lying in between them would have cold gas flowing through them.

Heat exchangers in which various corrugated sheets, or corrugated and plane sheets, are welded together at suitable points (Fig. 6-1, bottom) also belong to this group of recuperators. The numerous narrow ducts which are formed in this way usually produce a honeycombed cross-section through which more than two gases can flow. In low temperature technology, as has already been discussed on p. 139, this type of heat exchanger acquires great importance in the so-called Reversing Exchangers; see also Fig. 5-5.

The recuperator constructed of metal tubes is widely known and has already been mentioned many times;[1] see, for example, Figs. 5-3 and 5-4. A variation on metal tubes as component parts of heat exchangers are finned tubes which, as will be explained in more detail in Sec. 8-4, operate particularly favourably when a liquid is flowing through the inside of the tubes and a gas is flowing on the outside. In general finned tubes are only used in the middle temperature range of between $-50°C$ and $+500°C$. These are almost entirely produced in metal. Special designs also occur in ceramic materials, for example as heating elements for hot water heating.

However, not only the shape but also the dimensions of the flow cross-sections are dependent on the building material. Narrow cross-sections and thin walls are more easily produced from metal than from bricks. The fact that wider cross-sections are chosen for high temperatures, for example in the iron and steel industry, than for low temperatures, is not only a function of the building material but is based also on the fact that the pressures which occur are lower than, for example, in low temperature technology and thus only small pressure losses and thus low fluid velocities, can be permitted. Moreover, at very high temperatures, large volumes favour heat exchange through radiation.

Physical Properties and Temperature Resistance of Building Materials and their Resistance to Chemical Influences

The pressures existing in heat exchangers create physical stresses at most or at all parts of the heat exchanger. This is because the heat transferring walls, the casing

---

[1] Tubes made of graphite or similar materials have been suggested for specific particular purposes in the chemical industry; see for example [H211] and [K215, 216]. In addition, heat exchangers are occasionally produced in porcelain; see [K209].

and other parts of the heat exchanger have to withstand the pressure difference between the two gases, as well as a possible pressure difference between the outside gas and atmosphere. Moreover there is often the additional strain caused by thermal stresses which occur particularly at the onset of operation as the inner parts of a heat exchanger are heated or cooled much more quickly than, for example, the casing; see for example, [K206]. Correct heat exchanger measurements thus require stress calculations which it can be assumed can be undertaken and which will not be dealt with further. Care must further be taken, through particular, constructive measures, that the various parts of the heat exchanger, particularly at the joins, are tight in relation to the gas or liquid pressure.

In general it is more difficult to obtain adequate stability with *very high temperatures* than it is with low temperatures. This is because at the temperatures in question almost all building materials are under a great deal of stress because, as has already been mentioned, pressure resistance and tensile strength generally decrease with increasing temperature.

Finally the building materials must be sufficiently resistant to *corrosion* and other damage caused by chemical reactions. It is possible to protect against metal corrosion either by using corrosion resistant steel, for example alloy steel, or by providing the tubes with a protective coating, for example steel tubes coated with zinc or lead; see, for example, [K207]. Where it is possible corrosion can be precluded or at least minimized, by purifying or chemically pretreating the gases or liquids.

With heat resistant, high alloyed steels the upper temperature boundary for practical use is often determined by the build up of scale which is formed by the oxidation of the metal surface. Certain chrome-nickel steels show a particularly high resistance to scale because a protective oxide skin is formed on their surface from reaction with the oxygen.

Finally there is the possibility of severe *mechanical distortion of the construction material*. In addition to damage caused by strains which result from pressures which are too high, damage can also be caused by among other things, knocks and improper handling during transport and installation. Further, building materials can be mechanically pitted by broken solid particles which rub and grind against particular sections of the walls. Dust, ash, particles of ice, and the like, all have this effect.

HEAT AND "COLD" LOSSES IN RECUPERATORS

7-1 PROTECTION OF HEAT EXCHANGERS AGAINST HEAT AND "COLD" LOSSES

Precise recuperator calculations must, basically, also take into account the influence which heat and "cold" losses have on the temperature distribution and the required dimensions of a heat exchanger. Indeed, in what follows it will be shown that this influence is usually so small that it can be ignored without causing any major errors.

Where there are large temperature differences with respect to the surrounding area, this usually proves true only when the heat exchanger is insulated against heat or "cold" losses. On this basis the calculation of the losses will be premised by a short consideration of heat exchanger insulation. Details of this can be found in the literature and the comprehensive illustrations of heat and "cold" insulation.

There is no general rule as to when a heat exchanger should be insulated or what type of insulating material should be used, because, in addition to the magnitude of the losses, which increase with the temperature difference in relation to the surrounds (see the following paragraphs), the requirements of the exchanger are a determining factor as are, among other things, further considerations of the operation, feedback on the other parts of the apparatus and the ratio between the cost of insulation and the heat and "cold" saving achieved. It is significant, for estimating the efficiency of an insulating material, that the thermal conductivity λ_w of this material almost always increases with its gross density that is, density which is measured inclusive of the pores of the material. Thus loose insulating materials with a low gross density show the greatest insulating efficiency. However, as a rule, the lower the gross density, the lower is the resistance of the insulating material to high temperatures. At high temperatures a relatively dense insulating material must be used. Thus the insulating layer must be of a suitable thickness so that the effect of too small an insulation does not have to be tolerated. This is all the more important since the thermal conductivity increases with rising temperature.

The following tables give an overall view of the insulation effect of several heat and "cold" insulating materials; in the tables the thermal conductivity λ_w is given in W/(m K) for various gross densities ρ. They deal in average test values from which individual cases can deviate quite strongly.[1]

Table 7-1 Thermal insulation material for temperatures up to 120°C

Thermal insulation material	Gross density ρ kg/m^3	λ_w in W/(m K) at 20°C
Cork, peat or felt sheets	150	0.042
	300	0.058
	450	0.076
Peat litter, high water absorption		
dry	190	0.041
normal moisture	190	0.060

Table 7-2 Thermal insulation material for temperatures up to 600°C

Thermal insulation material	ρ kg/m^3	λ_w in W/(m K) at 100°C	300°C	500°C
Kieselguhr in powdered form	300	0.067	0.085	0.103
Burnt kieselguhr brick	300	0.087	0.118	0.145
(Plates, shells, shaped bricks)	400	0.097	0.127	0.157
	600	0.128	0.157	0.186
Mineral wool	250	0.050	0.079	0.125
Glass fibre, loose in sheets				
and as glass thread, on average	175	0.048	0.107	

Table 7-3 Thermal insulation material for high temperatures over 1000°C

Thermal insulation material	λ_w in W/(m K) at 500°C and with a gross density ρ of 700 kg/m^3	1300 kg/m^3	1900 kg/m^3	2500 kg/m^3
Silica bricks	0.35	0.70	1.40	2.20
Fireclay	0.43	0.55	1.05	2.05

[1] Further details can be found, for example, in Landolt-Börnstein [L201] Fig. IV/4b, pp. 417–43. and pp. 454–481 or VDI-Heat Atlas, 2nd edition. [1] pages Da 1 to Dc 27.

Table 7-4 "Cold" insulation materials[2]

Cold insulation materials	ρ kg/m^3	λ_w in W/(m K) at		
		$-200°C$	$-100°C$	$0°C$
Silk	100	0.022	0.032	0.043
Mineral wool	95	0.010	0.020	0.031
Cork grains, expanded, grains sizes about 3 mm	37	0.010	0.021	0.033
Polystyrene foam (styropor)	15		0.022	0.035
Moltoprene (polyurethane)	40		0.022	0.032
Glass fibre sheets, resin bonded	100		0.023	0.033

Table 7-5 "Cold" loss in the counterflow heat exchangers used in low temperature technology

(a) Loss to the surroundings

Type of counterflow heat exchanger	Average value of the gas flow rate kg/h	Dimensionless group $k_u F_u/C'$	Heat flowing in from the surroundings $\dot{Q}_u$ W	Cold loss $C \cdot \Delta\theta_2$ W	Increase in exit temperature at the cold end $\Delta\theta_2$
Straight tube assembly	6000	0.0016	200	86	0.06°
Straight tube assembly	32	0.0053	33	15	0.23°
Regenerator	3400	0.0020	180	120	0.13°
Short cross counterflow device with strong casing wall	360	0.0033	41	28	0.18°

(b) Loss due to longitudinal thermal conductivity

Type of counterflow heat exchanger	$(\dot{Q}_\lambda)_{lit}$ W	warm end $(\dot{Q}_\lambda)_0$	$(\dot{Q}_\lambda)_{ges}$	$\Delta\theta'_2$	$(\dot{Q}_\lambda)_{ges}/C'$
Straight tube assembly	110	69	102	0.02°	0.06°
Straight tube assembly	0.49	0.14	0.31	0.004°	0.007°
Regenerator	35	35	35	0.00°	0.04°
Short cross counterflow device with strong casing wall	181	66	136	0.58°	1.13°

[2] Lueger. Lexicon der Verfahrenstechnik [L208] key work "cold insulation material", p. 239–241.

The details on peat litter (German: *Torfmull*) in Table 7-1 indicate the strong influence of moisture. Thermal conductivity increases with moisture and thus the effect of insulation deteriorates at the same time. According to Cammerer [C201] the thermal conductivity of organic building and insulating materials increases by 1.25 per cent independent of the gross density, when the water content is increased by 1 per cent relative to the density of the insulation. This detrimental effect of moisture is noticeable, most of all, in insulation against "cold" where atmospheric moisture has the tendency to precipitate in the coldest parts of the insulation; see for example, [R202, 203] and [M205]. This danger can be largely avoided by protecting the surface of the insulating material by sealing it against penetration by the outside air. Such measures are not necessary with expanded polystyrene and moltoprene as the closed pores of these materials have a high resistance to water vapour diffusion.

Table 7-4 contains the thermal conductivity of some "cold" insulation materials in their "dry" state; see for example, [R202, M204] and more recent literature in [L201].

The number of insulating materials is so great that only a few can be mentioned. However, Alfol will be mentioned [S208]; this is an insulating material consisting of aluminium foils. These aluminium foils are arranged in several layers, each 1 cm apart and parallel to the surface being protected. The equivalent thermal conductivity of Alfol amounts to about $\lambda = 0.03$ at 20°C and to about $\lambda = 0.05$ W/(m K) at 200°C. The insulating effect of Alfol is founded in the low thermal conduction of the several latent layers of air lying one behind the other and, particularly at high temperatures, in the lowering of radiative heat transfer which is achieved through the subdivision of the temperature drop and the small radiation coefficient of aluminium. In place of smooth, parallel foils, crinkled foils can also be used where the crinkling guarantees a suitable average distance between the foils.

As heat exchangers are usually cylindrical on the outside, most of the insulation material is usually in the form of a cylindrical sheet fixed to the outside cylindrical casing. Thus the thermal flow through the insulation material can largely be calculated using Eqs. (5-11) or (5-12) which were derived for a cylindrical tube. The values for the insulation at the ends of the heat exchanger can also be calculated with Eq. (5-18), where the value of F_m can generally lie somewhere between the logarithmic and geometric mean of the inner and outer area of contact of the section of insulation concerned. In what follows, the total value of kF obtained will be denoted as $k_u F_u$ and is the quantity of heat flowing out into the surrounding area. If the average temperature difference between the gas flowing in the casing tube and the temperature of the surrounding area is found to be θ_u, the rate at which heat or "cold" $\dot{Q}_u$ is lost to the surroundings can be calculated using Eq. (5-20). $\dot{Q}_u$ can be obtained in the same way for non-insulated heat exchangers.

In the following paragraphs we will discuss how the heat or "cold" loss $\dot{Q}_u$ calculated affects the temperature distribution in a heat exchanger and thus also the heat transfer.

7-2 THE INFLUENCE EXERTED ON THE HEAT EXCHANGER BY THE LOSS OF HEAT OR "COLD" TO THE SURROUNDINGS

Basically, three types of heat or "cold" loss occur in heat exchangers. Compared with an ideal exchanger a loss takes place in the sense that the heat is only partly exchanged, that is, none of the gases is completely heated or cooled to the entry temperature of the other gas. A part of the heat or "cold" which could theoretically have been transferred, passes, unused, out of the heat exchanger with the gas that originally contained it. This exchange loss is known as the "exhaust gas loss" in the iron and steel industry* and combustion technology and is determined by the temperature difference between both gases at the entrance of the cold gas to the heat exchanger. In contrast, in low temperature technology it is the temperature difference at the warm end of the counter flow heat exchanger which is the determining factor for the exchange loss. Exchange loss has already been considered in Chapters Five and Six. It is either fixed by the prespecified entry and exit temperatures of the gases or can be determined from kF by calculating the exit temperatures using the process developed in Sec. 5-10.

It has already been mentioned in previous paragraphs, that a further loss occurs, this being that heat or "cold" which is lost to the surroundings through the outer casing and the end sections of the heat exchanger. This loss is often known as the "radiation loss", and in low temperature technology, occasionally as the "loss by absorption of radiant heat". However, particularly in insulated exchangers, as this loss occurs mainly by thermal conduction, partly through convection and only to a very small extent through radiation, this terminology is hardly the most appropriate.

A third loss can be attributed to thermal conduction in the construction material in the longitudinal direction of the heat exchanger.

Mathematical pursuance of the influence of these last two losses appears, at first, to be quite complicated. Fortunately, however, a relatively simple method of calculation can be developed to the end.

The Nesselmann Approach to the Precise Calculation of the Influence on the Exchange of Heat due to Thermal Losses to the Surroundings

Nesselmann [N203] has derived equations with which exact calculations of how the *losses of heat or "cold" to the surroundings* influence heat transfer in recuperators. His calculations relate to recuperators which are operated in either parallel flow or counterflow and assume the thermal capacities C and C' of the two gas flowrates to be constant.

The fundamental concept embodied in Nesselmann's considerations will be briefly explained in the case of counterflow with the assumption that the temperatures of the gases are higher than that of the surroundings. Let F_u be an average surface

* Editor's note: These are sometimes known as "stack losses" in British terminology.

within the insulation material through which heat is lost to the surroundings at temperature θ_u. Now df_u is an infinitely small part of f_u, for example between two neighbouring cross-sections of the recuperator at position x in Fig. 5-10. Further k_u is the standard heat transmission coefficient for heat loss. The heat loss through the end sections will either be ignored as unimportant or will be considered by selecting F_u to be somewhat larger than the actual outer surface of the casing. In addition df_u always needs to be considered to be proportional to the element df of the surface area F lying between the same cross-sections (see Sec. 5-3), this surface area being that which affects the desired exchange of heat between the gases. Thus, always, $df_u = \dfrac{F_u}{F} df$. Then the rate of heat conduction to the surroundings through df_u is

$$d\dot{Q}_u = k_u F_u (\theta' - \theta_u) \frac{df}{F} \qquad (7\text{-}1)$$

where, as previously, it is assumed that gas flows in the outer area at temperature θ'.

The influence of this loss on heat transfer can be calculated in the following way. According to Eqs. (5-28) and (5-32) the gas flowing inside through the surface df in unit time gives up the quantity of heat

$$-C\,d\theta = k(\theta - \theta')\,df \qquad (7\text{-}2)$$

to the gas flowing on the outside. However, according to Eq. (7-1) as this second gas loses the thermal quantity $d\dot{Q}_u$ to the surroundings, only the quantity

$$-C'\,d\theta' = k(\theta - \theta')\,df - \frac{k_u F_u}{F}(\theta' - \theta_u)\,df \qquad (7\text{-}3)$$

remains to be heated. The minus sign on the left-hand side of the equation takes into consideration that, according to Fig. 5-10, in counterflow θ' also decreases with increasing f. The differential Eqs. (7-4) and (7-5) are obtained for the unknown temperature differences $\theta - \theta_u$ and $\theta' - \theta_u$ from Eqs. (7-2) and (7-3).

$$\frac{d^2(\theta - \theta_u)}{df^2} + \left[\frac{k}{C} - \frac{k}{C'} - \frac{k_u F_u}{C'F}\right]\frac{d(\theta - \theta_u)}{df} - \frac{k}{C}\cdot\frac{k_u F_u}{C'F}(\theta - \theta_u) = 0 \qquad (7\text{-}4)$$

$$\frac{d^2(\theta' - \theta_u)}{df^2} + \left[\frac{k}{C} - \frac{k}{C'} - \frac{k_u F_u}{C'F}\right]\frac{d(\theta' - \theta_u)}{df} - \frac{k}{C}\cdot\frac{k_u F_u}{C'F}(\theta' - \theta_u) = 0 \qquad (7\text{-}5)$$

The temperature distribution in a heat exchanger which takes heat losses into consideration is determined from these differential equations.

The general solution of these equations takes the form:

$$\theta - \theta_u = B \cdot \exp(\beta f) + G \cdot \exp(\gamma f) \qquad (7\text{-}6)$$

$$\theta' - \theta_u = B' \cdot \exp(\beta f) + G' \cdot \exp(\gamma f) \qquad (7\text{-}7)$$

where B, G, B', G' denote constants which have yet to be determined and β and γ are given by

$$\beta F = -\frac{1}{2}\left(\frac{kF}{C} - \frac{kF}{C'} - \frac{k_u F_u}{C'}\right) + \sqrt{\frac{1}{4}\left(\frac{kF}{C} - \frac{kF}{C'} - \frac{k_u F_u}{C'}\right)^2 + \frac{kF}{C}\cdot\frac{k_u F_u}{C'}} \quad (7\text{-}8)$$

and

$$\gamma F = -\frac{1}{2}\left(\frac{kF}{C} - \frac{kF}{C'} - \frac{k_u F_u}{C'}\right) - \sqrt{\frac{1}{4}\left(\frac{kF}{C} - \frac{kF}{C'} - \frac{k_u F_u}{C'}\right)^2 + \frac{kF}{C}\cdot\frac{k_u F_u}{C'}} \quad (7\text{-}9)$$

The relationships (7-10) to (7-13) for the constants B, G, B', and G' are derived from the boundary conditions, in particular $\theta = \theta_1$ at $f = 0$ and $\theta' = \theta'_1$ at $f = F$ (see Fig. 5-10); subsequently Eqs. (7-6) and (7-7) are inserted into Eqs. (7-2) and (7-3).

$$B = \frac{(\theta_1 - \theta_u)\exp(\gamma F)\left(\beta F + \dfrac{kF}{C}\right) - (\theta'_1 - \theta_u)\dfrac{kF}{C}}{\exp(\gamma F)\left(\gamma F + \dfrac{kF}{C}\right) - \exp(\beta F)\left(\beta F + \dfrac{kF}{C}\right)} \quad (7\text{-}10)$$

$$G = -\frac{(\theta_1 - \theta_u)\exp(\beta F)\left(\beta F + \dfrac{kF}{C}\right) - (\theta'_1 - \theta_u)\dfrac{kF}{C}}{\exp(\gamma F)\left(\gamma F + \dfrac{kF}{C}\right) - \exp(\beta F)\left(\beta F + \dfrac{kF}{C}\right)} \quad (7\text{-}11)$$

$$B' = \frac{(\theta_1 - \theta_u)\exp(\gamma F)\cdot\dfrac{kF}{C} + (\theta'_1 - \theta_u)\left(\gamma F - \dfrac{kF}{C'} - \dfrac{k_u F_u}{C'}\right)}{\exp(\beta F)\left(\gamma F - \dfrac{kF}{C'} - \dfrac{k_u F_u}{C'}\right) - \exp(\gamma F)\left(\beta F - \dfrac{kF}{C'} - \dfrac{k_u F_u}{C'}\right)} \quad (7\text{-}12)$$

$$G' = -\frac{(\theta_1 - \theta_u)\exp(\beta F)\cdot\dfrac{kF}{C} + (\theta'_1 - \theta_u)\left(\beta F - \dfrac{kF}{C'} - \dfrac{k_u F_u}{C'}\right)}{\exp(\beta F)\left(\gamma F - \dfrac{kF}{C'} - \dfrac{k_u F_u}{C'}\right) - \exp(\gamma F)\left(\beta F - \dfrac{kF}{C'} - \dfrac{k_u F_u}{C'}\right)} \quad (7\text{-}13)$$

In these equations the dimensionless group $\dfrac{k_u F_u}{C'}$ expresses the influence of the loss of heat or cold to the surroundings.

Figure 7-1 illustrates an example calculated by Nesselmann from the above equations; in this example $C = C'$; $kF/C = 10$; $k_u F_u/C' = 0.172$; $\theta_1 = 100°$; $\theta'_1 = 25°$; $\theta_u = 10°$. The broken lines represent the temperature distribution which would occur without heat loss, the solid lines represent the temperature distribution which would appear under the influence of heat loss. Heat loss causes the curves to bend. In addition the drop in temperature and thus the emission of heat from the originally warm gas increases, while, on the other hand, the increase in temperature and the absorption of heat of the originally cold gas is reduced. The difference which results between heat emission and heat absorption is equal to the loss of heat $\dot{Q}_u$. The change in the exit temperatures θ_2 and θ'_2 occurs because the temperature difference between the two gases is increased at the warm end of the exchanger and decreased at the cold end.

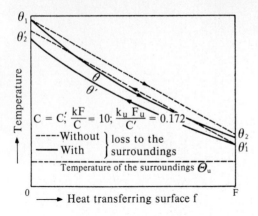

Figure 7-1 Temperature distribution in a counterflow heat exchanger with heat loss to the surroundings.

Derivation of a Simple Method of Calculation

In practice heat losses are usually less than in the example shown in Fig. 7-1 as exchangers are usually insulated to guard against larger losses.

In hot blast stoves for blastfurnaces, for example, the dimensionless group $k_u F_u/C'$ is of order of magnitude 0.03 or less. In the large counterflow heat exchangers of low temperature technology, which are usually embedded within the insulating material of the gas separation equipment, $k_u F_u/C'$ is usually smaller than 0.01. In such cases a considerably simpler process for calculating the influence of the loss to the surroundings can be derived from Nesselmann's equations in the following manner. Here the changes $\Delta\theta_2$ and $\Delta\theta'_2$ which are suffered by the exit temperatures as a result of the losses of heat and "cold" for fixed entry temperatures θ_1 and θ'_1 will be determined.

With small values of $k_u F_u/C'$ (7-14) can be used for the variations in $\Delta\theta_2$ and $\Delta\theta'_2$, with reasonable accuracy

$$\Delta\theta_2 = \left(\frac{d\theta_2}{d\dfrac{k_u F_u}{C'}}\right)_0 \cdot \frac{k_u F_u}{C'} \quad \text{and} \quad \Delta\theta'_2 = \left(\frac{d\theta'_2}{d\dfrac{k_u F_u}{C'}}\right)_0 \cdot \frac{k_u F_u}{C'} \quad (7\text{-}14)$$

where the differential quotients are developed for the limiting case of $k_u F_u/C' = 0$. From Eqs. (7-6) to (7-13)

$$\left(\frac{d\theta_2}{d\dfrac{k_u F_u}{C'}}\right)_0 = -a(\theta_1 - \theta_u) - b(\theta'_1 - \theta_u) \quad (7\text{-}15)$$

$$\left(\frac{d\theta'_2}{d\dfrac{k_u F_u}{C'}}\right)_0 = -a'(\theta_1 - \theta_u) - b'(\theta'_1 - \theta_u) \quad (7\text{-}16)$$

are obtained for these differential quotients, where the constants a, b, a' and b' are determined by equations

$$a = \cfrac{2\left\{1-\exp\left[\left(\dfrac{C}{C'}-1\right)\dfrac{kF}{C}\right]\right\}+\left(\dfrac{C}{C'}-1\right)\dfrac{kF}{C}\left\{1+\exp\left[\left(\dfrac{C}{C'}-1\right)\dfrac{kF}{C}\right]\right\}}{\dfrac{kF}{C}\left(\dfrac{C}{C'}-1\right)\left\{\dfrac{C}{C'}\exp\left[\left(\dfrac{C}{C'}-1\right)\dfrac{kF}{C}\right]-1\right\}^2}\cdot\dfrac{C}{C'}$$

$$\times\exp\left[\left(\frac{C}{C'}-1\right)\frac{kF}{C}\right]$$

(7-17)

$$b = \cfrac{\dfrac{C}{C'}\exp\left[2\left(\dfrac{C}{C'}-1\right)\dfrac{kF}{C}\right]-1-\left(\dfrac{C}{C'}-1\right)\left[1+\left(\dfrac{C}{C'}+1\right)\dfrac{kF}{C}\right]\exp\left[\left(\dfrac{C}{C'}-1\right)\dfrac{kF}{C}\right]}{\dfrac{kF}{C}\left(\dfrac{C}{C'}-1\right)\left\{\dfrac{C}{C'}\exp\left[\left(\dfrac{C}{C'}-1\right)\dfrac{kF}{C}\right]-1\right\}^2}$$

(7-18)

$$a' = \cfrac{\dfrac{C}{C'}\exp\left[2\left(\dfrac{C}{C'}-1\right)\dfrac{kF}{C}\right]-\left(\dfrac{C}{C'}-1\right)\left[\left(\dfrac{C}{C'}+1\right)\dfrac{kF}{C}+1\right]\exp\left[\left(\dfrac{C}{C'}-1\right)\dfrac{kF}{C}\right]-1}{\dfrac{kF}{C}\left(\dfrac{C}{C'}-1\right)\left\{\dfrac{C}{C'}\exp\left[\left(\dfrac{C}{C'}-1\right)\dfrac{kF}{C}\right]-1\right\}^2}$$

$$\times\frac{C}{C'}$$

(7-19)

$$b' = \cfrac{\dfrac{C}{C'}\left[\left(\dfrac{C}{C'}-1\right)\dfrac{C}{C'}\cdot\dfrac{kF}{C}-2\right]\exp\left[\left(\dfrac{C}{C'}-1\right)\dfrac{kF}{C}\right]+\left(\dfrac{C}{C'}-1\right)\dfrac{kF}{C}+2\dfrac{C}{C'}}{\dfrac{kF}{C}\left(\dfrac{C}{C'}-1\right)\left\{\dfrac{C}{C'}\exp\left[\left(\dfrac{C}{C'}-1\right)\dfrac{kF}{C}\right]-1\right\}^2}$$

(7-20)

where it will be seen that a, b, a' and b' are dependent only on kF/C and C/C'.

The relationships for a, b, a' and b' appear, at first, to be very complicated. In order to make the calculation of these quantities simpler, the average temperature

$$\theta'_M = \frac{1}{F}\int_0^F \theta'\,df$$

(7-21)

of the exit gases in the recuperator is introduced. It is sufficient to calculate this for the temperature curve which is not influenced by the heat losses. By integrating the second Eq. (5-44)

$$\theta'_2 - \theta'_M = (\theta_1 - \theta'_2)\frac{C}{C'-C}\left\{1-\frac{\exp\left[\left(\dfrac{1}{C'}-\dfrac{1}{C}\right)kF\right]-1}{\left(\dfrac{1}{C'}-\dfrac{1}{C}\right)kF}\right\}$$

is obtained for θ'_M. In order to rearrange this equation still further the ratio τ of the temperature differences $\Delta\theta_a$ and $\Delta\theta_b$ at both ends of the heat exchanger is used and from Eq. (5-42) is yielded

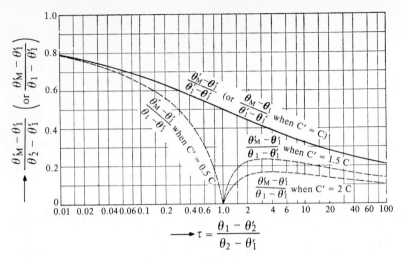

Figure 7-2 Diagram for determining the average temperature θ'_M of the exit gas in a heat exchanger.

$$\tau = \frac{\Delta\theta_a}{\Delta\theta_b} = \frac{\theta_1 - \theta'_2}{\theta_2 - \theta'_1} = \exp\left[-\left(\frac{1}{C'} - \frac{1}{C}\right)kF\right] \tag{7-22}$$

By setting $\dfrac{C'}{C} = \dfrac{\theta_1 - \theta'_2}{\theta_2 - \theta'_1}$ this last equation for θ'_M can be put into the simple form:

$$\frac{\theta'_M - \theta'_1}{\theta'_2 - \theta'_1} = \frac{1}{\ln\tau} - \frac{1}{\tau - 1} \tag{7-23}$$

It follows that $\dfrac{\theta'_M - \theta'_1}{\theta'_2 - \theta'_1}$ and thus θ_M can be very quickly determined from this equation. However the illustration which will be discussed later (top curve in Fig. 7-2) can also be used for this.

With θ'_M the rate of loss of heat $\dot{Q}_u$ of the total heat exchanger becomes

$$\dot{Q}_u = k_u F_u (\theta'_M - \theta_u) \tag{7-24}$$

Employing this relationship Eqs. (7-25) and (7-26) are obtained from Eqs. (7-14), (7-15) and (7-16) for small values of $k_u F_u$.

$$\frac{-C\,\Delta\theta_2}{\dot{Q}_u} = \frac{C}{C'}\left[a\frac{\theta_1 - \theta_u}{\theta'_M - \theta_u} + b\frac{\theta'_1 - \theta_u}{\theta'_M - \theta_u}\right] \tag{7-25}$$

$$\frac{-C'\,\Delta\theta'_2}{\dot{Q}_u} = a'\frac{\theta_1 - \theta_u}{\theta'_M - \theta_u} + b'\frac{\theta'_1 - \theta_u}{\theta'_M - \theta_u} \tag{7-26}$$

Equation (7-25) can then be put in the form

$$\frac{-C'\,\Delta\theta_2}{\dot{Q}_u} = \frac{\theta'_2 - \theta'_1}{\theta_1 - \theta'_1}\left[\frac{C}{C'}(a+b)\frac{\theta_1 - \theta'_1}{\theta'_2 - \theta'_1} - E\frac{\theta'_2 - \theta'_1}{\theta'_M - \theta_u}\right] \tag{7-27}$$

with

$$E = \frac{\theta_1 - \theta_1'}{\theta_2' - \theta_1'} \left[\frac{C}{C'} b \frac{\theta_M' - \theta_1'}{\theta_2' - \theta_1'} - \frac{C}{C'} a \left(\frac{\theta_1 - \theta_1'}{\theta_2' - \theta_1'} - \frac{\theta_M' - \theta_1'}{\theta_2' - \theta_1'} \right) \right] \qquad (7\text{-}28)$$

Finally, from Eqs. (5-46) and (7-22) is obtained

$$\frac{\theta_2' - \theta_1'}{\theta_1 - \theta_1'} = \frac{\tau - 1}{\dfrac{C'}{C} \tau - 1} \qquad (7\text{-}29)$$

and thence Eqs. (7-30) and (7-31) can be developed from Eqs. (7-17) to (7-20) taking note of Eqs. (7-22) and (7-23)

$$\frac{C}{C'}(a+b)\frac{\theta_1 - \theta_1'}{\theta_2' - \theta_1'} = \frac{1}{\ln \tau} - \frac{1}{\tau - 1} = \frac{\theta_M' - \theta_1'}{\theta_2' - \theta_1'} \qquad (7\text{-}30)$$

$$E = \frac{1}{(\ln \tau)^2} - \frac{\tau}{(\tau - 1)^2} \qquad (7\text{-}31)$$

Finally, using Eq. (7-30), one can write

$$\frac{-C\Delta\theta_2}{\dot{Q}_u} = \frac{\theta_2' - \theta_1'}{\theta_1 - \theta_1'} \left[\frac{\theta_M' - \theta_1'}{\theta_2' - \theta_1'} - E \frac{\theta_2' - \theta_1'}{\theta_M' - \theta_u'} \right] \qquad (7\text{-}32)$$

instead of Eq. (7-27). Similarly Eq. (7-26) can also be rearranged, giving

$$\frac{-C'\Delta\theta_2'}{\dot{Q}_u} = 1 - \frac{-C\Delta\theta_2'}{\dot{Q}_u} \qquad (7\text{-}33)$$

The latter notation is obtained immediately by considering that $\dot{Q}_u$ splits into two parts, namely $-C\Delta\theta_2$ and $-C'\Delta\theta_2'$. Then $-C\Delta\theta_2$ represents the increased thermal emission of the warmer gas and $-C'\Delta\theta_2'$ represents the decreased heat absorption of the colder gas as a result of heat loss to the surroundings.

The required values of $\Delta\theta_2$ and $\Delta\theta_2'$ can be calculated comparatively simply from Eqs. (7-23), (7-24), (7-31), (7-32), and (7-33); this will be explained more fully later.

Application of the Simplified Process for Calculating the Influence of the Loss of Heat to the Surroundings

If we again briefly put together the sense and purpose of the previous derivation we are able to state the following: according to Fig. 7-1 a loss of heat to the surroundings results in the temperature curve dipping down in contrast to the temperature curve for a loss-free heat exchanger. In addition, the temperature difference between the two gases is increased at the warm end of the exchanger and decreased at the cold end. This is equivalent to lowering the exit temperatures θ_2 and θ_2' of the two gases by specific amounts $-\Delta\theta_2$ and $-\Delta\theta_2'$. Here it is assumed that the entry temperatures θ_1 and θ_1' remain unchanged.

In almost all practical cases it is sufficient to determine only the changes $\Delta\theta_2$

and $\Delta\theta_2'$ in the exit temperatures with the assumption that the loss of heat to the surroundings is small. These variations can be calculated from the previously derived Eqs. (7-23), (7-24), (7-31), (7-32), and (7-33) in the following way. In addition to the entry temperatures θ_1 and θ_1' let the exit temperatures θ_2 and θ_2' also be known, as calculated for a loss-free recuperator. Gas at temperature θ' flows in the outer area of the recuperator. First the average temperature θ_M' of this gas and the rate of heat loss $\dot{Q}_u$ are determined. Further the ratio $\tau = \Delta\theta_a/\Delta\theta_b$ of the temperature differences $\Delta\theta_a = \theta_1 - \theta_2'$ and $\Delta\theta_b = \theta_2 - \theta_1'$ at the ends of the recuperator is determined. Given the value of τ so determined, Eq. (7-23) gives the ratio $\dfrac{\theta_M' - \theta_1'}{\theta_2' - \theta_1'}$, from which θ_M' can be obtained directly. The rate $\dot{Q}_u$ at which heat passes into the surroundings can thus be obtained from Eq. (7-24) in which θ_u denotes the temperature of the surroundings, F_u denotes the average surface area of the insulation through which the heat loss $\dot{Q}_u$ flows and k_u denotes the corresponding heat transmission coefficient. Finally, upon determining the quantity E (introduced as an abbreviation of a longer expression) from Eq. (7-31) with the help of τ, the required values $\Delta\theta_2$ and $\Delta\theta_2'$ can then be calculated from Eqs. (7-32) and (7-33).

The calculation described can be shortened still further by taking $\dfrac{\theta_M' - \theta_1'}{\theta_2' - \theta_1'}$ and E from Figs. 7-2 and 7-3 where these quantities are shown as the top solid lines and dependent on τ. If, as in most cases, τ lies between 0.1 and 10, the first of these two expressions can be calculated using the simple equation

$$\frac{\theta_M' - \theta_1'}{\theta_2' - \theta_1'} = 0.5 - 0.18 \lg \tau$$

In place of (7-32) and (7-33) the following equations can be used. Using the new abbreviation

$$E \cdot \left(\frac{\theta_2 - \theta_1'}{\theta_1 - \theta_1'}\right)^2 = D \qquad (7\text{-}34)$$

Eqs. (7-32) and (7-33) become

$$\frac{-C \Delta\theta_2}{\dot{Q}_u} = \frac{\theta_M' - \theta_1'}{\theta_1 - \theta_1'} - D \frac{\theta_1 - \theta_1'}{\theta_M' - \theta_u} \qquad (7\text{-}35)$$

$$\frac{-C' \Delta\theta_2'}{\dot{Q}_u} = 1 - \frac{\theta_M' - \theta_1'}{\theta_1 - \theta_1'} + D \frac{\theta_1 - \theta_1'}{\theta_M' - \theta_u} \qquad (7\text{-}36)$$

Calculation of $\Delta\theta_2$ and $\Delta\theta_2'$ is simpler using these equations than it is using Eqs. (7-32) and (7-33) provided the graphs for $\dfrac{\theta_M' - \theta_1'}{\theta_1 - \theta_1'}$ and D as a function of τ are available. A disadvantage is that C/C' occurs in these diagrams as a parameter so that sets of curves are obtained instead of the simple lines described. Thus, the solid lines in Figs. 7-2 and 7-3 are only valid for $C = C'$. In Figs. 7-2 and 7-3 the curves

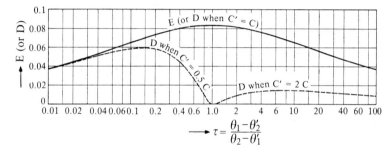

Figure 7-3 Variable E or D for determining the temperature curve in a heat exchanger with heat loss to the surroundings.

for $C' = 2C$ and $C' = 0.5C$ and a curve for $C' = 1.5C$ are shown as dashed lines in order to give some idea of the complete diagrams for $\dfrac{\theta'_M - \theta_1}{\theta_1 - \theta'_1}$ and D.

If the methods of calculation described are applied, for example, to modern blast furnaces figures of the order of magnitude of 10°C emerge for $\Delta\theta_2$ and $\Delta\theta'_2$. In contrast, in low temperature technology and in regenerators as well as in large tube counterflow heat exchangers with an insulation thickness of about 30 cm, the exit temperature at the warm end is increased by less than 0.1°C because of cold loss.

The above values of $k_u F_u / C'$ (0.03 or 0.01) (p. 202) form the basis for these calculations. The calculated perturbations of θ_2 and θ'_2 are frequently within the range of accuracy with which the heat exchanger calculations are generally performed. Thus the consideration of the loss to the surroundings can often be neglected. However, for cases where this loss cannot be ignored, Fig. 7-4 shows that its influence can be calculated very accurately with the help of Eqs. (7-23), (7-24) and (7-31) to (7-33) and Figs. 7-2 and 7-3. In these diagrams the temperature differences $\Delta\theta_a = \theta_1 - \theta'_2$ and $\Delta\theta_b = \theta_2 - \theta'_1$ at the ends of the heat exchanger,

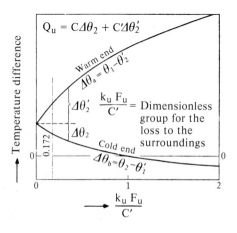

Figure 7-4 Temperature differences $\Delta\theta_a$ and $\Delta\theta_b$ at the ends of a counterflow heat exchanger dependent on the dimensionless group $k_u F_u / C'$ for heat loss to the surroundings.

accurately calculated employing Nesselmann's Eqs. (7-6) to (7-13), are illustrated by the unbroken curves as a function of $k_u F_u/C'$. Again the same relationships are used as in Fig. 7-1; only $k_u F_u/C'$ is assumed to be variable and drawn as the abscissa. $\Delta\theta_2$ and $\Delta\theta_2'$ are equal to the differences between the ordinates at the required value of $k_u F_u/C'$ and the ordinate value at $k_u F_u/C' = 0$ as shown in the diagram. It will be seen that the exact solid lines for $k_u F_u/C' < 0.1$ can be replaced quite adequately by their tangents at the origin ($k_u F_u/C' = 0$); this corresponds to the basic thinking behind the approximation process described above.

Application of the Simplified Calculation Process to General Cases

It is evident that the equations and diagrams which have just been described can also be applied, with sufficient accuracy, to the temperature dependent thermal capacities of gases, by choosing suitable average values for C and C'. The simplified calculation process as well as the more precise method of calculation is not only valid for recuperators with tube assembly arrangements; obviously it can also be extended to all types of heat exchanger which operate, mainly, in counterflow but it can be applied to cross counterflow heat exchangers (see the next chapter) and regenerators (see Part Four), which will be dealt with later.

Furthermore, the influence of the loss of heat to the surroundings depends only slightly on whether the exit gas is warm or cold. If the gas is warm the rate of loss of heat $\dot{Q}_u$ is somewhat larger than in the case dealt with so far, because of the greater difference between the average temperature θ_M of this gas and the temperature of the surroundings, θ_u. The influence on the heat exchanged however is on the contrary, reduced since the heat lost is directly transferred to the surroundings, and does not appear as an additional load upon the heating surface area F. On the whole this difference is so small that the method of calculation presented here is sufficiently accurate even when the exit gas is warm. The temperature distribution, for example as shown in Fig. 7-1, will also remain almost exactly the same.

The considerations and the methods of calculation developed previously can be applied, without any difficulty, to heat exchangers operating below the ambient temperature.

Estimation of the Temperature Distribution in the Counterflow Recuperator when Heat Losses to the Surroundings Take Place

Equations (7-6) to (7-13) are sufficiently accurate to enable the temperature distribution to be calculated for a counterflow recuperator for circumstances where the influence of the heat loss $\dot{Q}_u$ to the surroundings must be considered. The following method of estimation enables this temperature distribution to be obtained more simply and more quickly provided $\Delta\theta_2$ and $\Delta\theta_2'$ have been calculated previously.

As has been mentioned, the increase in the cooling of the warmer gas will be considerably greater than the decrease in the heating of the cooler gas when such heat losses take place as compared with the situation where no heat losses occur, (see Fig. 7-1). The same changes in the exit temperatures can also be obtained in a loss-free heat exchanger by suitably increasing the ratio C'/C of the thermal capacities of both gases. The value that this ratio must assume can be found from Eq. (5-31) by inserting the exit temperatures into this equation in place of θ_2 and θ'_2; these exit temperatures can be calculated by considering the heat loss and using one of the methods of calculation described previously.

It would appear that the total temperature distribution in the recuperator under the influence of heat loss $\dot{Q}_u$ is approximately the same as that in a loss-free heat exchanger with the increased value for the ratio C'/C. With this value of C'/C the required temperature curve can be calculated fairly accurately using Eqs. (5-43) and (5-44) in Sec. 5-6. That this process is adequate in almost all practical cases can be shown by applying it, for example, to the data in Fig. 7-1. The temperature distribution determined by this process deviates only very slightly from the curves in Fig. 7-1 calculated accurately.

The Effect of Heat Loss to the Surroundings on the Thermal Performance of a Recuperator

Finally, consideration must be given to how much of a real loss to the heat required to be exchanged is caused by the rate $\dot{Q}_u$ at which heat is given up to the surroundings. When the heat exchanger is operating at a temperature higher than that of the surroundings it is, as a rule, being used to heat the originally colder gas to a specific end temperature θ_2. If, as a result of the heat being given up to the surroundings, an end temperature results which is lower by $\Delta\theta_2$, the quantity of heat $-C'\Delta\theta_2$ not taken up represents a real loss which is called the *thermal loss*. However, $\dot{Q}_u$ consists of two parts, namely $-C'\Delta\theta_2$ and $-C\Delta\theta_2$ (see, for example, Eq. (7-32)). The second quantity refers to the fact that, as can be seen from Fig. 5-20, the originally warmer gas is cooled by $-C\Delta\theta_2$ more than in a loss-free exchanger. It follows that $-C\Delta\theta_2$ represents no real loss to the exhange of heat process.

The thermal loss $-C'\Delta\theta_2$, which forms that fraction of the total quantity of heat which is transferred to the surroundings and which can be recorded as a loss, can be determined using Eqs. (7-23) and (7-33) or (7-36). If C is approximately equal to C' the fraction $\dfrac{-C'\Delta\theta_2}{\dot{Q}_u}$ can easily be estimated. When $C = C'$, $\Theta = 1$ and thus, according to Figs. 7-2 and 7-3, $\dfrac{\theta'_M - \theta'_1}{\theta_1 - \theta'_1} = 0.5$ and $E = 0.083$. Since, as a first approximation, we can set $\dfrac{\theta'_2 - \theta'_1}{\theta_1 - \theta'_1} = 1$, and $\dfrac{\theta'_2 - \theta'_1}{\theta'_M - \theta_M}$ (when $C = C'$) generally only fluctuates between 0 and 2, according to Eq. (7-32) $\dfrac{-C\Delta\theta_2}{\dot{Q}_u}$ lies between 0.5 and 0.33 and thus, according to Eq. (7-33), $\dfrac{-C'\Delta\theta_2}{\dot{Q}_u}$ lies between 0.5 and 0.67. Thus in such, and similar, cases the thermal loss $-C'\Delta\theta_2$ is, on average, somewhat greater than half the total loss $\dot{Q}_u$.

7-3 INFLUENCE OF THERMAL CONDUCTION IN THE BUILDING MATERIALS IN THE LONGITUDINAL DIRECTION OF THE HEAT EXCHANGER

A further heat loss occurs as already mentioned, through the conduction of heat in the longitudinal direction through the building material used to construct the partition walls and the casing of the heat exchanger. To calculate this thermal quantity $\dot{Q}_\lambda$ it is necessary to know the *temperature drop in the longitudinal direction* at all positions in the *heat transferring walls*. As this heat conduction loss only represents a correction, it is sufficient when determining the temperature drop, to start from the temperature distribution which is uninfluenced by heat loss. Again, this discussion is limited to the case of counterflow. This limiting of the discussion to counterflow heat exchangers is all the more appropriate because in parallel flow the longitudinal thermal conduction loss is, as a rule, small enough to be disregarded, because of the almost uniform temperature of the tube wall.

It follows from Eq. (5-52) in Sec. 5-7 that the average wall temperature $\Theta_m = \dfrac{\Theta_0 + \Theta'_0}{2}$ can be calculated using

$$\Theta_m = \frac{1}{2}\left(1 - \frac{k}{\alpha} + \frac{k}{\alpha'}\right)\theta + \frac{1}{2}\left(1 - \frac{k}{\alpha'} + \frac{k}{\alpha}\right)\theta' \tag{7-37}$$

if, for the purposes of simplification, the differences between F_i, F_a and F are ignored. From this

$$\frac{d\Theta_m}{dx} = \frac{1}{2}\left(1 - \frac{k}{\alpha} + \frac{k}{\alpha'}\right)\frac{d\theta}{df}\cdot\frac{df}{dx} + \frac{1}{2}\left(1 - \frac{k}{\alpha'} + \frac{k}{\alpha}\right)\frac{d\theta'}{df}\cdot\frac{df}{dx} \tag{7-38}$$

provides the drop in temperature in the longitudinal direction, where, as in Fig. 5-10, x represents distance from the point of entry of the warm gas. x and f will be treated as positive in the direction of decreasing temperatures θ and θ'. Furthermore

$$\frac{d\theta}{df} = -\frac{k}{C}(\theta - \theta') \tag{7-39}$$

$$\frac{d\theta'}{df} = -\frac{k}{C'}(\theta - \theta') \tag{7-40}$$

are obtained from Eqs. (5-28) and (5-32). If these are then inserted into Eq. (7-38)

$$\frac{d\Theta_m}{dx} = -\frac{uk}{2}(\theta - \theta')\left[\frac{1}{C} + \frac{1}{C'} - \left(\frac{k}{\alpha} - \frac{k}{\alpha'}\right)\left(\frac{1}{C} - \frac{1}{C'}\right)\right] \tag{7-41}$$

follows. Finally, if λ_s is the thermal conductivity of the building material, δ is the thickness of the heat transferring walls, u is the average circumference of the tubes and thus $u \cdot \delta$ are their cross-sectional surface areas, the quantity of heat

$$(\dot{Q}_\lambda)_W = -\lambda_s \cdot \frac{d\Theta_m}{dx} \cdot u\,\delta \tag{7-42}$$

flows in the longitudinal direction in unit time at position x, where $d\Theta_m/dx$ is determined using Eq. (7-41).

In addition to this, there is a quantity of heat $(\dot{Q}_\lambda)_M$ which is conducted in the longitudinal direction of the heat exchanger by the casing tube. If this casing tube is insulated it can be assumed, as a first approximation, that it has this same temperature θ' as the exit gas. Using Eq. (7-40) the temperature drop in the outer casing is then determined by

$$\frac{d\theta'}{dx} = \frac{d\theta'}{df}\cdot\frac{df}{dx} = -\frac{uk}{C'}(\theta-\theta') \qquad (7\text{-}43)$$

Finally, if λ_M is the thermal conductivity, u_M is the average circumference and δ_M is the wall thickness of the casing tube, the rate at which heat flows in the longitudinal direction is given by

$$(\dot{Q}_\lambda)_M = -\lambda_M\frac{d\theta'}{dx}u_M\,\delta_M \qquad (7\text{-}44)$$

Using Eqs. (7-41) to (7-44)

$$\dot{Q}_\lambda = \left[\lambda_s u\,\delta\frac{1}{2}\left\{\frac{1}{C}+\frac{1}{C'}-\left(\frac{k}{\alpha}-\frac{k}{\alpha'}\right)\left(\frac{1}{C}-\frac{1}{C'}\right)\right\}+\lambda_M u_M\,\delta_M\frac{1}{C'}\right]uk(\theta-\theta') \qquad (7\text{-}45)$$

gives the total rate $\dot{Q}_\lambda = (\dot{Q}_\lambda)_W+(\dot{Q}_\lambda)_M$ at which heat flows in the longitudinal direction of the heat exchanger.

For the sake of completeness the corresponding equation *for regenerators* will also be given here. As will follow from considerations in Part Four, well insulated regenerators have an enclosing tube which, on average, has the same temperature Θ_m as the heat-storing mass. Thus $d\theta'/dx$ is substituted for $d\Theta_m/dx$ in Eq. (7-44). It follows, for regenerators that

$$\dot{Q}_\lambda = (\lambda_s u\,\delta+\lambda_M u_M\,\delta_M)\frac{1}{2}\left[\frac{1}{C}+\frac{1}{C'}-\left(\frac{k}{\alpha}-\frac{k}{\alpha'}\right)\left(\frac{1}{C}-\frac{1}{C'}\right)\right]uk(\theta-\theta') \qquad (7\text{-}46)$$

If, as for example in regenerators in low temperature technology, the heat-storing mass is often broken into sections in the longitudinal direction, only a small amount of heat will be conducted in this direction. This can be allowed for in Eq. (7-46) by choosing a very small value for λ_s or by ignoring it, this will usually give sufficient accuracy.

It is noticeable that, according to the very general Eqs. (7-45) and (7-46), the heat flow $\dot{Q}_\lambda$ is always proportional to the temperature difference $\theta-\theta'$ at the part of the exchanger being considered. Of course, if C, C', α, α', k, λ_B, etc., vary, the constant of proportionality itself can depend upon position x or f. In general, however, these variations are only very slight.

Development of a Simple Rule

If the quantities mentioned are constant so that only θ and θ' vary on the right hand side of Eqs. (7-45) and (7-46), a simple rule can be derived for calculating $\dot{Q}_\lambda$.

If $(\Theta_m)_a$ and $(\Theta_m)_b$ are the values of the average wall temperature Θ_m at $x = 0$ and $x = L$, that is, as seen in Fig. 5-10, at the ends of the heat exchanger, there follows from Eq. (5-52) with $F = F_i = F_a$

$$\frac{(\Theta_m)_b - (\Theta_m)_a}{L} = -\frac{F}{2L}\left[\left(1 - \frac{k}{\alpha} + \frac{k}{\alpha'}\right)\cdot\frac{\theta_1 - \theta_2}{F} + \left(1 - \frac{k}{\alpha'} + \frac{k}{\alpha}\right)\frac{\theta'_2 - \theta'_1}{F}\right] \quad (7\text{-}47)$$

Furthermore, for the average temperature difference $\Delta\theta_M$ between the two gases,

$$\frac{\theta_1 - \theta_2}{F} = \frac{k}{C}\Delta\theta_M \quad \text{and} \quad \frac{\theta'_2 - \theta'_1}{F} = \frac{k}{C'}\Delta\theta_M$$

it follows from Eqs. (5-20) and (5-31) that Eq. (7-47) can be converted, using $\dfrac{F}{L} = \dfrac{df}{dx}u$, to

$$\frac{(\Theta_m)_b - (\Theta_m)_a}{L} = -\frac{uk}{2}\Delta\theta_M\left[\frac{1}{C} + \frac{1}{C'} - \left(\frac{k}{\alpha} - \frac{k}{\alpha'}\right)\left(\frac{1}{C} - \frac{1}{C'}\right)\right] \quad (7\text{-}48)$$

If the temperature varies linearly between $(\Theta_m)_a$ and $(\Theta_m)_b$ the heat flow rate through the wall amounts to

$$(\dot{Q}_\lambda)_{W, lin} = -\lambda_s\frac{(\Theta_m)_b - (\Theta_m)_a}{L}u\,\delta \quad (7\text{-}49)$$

Thus, using Eq. (7-42) and then taking Eqs. (7-41) and (7-48) into consideration, we obtain:

$$(\dot{Q}_\lambda)_W = (\dot{Q}_\lambda)_{W, lin}\cdot\frac{\theta - \theta'}{\Delta\theta_M} \quad (7\text{-}50)$$

The same consideration also applies to heat flow in the casing of the heat exchanger and it follows that a relationship corresponding to Eq. (7-50) can be obtained both for recuperators and regenerators, namely

$$\dot{Q}_\lambda = (\dot{Q}_\lambda)_{lin}\cdot\frac{\theta - \theta'}{\Delta\theta_M} \quad (7\text{-}51)$$

This is applicable to the total flow of heat in the longitudinal direction of the heat exchanger.

The following simple procedure therefore emerges from this as a method of calculation of $\dot{Q}_\lambda$. First the flow of heat $(\dot{Q}_\lambda)_{lin}$ is calculated, as that which is conducted in the heat transferring walls and the casing of the heat exchanger for a linear temperature distribution. The actual heat flow $\dot{Q}_\lambda$ at an arbitrary position x is then obtained by multiplying $(\dot{Q}_\lambda)_{lin}$ by the ratio of the temperature difference $\theta - \theta'$ at the position under consideration to the average temperature difference $\Delta\theta_M$. Although this procedure is really only applicable when the values of C, C', α, α', k, λ_B, etc., are constant, it can also be applied in almost all other cases and can produce a result which is sufficiently accurate.

Influence of Longitudinal Thermal Conductivity on the Temperature Distribution

The influence of thermal conductivity in the longitudinal direction of the heat exchanger not only manifests itself in the quantity $\dot{Q}_\lambda$ of the heat which is conducted in this direction but also in the disturbance of the temperature distribution of the gases, and thus their exit temperatures θ_2 and θ'_2. Indeed, if it is assumed that the temperature difference $\theta - \theta'$ between the two gases varies along the length of the recuperator, it may be seen from Eq. (7-51) that the same variation applies to the heat conducted in the walling $\dot{Q}_\lambda$. For example, $\dot{Q}_\lambda$ may become larger with increasing f. In a very small section of the heat exchanger with the heat-transferring surface df, a small part $d\dot{q}$ of the heat given up by the warm gas serves to increase $\dot{Q}_\lambda$ and only the remainder $d\dot{q}' = d\dot{q} - d\dot{Q}_\lambda$ is then transferred to the cold gas.

If the differences between F_i, F_a and F are neglected, the following relationships for $d\dot{q}$ and $d\dot{q}'$ can be developed:

$$d\dot{q} = \alpha(\theta - \Theta_0)\,df \qquad (7\text{-}52)$$

$$d\dot{q}' = \alpha'(\Theta'_0 - \theta')\,df \qquad (7\text{-}53)$$

The average value of $d\dot{q}$ and $d\dot{q}'$ can be set equal to the rate of heat conduction through the wall and this is displayed in Eq. (7-54).

$$\frac{1}{2}(d\dot{q} + d\dot{q}') = \frac{\lambda_s}{\delta}(\Theta_0 - \Theta'_0)\,df \qquad (7\text{-}54)$$

If these three equations are combined using the temperature differences $\theta - \Theta_0$, etc., and then taking Eq. (5-2) into consideration, Eq. (7-55) is obtained:

$$\theta - \theta' = \frac{1}{k}\cdot\frac{d\dot{q}}{df} + \left(\frac{1}{\alpha'} + \frac{\delta}{2\lambda_s}\right)\frac{d\dot{q}' - d\dot{q}}{df} \qquad (7\text{-}55)$$

The abbreviation (7-56) is first introduced.

$$\frac{(\dot{Q}_\lambda)_{lin}}{\Delta\theta_M} = A \qquad (7\text{-}56)$$

Then using Eq. (7-51), with this abbreviation, namely

$$d\dot{q} - d\dot{q}' = d\dot{Q}_\lambda = A\,d(\theta - \theta')$$

Eq. (7-57) can be readily developed from Eq. (7-55).

$$\frac{d\dot{q}}{\theta - \theta'} - \left(\frac{k}{\alpha'} + \frac{k\delta}{2\lambda_s}\right)A\,\frac{d(\theta - \theta')}{\theta - \theta'} = k\,df \qquad (7\text{-}57)$$

Equation (7-59) follows from the heat-balance equation, which is:

$$\dot{q} = C(\theta_1 - \theta) = C'(\theta'_2 - \theta') + \dot{Q}_\lambda - \dot{Q}_{\lambda 0} = C'(\theta'_2 - \theta') + A(\theta - \theta') - A(\theta_1 - \theta_2) \qquad (7\text{-}58)$$

where $\dot{Q}_{\lambda 0}$ denotes the value of $\dot{Q}_\lambda$ when $x = 0$.

$$\theta_1 - \theta = \frac{C' + A}{C - C'} [(\theta - \theta') - (\theta_1 - \theta_2')] \tag{7-59}$$

Forming first $d\dot{q} = C\, d(\theta_1 - \theta)$, Eq. (7-60) is obtained by the insertion of this expression into Eq. (7-57) and by subsequent integration.

$$\theta - \theta' = (\theta_1 - \theta_2') \cdot \exp\left(\frac{kf}{\dfrac{C}{C - C'}(C' + A) - \left(\dfrac{k}{\alpha'} + \dfrac{k\,\delta}{2\lambda_s}\right)A}\right) \tag{7-60}$$

Equation (7-59) can then be converted into the following form:

$$\theta_1 - \theta = (\theta_1 - \theta_2')\frac{C' + A}{C - C'}\left[\exp\left(\frac{kf}{\dfrac{C}{C - C'}(C' + A) - \left(\dfrac{k}{\alpha'} + \dfrac{k\,\delta}{2\lambda_s}\right)A}\right) - 1\right] \tag{7-61}$$

The temperature distribution in a heat exchanger under the influence of longitudinal thermal conduction $\dot{Q}_\lambda$ is determined by Eqs. (7-60) and (7-61), provided that C, C', α, α', λ_s and k are constant and the entry temperatures θ_1 and θ_1' are known. The exit temperatures θ_2 and θ_2' of the gases which occur when the influence of longitudinal conduction cannot be neglected are also determined by these equations. The value of θ_2 is obtained directly by using the following relationship for $f = F$ from Eqs. (7-60) and (7-61) namely:

$$\theta_2 - \theta_1' = (\theta_1 - \theta_1') \cdot \frac{\exp\left(\dfrac{kF}{\dfrac{C}{C - C'}(C' + A) - \left(\dfrac{k}{\alpha'} + \dfrac{k\,\delta}{2\lambda_s}\right)A}\right)}{\dfrac{C + A}{C - C'} \cdot \exp\left(\dfrac{kF}{\dfrac{C}{C - C'}(C' + A) - \left(\dfrac{k}{\alpha'} + \dfrac{k\,\delta}{2\lambda_s}\right)A}\right) - \dfrac{C' + A}{C - C'}} \tag{7-62}$$

It is evident that the temperature distribution is influenced by $\dot{Q}_\lambda$ as seen in Eqs. (7-60) and (7-61) and that this corresponds exactly to the loss-free case with constant, but somewhat different values of C and C'. Equations (5-42) and (5-43) may be converted into Eqs. (7-60) and (7-61) if, upon taking Eq. (5-2) into consideration,

$$C \text{ is set equal to } C + \frac{C' - C}{C' + A}\left(\frac{k}{\alpha'} + \frac{k\,\delta}{2\lambda_s}\right)A$$

and

$$C' \text{ is set equal to } C' - \frac{C' - C}{C + A}\left(\frac{k}{\alpha} + \frac{k\,\delta}{2\lambda_s}\right)A \tag{7-63}$$

It can be seen from this that, in the case of $C' > C$ the thermal capacity C of the warmer gas is perceptibly increased by the effect of $\dot{Q}_\lambda$ and the thermal capacity of C' of the colder gas becomes noticeably smaller. The difference in the temperature distributions of θ and θ' which arise from this is illustrated in Fig. 7-5, in which the dashed lines represent the temperature curve when there is no longitudinal

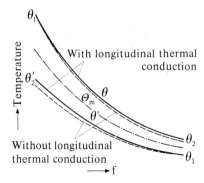

Figure 7-5 Temperature distribution in a counterflow heat exchanger with $C' > C$ under the influence of thermal conductivity through the walls in the longitudinal direction.

conduction, and the solid lines represent the temperature distribution under the influence of $\dot{Q}_\lambda$. It may be seen that in this case, where the entry temperatures θ_1 and θ_1' are constant, the temperatures θ and θ' of the gases will be increased elsewhere in the heat exchanger. Consequently the average temperature Θ_m of the partition walls must also increase. These phenomena can be explained in the following way:

It will be assumed, as in previous derivations, that the heat $\dot{Q}_\lambda$ conducted in the longitudinal direction leaves unhindered from the outside ends of the heat exchanger. Then, in the example illustrated, $\dot{Q}_\lambda$ will decrease steadily with increasing f, because in Fig. 5-20 the temperature drop $-d\Theta_m/dx$ is smaller the further towards the right one goes. Thus, when $f = 0$, a considerably larger quantity of heat $(\dot{Q}_\lambda)_0$ will be supplied to the heat exchanger from the outside than the heat $(\dot{Q}_\lambda)_F$ which is given up to the outside when $f = F$. Part of the difference $(\dot{Q}_\lambda)_0 - (\dot{Q}_\lambda)_F$ will be transferred to the colder gas which will thus be heated more than in the uninterrupted exchange of heat between the gases. On the other hand there occurs a decreased thermal emission from the warmer gas, so that, on the whole, this will be cooled less. Since there is an increase in the temperatures of both the gases, the average temperature Θ_m of the heat transferring walls will also increase.

Increase in Thermal Loss Through Temperature Variations in the Gases

The extent to which the temperature variations described constitute a loss, can be recognized if, in Fig. 7-5, the left-hand side of the heat exchanger is imagined to be coupled to an apparatus in which some sort of process, for example a chemical reaction, is carried out at high temperatures. The conduction heat $(\dot{Q}_\lambda)_0$ originates from this apparatus. Nevertheless, however, only the considerably smaller amount $(\dot{Q}_\lambda)_F$ at the cold end of the recuperator is directly lost through conduction. However an additional loss takes place because the originally warmer gas leaves at a higher temperature θ_2 and thus takes more heat out with it. The exchange loss is increased by this. If the exit temperature is increased by $\Delta_\lambda\theta_2$ the additional loss amounts to $C \cdot \Delta_\lambda\theta_2$.

An exception to this is the special case $C = C'$ where, because the temperature distribution is linear, $\dot{Q}_\lambda = (\dot{Q}_\lambda)_0 = (\dot{Q}_\lambda)_F = \text{const}$. Here the heat $\dot{Q}_\lambda$ flows longitudinally along the partition wall without influencing the exchange of heat taking place between the two gases perpendicular to the walls. It follows from this that the exit temperatures θ_2 and θ'_2 remain unchanged and $(\dot{Q}_\lambda)_F$ represents the total loss of heat due to thermal conduction in the longitudinal direction. The additional heat loss $C \cdot \Delta_\lambda \theta_2$ only takes place with a curved temperature distribution.

Only slight variations to this occur when the quantities of heat $(\dot{Q}_\lambda)_0$ and $(\dot{Q}_\lambda)_F$ neither enter from nor escape to the outside. If, for example, the extreme case is assumed where the heat transferring walls are completely insulated at the ends of the heat exchanger, the main direction of the temperature distribution will remain as in Fig. 7-5. However the curve for Θ_m will bend at the ends of the exchanger in such a way that the temperature gradient $-d\Theta_m/dx = 0$ when $f = 0$ and $f = F$, thus the curve for Θ_m becomes horizontal at the exits (see Fig. 7-6). Because of this the temperature difference between the wall and the warmer gas will be greater when $f = 0$ while the temperature difference between the wall and the colder gas will be greater when $f = F$. Essentially this has the effect that the partition walls at one end additionally absorb the quantity of heat $(\dot{Q}_\lambda)_0$ from the warmer gas and then finally, at the other end, release this additional heat $(\dot{Q}_\lambda)_F$ to the colder gas. Thus both gases undergo a somewhat increased temperature change immediately after entry while, conversely, the temperature change shortly before exit is somewhat decelerated. Thus the temperatures θ and θ' strive to come together in the centre part of the heat exchanger. However, an unchanged value of $\Delta\theta_M$ can only remain when the exit temperature θ_2, in compensation, is somewhat increased, contrary to Fig. 7-6, and the exit temperature θ'_2 is somewhat decreased.

The precise calculation of this temperature is rather complicated and is therefore not discussed here. However, as the differences between the pairs of temperature curves in Fig. 7-5 are only very slight overall, it is to be expected that the total loss $(\dot{Q}_\lambda)_{ges}$ due to longitudinal conduction will be approximately the same in both cases. However, the following difference still exists. While, in the first case (Fig. 7-5), $(\dot{Q}_\lambda)_{ges}$ is a factor of $(\dot{Q}_\lambda)_F$ and $C \cdot \Delta_\lambda \theta_2$, in the second case (Fig. 7-6) $(\dot{Q}_\lambda)_{ges}$ only consists of the exchange loss $C \cdot \Delta_\lambda \theta_2$ which results in $\Delta_\lambda \theta_2$ being correspondingly larger.

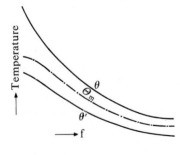

Figure 7-6 Temperature distribution in a counterflow heat exchanger under the influence of longitudinal thermal conduction where the tube walls are insulated at the ends.

In the first case $(\dot{Q}_\lambda)_F$ and $\Delta_\lambda\theta_2$ can be calculated using Eqs. (7-51), (7-56) and (7-62) from which $(\dot{Q}_\lambda)_{ges} = (\dot{Q}_\lambda)_F + C\cdot\Delta_\lambda\theta_2$ can then be calculated. $\Delta_\lambda\theta_2$ is then obtained by deducting the value for $A = 0$ from the value for θ_2 determined using Eq. (7-62). However if, as in almost all practical cases, the longitudinal thermal conduction loss is small, $\Delta_\lambda\theta_2$ and thus also $(\dot{Q}_\lambda)_{ges}$ can be found more simply by using the following process.

Simplified Calculation of the Influence of Longitudinal Thermal Conductivity upon the Exit Temperature of the Warmer Gas

It is assumed that the rate $\dot{Q}_\lambda$ at which heat is conducted in the longitudinal direction, and thus also $A = (\dot{Q}_\lambda)_{lin}/\Delta\theta_M$, is small. The critical case of $A = 0$ is taken as the starting point. If Eq. (7-62) is differentiated with respect to A and then when the limit as $A \to 0$ is taken Eq. (7-64) will be obtained.

$$\left(\frac{d\theta_2}{dA}\right)_{A=0}$$
$$= (\theta_2 - \theta_1')\frac{1-\left\{1+\left(\dfrac{1}{C'}-\dfrac{1}{C}\right)\left[1+\dfrac{C'-C}{C}\left(\dfrac{k}{\alpha'}+\dfrac{k\delta}{2\lambda_s}\right)\right]kF\right\}\exp\left[-\left(\dfrac{1}{C'}-\dfrac{1}{C}\right)kF\right]}{C'\cdot\exp\left[-\left(\dfrac{1}{C'}-\dfrac{1}{C}\right)kF\right]-C}$$

$$(7\text{-}64)$$

Because the value of A is small this derivative can be set equal to $\Delta_\lambda\theta_2/A$. Further, following from Eqs. (7-51) and (7-56), and noting that $(\dot{Q}_\lambda)_F = A(\theta_2 - \theta_1')$, Eq. (7-65) is obtained, using Eq. (7-22)

$$\frac{C\cdot\Delta_\lambda\theta_2}{(\dot{Q}_\lambda)_F} = \frac{1+\left\{\left[1+\dfrac{C'-C}{C}\left(\dfrac{k}{\alpha'}+\dfrac{k\delta}{2\lambda_s}\right)\right]\ln\tau-1\right\}\tau}{\dfrac{C'}{C}\tau-1}$$

$$(7\text{-}65)$$

Equation (7-66) can then be developed for the total loss due to thermal conduction $(\dot{Q}_\lambda)_{ges} = (\dot{Q}_\lambda)_F + C\,\Delta_\lambda\theta_2$:

$$\frac{(\dot{Q}_\lambda)_{ges}}{(\dot{Q}_\lambda)_F} = 1 + \frac{C\cdot\Delta_\lambda\theta_2}{(\dot{Q}_\lambda)_F} = \frac{\tau\left\{\left[1+\dfrac{C'-C}{C}\left(\dfrac{k}{\alpha'}+\dfrac{k\delta}{2\lambda_s}\right)\right]\ln\tau + \dfrac{C'-C}{C}\right\}}{\dfrac{C'}{C}\tau-1}$$

$$(7\text{-}66)$$

In the special case where $C = C'$, this equation takes on the following remarkably simple form, once Eqs. (7-22) and (5-56) have been considered:

$$\frac{(\dot{Q}_\lambda)_{ges}}{(\dot{Q}_\lambda)_F} \cdot \frac{\tau\ln\tau}{\tau-1} = \frac{\theta_1-\theta_2}{\Delta\theta_M} \qquad \text{(with } C = C'\text{)} \qquad (7\text{-}67)$$

In addition, if it is observed that according to Eq. (5-95),

$$(\dot{Q}_\lambda)_F = (\dot{Q}_\lambda)_{lin} \cdot \frac{\theta_2 - \theta_1'}{\Delta\theta_M} = (\dot{Q}_\lambda)_{lin} \cdot \frac{\ln \tau}{\tau - 1}$$

Eq. (7-68) can be obtained as the final result, utilizing Eqs. (7-66) and (7-67)

$$\frac{(\dot{Q}_\lambda)_{ges}}{(\dot{Q}_\lambda)_{lin}} = \frac{\tau \ln \tau}{(\tau - 1)\left(\dfrac{C'}{C}\tau - 1\right)} \left\{ \left[1 + \frac{C' - C}{C}\left(\frac{k}{\alpha'} + \frac{k\delta}{2\lambda_s}\right) \right] \ln \tau + \frac{C' - C}{C} \right\} \quad (7\text{-}68)$$

and in the special case where $C = C'$

$$\frac{(\dot{Q}_\lambda)_{ges}}{(\dot{Q}_\lambda)_{lin}} = \frac{(\theta_1 - \theta_2')(\theta_2 - \theta_1')}{(\Delta\theta_M)^2} \qquad \text{(with } C = C') \qquad (7\text{-}69)$$

If it is required to use Eq. (7-69) for the development of the case where $C = C'$ only, the expression on the right-hand side can be set equal to 1 since in this case kF is fixed in advance and the temperature differences $\theta_1 - \theta_2'$, $\theta_2 - \theta_1'$ and $\Delta\theta_M$ are equal to one another. This is in agreement with an earlier finding that when $C = C'$ no loss occurs due to temperature variations in the gases and thus $(\dot{Q}_\lambda)_{ges} = (\dot{Q}_\lambda)_F = (\dot{Q}_\lambda)_0 = (\dot{Q}_\lambda)_{lin}$. The importance of Eq. (7-69) lies in the fact that it also provides good approximations even when $C \neq C'$.

Diagram and Simple Rule for Determining the Total Loss due to Longitudinal Thermal Conductivity

The results obtained so far can be summarized in the following way. The total rate of heat loss $(\dot{Q}_\lambda)_{ges}$ which is produced by longitudinal thermal conductivity in the heat transferring walls and the casing tube, does not consist solely of the quantity $(\dot{Q}_\lambda)_F$ which is produced indirectly from the longitudinal thermal conduction in the recuperator. Rather, longitudinal thermal conduction influences the whole temperature distribution in the recuperator. An increase in exchange loss can result from this as is illustrated in Fig. 7-5; here the originally warmer gas leaves the recuperator at a higher temperature, than it would without thermal conduction. Nevertheless this exchange loss can also be negative, that is when the temperature distributions for $C' < C$ are curved in the opposite direction to that shown in Fig. 7-5. This additional loss must be considered, together with its plus or minus sign, in the value of the total loss $(\dot{Q}_\lambda)_{ges}$.

It has been indicated that it is easier to calculate $(\dot{Q}_\lambda)_{ges}$ once the thermal quantity $(\dot{Q}_\lambda)_{lin}$ has first been determined; in the case of a linear temperature distribution this would be the quantity of heat $(\dot{Q}_\lambda)_{lin}$ conducted between the actual temperatures at the ends of the recuperator. The ratio $(\dot{Q}_\lambda)_{ges}/(\dot{Q}_\lambda)_{lin}$ is then determined utilizing Eq. (7-68) or Eq. (7-69). Essentially its value is dependent only on the ratio $\tau = \Delta\theta_a/\Delta\theta_b$ of the temperature differences of the gases at the ends of the heat exchanger and the ratio C'/C of the thermal capacities of the two gases. The influence of k, α', δ and λ_s in Eq. (7-68) on the other hand, is only very slight.

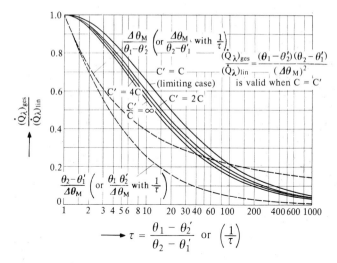

Figure 7-7 The ratio of the total loss $(\dot{Q}_\lambda)_{ges}$ due to longitudinal thermal conductivity and $(\dot{Q}_\lambda)_{lin}$ as a function of $\tau = \Delta\theta_a/\Delta\theta_b$ with $\alpha = \alpha'$.

It is possible to determine the value of $(\dot{Q}_\lambda)_{ges}/(\dot{Q}_\lambda)_{lin}$ very rapidly using Fig. 7-7 in which this ratio is plotted (using Eq. (7-68)) as a function of τ for various values of C'/C. α is set equal to α' so that $\dfrac{k}{\alpha'} + \dfrac{k\delta}{2\lambda_\beta} = \dfrac{1}{2}$. It is assumed that $\tau > 1$ and $C'/C > 1$. The curves shown in Fig. 7-7 can be used for $\tau < 1$, and also for $C'/C < 1$ if $1/\tau$ is chosen as the abscissa and C/C' is chosen as the parameter. The use of these curves is made very simple since the abscissa deals only with values of $\tau \geqslant 1$; similarly the values C'/C are all greater than or equal to one. It can be seen from the curves that the influence of C'/C is only very small. It can usually be neglected as, in the case of $(\dot{Q}_\lambda)_{ges}$ it is only used as a correction factor. Thus $(\dot{Q}_\lambda)_{ges}/(\dot{Q}_\lambda)_{lin}$ can almost always be calculated approximately using the simple Eq. (7-69), which corresponds to the topmost curve in Fig. 7-7. It follows from this equation that a simple rule can be applied in order to calculate the total loss $(\dot{Q}_\lambda)_{ges}$ due to longitudinal thermal conductivity. The heat flow $(\dot{Q}_\lambda)_{lin}$ is first determined as the rate at which heat is conducted in the longitudinal direction of the heat exchanger for a linear temperature distribution. The value thus obtained is then multiplied by the product of the two temperature differences of the gases at the ends of the heat exchanger and finally divided by the square of the average temperature difference $\Delta\theta_M$.

Further $\dfrac{\theta_2 - \theta'_1}{\Delta\theta_M}$ and $\dfrac{\Delta\theta_M}{\theta_1 - \theta'_2}$ are shown as functions of τ in Fig. 7-7. The first of these curves can be used, utilizing Eq. (7-51), for the rapid determination of the heat loss $(\dot{Q}_\lambda)_F$ which is itself indirectly produced by longitudinal thermal conduction; the second curve can be used for determining the average temperature $\Delta\theta_M$ from the end temperature differences $\Delta\theta_a$ and $\Delta\theta_b$, (also see Fig. 5-18). The quotient of the quantities represented by these two curves again yields the expression on the right-hand side of Eq. (7-69), but it is more simple to read it off from the topmost curve mentioned initially.

Magnitude of the Losses in Low Temperature Technology

In order to obtain an overall picture of the magnitude of the losses due to heat or "cold" emissions to the surroundings and due to longitudinal thermal conductivity, some of the values calculated for a counterflow heat exchanger used in low temperature technology are given below. The influence of these losses on the temperature distribution can be recognized if Figs. 7-1 and 7-5 are imagined to be mirrored across a horizontal line corresponding to the temperature of the surroundings θ_u, so that all temperatures now lie below θ_u. It follows from Fig. 7-1 that this loss of cold $\dot{Q}_u$ to the surroundings has the effect that the originally warmer gas will be cooled less and the originally cooler gas will be heated more than would be the case without this loss. It will be seen from Fig. 7-5 that the longitudinal thermal conduction creates an increase in the temperature difference between the gases at the warm end and thus increases the exchange loss.

The following table contains the results of a calculation, performed previously, of the "cold" loss in four counterflow heat exchangers which have been chosen to be as different as possible the one from the other. Two are cases of recuperators constructed of an assembly of straight tubes (see Fig. 5-10), one is a regenerator (see Part Four) and one is an example of a cross-counterflow heat exchanger (see Sec. 8-3). The cross-counterflow device is very short and has a heavily constructed casing tube to withstand the high pressure which exists even in the outer area.

The third column of the table at the top shows that the dimensionless group $k_u F_u / C'$, which relates to the loss to the surroundings, is very small. $\dot{Q}_u$ is the rate at which heat flows into the heat exchanger from the surroundings. As mentioned in earlier discussions only part of this, $C \cdot \Delta\theta_2$, represents an actual loss of "cold", where C is the thermal capacity rate of the originally warmer gas and $\Delta\theta_2$ is the increase in the exit temperature produced by the loss to the surroundings. In spite of the large differences between the counterflow devices, $\Delta\theta_2$ is of the same order of magnitude in all four cases; it mostly lies between $0.1°$ and $0.2°$ and, in general, increases with decreasing performance.

The bottom table shows the influence of the loss due to longitudinal thermal conduction for the same counterflow devices. The second column gives the quantity of heat $(\dot{Q}_\lambda)_{lin}$ flowing in the longitudinal direction through the building materials of the walls and the casing tube when a linear temperature distribution is assumed. In the straight tube assemblies and the regenerator, $(\dot{Q}_\lambda)_{lin}$ is roughly in proportion to the gas flowrate. In contrast $(\dot{Q}_\lambda)_{lin}$ is considerably larger in the cross-counterflow heat exchanger because this is short and has a strong-walled casing tube made of copper. $(\dot{Q}_\lambda)_0$ gives the quantity of heat actually flowing in at the warm end of the heat exchanger, by means of longitudinal thermal conduction, (when $f = 0$). Moreover the longitudinal thermal conduction causes a lowering of the exit temperature θ'_2 of the originally colder gas so that additional "cold" is carried out. In the next column $(\dot{Q}_\lambda)_{ges}$ is the total "cold" loss due to longitudinal thermal conductivity and consists of the two quantities mentioned previously. The penultimate column gives the decrease $\Delta\theta_2'$ of the exit temperature. In the last column $(\dot{Q}_\lambda)_{ges}$ is represented in a corresponding way as a variation of the exit temperature.

In estimating the influence of the calculated losses it must be remembered that the total temperature difference at the warm end of the counterflow heat exchanger lies between 2° and 5°.

EIGHT

RECUPERATORS OPERATED IN CROSS FLOW[1]

8-1 VARIOUS ARRANGEMENTS FOR HEAT TRANSFER IN CROSS FLOW

In cross flow the directions of flow of the two gases cross one another. One gas flows, as shown for example in Fig. 8-1, inside the tubes which are joined together in a tube assembly; as a rule the other gas impinges perpendicularly, or approximately so, on the tubes and flows past them in this direction. *Pure cross flow* occurs when the tubes are straight and, apart from any minor deviations caused by the tubes themselves, the gases follow a straight path.

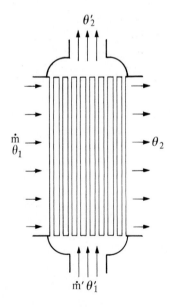

Figure 8-1 Pure cross flow.

[1] See also Kühl's detailed description [K217].

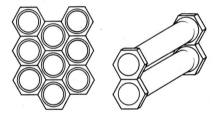

Figure 8-2 Air tube radiator for motor vehicles (honeycomb cooler).

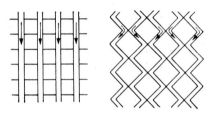

Figure 8-3 Water tube radiator for motor vehicles.

In this way, cross flow is often used for preheating the feed water for boiler plants in such a way that, for example, the water flows through vertical tubes while the boiler gas flows outside the tubes in a horizontal direction.

A very well known arrangement for pure cross flow is represented by air tube radiators in motor vehicles (see Fig. 8-2). Here the air flows through numerous, parallel, horizontal tubes which are widened at the ends, for example, and then expanded and fabricated together in a honeycomb formation. The water to be cooled flows from top to bottom through the narrow spaces which exist between the tubes.* Pure cross flow can also be realized in water tube radiators as in Fig. 8-3. Here the water flows through vertical smooth, or zig-zag tubes while the air is driven horizontally through the spaces between the tubes.

Pure cross flow can also be realized in the laminar arrangement shown in Fig. 6-1, bottom left, where the two gases flow across one another in the spaces between the parallel plates. Here the first, third, fifth, etc., intersections are for one gas while the second, fourth, sixth, etc., are for the other gas. Arrangements such as those shown in Fig. 6-1, bottom, middle, and Fig. 5-5 permit pure cross flow operation.

Cross flow is more usually combined with *parallel flow* or *counterflow* rather than in straightforward cross flow. Arrangements of this type are often known as *mixed switching.*† Such switching is obtained when, as shown in Fig. 8-4, for example, numerous plates are arranged one above the other, see Fig. 6-1, bottom, left. While, as before, each individual element works as in cross flow, the main movement of air is from bottom to top, in the opposite direction to the flow of the exhaust gas. In this sense, this is therefore counterflow. A basically similar type of mixed switching is often employed in superheated boilers and feed water preheaters the arrangements of which are shown in Fig. 8-5. The steam or water flows in numerous tube "coils", lying one behind another; each "coil" consists for the most part, of horizontal, straight pieces of tube. The exhaust gases impinge on the pieces of tube, for example at right angles to them, from the top, and move in counterflow or parallel flow, along the main direction of flow of the steam or water. More complex switching, an example of which is a superheater, is illustrated in Fig. 8-6.

---

* Editor's note: In many motor vehicles, the water flows from bottom to top employing the thermo-syphon effect as well as being pumped through the radiator in this direction.

† Editor's note: In the American literature, this arrangement is known as *Cross flow with two, three or multipass counterflow/parallel flow.*

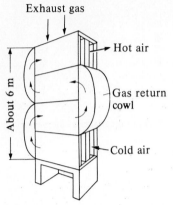

Exhaust gas

Hot air

Gas return cowl

Cold air

About 6 m

Figure 8-4 A plate air warmer with "mixed switching". The individual elements consist of parallel, smooth plates.

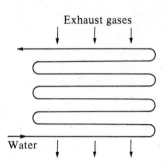

Exhaust gases

Water

Figure 8-5 Flow of water in a feed water preheater.

The direction of flow of the combustion gases, which is indicated by the solid arrows, is counter to the main direction of flow of the steam in the upper part while, by contrast, in the lower part which is exposed to the highest temperatures, the steam flows in the same direction as the combustion gases. Parallel flow has the effect that the tubes of the superheater do not become too hot and thus damage to the material of which they are constructed, is avoided. Switching can be even more

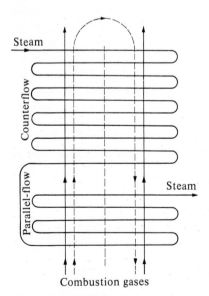

Steam

Counterflow

Parallel-flow

Steam

Combustion gases

Figure 8-6 Superheater where the steam and exhaust gas flow in counterflow and parallel flow with cross flow also taking part.

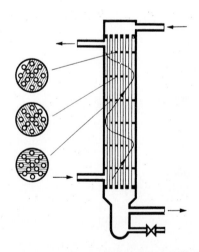

Figure 8-7 Counterflow heat exchanger with cross flow directed by means of baffles.

complicated when the outer area is divided by a vertical wall so that the combustion gases have to follow the path indicated by dashed lines in Fig. 8-6.

Finally the tube assembly arrangement with guideplates shown in Fig. 8-7 represents a combination of cross flow and counterflow. Here, also, the gas flowing on the outside is guided in cross flow against the straight tube assemblies but the cross flow direction changes periodically down the length of the heat exchanger.

The steel tube hot blast stove, for example, is built according to this principle; this has been developed by Schack [S202, J203] for the purpose of preheating the air needed in blast-furnaces and is intended to replace the hot blast stoves with ceramic chequerwork which have, hitherto, been used exclusively.

The most complete combination of cross flow and counterflow is to be found in the "cross-counterflow" heat exchanger which has been developed in low temperature technology. This type of cross-counterflow device is basically composed of numerous helically wound tubes, arranged in several layers, one on top of another; Fig. 8-8 shows this without its casing tube. The tubes are alternately coiled to the left and to the right. This allows all the tubes to be approximately the same length and if a greater diameter is required, the number of tubes can be increased. The gas flowing on the outside in the direction of the axis impinges approximately, but not exactly, perpendicularly on the tubes.

As long as tubes are used as the structural elements, the advantage of cross flow in all the arrangements described lies in the fact that the heat transmission coefficient is considerably higher on the outside of the tubes in cross flow than it is in parallel flow. With cross-counterflow, i.e., with mixed switching, this gain is combined with the previously mentioned advantages of counterflow; together they give a particularly favourable effect (see Sec. 8-3).

A further increase in heat transfer on the outside can be achieved using *finned tubes* where it is generally assumed that cross flow is employed for each individual tube. Finned tubes can be used both in pure cross flow, for example in the arrangements as shown in Fig. 8-1, and in mixed switching, particularly in the forms shown in Figs. 8-5 and 8-6.

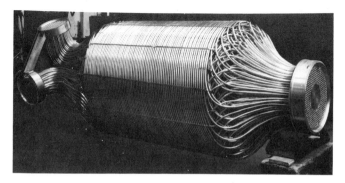

Figure 8-8 Cross-counterflow heat exchanger for low temperatures in the course of construction. (Industrial photography by Linde AG, Höllriegelskreuth, Munich).

8-2 TEMPERATURE DISTRIBUTION AND HEAT TRANSFER WITH PURE CROSS FLOW

The Differential Equations

The gas temperature distribution which appears in pure cross flow can be calculated using the theories first given by Nusselt [N206]. In order to establish the relevant differential equations, it will be assumed, as in Fig. 8-1, that the originally warmer gas flows from left to right between the vertical tubes of the cross flow recuperator, while the originally colder gas flows from the bottom to the top of the exchanger, inside the tubes. In order to simplify the calculation all the tubes will be thought of as having been split along their length and bent to form a flat sheet (Fig. 8-9), the surface area F of which is equal to the total heating surface area of all the tubes. One gas flows in front of the sheet from left to right and the other gas flows behind the sheet from the bottom to the top giving pure cross flow. A particular place of the sheet will be characterized by the coordinates x and x', the lengths of the edges of the sheet will be denoted as L and L' (Fig. 8-9). If θ and θ' are the temperatures of the gases at position x, x' the rate

$$d\dot{q} = k\,dx\,dx'(\theta - \theta')$$

at which heat will be transferred through a surface element $df = dx\,dx'$ at this position is given by this expression. The quantity $\dot{m} \cdot \dfrac{dx'}{L'}$ of gas I flows over df in unit time and is cooled by $\dfrac{\partial\theta}{\partial x}dx$; at the same time, the quantity $\dot{m}' \cdot \dfrac{dx}{L}$ of gas II flows past the same element and is heated by $\dfrac{\partial\theta'}{\partial x'}dx'$. If C and C' again denote

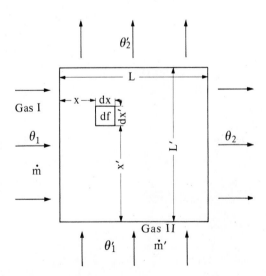

Figure 8-9 The exchange of heat in cross flow on a flat sheet.

the thermal capacity rates of the gases, the quantity of heat passing through df can be expressed by

$$d\dot{q} = -C\frac{dx'}{L'}\frac{\partial\theta}{\partial x}dx = C'\frac{dx}{L}\frac{\partial\theta'}{\partial x'}dx'$$

By comparison with the previous expression for $d\dot{q}$ the differential Eqs. (8-1) and (8-2) are obtained

$$\frac{C}{kL'}\frac{\partial\theta}{\partial x} = \theta' - \theta \qquad (8\text{-}1)$$

$$\frac{C'}{kL}\frac{\partial\theta'}{\partial x'} = \theta - \theta' \qquad (8\text{-}2)$$

If, in addition, the dimensionless variables

$$\xi = \frac{kL'}{C}x \qquad \text{and} \qquad \xi' = \frac{kL}{C'}x' \qquad (8\text{-}3)$$

are introduced in place of x and x', Eqs. (8-1) and (8-2) take the form

$$\partial\theta/\partial\xi = \theta' - \theta \qquad (8\text{-}4)$$

$$\partial\theta'/\partial\xi' = \theta - \theta' \qquad (8\text{-}5)$$

These relationships are the required differential equations for pure cross flow.

It is to be noted that Eqs. (8-4) and (8-5) are the same differential equations for the temperature distribution in regenerators, which will be dealt with in Part Four (Sec. 13-3, Eqs. (13-26) and (13-27)); θ' takes the place of the temperature Θ of the heat storing mass and ξ' takes the place of the reduced time η (see Sec. 13-3, Eq. (13-24)). The apparently complicated solution of these differential equations will be dealt with in detail in Part Four. The method of solution suitable for calculating pure cross flow in recuperators will thus also be found there. The solution developed in Sec. 13-4 for the preliminary heating of a heat-storing mass is applicable to pure cross flow (Eqs. (13-37) to (13-39)); it is assumed there that the heat-storing mass originally has a uniform temperature and that, at a particular instant, a gas at a higher constant entry temperature flows through it. The approximate methods of calculation developed in Sec. 15-1 to Sec. 16-5 in Part Four are also suitable for solving differential Eqs. (8-4) and (8-5) and are thus also suitable for calculating the temperature distribution in pure cross flow. It is thus not necessary to go any further into the solutions of Eqs. (8-4) and (8-5) at this stage.

A solution developed by Nusselt [N207] is mentioned here as this will not be discussed later; this is derived with the help of an integral equation and is set out below in simplified form;

$$\frac{\theta - \theta'_1}{\theta_1 - \theta'_1} = 1 - e^{-(\xi + \xi')}\left[\xi + \frac{\xi^2}{2!}(1+\xi') + \frac{\xi^3}{3!}\left(1+\xi' + \frac{\xi'^2}{2!}\right) + \cdots\right.$$

$$\left. + \frac{\xi^n}{n!}\left(1+\xi' + \frac{\xi'^2}{2!} + \cdots + \frac{\xi'^{n-1}}{(n-1)!}\right) + \cdots\right], \qquad (8\text{-}6)$$

$$\frac{\theta' - \theta'_1}{\theta_1 - \theta'_1} = 1 - e^{-(\xi + \xi')}\left[1 + \xi(1 + \xi') + \frac{\xi^2}{2!}\left(1 + \xi' + \frac{\xi'^2}{2!}\right) + \cdots \right.$$

$$\left. + \frac{\xi^n}{n!}\left(1 + \xi' + \frac{\xi'^2}{2!} + \cdots + \frac{\xi'^n}{n!}\right) + \cdots \right] \qquad (8\text{-}7)$$

These series can be developed in a somewhat simpler manner using the method of Anzelius [A304] consideration of which will be given in Sec. 13-4.

It will be noted that, according to Eq. (8-3), because $LL' = F$, when $x = L$, ξ becomes kF/C and when $x' = L'$, ξ' becomes equal to kF/C'. Accordingly the dimensions of the planes which represent the tube walls in the mathematical model and illustrated in Fig. 8-9, are represented in dimensionless form by the quantities kF/C and kF/C'. On the other hand these two dimensionless groups can be calculated in a straightforward manner from the dimensions and the fluid velocities for the recuperator working in cross flow under consideration, using the method described in Sec. 6-2. In the case of tube assemblies it is necessary to base the heat transfer coefficient appropriate to cross flow in the outer area, using the equations given in Sec. 2-11.

Temperature Distribution and Heat Transfer in Pure Cross Flow

Illustrated in Fig. 8-10 is the temperature distribution in pure cross flow, as calculated using the approximate method described in Sec. 16-3 to Sec. 16-5. It is assumed here that both gases, related to unit time, have the same thermal capacities $C = C'$, and that the quantities involved in determining the heat transfer yield kF/C and $k'F/C' = 10$. It is simplest to imagine that the gases flow past both sides of a plane as shown in Fig. 8-9.

The dimensionless longitudinal coordinate ξ is plotted on the horizontal axis, ξ' is plotted as the third dimension and the temperatures θ and θ' of the gases are

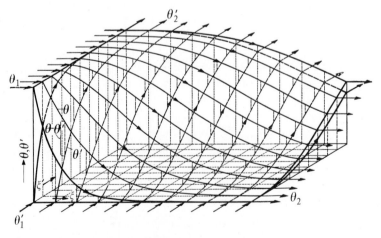

Figure 8-10 Temperature distribution for pure cross flow.

plotted on the vertical axis. The lines with arrows represent the temperature variation of a specific section of gas on its way through the heat exchanger. The arrows indicate the direction of flow. Thus gas I flows from left to right and gas II flows from the front to the back.

Temperature differences $\theta - \theta'$ are indicated as dashed, vertical lines. This diagram clearly illustrates the physical peculiarity of cross flow, when the following is taken into account. In parallel flow and counterflow it is known that all sections of gas undergo the same temperature changes on their parallel routes through the heat exchanger. In contrast to this, Fig. 8-10 shows that, in cross flow, sections of gas II which pass at various positions over the front edge of the partition wall, have different temperatures to those of gas I and are thus heated at different speeds and attain different final temperatures. The section flowing at $\xi = 0$ is heated the most; this encounters in its entirety, the high entry temperature θ_1 of gas I. The smallest temperature change occurs at $\xi = kF/C = 10$. The individual sections of gas I all undergo the same, differing temperature changes but are cooled down rather than heated up.

Heat Transfer and Efficiency in Cross Flow

It can be recognized from Fig. 8-10 that more heat will be transferred in the case illustrated than in parallel flow; this is because on average most of gas I is cooled below the mean value $\dfrac{\theta_1 + \theta'_1}{2}$ of the entry temperatures and similarly most of gas II is heated above this same mean value. The rate $\dot{Q}$ at which heat is exchanged using Eq. (5-31), is determined if the average exit temperature θ_2 or θ'_2 is computed from the exit temperatures for all section of gas I or gas II using the equations

$$\theta_2 = \frac{C'}{kF} \int_0^{kF/C'} \theta \, d\xi' \qquad \text{when } \xi = \frac{kF}{C} \qquad \text{and}$$

$$\theta'_2 = \frac{C}{kF} \int_0^{kF/C} \theta' \, d\xi \qquad \text{when } \xi' = \frac{kF}{C'}$$

This averaging process can be carried out by simple integration of Eq. (8-6) or (8-7). However, if, as in Fig. 8-10, the temperature distribution has been determined using one of the methods described in Sec. 15-1 to Sec. 16-5 in Part Four, the average exit temperatures can be obtained by numerical quadrature, for example by means of the Simpson rule.

Equations (5-66) and (5-67) which were developed for parallel flow and counterflow, remain valid, unchanged, for the efficiency η_w of the exchange of heat in cross flow, as long as C and C' are again assumed to be constant. Thus the efficiency function

$$\eta^* = \frac{\theta_1 - \theta_2}{\theta_1 - \theta'_1} \tag{8-8}$$

can also be determined for cross flow so that, in this case,

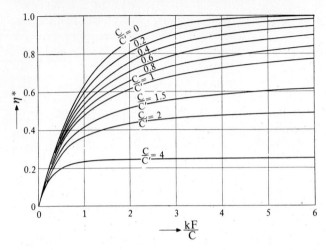

Figure 8-11 Efficiency function η^* in pure cross flow.

$$\eta_w = \eta^* \text{ is applicable when } C \leqslant C' \qquad (8\text{-}9)$$

$$\eta'_w = \frac{C}{C'} \eta^* \text{ is applicable when } C \geqslant C' \qquad (8\text{-}10)$$

In Fig. 8-11 the efficiency function η^* for cross flow is plotted as a function of kF/C for various ratios of C/C'. This diagram can similarly be used for calculating the heat exchanged, as has already been explained for parallel flow and counterflow in Figs. 5-20 and 5-21.

The average temperature difference $\Delta\theta_M$ can also be determined for pure cross flow so that Eq. (5-20) is satisfied

$$\dot{Q} = kF\,\Delta\theta_M \qquad (8\text{-}11)$$

In contrast to parallel flow and counterflow, $\Delta\theta_M$ cannot be calculated only from $\Delta\theta_a$ and $\Delta\theta_b$ (see Eq. (5-56)), because, in cross flow, $\Delta\theta_M$ is also dependent upon the difference $\theta_1 - \theta'_1$ of the entry temperatures. However, since on the basis of Eqs. (8-11), (8-8) and (5-31), the relationship

$$\frac{\Delta\theta_M}{\theta_1 - \theta'_1} = \frac{\eta^*}{\dfrac{kF}{C}} \qquad (8\text{-}12)$$

can be applied, unchanged, to cross flow, Fig. 8-11 can be used for determining the average temperature difference $\Delta\theta_M$ in cross flow.

Kühne [K219, 221] has presented a graph for directly determining $\Delta\theta_M$ from the entry and exit temperatures. Figure 8-12 shows this diagram derived from precise calculations undertaken by Rötzel [R207]. Here $\dfrac{\theta'_2 - \theta'_1}{\theta_1 - \theta'_1}$ is drawn as the

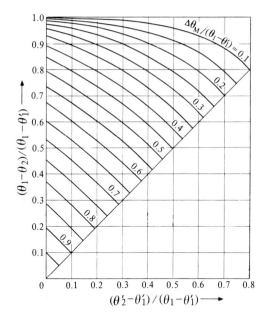

Figure 8-12 The Kühne diagram for determining the average temperature difference $\Delta\theta_M$ in pure cross flow, based on the precise calculations of Roetzel.

abscissa and $\dfrac{\theta_1-\theta_2}{\theta_1-\theta_1'}$ is drawn as the ordinate; θ_2 and θ_2' denote the average exit temperatures of the gases. It is assumed that $C' \geqslant C$. It will be seen that $\Delta\theta_M/(\theta_1-\theta_1')$ can be read off from the curves and the value of $\Delta\theta_M$ can easily be determined from this.

Comparisons Between Parallel Flow, Counterflow, and Pure Cross Flow

In Fig. 8-13 the efficiency $\eta_w(=\eta^*)$ when $C = C'$ is shown as a function of kF/C for pure cross flow, parallel flow and counterflow in order to compare the effectiveness

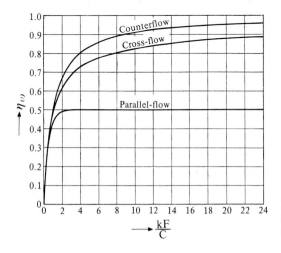

Figure 8-13 Efficiency η_w of a heat exchanger in counterflow, parallel flow and cross flow when $C = C'$.

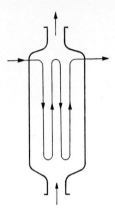

Figure 8-14 Cross flow with parallel flow as a separate process.

of these three types of flow. This confirms the outcome of earlier discussions where, for a given value of kF/C, counterflow always emerges as the most favourable mode of operation for heat transfer. However, pure cross flow is superior to parallel flow: this is because with cross flow, the efficiency tends to the limit, 1, with increasing value of kF/C, but the rate of convergence to this limit is still slower than in the case of counterflow. The fact that a considerable difference exists between cross flow and counterflow can be recognized, for example, from the fact that, according to Fig. 8-13, for an efficiency of 90 per cent the value of kF/C must be three times larger for pure cross flow than for counterflow. Therefore, with specified values of k and C the heating surface area F must be three times larger for cross flow. With tube assemblies this difference is, in practice, smaller, because, as has already been mentioned, higher heat transfer coefficients arise on the outside of the tubes for cross flow than flow parallel to the axes of the tubes.

8-3 CROSS FLOW COMBINED WITH PARALLEL FLOW IN THE CROSS-COUNTERFLOW HEAT EXCHANGER[2]

As has already been discussed in Sec. 8-1, parallel flow and cross flow often occur together. Here, either cross flow represents the main process and parallel flow a separate process, or vice-versa. In Fig. 8-14 one gas flows through a tube spiral which, in the main, consists of straight pieces.

These straight pieces lie in the direction of the flow of the outside gas. As all the individual paths of the gas flowing inside the tubes combine to form a total path from left to right, the main flow of the gas corresponds to pure cross flow. Meanwhile, flow parallel to the direction of the outside gas takes place in the individual straight pieces of the tube spiral in the form alternately, of parallel flow and counterflow. The efficiency of such an arrangement generally lies somewhere

[2] The author's considerations of this matter and the relevant calculations were first published in the first edition of this book.

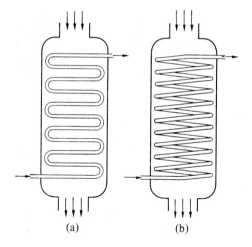

Figure 8-15a and b Counterflow with cross flow as a separate process. (a) With opposing directions. (b) With the same directions.

(a)   (b)

between that of pure cross flow and that of parallel flow. Heat exchangers of this type represent a border-line case for recuperators consisting of several passages joined in series, which are discussed in Sec. 9-1. The theories relating to this type of recuperator will therefore be discussed in this section (as a limiting case for an infinite number of passages).

Cross-counterflow Heat Exchangers

The combination of cross and counterflow is most efficient when the individual parts of the exchanger operate in cross flow while the overall process is counterflow. This occurs most often in the mixed switching discussed in Sec. 8-1. When such heat exchangers are built of tubes, the significant increase in heat transfer, which has already been mentioned several times, occurs on the outside of the tubes. Two typical arrangements are illustrated in Fig. 8-15.

The gas flowing inside, moves for example, through a tube spiral, all the sections of which, up to the tube bends, lie perpendicularly, or almost so, to the direction of flow of the outside gas. Although essentially, cross flow exists at each individual position, the resulting total movement of the gas flowing inside is parallel and opposite to the fluid direction of the gas in the outer area. The difference between the two arrangements in Fig. 8-15 (a) and (b) results from the following considerations.

Counter Directed and Unidirected Flow in Cross-Counterflow*

If the straight parts of the tube spiral are arranged in a plane, as in Fig. 8-15(a), the gas flowing inside the tubes alters its direction from one section of tube to another. Thus, "counter-directional flow" is said to be manifest. This also applies when

---

* Editor's note: This is an attempted translation of "Gegensinnige und gleichsinnige Führung bei Kreuz-Gegenstrom".

several such tube spirals lie one behind the other. Examples of counter-directional flow are shown in Figs. 8-4, 8-5 and 8-6. The arrangement with baffles shown in Fig. 8-7 also works in a basically similar manner. Here the outside gas is directed in cross flow, against the straight tube assembly, whereby the direction of cross flow changes periodically. However, if, as is shown in the arrangement in Fig. 8-15(b), the individual tube spirals are wound helically, the gas in all the vertical tube sections lying one above another will flow in the same direction. This is then said to be "unidirectional flow". The most important, and indeed the most complete example of this is given by the cross-counterflow heat exchanger, which is most of all used in low temperature technology; see Fig. 8-8.

The difference mentioned between unidirectional and counter-directional flow corresponds to the difference between unidirectional and counter-directional liquid flow in rectification. This is because in a rectifying column the main process always consists of counterflow while the particular process on the plates consists of the cross flow conduction of liquid and steam; see [H206].

It turns out that unidirectional flow is superior to counter-directional flow although the differences are usually only very slight.

Cross-Counterflow Heat Exchanger with a Large Number of Coils

If with the aid of Fig. 8-15, for example, the temperature changes in the outside gas, which it is assumed is the colder gas, were to be followed, the following picture would emerge. The temperature difference between this gas and the warmer gas flowing inside is at its greatest each time the gas impinges on a tube. This temperature difference is then reduced by the exchange of heat which then takes place. When the outside gas flows past a tube its temperature θ' increases rapidly at first, and then more slowly.[3]

The temperature θ' remains unchanged over the short path to the next section of tube. The temperature of the gas flowing inside can, on the contrary, be considered reasonably accurately to be constant within a tube cross-section. When plotted as a function of the flow route, the temperature curve which results is composed of steps, as is shown in Fig. 8-16.

If the number of tube coils against which the outside gas impinges is large, the individual steps are only very small. In this case, the lines for θ' and θ in Fig. 8-16 can be replaced by smooth curves, fairly accurately, as for parallel flow or counterflow (see Figs. 5-15 and 5-16). From a practical viewpoint the relationships for pure parallel flow are applicable and the differences between unidirectional and counter-directional flow disappear. Cross-counterflow heat exchangers with a large number of tube coils can thus be calculated using essentially the same process that was developed in Chapters Five and Six for pure counterflow. Care must merely be taken that the relationships given for cross flow in Sec. 2-11 and Sec. 4-4 are used for the heat transfer coefficients and pressure drop in the outer area.

---

[3] At least when α' is uniformly equal, which does not in fact occur exactly.

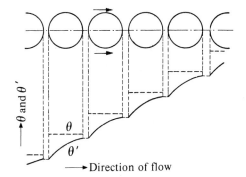

θ and θ'

θ

θ'

——→ Direction of flow

Figure 8-16 Curve of the temperature θ' of an outside gas in a cross-counterflow heat exchanger.

Cross-Counterflow Heat Exchangers with a Small Number of Tube Coils

When the number of tube coils is relatively small, the change in temperature of the outside gas at a coil in the tube can be of the same order of magnitude as the temperature difference which originally existed between the two gases at the point of the tube coil under consideration. At the position indicated by the arrow in Fig. 8-17 the outside gas again has the temperature θ' before impinging on the tube. If a complete exchange of heat could be achieved by flowing past the tube, the outside gas would assume the temperature θ of the inside gas. In a theoretically ideal case, according to this, the outside gas would undergo an increase in temperature of $\theta - \theta'$.

However, this optimum exchange of heat can only be achieved if either the heat transmission coefficient k is infinitely large or each tube coil has an infinitely large heating surface area ΔF. In pure counterflow, on the contrary, a finite value of $k\,\Delta F$ is sufficient for the same temperature change. In this context cross flow becomes more inferior to counterflow the more the temperature change at a tube coil approaches the temperature difference $\theta - \theta'$.

The temperature distribution in a tube coil in a cross-counterflow heat exchanger will be calculated next in order to pursue, numerically, this difference between cross-counterflow and pure counterflow.

Temperature Distribution in a Tube Coil in a Cross-Counterflow Heat Exchanger

Figure 8-18 represents the nth coil, (the top coil being the 1st), which, in the case of the helical arrangement as shown in Fig. 8-15(b), is thought of as a curved plane. Even when, as in Fig. 8-15(a), there are straight tube sections these will be spoken of as "tube coils" to maintain consistency. The originally warm gas enters the tube

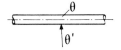

θ

θ'

Figure 8-17 Section of tube in cross-counterflow.

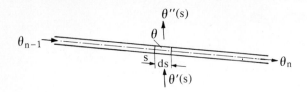

Figure 8-18 Heat transfer at a tube coil in cross-counterflow.

coil from the left at the temperature θ_{n-1}, and leaves on the right at the temperature θ_n. A position on the tube will now be considered which is at a distance s from the topmost end of the total coiled tube; this is measured along the length of the tube. The length of a tube coil will therefore be chosen as the unit for s. Thus s has a value of $n-1$ at the left-hand end of the tube coil illustrated in Fig. 8-18, and has a value n at the right-hand end. The temperature θ' of the exit gas is also, in general, dependent on s. It is denoted by $\theta'(s)$ on the diagram just below the coil at the position s being considered.

The average value of all temperatures $\theta'(s)$ below the coil being considered is equal to the average temperature of the total quantity of rising gas below this coil. This average temperature is θ'_n. Thus

$$\int_{n-1}^{n} \theta'(s)\,ds = \theta'_n \tag{8-13}$$

Furthermore, if C' is again the thermal capacity of the quantity of gas flowing past outside the coil in unit time, the thermal capacity of the quantity of gas ascending between s and $s+ds$ totals $C'\,ds$. This quantity of gas with an initial temperature of $\theta'(s)$ has the temperature $\theta''(s)$ when it has completed its exchange of heat with the tube coil. The gas flowing inside with the thermal capacity C is thus cooled by $d\theta$. According to this, the relationship

$$C'\,ds[\theta''(s)-\theta'(s)] = -C\,d\theta \tag{8-14}$$

represents the rate at which heat is transferred between s and $s+ds$. If a complete exchange of heat is to be achieved, the gas flowing outside must be fully heated to the temperature θ of the gas flowing inside. The ratio ε of the temperature change which is actually achieved to its theoretically highest value:

$$\varepsilon = \frac{\theta''(s)-\theta'(s)}{\theta-\theta'(s)} \tag{8-15}$$

is known as the efficiency per unit of exchange. The value of ε can generally be considered to be constant for all positions on the tube, as long as each of the heat transfer coefficients α and α' on the inner and outer sides of the tube has the same value overall. If $\theta''(s)-\theta'(s)$ from Eq. (8-15) is inserted into the previous equation, the following *differential equation is obtained for the temperature θ of the gas flowing inside the tube:*

$$\frac{d\theta}{ds} + \frac{C'}{C}\varepsilon[\theta-\theta'(s)] = 0 \tag{8-16}$$

Boundary Conditions

The differential Eq. (8-16) can only be solved if the temperature $\theta'(s)$ is known at all positions s below the tube coil being considered. For the bottommost tube coil $\theta'(s)$ is equal at all times to the temperature θ'_1, at which the outside gas enters the heat exchanger. In this case the solution

$$\theta(s) - \theta_2 = (\theta_2 - \theta_1)\left\{\exp\left[\varepsilon \frac{C'}{C}(n - s)\right] - 1\right\} \tag{8-17}$$

is obtained where the average temperature θ'_n of the gas ascending outside is assumed to be equal to θ'_1 and the end temperature θ_n of the gas flowing inside is assumed to be equal to the exit temperature θ_2. In this case n is equal to the number of the bottommost coil and is thus equal to the total number N of all coils.

On the other hand $\theta'(s)$ is unknown at first for the other coils and must therefore be determined for a particular boundary condition. This follows from the consideration that, for each position on a tube coil, the end temperature $\theta''(s)$ of the gas flowing outside is equal to the initial temperature for the next tube up. The relationship which results from this varies, depending on whether the gas flowing inside the successive tube coils flows in the same or the opposite direction. As can be seen from Figs. 8-19 and 8-20 the position $s - 1$ lies above the position s in the nth coil in the first case and position $2(n - 1) - s$ lies above position s in the second case.

If the temperature of the gas immediately before it impinges against such a higher position is denoted by $\theta'(s - 1)$ or $\theta'(2(n - 1) - s)$, we obtain;

for unidirectional flow

$$\theta''(s) = \theta'(s - 1) \tag{8-18}$$

for counter-directional flow

$$\theta''(s) = \theta'(2(n - 1) - s) \tag{8-19}$$

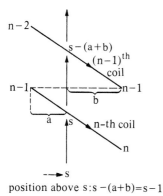

position above $s : s - (a + b) = s - 1$

Figure 8-19 Two tube coils lying one above the other with unidirectional flow.

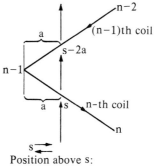

Position above s:
$s - 2a = 2(n - 1) - s$

Figure 8-20 Two tube coils lying one above the other with counter-directional flow.

Using Eq. (8-15) this *boundary condition* can be used the following way:

for unidirectional flow:

$$\theta'(s-1) = \varepsilon\theta(s) + (1-\varepsilon)\theta'(s) \tag{8-20}$$

for counter-directional flow:

$$\theta'(2(n-1)-s) = \varepsilon\theta(s) + (1-\varepsilon)\theta'(s) \tag{8-21}$$

Solutions of the Differential Equations

The temperature distribution in the various tube coils of the cross-counterflow heat exchanger can basically be calculated in such a way that, beginning with the bottommost coil, corresponding to each coil the solution of the differential Eq. (8-16) is sought and is then linked, using the boundary condition (8-20) or (8-21), to the next coil higher up. It would appear here that the temperature θ'_1 which is independent of s, and which is the temperature of the gas flowing on the outside below the bottommost coil, will indeed influence the temperature distribution in the next coils up but that this influence will very quickly fade. Thus, after only a few coils, a temperature distribution develops which is, on the whole, fairly stable but which still varies from coil to coil but which is no longer dependent on the temperature distribution beneath the bottommost coil, i.e., no longer dependent on what has gone before. A solution of the differential Eq. (8-16) can be immediately found for this component of the temperature distribution by utilizing the boundary condition (8-20) or (8-21).[4] In what follows this solution will only be considered because it alone almost totally characterizes the complete behavior of the cross-counterflow heat exchanger while the behavior of the bottommost coil, which deviates from this, can be neglected as unimportant.

If, for example, the temperature distribution in the higher coils with constant values of C and C' were to be considered diagrammatically, one could quickly surmise that the processes in the successive coils are similar to one another. This means that the ratio of the corresponding temperature differences for two coils lying one above another is constant. Thus the statement

$$\frac{\theta'(s-1)-\theta'_{n-1}}{\theta'(s)-\theta'_n} = \frac{\theta_{n-1}-\theta'_{n-1}}{\theta_n-\theta'_n} = 1+\lambda \tag{8-22}$$

can be made about the nth coil and the $(n-1)$th coil lying above it, where the numerical value λ is considered to be constant and θ_n and θ_{n-1} again denote the average temperatures of the outside gas below the nth or $(n-1)$th coils. Equation (8-23) is obtained by rearranging Eq. (8-22):

$$\theta'(s-1) = \theta'_{n-1} + (1+\lambda)(\theta'(s)-\theta'_n) \tag{8-23}$$

---

[4] The derivations of the relationships obtained are more or less the same as those for rectification in "linear unbalancing". See H. Hausen [H206].

If the heat balance equation for the nth coil is considered,

$$C'(\theta'_{n-1} - \theta'_n) = C(\theta_{n-1} - \theta_n) \tag{8-24}$$

the differential Eq. (8-16) can be integrated as follows using the boundary conditions (8-20) or (8-21) and Eqs. (8-23) and (8-24).

1. Unidirectional Flow

If, first of all, a comparison of Eqs. (8-20) and (8-23) is made, we obtain:

$$\theta'(s) = \frac{(1 + \lambda)\theta'_n - \theta'_{n-1} + \varepsilon\theta(s)}{\lambda + \varepsilon} \tag{8-25}$$

If this is inserted into the differential Eq. (8-16), by integrating between s and n or between $\theta = \theta(s)$ and θ_n, the solution

$$\theta - \theta_n = (\theta_n - \theta'_n)\frac{C'}{C' - C}\left\{\exp\left[\frac{C'}{C} \cdot \frac{\lambda\varepsilon}{\lambda + \varepsilon}(n - s)\right] - 1\right\} \tag{8-26}$$

is obtained. λ can be determined by using this equation for $s = n - 1$ and correspondingly for $\theta = \theta_{n-1}$. Taking into consideration the second relationship (8-22) and Eq. (8-24) the following equation in λ is yielded:

$$\frac{C'}{C} = \frac{\lambda + \varepsilon}{\lambda\varepsilon}\ln(1 + \lambda) \tag{8-27}$$

This equation can also be derived if, upon taking into consideration Eq. (8-26), Eq. (8-25) is inserted into Eq. (8-13). Moreover, in this way the correctness of the expression (8-22) can be checked. Given ε and C/C', λ can be determined iteratively from Eq. (8-27). It follows from Eqs. (8-26) and (8-25), that the temperature distribution in and immediately below the nth coil can be determined using this value of λ.

2. Counter-Directional Flow

If s is replaced by $2n - 1 - s$ in the boundary condition (8-21), and a comparison is made with Eq. (8-23) the following expression is obtained:

$$\varepsilon\theta(2n - 1 - s) + (1 - \varepsilon)\theta'(2n - 1 - s) = \theta'_{n-1} + (1 + \lambda)(\theta'(s) - \theta'_n)$$

If, again, s is replaced here by $2n - 1 - s$ and $\theta'(2n - 1 - s)$ is eliminated from the previous equation and the new equation, the relationship (8-28) results:

$$\theta'(s) = \frac{\varepsilon\theta_n - \theta'_{n-1} + (1 + \lambda)\theta'_n}{\lambda + \varepsilon} + \frac{\varepsilon(1 - \varepsilon)}{(1 + \lambda)^2 - (1 - \varepsilon)^2}(\theta(s) - \theta_n)$$

$$+ \frac{\varepsilon(1 + \lambda)}{(1 + \lambda)^2 - (1 - \varepsilon)^2}(\theta(2n - 1 - s) - \theta_n) \tag{8-28}$$

With this the differential Eq. (8-16) takes the form:

$$\frac{C}{C'}\frac{\lambda+\varepsilon}{\varepsilon}\frac{d(\theta(s)-\theta_n)}{ds}+\frac{C'\lambda}{C'-C}(\theta_n-\theta'_n)+\frac{(1+\lambda)^2-(1-\varepsilon)}{2+\lambda-\varepsilon}(\theta(s)-\theta_n)$$
$$-\frac{(1+\lambda)\varepsilon}{2+\lambda-\varepsilon}(\theta(2n-1-s)-\theta_n)=0 \tag{8-29}$$

This equation is solved using the expression

$$\theta(s)-\theta_n = A_1\exp(rs)+A_2\exp[r(2n-1-s)]+A_3$$

where A_1, A_2, A_3 and r are constants. With a slight transformation this can also be expressed using three other constants $B_1, B_2,$ and B_3:

$$\theta(s)-\theta_n = B_1\sinh r(s-n+\tfrac{1}{2})+B_2\cosh r(s-n+\tfrac{1}{2})+B_3$$

Thus Eq. (8-30) is obtained as a solution of the differential Eq. (8-29) where:

$$\theta-\theta_n = (\theta_n-\theta'_n)\frac{C'}{C'-C}\left[\frac{\sinh r\left(s-n+\dfrac{1}{2}\right)-\dfrac{\lambda+\varepsilon}{\lambda\varepsilon}\dfrac{C}{C'}r\cosh r\left(s-n+\dfrac{1}{2}\right)}{\sinh\dfrac{r}{2}-\dfrac{\lambda+\varepsilon}{\lambda\varepsilon}\dfrac{C}{C'}r\cosh\dfrac{r}{2}}-1\right]$$

with
$$\tag{8-30}$$

$$r = \pm\frac{C'}{C}\varepsilon\sqrt{\frac{(1+\lambda)^2-1}{(1+\lambda)^2-(1-\varepsilon)^2}} \tag{8-31}$$

In this way the constant of integration is so determined that, when $s = n$, $\theta = \theta(s)$ becomes θ_n. Further if $s = n-1$ and $\theta = \theta_{n-1}$ are substituted into Eq. (8-30), when $C'/C > 1(\lambda > 0)$, the non-linear equation for λ (8-32) can be evolved.

$$\frac{C'}{C} = \frac{\operatorname{artanh}\left(\dfrac{\lambda+\varepsilon}{\lambda+2}\sqrt{\dfrac{(1+\lambda)^2-1}{(1+\lambda)^2-(1-\varepsilon)^2}}\right)}{\dfrac{\varepsilon}{2}\sqrt{\dfrac{(1+\lambda)^2-1}{(1+\lambda)^2-(1-\varepsilon)^2}}} \tag{8-32}$$

This relationship can be derived from Eq. (8-13). λ can again be determined iteratively using Eq. (8-32). Using this value of λ Eqs. (8-30) and (8-28) provide the required temperature distribution.

The same relationships are valid when $C'/C < 1$ ($\lambda < 0$) if, in Eqs. (8-31) and (8-32), the imaginary root is replaced by its absolute value, if in Eq. (8-30) sinh and cosh are replaced by sin and cos and if in Eq. (8-32) artanh is replaced by arctan.

Special Case $C = C'$

If $C = C'$, $\lambda = 0$ both in unidirectional and counter-current gas flow. In this case the equations developed so far are considerably simplified and this can best be

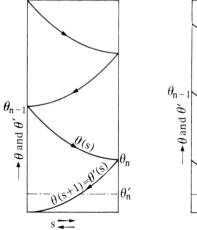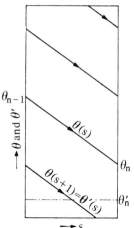

Figure 8-21 Temperature distribution in a cross-counterflow heat exchanger with $C = C'$ and $\varepsilon = 1$. Left: counter-directional flow; right: unidirectional flow.

illustrated by expanding the differential Eq. (8-16) for $\lambda = 0$. Equation (8-33) is obtained in place of Eq. (8-26) for unidirectional flow:

$$\theta - \theta_n = (\theta_n - \theta'_n) \frac{2\varepsilon}{2 - \varepsilon} (n - s) \tag{8-33}$$

and Eq. (8-34) is obtained in place of Eq. (8-30) for counter-directional flow:

$$\theta - \theta_n = (\theta_n - \theta'_n) \frac{3}{2} \frac{\varepsilon}{1 + (1 - \varepsilon)(2 - \varepsilon)} (2(1 - \varepsilon)(n - s) + \varepsilon(n - s)^2) \tag{8-34}$$

The values of $\theta'(s)$ can be derived from Eq. (8-25) or (8-28) with $\lambda = 0$.

Figure 8-21 illustrates the temperature distribution for $C = C'$ and $\varepsilon = 1$ as a function of s for unidirectional and counter-directional flow in several tube coils lying one above another. The ordinate not only gives the temperature θ of the inside gas but also gives the temperature $\theta''(s)$ for $\varepsilon = 1$, that is the temperature taken up by the outside gas after the exchange of heat with the tube coil. The straight lines with unidirectional flow are remarkable; however, with $C \neq C'$, these lines become curved although they are of a similar form.

It can be seen that the inside and thus also the outside gas in or at each tube coil, undergoes a greater temperature change in unidirectional flow than in counter-directional flow. This superiority of unidirectional flow is essentially due to the fact that with this type of flow when $C = C'$ the temperature difference between two tube coils lying one above another is of the same overall size, while with counter-directional flow the difference is small in some parts of the coil and towards one end, it even reduces to zero. It follows that with counter-directional flow a smaller temperature difference is available, on average, for heat transfer, than with unidirectional flow.

Comparison of Cross-Counterflow with Pure Counterflow

It has already been indicated that pure counterflow is basically superior to cross flow insofar as with counterflow a specific rate of heat transfer $\Delta \dot{Q}$ can be achieved with a smaller value of $k\,\Delta F$ than with cross-counterflow. This can now also be demonstrated numerically on the basis of the relationships which have been obtained. To this end it is necessary to calculate the value of $(k\,\Delta F)_K$ for a tube coil in a cross-counterflow heat exchanger and to compare it with the value of $(k\,\Delta F)_G$ with which, with pure counterflow but otherwise the same operating conditions, the same transfer of heat can be achieved.

At a position lying between s and $s + ds$ on the cross flow coil in Fig. 8-18, the rate of transfer of heat is given by Eq. (8-35) namely:

$$C'\,ds[\theta''(s) - \theta'(s)] = (k\,\Delta F)_K \cdot ds \cdot \Delta \theta_M(s) \tag{8-35}$$

(see also Eq. (8-14)), where $(k\,\Delta F)_K\,ds$ denotes the product of the heat transmission coefficient and the heating surface area for the element ds of the tube coil and $\Delta \theta_M(s)$ denotes the average temperature difference between both the gases at position s in the temperature range from $\theta'(s)$ to $\theta''(s)$. As the initial temperature difference at the position under consideration is equal to $\theta - \theta'(s)$, the temperature difference after the exchange of heat is equal to $\theta - \theta''(s)$; using Eq. (5-56), Eq. (8-36) is obtained for the average temperature difference at position s:

$$\Delta \theta_M(s) = \frac{\theta''(s) - \theta'(s)}{\ln \dfrac{\theta - \theta'(s)}{\theta - \theta''(s)}} \tag{8-36}$$

From Eq. (8-15)

$$\Delta \theta_M(s) = \frac{\theta''(s) - \theta'(s)}{-\ln(1 - \varepsilon)}$$

can also be developed. With this, Eq. (8-35) is converted into

$$\frac{(k\,\Delta F)_K}{C'} = -\ln(1 - \varepsilon) \tag{8-37}$$

Thus, with cross-counterflow, if ε, C and C' are given, $(k\,\Delta F)_K$ can also be determined using Eq. (8-37).

In addition it is necessary to calculate the value $(k\Delta F)_G$ for pure counterflow which gives rise to the same temperature variation $\theta'_{n-1} - \theta'_n$ as the cross flow coil under consideration. Equation (8-38) represents the exchange of heat with counterflow

$$C'(\theta'_{n-1} - \theta'_n) = (k\,\Delta F)_G \cdot (\Delta \theta_M)_n^{n-1} \tag{8-38}$$

where $(\Delta \theta_M)_n^{n-1}$ denotes the average temperature difference of both gases between the temperatures θ'_n and θ'_{n-1}. While, with cross flow, θ'_n and θ'_{n-1} denote average temperatures corresponding to Eq. (8-13), with counterflow they represent the total gas temperatures because in this case no temperature differences exist between the particles of gas in the same cross-section. It follows from Eq. (5-56) that:

$$(\Delta\theta_M)_n^{n-1} = \frac{(\theta_n - \theta'_n) - (\theta_{n-1} - \theta'_{n-1})}{\ln \dfrac{\theta_n - \theta'_n}{\theta_{n-1} - \theta'_{n-1}}} \tag{8-39}$$

Equation (8-40) follows from Eq. (8-38) upon taking into account Eqs. (8-22) and (8-24):

$$\frac{(k\,\Delta F)_G}{C'} = \frac{\ln(1+\lambda)}{\dfrac{C'}{C} - 1} \tag{8-40}$$

The required value of $(k\,\Delta F)_G$ can then be determined using this equation.

In the special case where $C = C'$, taking into account Eqs. (8-22) and (8-24), Eq. (8-40) is converted into Eq. (8-41) as λ becomes infinitely small:

$$\frac{(k\,\Delta F)_G}{C'} = \frac{\lambda}{\dfrac{C'}{C} - 1} = \frac{\theta_{n-1} - \theta_n}{\theta_n - \theta'_n} \tag{8-41}$$

If the temperature difference ratio which occurs here is calculated using Eq. (8-33) or (8-34), Eq. (8-42) will be obtained for co-current directional flow from Eq. (8-41)

$$\frac{(k\,\Delta F)_G}{C'} = \frac{2\varepsilon}{2-\varepsilon} \tag{8-42}$$

and Eq. (8-43) will be obtained for counter-current directional flow.

$$\frac{(k\,\Delta F)_G}{C'} = \frac{3}{2} \frac{\varepsilon(2-\varepsilon)}{1+(1-\varepsilon)(2-\varepsilon)} \qquad \text{(with } C = C') \tag{8-43}$$

Limiting Cases Where $\varepsilon = 0$ and $\varepsilon = 1$

When ε is *infinitely small*, Eqs. (8-27) and (8-32) can only be satisfied if λ is also assumed to be infinitely small. In this case both equations adopt the same form

$$\frac{C'}{C} = \frac{\lambda + \varepsilon}{\varepsilon}$$

from which follows:

$$\lambda = \varepsilon\left(\frac{C'}{C} - 1\right)$$

By inserting this into Eq. (8-40) with an infinitely small ε, and noting that $\lim_{\lambda=0} \ln(1+\lambda) = \lambda$, we obtain:

$$\frac{(k\,\Delta F)_G}{C'} = \varepsilon \qquad \text{(for } \lim \varepsilon = 0) \tag{8-44}$$

Correspondingly, there follows from Eq. (8-37) for an infinitely small ε:

$$\frac{(k\,\Delta F)_K}{C'} = \varepsilon \qquad \text{(for lim } \varepsilon = 0) \qquad (8\text{-}45)$$

Thus $(k\,\Delta F)_G$ and $(k\,\Delta F)_K$ are equal to one another for very small values of ε. Accordingly cross-counterflow and pure counterflow are identical to one another so long as the temperature change of the outside gas at each position on the tube coil is small compared with the temperature difference between the two gases, in which case, the coil coefficient is very large.

According to Eq. (8-37), $(k\,\Delta F)_K$ is infinitely large in the limiting case where $\varepsilon = 1$. This is physically significant in that, as has already been mentioned, only one finite temperature can be reached in a cross-counterflow heat exchanger coil, even if the heating surface area of the coil or the heat transmission coefficient is made infinitely large. In contrast, in counterflow the same rate of heat transfer can be achieved with a finite value of $(k\,\Delta F)_G$.

Efficiency Factor for Cross-Counterflow

In order to examine numerically the relationship between cross-counterflow and pure counterflow, even for the most general cases, the cross-counterflow factor

$$\eta_K = \frac{(k\,\Delta F)_G}{(k\,\Delta F)_K} \qquad (8\text{-}46)$$

is introduced. By combining this factor with Eqs. (8-37) and (8-40), the essential conclusion to be drawn from this equation is that the efficiency factor has the same value for all coils, given constant values of C', C and ε and thus also λ, provided these have the same heat transmission coefficient k and the same heating surface area ΔF_K. Thus, instead of relating the factor to ΔF_K and ΔF_G it can also be expressed in terms of the total heating surface area of the cross-counterflow heat exchanger or the equivalent pure counterflow device and can be written

$$\eta_K = \frac{(kF)_G}{(kF)_K} \qquad (8\text{-}47)$$

The practical significance of this efficiency factor lies in the fact that, once its value is known, a cross-counterflow heat exchanger can be calculated in the same manner as a pure counterflow device. Subsequently all that is necessary is simply to divide the value of $(kF)_G$ obtained, by η_K in order to find $(kF)_K$ for the cross-counterflow heat exchanger.

The efficiency factor η_K can be developed somewhat differently on the basis of the following considerations. The exchange of heat in a pure counterflow heat exchanger is determined using Eq. (5-20). For the sake of clarity it is now necessary to write $(\Delta\theta_M)_G$ instead of $\Delta\theta_M$ in this equation for the average temperature difference in the total heat exchanger. Correspondingly an average temperature difference $(\Delta\theta_M)_K$ for the cross-counterflow exchanger is defined by the expression

$$\dot{Q} = (kF)_K \cdot (\Delta\theta_M)_K \tag{8-48}$$

where $\dot{Q}$ should have the same value as for pure counterflow. Equation (8-49) is thus obtained from Eq. (8-47) together with Eq. (5-20) and Eq. (8-48):

$$\eta_K = \frac{(\Delta\theta_M)_K}{(\Delta\theta_M)_G} \tag{8-49}$$

Thus in cross-counterflow the average temperature difference is smaller relative to the efficiency factor η_K than in pure counterflow. The large value of $kF = (kF)_K$ in cross-counterflow, as calculated using Eq. (8-47), is thus required to balance the small value of the average temperature difference.

In calculating the efficiency factor η_K it is most important to note that according to Eq. (8-46) and the relationships obtained for $(k\,\Delta F)_G$ and $(k\,\Delta F)_K$, namely (8-37) and (8-40), η_K is a function of ε (or λ) and C'/C. However as ε cannot be estimated directly, a value ψ can be introduced in place of ε and this value can be determined easily from the entry and exit temperatures of the gases. ψ is the ratio of the temperature change $\theta'_{n-1} - \theta'_n$ of the outside gas at one coil of the cross-counterflow heat exchanger to the average temperature difference $(\Delta\theta_M)_n^{n-1}$ which exits there, and this temperature difference can be calculated as in counterflow, by utilizing Eq. (8-39). The following equation is thus derived:

$$\psi = \frac{\theta'_{n-1} - \theta'_n}{(\Delta\theta_M)_n^{n-1}} \tag{8-50}$$

Using Eq. (8-38), (8-51) is obtained:

$$\psi = \frac{(k\,\Delta F)_G}{C'} \tag{8-51}$$

However this expression needs to be brought into a more convenient form for calculation purposes. Let the number of all the tube coils lying one on top of another (Fig. 8-15) be N. As $kF = (kF)_G = N \cdot (k\,\Delta F)_G$ for the equivalent pure counterflow exchanger, because all the tube coils are assumed to be identical, using Eqs. (5-20) and (5-31) the rate of transfer of heat is given by

$$\dot{Q} = C'(\theta'_2 - \theta'_1) = N(k\,\Delta F)_G \cdot (\Delta\theta_M)_G \tag{8-52}$$

Using this Eq. (8-51) can be manipulated into the form:

$$\psi = \frac{\theta'_2 - \theta'_1}{N(\Delta\theta_M)_G} \tag{8-53}$$

Thus, it follows from these equations that ψ can be determined simply by dividing the total temperature change $\theta'_2 - \theta'_1$ of the outside gas by the average temperature difference $(\Delta\theta_M)_G$ which is obtained for counterflow using Eq. (5-56) and by the number N of cross flow coils lying one above another.

Now, in order to represent the efficiency factor η_K of the cross-counterflow heat exchanger as a function of ψ and C/C', it must first be noted that, according to Eqs.

(8-51) and (8-40) the relationship (8-54) also applies to ψ namely:

$$\psi = \frac{\ln(1+\lambda)}{\dfrac{C'}{C} - 1} \tag{8-54}$$

If, upon utilizing Eq. (8-51), $(k\,\Delta F)_G$ is replaced by $C'\psi$, Eq. (8-55) can be obtained from Eq. (8-46), if at the same time note is taken of Eq. (8-37).

$$\eta_K = \frac{\psi}{\ln\dfrac{1}{1-\varepsilon}} \tag{8-55}$$

The efficiency factor of the cross-counterflow heat exchanger can be calculated for each pair of values of ψ and C/C' from Eqs. (8-54) and (8-55). For this purpose Eq. (8-54) is first solved for λ and with this, the value of ε is determined from Eq. (8-27) or (8-32). Finally this inserted into Eq. (8-55).

The complete expression which follows for the efficiency factor η_K as a function of ψ and C'/C is found, in this manner, for unidirectional flow

$$\frac{1}{\eta_K} = \frac{1}{\psi} \cdot \ln \frac{\dfrac{C'}{C}\exp\left[\left(\dfrac{C'}{C} - 1\right)\psi\right] - \left(\dfrac{C'}{C} - 1\right)\psi - \dfrac{C'}{C}}{\left[\dfrac{C'}{C} - \left(\dfrac{C'}{C} - 1\right)\psi\right]\exp\left[\left(\dfrac{C'}{C} - 1\right)\psi\right] - \dfrac{C'}{C}} \tag{8-56}$$

In contrast a corresponding expression for η_K is not given for *counter-directional flow*, except for the special case where $C = C'$, because ε can only be obtained from Eq. (8-32) iteratively.

In this special case where $C = C'$ the expressions on the right-hand side of Eqs. (8-42) and (8-43) can be applied directly to Eq. (8-51) for ψ. If these equations are solved for ε, Eqs. (8-57) and (8-58) are obtained by insertion into Eq. (8-55):

for unidirectional flow

$$\frac{1}{\eta_K} = \frac{1}{\psi}\ln\frac{2+\psi}{2-\psi} \qquad (C = C') \tag{8-57}$$

for counter-directional flow

$$\frac{1}{\eta_K} = \frac{1}{\psi}\ln\frac{3+2\psi}{\sqrt{9-3\psi^2}-\psi} \qquad (C = C') \tag{8-58}$$

The Results of this Calculation

The relatively simple results obtained above can be summarized in the following way. If, given the gas temperatures, a specific rate of heat transfer $\dot{Q}$ is required, the product $kF = (kF)_K$ should be $1/\eta_K$ times larger for a cross-counterflow heat exchanger than for a pure counterflow device built for the same performance,

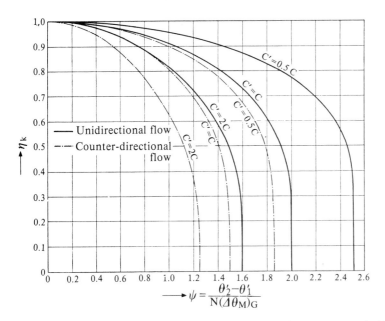

Figure 8-22 Efficiency factor η_K of cross-counterflow heat exchangers compared with pure counterflow for unidirectional and counter-directional flow. $\theta_2' - \theta_1'$ is the total temperature variation of the outside gas. $(\Delta\theta_M)_G$ is the average temperature difference between the two gases calculated for pure counterflow and N is the number of coils.

where η_K is the efficiency factor of the cross-counterflow heat exchanger. η_K is different for unidirectional flow and counter-directional flow. In addition η_K is a function of the ratio C'/C of the thermal capacities of the outside and inside gases as well as of the temperature ratio ψ mentioned above which is obtained from Eq. (8-53). Thus the value of ψ can easily be calculated from the total temperature difference $\theta_2' - \theta_1'$ of the outside gas, from the number N of the tube coils or tube pieces lying one behind another in the direction of flow, and from the average temperature difference $(\Delta\theta_M)_G$ between two gases which is determined for pure counterflow using Eq. (5-56). As η_K is dependent on C'/C and ψ, it is determined using Eqs. (8-54) to (8-58) which are derived above, as well as by Eqs. (8-27) to (8-32).

In Fig. 8-22 the values of the efficiency factor η_K calculated using these equations are shown as a function of ψ for different values of C'/C. The solid lines apply to unidirectional flow, the dot-dash lines apply to counter-directional flow. The efficiency factor η_K can thus be read off from a diagram of this type for practical calculations. According to Fig. 8-22 the efficiency factor is very close to unity when ψ is very small, as often occurs when there are a very high number of coils N. Thus, as happens very frequently, a cross-counterflow heat exchanger can be calculated accurately by treating it as a pure counterflow heat exchanger. Hence it is only necessary to take into account the fact that higher heat transfer coefficients occur on the outside of the tubes than is the case where the outside gas flows parallel to the tubes.

η_K decreases slowly at first and then more and more rapidly as ψ increases. Up to about $\psi = 1$ the decrease for counter-directional flow is roughly twice as large as it is for unidirectional flow. The following empirically based approximating equation can be used up to about $\psi = 1$:

for unidirectional flow:

$$\eta_K = 1 - \frac{C'}{C} \cdot \frac{\psi^2}{12},$$

(8-59)

with $\psi < 1$

for counter-directional flow:

$$\eta_K = 1 - \frac{C'}{C} \cdot \frac{\psi^2}{6}.$$

(8-60)

It will be seen in Fig. 8-18 that when $\psi = 1$ and $C = C'$, η_K takes the value 0.91 for unidirectional flow and 0.81 for counter-directional flow. The deterioration compared with pure counterflow expressed by these numerical values is only relatively slight if it is remembered that 1 is a very undesirable value for ψ and rarely occurs in practice. Above all it must be noted that, compared with where the gas flows parallel to the tubes, cross flow generally doubles or triples the heat transfer coefficient on the outside of the tube. The increase in k caused by this usually far outweighs the unfavourable influence of η_K.

8-4 HEAT TRANSFER THROUGH FINNED TUBES

Often the heat transfer coefficient on one side of a heat transferring wall is considerably smaller than that on the other side, for example when a gas is flowing on the one side and a liquid is flowing on the other. The difference can, in part, be counter balanced, as mentioned in Sec. 2-11 and Sec. 8-2, by directing the gas in cross flow onto the outside of the tubes. It is even more efficient if fins are arranged on the side of the wall which has the lower heat transfer coefficient so that the surface area on this side is considerably enlarged. Straight fins occur mostly on vertical surfaces, for example on air-cooled containers or on certain types of radiators. Circular fins are more important and these are equally spaced apart.

The heat which is taken from the gas by such a fin must flow into the inside of the fin and finally into the root of the fin or vice versa. On average, as a result of the temperature gradient within the fin, there is a smaller temperature difference between the surrounding gas and the surface of the fin than with the root of the fin. The ratio of these two temperature differences is called the fin efficiency factor η_R.

It will now be shown how the temperature gradient in the fin, and thus the fin efficiency factor can be calculated if a constant heat transfer coefficient is assumed at the fin surface area. How correct is this assumption and which heat transfer coefficient value can be expected, on the basis of experimental considerations, to be applicable, has already been discussed in Sec. 2-12 in Part Two of this book.

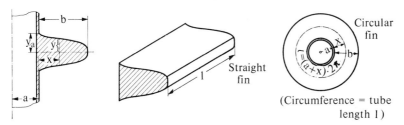

(Circumference = tube
length 1)

Figure 8-23 Cross-sections and views of fins.

Differential Equation for the Temperature Distribution in the Fin

The differential equation should be derived in a general form so that it is valid both for circular and straight fins. Helically wound fins will be treated as circular fins. Figure 8-23, left, shows the cross-section of a fin; in order to make things clear this is shown broader than in reality. x is the distance of a point under consideration from the fin root, y is the semi-thickness of the fin at position x; b is the distance of the outer fin edge from the fin root, a is the outer radius of the tube and thus, also, the distance of the fin root from the tube axis. The length l of a straight fin is understood to be its size in a direction perpendicular to the cross-section under consideration (Fig. 8-23, centre). In circular fins it is appropriate that the length l be measured as the circumference of a circle which is positioned at a distance x from the fin root and is concentric with the tube axis (dashed line in Fig. 8-23, right). It follows from this that

$$l = 2\pi(a+x) \qquad (8\text{-}61)$$

Thus l is dependent on x in finned tubes. It will be seen in Fig. 8-23, left, that the cylindrical surface area $l \cdot y$ represents half the area perpendicular to the x direction through which the heat within the fin must flow essentially in a radial direction. Let λ_s be the thermal conductivity of the material of which the fin is constructed and let Θ be the temperature difference between the fin at position x and the gas on the outside, the temperature of which has the same value at all positions. Then the rate of transfer of heat which flows through the surface area ly is given by:

$$\dot{Q} = -\lambda_s ly \frac{d\Theta}{dx} \qquad (8\text{-}62)$$

Furthermore the fin half under consideration is in contact with the gas between x and $x+dx$ over a surface area very approximately equal to $l \cdot dx$. The rate at which this surface gives up heat is given by:

$$-d\dot{Q} = \alpha_R \Theta l \, dx \qquad (8\text{-}63)$$

where α_R denotes the effective heat-transfer coefficient between the fin and the gas. For this calculation let α_R be assumed to be constant or only to be dependent on x.

If Eq. (8-62) is differentiated with respect to x the following differential equation for the temperature distribution in fins is obtained through substitution into

Eq. (8-63) and by taking Eq. (8-61) into consideration:

$$\frac{d^2\Theta}{dx^2} + \left(\frac{1}{y}\frac{dy}{dx} + \frac{1}{a+x}\right)\frac{d\Theta}{dx} - \frac{\alpha_R}{\lambda_s y}\Theta = 0 \tag{8-64}$$

In straight fins a will be infinitely large and thus $\dfrac{1}{a+x}$ will be equal to 0. This differential equation for the temperature distribution in fins was first formulated by E. Schmidt [S207] but, in contrast to Eq. (8-64) above, Schmidt used two equations, one for straight fins and one for circular fins.

Solutions to the Differential Equations Developed by E. Schmidt

Schmidt [S207] has derived the following complete solutions of the differential equation for the most important special case where $\alpha_R = \text{const}$. Let y_a be the semi-thickness of the fin at the fin root and Θ_a be the temperature difference between the fin and the gas at that point. Furthermore, in circular fins, let r be the distance from the position x under consideration from the tube axis and let R be the outer fin radius, so that r will be equal to $a+x$ and R will be equal to $a+b$. Moreover, if the abbreviation

$$\mu = \sqrt{\frac{\alpha_R}{\lambda_s y_a}} \tag{8-65}$$

is introduced, Schmidt's solutions take the following forms:

1. for straight fins with a right-angled cross-section
   $(y = y_a = \text{const}; \text{see Fig. 8-25})$;

$$\frac{\Theta}{\Theta_a} = \frac{\cosh\mu(b-x)}{\cosh\mu b}; \tag{8-66}$$

2. for straight fins with a triangular cross-section
   $(y = 0 \text{ for } x = b; \text{see Fig. 8-25})$;

$$\frac{\Theta}{\Theta_a} = \frac{J_0(2i\mu\sqrt{b(b-x)})}{J_0(2i\mu b)};^5 \tag{8-67}$$

3. for circular fins with a right-angled cross-section
   (Fig. 8-25, bottom);

$$\frac{\Theta}{\Theta_a} = \frac{J_0(i\mu r)H_1(i\mu R)+iH_0(i\mu r)\cdot iJ_1(i\mu R)}{J_0(i\mu a)H_1(i\mu R)+iH_0(i\mu a)\cdot iJ_1(i\mu R)} \tag{8-68}$$

where J_0, J_1 and H_0, H_1 are the Bessel and Hankel functions respectively of zero or first order. The excess temperature t_b, relative to the gas temperature at the fin edge, is obtained from these equations by putting $x = b$ or $r = R$.

---

[5] It has been common of late to write $I(x)$ in place of $i^n \cdot J_n(ix)$.

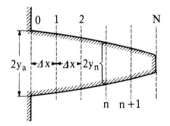

Figure 8-24 Division of a fin into N sections of equal width Δx.

A decision regarding the effect of a fin of given dimensions can be made very quickly using the following small development of the approach of Schmidt to this problem; see [H209]. A fin efficiency factor[6] is introduced (8-69) which represents the ratio of the rate $\dot{Q}$ at which heat is transferred by half the fin to the rate $\dot{Q}_a$ at which heat would be transferred if the fin had the same excess temperature Θ_a overall relative to the gas temperature it has at the fin root.

$$\eta_R = \dot{Q}/\dot{Q}_a \qquad (8\text{-}69)$$

Using the Schmidt [S207] equations developed for Θ/Θ_a the following relationships can be developed for the fin efficiency factor:

1. for straight fins with a right-angled cross-section

$$\eta_R = \frac{\tanh(\mu b)}{\mu b} \qquad (8\text{-}70)$$

2. for straight fins with a triangular cross-section

$$\eta_R = \frac{-iJ_1(2i\mu b)}{\mu b J_0(2i\mu b)} \qquad (8\text{-}71)$$

3. for circular fins with a right-angled cross-section

$$\eta_R = \frac{-2a}{i\mu(R^2 - a^2)} \cdot \frac{J_1(i\mu a)H_1(i\mu R) - H_1(i\mu a)J_1(i\mu R)}{J_0(i\mu a)H_1(i\mu R) - H_0(i\mu a)J_1(i\mu R)} \qquad (8\text{-}72)$$

Numerical Method for Calculating Fin Performance with Arbitrary Cross-Section

The following approximation process [H209] is suitable for calculating the temperature distribution in fins with an arbitrarily shaped cross-section.

The fins are assumed to be divided, in the x direction, into an arbitrary number N of sections of equal width Δx, where the sectors from the fin root to the fin edge are numbered consecutively 0, 1, 2, ... up to N (see Fig. 8-24). The excess temperature relative to the gas temperature at position n is denoted by Θ_n, whose location is

---

[6] It is sometimes considered that it is better to use the term "coefficient of excellence" or a "quality coefficient" rather than an "efficiency factor" in this particular case.

determined by $x = n\,\Delta x$. Correspondingly at positions $n-1$ or $n+1$ the excess temperatures are Θ_{n-1} or Θ_{n+1}.

At position n, if the following approximations [Eqs. (8-73) and (8-74)] are used

$$\frac{d\Theta}{dx} = \frac{\Theta_{n+1}-\Theta_{n-1}}{2\Delta x} \tag{8-73}$$

and

$$\frac{d^2\Theta}{dx^2} = \frac{1}{\Delta x}\left(\frac{\Theta_{n+1}-\Theta_n}{\Delta x} - \frac{\Theta_n-\Theta_{n-1}}{\Delta x}\right) = \frac{\Theta_{n-1}-2\Theta_n+\Theta_{n+1}}{\Delta x^2} \tag{8-74}$$

the differential Eq. (8-64) is converted into the following difference equation

$$\frac{\Theta_{n-1}}{\Theta_b} = 2\cdot\frac{2+\dfrac{\alpha_{Rn}}{\lambda_{sn}y_n}\Delta x^2}{2-\left(\dfrac{1\,dy}{y\,dx}+\dfrac{1}{a+x}\right)_n\Delta x}\cdot\frac{\Theta_n}{\Theta_b} - \frac{2+\left(\dfrac{1\,dy}{y\,dx}+\dfrac{1}{a+x}\right)_n\Delta x}{2-\left(\dfrac{1\,dy}{y\,dx}+\dfrac{1}{a+x}\right)_n\Delta x}\cdot\frac{\Theta_{n+1}}{\Theta_b} \tag{8-75}$$

in which both sides are still divided by the excess temperature Θ_b at the fin edge. α_{Rn}, λ_{sn}, y_n and $\left(\dfrac{1\,dy}{y\,dx}+\dfrac{1}{a+x}\right)_n$ denote the values α_R, λ_R, etc., at position n which are assumed to vary with x. Using Eq. (8-75) Θ_{n-1}/Θ_b can be calculated progressively, step by step in a direction from the fin edge to the fin root.*

It is only at the outermost step at the fin edge, where the calculation must begin with $\Theta_N/\Theta_b = 1$, that Eq. (8-75) cannot be applied directly. Thus the following special expression is formed for the excess temperature at this position:

$$\Theta = \Theta_b+\left(\frac{d\Theta}{dx}\right)_b(x-b) + \frac{1}{2}\left(\frac{d^2\Theta}{dx^2}\right)_b(x-b)^2 \tag{8-76}$$

It follows from Eq. (8-64), that at position $x = b$:

$$\left(\frac{d^2\Theta}{dx^2}\right)_b = \frac{\alpha_{Rb}}{\lambda_{sb}y_b}\Theta_b - \left[\frac{1}{y_b}\left(\frac{dy}{dx}\right)_b + \frac{1}{a+b}\right]\left(\frac{d\Theta}{dx}\right)_b \tag{8-77}$$

It is now necessary to decide whether $y_b = 0$, or $y_b > 0$ as for the triangular fin. Equation (8-78) is developed from Eq. (8-77), by multiplying both sides of this equation by y_b, when $y_b = 0$.

$$\left(\frac{d\Theta}{dx}\right)_b = \frac{\alpha_{Rb}}{\lambda_{sb}}\cdot\frac{\Theta_b}{\left(\dfrac{dy}{dx}\right)_b} \tag{8-78}$$

It is sufficiently accurate to set the value of $\left(\dfrac{d^2\Theta}{dx^2}\right)_b$, which thus remains

---

* Editor's note: This is an explicit method for solving Eq. (8-64). More modern numerical methods are likely to be used nowadays.

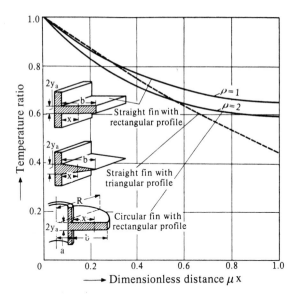

Figure 8-25 Temperature distribution in fins with $\mu \cdot b = 1$.

$$\mu = \sqrt{\frac{\alpha_R}{\lambda_s \cdot y_a}}$$

undetermined, equal to 0. This, for example, can be seen from the almost straight curve of the dashed line in Fig. 8-25 which has yet to be discussed. Then Eq. (8-79) is obtained for position $N-1$ by inserting Eq. (8-78) into Eq. (8-76):

$$\frac{\Theta_{N-1}}{\Theta_b} = 1 - \frac{\alpha_{Rb}}{\lambda_{sb}} \cdot \frac{\Delta x}{\left(\dfrac{dy}{dx}\right)_b} \qquad \text{(with } y_b = 0) \tag{8-79}$$

If, on the other hand, y_b is not equal to 0, heat will also be transferred through the thin, cylindrical fin edge. However, as the quantity of heat involved is relatively small it can be neglected, as it has been by E. Schmidt [S207]; reasonable accuracy can be obtained if again we set $\left(\dfrac{d\Theta}{dx}\right)_b = 0$. With this, Eq. (8-80) can be derived for the temperature at position $N-1$ using Eqs. (8-76) and (8-77):

$$\frac{\Theta_{N-1}}{\Theta_b} = 1 + \frac{\alpha_{Rb}}{\lambda_{sb}} \cdot \frac{(\Delta x)^2}{2 y_b} \qquad \left(\text{with} \quad y_b > 0 \quad \text{and} \quad \left(\frac{d\Theta}{dx}\right)_b = 0 \right) \tag{8-80}$$

Now, using Eq. (8-79) or (8-80) it is possible to determine the value of Θ_{N-1}/Θ_b without actually knowing Θ_b. The following values Θ_{N-2}/Θ_b, etc., can then be calculated progressively step by step using Eq. (8-75). Θ_a/Θ_b is obtained as the final value. Thus, if the temperature Θ_a is known at the fin root Θ_b can also now be determined in this way at the fin edge and thus, on the basis of this step by step calculation, the total temperature distribution is determined. If the fin thickness $2y$ does not vary a great deal as few as four to five steps can be used to give sufficiently accurate results.

Calculation of the Fin Efficiency Factor Using the Step by Step Process

Having determined the temperature curve using the step by step process it is then possible, by means of a simple calculation, to determine also the rate at which heat is transferred through the fin and thus, using Eq. (8-69), to determine the fin efficiency factor.

As Θ denotes the excess temperature relative to the outside gas, the rate at which heat is transferred through the fin surface is given by

$$\dot{Q} = \int_0^b \alpha_R \Theta l \, dx \tag{8-81}$$

where l is determined by Eq. (8-61).

This integral can be evaluated using numerical quadrature, for example, employing Simpson's Rule, as soon as the values of Θ_a, Θ_1, Θ_2, etc., up to Θ_b are known from the previous calculation.

In contrast, the rate at which heat $\dot{Q}_a$ would be transferred if the surface area had the same excess temperature Θ_a relative to the gas temperature at all positions is obtained from Eq. (8-81) if one sets $\Theta = \Theta_a = $ constant. If, for example, α_R is constant, Eq. (8-82) applies:

$$\dot{Q}_a = \alpha_R (2a+b) b \pi \Theta_a \tag{8-82}$$

Thus the fin efficiency factor is obtained by inserting $\dot{Q}$ from Eq. (8-81) and $\dot{Q}_a$ from Eq. (8-82) into Eq. (8-69).

Calculation of Fins with Varying Heat Transfer Coefficient

The step by step process described can be applied to the case where the heat transfer coefficient α_R is dependent on the distance x from the fin root. Indeed it follows from the experimental results described in Sec. 2-12 (Fig. 2-19) that with circular fins α_R often also varies with the angle of the fin radius being considered relative to the direction of flow. Fins of this type can be divided into sections such that within each section the angle dependence of α_R can be neglected without causing too much error. The calculation can be carried out separately for each section. This calculation could perhaps be improved if the equation was integrated numerically along the line of flow of the heat rather than in a radial direction.

Results of the Calculation

Figures 8-25, 8-26 and 8-27 illustrate the result of calculations carried out, in the main, using E. Schmidt's Eqs. (8-66) and (8-68) as well as Eqs. (8-70) to (8-72). The dimensionless group μx or μb is chosen as the abscissa where μ is determined by Eq. (8-65) and α_R and λ_s are assumed to be constant.

Figure 8-25 illustrates the temperature distribution in fins of width b such that $\mu b = 1$. Thus the ratio Θ/Θ_a of the excess temperature Θ at position x to the excess temperature Θ_a at the fin root is shown as a function of μx. The solid lines represent

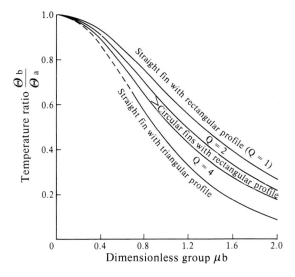

Figure 8-26 Temperature Θ_b at the fin edge.

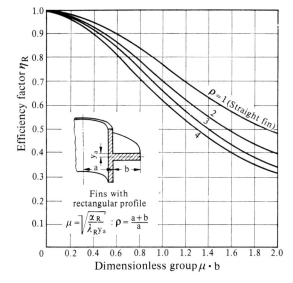

Figure 8-27 Efficiency factor η_R of fins with a rectangular profile.

fins with a rectangular cross-section. The top curve refers to a straight fin[7], the bottom curve refers to a circular fin with a radius ratio $\rho = R/a = 2$, where thus the outer fin radius R is twice as big as the outer tube radius a. As the transfer of heat at the thin cylindrical edge surface is neglected, the two curves at the fin edge have a horizontal tangent. In addition the temperature distribution for a straight fin with a triangular cross-section is shown as a dashed line. The decending temperature

---

[7] According to E. Hofmann [H215], helical fins which are curved at the fin root can be treated as though they were straight fins because the fin surface and the cross-sections for the internal flow of heat are the same in both cases.

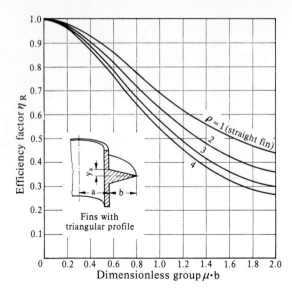

Figure 8-28 Efficiency factor η_R of fins with a triangular profile.

curve in this case is almost a straight line up to the fin edge, as the cross-section of flow is reduced in the same proportion as is the heat flow in the outer parts of the fin.

Figure 8-26 illustrates the ratio Θ_b/Θ_a of the excess temperature relative to the gas temperature at the fin edge to the excess temperature at the fin root as a function of μb. The solid curves again represent fins with a rectangular cross-section and the dashed lines represent fins with a triangular cross-section.

Figures 8-27 and 8-28 show the fin efficiency factor η_R as a function of μb for these fins as well as for other fins with a triangular cross-section. Thus, if the dimensions b, y_a and $\rho = R/a$ (Fig. 8-23), and the values of α_R and λ_s are known for a fin, η_R can be read off directly from Fig. 8-27 or 8-28. Calculation of the heat emitted by a fin is very simple in cases where the heat transfer coefficient α_R can be assumed to be constant over the entire fin surface area.

If the values determined by Th. E. Schmidt, expressed by Eq. (2-64) in Sec. 2-12, which are applicable to the tube surface areas lying between the fins, are also used for α_R, the rate at which heat is emitted or absorbed by a finned tube will be obtained using Eq. (2-63) in Sec. 2-12.

NINE

RECUPERATORS WITH SEVERAL PASSAGES

9-1 PASSAGES JOINED IN SERIES

In a recuperator constructed from tubes, the tube assembly is often divided into several groups of tubes lying close together which are joined in series (see Fig. 9-1). Thus there is a bundle of inner passages through which one gas flows with several changes of direction while the other gas moves in the outer area. This type of heat exchanger will be called the "multiple" heat exchanger.[1] In the following, "passage" is taken to be the entire inner area of all the tubes in a tube assembly, unless otherwise stated.

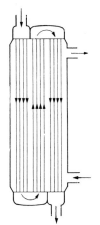

Figure 9-1 Recuperator with three inner passages arranged in series.

[1] In America these heat exchangers are known as "multipass heat exchangers".

Multipass heat exchangers* are used, above all, when the heat transfer coefficient is higher in the outer area than it is inside the tubes. This higher heat transfer coefficient can be caused, for example, by an externally flowing liquid, by condensation of steam, by an externally flowing gas especially where cross flow is enforced by means of baffles as in Fig. 5-4. By dividing the tubes into several passages the fluid velocity of the material flowing within the tube, and thus its heat transfer coefficient, is increased.

When there is an even number of passages, for example two or four, the gas flowing inside the tubes just as often moves in parallel flow as in counterflow relative to the gas flowing outside the tubes. When the number of passages is uneven the movement is selected in such a way that counterflow occurs once more than parallel flow. Figure 9-1 illustrates an example of a recuperator arranged in this way with three passages.

Sometimes the outer area is also divided by partition walls arranged parallel to the axis of the recuperator in order to increase heat transmission. As will be discussed in more detail, this has essentially the same effect as recuperators divided into several parts and arranged in series with a correspondingly smaller number of tubes.

The Calculation of Heat Exchangers with Several Passages

Methods of calculation for these types of heat exchanger have been developed by Davis [D201], Nagle [N202], Underwood [U202], Bowman [B210] and Fischer [F204].[2] As it has always been assumed, in the development of these methods, that the originally warmer gas is flowing outside, in what follows, the temperature of the gas flowing outside will be denoted by θ and that of the gas flowing inside the tubes will be denoted by θ', contrary to previous conventions. θ has the same value at all positions on a heat exchanger cross-section.

The first case to be considered is that where the outer area is not divided and the number of internal passages arranged in series is even. In this case Davis [D201] and Nagle [N202] found that if it is assumed that the heating surface areas, the heat transfer coefficients, etc., of all tube groups are the same, the heat exchanged is independent of whether the number of passages is two, four, six or however many more. This can be stated more explicitly as follows. If the entry temperatures θ_1 and θ'_1 and the exit temperatures θ_2 and θ'_2 of both the gases are given for these conditions, the average temperature difference $\Delta\theta_M$ for the total recuperator defined by Eq. (5-20) has almost always the same value, irrespective of how many passages, presumably an even number, there are. It is obvious that the size of this average temperature difference lies somewhere between the values for parallel flow and counterflow as counterflow and parallel flow alternate in the individual groups of tubes.

While Davis has only provided curves for determining $\Delta\theta_M$, and has given no

---

* Editor's note: The American terminology will be adopted here.

[2] See also VDI Heat Atlas, 2nd edition [1], pp. Ca8 to Ca11.

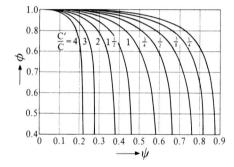

Figure 9-2 Correction factor for two-, four-, six- and more passage multipass recuperators.

further background reasoning for them, and while Nagle has been able to derive equations which can only be solved iteratively,* Underwood [U202] has developed the following closed equation for the average temperature difference with two passages joined in series;

$$\Delta\theta_M = \frac{\sqrt{(\theta_1-\theta_2)^2+(\theta_2'-\theta_1')^2}}{\ln\dfrac{\theta_1+\theta_2-\theta_1'-\theta_2'+\sqrt{(\theta_1-\theta_2)^2+(\theta_2'-\theta_1')^2}}{\theta_1+\theta_2-\theta_1'-\theta_2'-\sqrt{(\theta_1-\theta_2)^2+(\theta_2'-\theta_1')^2}}} \qquad (9\text{-}1)$$

This relationship will be derived later so that considerations which are more concerned with the results and the application of the theory, can be pursued immediately. A much more complicated equation which can only be solved iteratively has been obtained by Underwood [U202] for four passages joined in series. A corresponding calculation for three passages can be found in the work of Fischer [F204].

The Results of the Calculations

The efficiency of the exchange of heat can be assessed by comparing the average temperature difference $\Delta\theta_M$ of multipass heat exchangers with the average temperature difference $(\Delta\theta_M)_G$ of a pure counterflow heat exchanger, using Eq. (5-56). The relationship

$$\phi = \frac{\Delta\theta_M}{(\Delta\theta_M)_G} \qquad (9\text{-}2)$$

represents a correction factor with which $(\Delta\theta_M)_G$ must be multiplied in order to obtain the true average temperature difference $\Delta\theta_M$ for the multipass heat exchanger.[3]

Figure 9-2 displays this correction factor for recuperators with two, four, six, etc.,

---

* Editor's note: The lack of computing facilities when this book was first written is reflected here in an apparent distaste for solving non-linear algebraic equations iteratively. Clearly any misgivings can be displaced if it is planned to program such iterative methods for a computer.

[3] This correction factor corresponds to the efficiency factor η_K of a cross-counterflow heat exchanger (see Sec. 8-3, Eq. (8-49)).

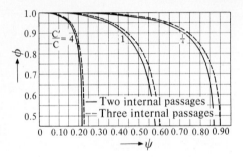

Figure 9-3 Comparison of recuperators with two and three inner passages.

passages for different ratios C'/C of the thermal capacities of the inside and outside gases. The dimensionless group shown in Eq. (9-3) is chosen as the abscissa:

$$\psi = \frac{\theta'_2 - \theta'_1}{\theta_1 - \theta'_1} \tag{9-3}$$

This represents the ratio of the temperature increase in the inside gas to the difference between the entry temperatures of both gases. Following from Eqs. (5-67) and (5-31) when $C' \leqslant C$, ψ is equal to the efficiency factor for the exchange of heat. Equations (9-25) and (9-26), which are derived later on for two passages, are used for the calculation of the curves in Fig. 9-2. However, these curves in Fig. 9-2 are also applicable for four, six, etc., passages, as has already been mentioned, and provide a high degree of accuracy.

In recuperators with three passages joined in series where two operate in counterflow and one operates in parallel flow (see Fig. 9-3), a somewhat higher average temperature difference $\Delta\theta_M$ occurs, according to Fischer's calculations, than in heat exchangers with an even number of passages; this corresponds rather more closely to pure counterflow. Thus, as Figs. 9-3 and 9-4 show, the correction factor ϕ takes a somewhat higher value, given otherwise the same operating conditions.

The performance of a multipass heat exchanger can be calculated in the manner described almost as easily as a pure counterflow recuperator, using Figs. 9-2 and 9-4, in the cases where the entry and exit temperatures θ_1, θ'_1, θ_2 and θ'_2 and the thermal capacity rates C and C' of the two gases are given. The average temperature difference $(\Delta\theta_M)_G$ for counterflow and the dimensionless quantity ψ

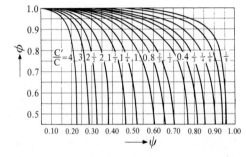

Figure 9-4 Correction factor for three passage recuperators.

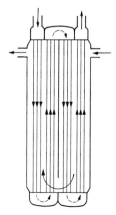

Figure 9-5 Recuperator with four internal and two external passages.

are first determined using Eq. (9-3). Using the correction factor obtained from Fig. 9-2 or 9-4 the true average temperature difference $\Delta\theta_M$ for the multipass heat exchanger is then obtained from Eq. (9-2).

Recuperators with Several External Passages

Nagle [N202], Bowman [B210] and Fischer [F204] have also dealt numerically with recuperators which have several outside passages (see Fig. 9-5). Thus several, usually two or four, internal passages meet at each external passage. The external passages are joined in series, as are all the internal passages. Heat exchangers of this type are intended more than anything else to achieve a behavior which approximates more closely to pure counterflow, and thus achieve a better exchange of heat, than would be possible for heat exchangers with only one external passage. In addition, the heat transmission coefficient is enlarged since fluid velocity is increased.

In their calculations, Nagle, Bowman and Fischer all assume that each external passage has the same proportion of the total heating surface area F and thus with N such passages, the area F/N is apportioned to each one. Further it is assumed that the heat transmission coefficient k has the same value for all passages. They arrive at the result that in these cases the correction factor ϕ has the same value for all the external passages and is equal to the correction factor for the total heat exchanger. They develop equations which describe this system in a rather intricate manner. The same result can be obtained considerably more simply on the following basis.[4]

Figures 9-2 and 9-4 apply to each individual external passage if ψ is determined from the temperatures at the beginning and end of the passage concerned. However, if only the entry and exit temperatures θ_1, θ_1' and θ_2 and θ_2' for the total heat exchanger are known, ψ can be determined for an individual passage using a transformation of the following form.

[4] From H. Hausen, first published in the first edition of this book.

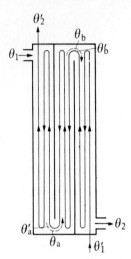

Figure 9-6 Temperatures in a recuperator with several external passages.

If one was to examine the external passages arranged in series, for example as shown in Fig. 9-6, following the movement of the gas flowing externally, which enters at a temperature θ_1, the temperature of this gas at the end of the first passage could be denoted by θ_a and by θ_b at the end of the second passage. At the end of the last passage the gas reaches the exit temperature θ_2.

Correspondingly the gas flowing internally has the temperature θ'_a at the transition point from the second to the first of these passages and the temperature θ'_b at the transition from the third to the second of these passages, and so on. Furthermore, if it is assumed that F/N, k, and even C'/C have the same value in all external passages, these passages are equal to one another in the sense that the ratio of the two temperature differences at the ends of the individual external passages are always the same size. When there is only one internal passage correspondingly to each external passage, the use of F/N instead of F emerges from the second Eq. (5-46). Thus the following must apply (see Fig. 9-6).

$$\frac{\theta_a - \theta'_a}{\theta_1 - \theta'_2} = \frac{\theta_b - \theta'_b}{\theta_a - \theta'_a} = \frac{\theta_c - \theta'_c}{\theta_b - \theta'_b}, \quad \text{etc.}$$

From this is yielded:

$$\frac{\theta_b - \theta'_b}{\theta_1 - \theta'_2} = \left(\frac{\theta_a - \theta'_a}{\theta_1 - \theta'_2}\right)^2; \quad \frac{\theta_c - \theta'_c}{\theta_1 - \theta'_2} = \left(\frac{\theta_a - \theta'_a}{\theta_1 - \theta'_2}\right)^3$$

Taking all N external passages into account, we obtain

$$\frac{\theta_2 - \theta'_1}{\theta_1 - \theta'_2} = \left(\frac{\theta_a - \theta'_a}{\theta_1 - \theta'_2}\right)^N$$

or

$$\frac{\theta_a - \theta'_a}{\theta_1 - \theta'_2} = N\sqrt{\frac{\theta_2 - \theta'_1}{\theta_1 - \theta'_2}} \qquad (9\text{-}4)$$

Further the heat balance Eq. (9-5) can be applied to the first external passage

$$C(\theta_1 - \theta_a) = C'(\theta'_2 - \theta'_a) \tag{9-5}$$

By inserting θ_a from Eq. (9-4) into Eq. (9-5), Eq. (9-6) is obtained.

$$\theta'_2 - \theta'_a = (\theta_1 - \theta'_2)\frac{C}{C' - C}\left(1 - N\sqrt{\frac{\theta_2 - \theta'_1}{\theta_1 - \theta'_2}}\right) \tag{9-6}$$

If the definition of ψ (see Eq. (9-3)) is used at the first external passage, then we can develop Eq. (9-7) below

$$\psi = \frac{\theta'_2 - \theta'_a}{\theta_1 - \theta'_a} \tag{9-7}$$

Equation (9-8) is obtained by substituting for θ'_a using Eq. (9-6)

$$\psi = \frac{1 - N\sqrt{\dfrac{\theta_2 - \theta'_1}{\theta_1 - \theta'_2}}}{\dfrac{C'}{C} - N\sqrt{\dfrac{\theta_2 - \theta'_1}{\theta_1 - \theta'_2}}} \tag{9-8}$$

This provides the necessary equation for ψ in which, in addition to C and C', only the entry and exit temperatures θ_1, θ'_1, θ_2 and θ'_2 of the two gases appear as they relate to the total heat exchanger. As can be concluded from the equivalence, mentioned above, of the external passages or from a particular derivation for any other arbitrary passage, Eq. (9-8) is equally applicable to all passages. Thus, using the value of ψ calculated from Eq. (9-8), the correction factor ϕ can be determined, using Fig. 9-2 or 9-4, for each arbitrary external passage and thus for the total heat exchanger.

Derivation of Eq. (9-1) for a Dual Passage Heat Exchanger

The derivation of Eq. (9-1), which has been delayed up to now, is here presented in order to explain the developments made by Underwood. As has already been mentioned Underwood [U202] developed Eq. (9-1) for the case where two internal passages are joined in series and have the same number of tubes and thus the same heating surface area and where, by contrast, the outer area is not subdivided (Fig. 9-7). Here C, C' and the heat transmission coefficient k are taken to be constant. In addition it is assumed that the temperature θ in a specific cross-section of the outer area has the same value overall which is what happens with baffles present in the outer area or when there is a strong intermixing of the gas. The longitudinal coordinates f of the heat exchanger will be treated as increasing positively in the direction of increasing temperature θ of the outside gas; thus in Fig. 9-7 this will be from the top to the bottom. The gas at temperature θ' which is to be heated first flows into the left-hand half of the tube assembly in an upwards direction and then flows downwards in the right-hand half.

In an arbitrary cross-section $A-B$ the gas is at the temperature θ'_l on the

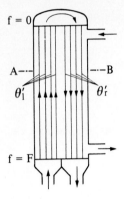

Figure 9-7 Recuperator with two internal passages.

left-hand side and is at the temperature θ'_r on the right-hand side. The heating surface area df, related to all tubes, lies between this cross-section and a cross-section lying infinitely close to it. The rate at which heat is transferred through this heating surface area is given by

$$d\dot{q} = -C\,d\theta = -C'\,d\theta'_l + C'\,d\theta'_r \tag{9-9}$$

The minus and plus signs indicate that θ and θ'_l fall in the direction of increasing f. In addition (9-10) and (9-11) represent the transfer of heat through the heating surface area $df/2$ in each of the two passages:

$$d\dot{q}_l = -C'\,d\theta'_l = k\frac{df}{2}(\theta - \theta'_l) \tag{9-10}$$

$$d\dot{q}_r = +C'\,d\theta'_r = k\frac{df}{2}(\theta - \theta'_r) \tag{9-11}$$

Thus the following three simultaneous equations are obtained for the temperature distributions

$$-C\frac{d\theta}{df} = C'\left(-\frac{d\theta'_l}{df} + \frac{d\theta'_r}{df}\right) \tag{9-12}$$

$$-\frac{d\theta'_l}{df} = \frac{k}{2C'}(\theta - \theta'_l) \tag{9-13}$$

$$+\frac{d\theta'_r}{df} = \frac{k}{2C'}(\theta - \theta'_r) \tag{9-14}$$

The solutions to these equations are set out below

$$\theta = A + B\exp(\beta f) + D\exp(\gamma f) \tag{9-15}$$

$$\theta'_l = A + \frac{k}{k - 2C'\beta}B\exp(\beta f) + \frac{k}{k - 2C'\gamma}D\exp(\gamma f) \tag{9-16}$$

$$\theta'_r = A + \frac{k}{k + 2C'\beta}B\exp(\beta f) + \frac{k}{k + 2C'\gamma}D\exp(\gamma f) \tag{9-17}$$

where A, B and D denote constants which have still to be determined and where β and γ are determined in the following way

$$\beta = \frac{k}{2C}\left[-1 + \sqrt{1 + \left(\frac{C}{C'}\right)^2}\right] \tag{9-18}$$

$$\gamma = \frac{k}{2C}\left[-1 - \sqrt{1 + \left(\frac{C}{C'}\right)^2}\right] \tag{9-19}$$

The requirement that $\theta_l' = \theta_r'$ when $f = 0$ (see Fig. 9-7) gives

$$D = -B$$

which can be confirmed most easily by considering that, according to Eqs. (9-18) and (9-19), β and γ are the two roots of the equation

$$\frac{k}{k - 2C'x} - \frac{k}{k + 2C'x} = \frac{C}{C'}$$

where x is the unknown variable. The following is obtained from the further considerations that $\theta = \theta_1$ when $f = 0$ and $\theta = \theta_2$, $\theta_l' = \theta_1'$ and $\theta_r' = \theta_2'$ when $f = F$:

$$\theta_1 - \theta_2 = B[\exp(\gamma F) - \exp(\beta F)]$$

and

$$\theta_1 - \theta_2' = B\left[\frac{k}{k + 2C'\gamma}\exp(\gamma F) - \frac{k}{k + 2C'\beta}\exp(\beta F)\right]$$

If B is eliminated from these two equations, after slight rearrangement, the following results

$$\exp[(\beta - \gamma)F] = \frac{(\theta_1 - \theta_2)\dfrac{k}{k + 2C'\gamma} - (\theta_1 - \theta_2')}{(\theta_1 - \theta_2)\dfrac{k}{k + 2C'\beta} - (\theta_1 - \theta_2')}$$

By combining this with Eqs. (9-18) and (9-19), we yield Eq. (9-20)

$$\frac{kF}{C}\sqrt{1 + \left(\frac{C}{C'}\right)^2} = \ln\frac{(\theta_1 - \theta_2)\left[1 - \dfrac{C'}{C} + \sqrt{1 + \left(\dfrac{C'}{C}\right)^2}\right] + 2\dfrac{C'}{C}(\theta_1 - \theta_2')}{(\theta_1 - \theta_2)\left[1 - \dfrac{C'}{C} - \sqrt{1 + \left(\dfrac{C'}{C}\right)^2}\right] + 2\dfrac{C'}{C}(\theta_1 - \theta_2')} \tag{9-20}$$

This relationship can be finally converted to the required Eq. (9-1) when the following are formed using Eqs. (5-20) and (5-31)

$$\frac{kF}{C} = \frac{\theta_1 - \theta_2}{\Delta\theta_M} \tag{9-21}$$

and using Eq. (5-31)

$$\frac{C'}{C} = \frac{\theta_1 - \theta_2}{\theta_2' - \theta_1'} \tag{9-22}$$

Underwood [U202] indicated that Eq. (9-1) is also valid, unchanged, for the situation where the inlet gas in two internal passages arranged in series first moves in parallel flow and then in counterflow to the exit gas.

In order to obtain ϕ as a function of ψ (see Figs. 9-2, 9-3 and 9-4), Eqs. (9-20) and (5-56) are rearranged with the help of Eqs. (9-21) and (9-3) in the following way

$$\frac{\theta_1 - \theta_2}{\Delta\theta_M}\sqrt{1 + \left(\frac{C}{C'}\right)^2} = \ln\frac{2 - \psi\left[1 + \dfrac{C'}{C} - \sqrt{1 + \left(\dfrac{C'}{C}\right)^2}\right]}{2 - \psi\left[1 + \dfrac{C'}{C} + \sqrt{1 + \left(\dfrac{C'}{C}\right)^2}\right]} \qquad (9\text{-}23)$$

$$(\Delta\theta_M)_G = \frac{(\theta_1 - \theta_2)\left(1 - \dfrac{C}{C'}\right)}{\ln\dfrac{1 - \psi}{1 - \dfrac{C'}{C}\psi}} \qquad (9\text{-}24)$$

Thus using Eq. (9-2) we obtain

$$\phi = \frac{\Delta\theta_M}{(\Delta\theta_M)_G} = \frac{\sqrt{1 + \left(\dfrac{C'}{C}\right)^2}}{\dfrac{C'}{C} - 1} \cdot \frac{\ln\dfrac{1 - \psi}{1 - \dfrac{C'}{C}\psi}}{\ln\dfrac{2 - \psi\left[1 + \dfrac{C'}{C} - \sqrt{1 + \left(\dfrac{C'}{C}\right)^2}\right]}{2 - \psi\left[1 + \dfrac{C'}{C} + \sqrt{1 + \left(\dfrac{C'}{C}\right)^2}\right]}} \qquad (9\text{-}25)$$

When $C = C'$ this equation is simplified by noting that

$$\lim_{C \to C'}\left(\frac{C'}{C} - 1\right) = 0$$

and becomes

$$\phi = \frac{\psi}{1 - \psi} \cdot \frac{\sqrt{2}}{\ln\dfrac{2 - \psi(2 - \sqrt{2})}{2 - \psi(2 + \sqrt{2})}} \qquad (9\text{-}26)$$

Limiting Case of an Infinite Number of Passages

The method of calculation for more than two internal passages proposed by Underwood [U202] and others, leads, as mentioned at the outset, to very complex equations which can only be solved iteratively. Repetition of these equations will therefore be abandoned. Moreover as it appears that there are only very slight differences between two, four, six or more internal passages, it is sufficient to deal with the limiting case where there are an infinite number of passages. This is relatively easy to deal with as a consequence of the following considerations.[5]

A recuperator with an infinite number of internal passages corresponds to the arrangement shown in Fig. 8-14 if it is assumed that the tube coil illustrated there consists of an infinite number of vertical tube sections. It follows that the change in the temperature θ' of the inside gas is only infinitely small in an individual passage or tube section. Thus the temperature of this gas only varies very slowly with the number of passages through which the gas is passing and therefore, as will be seen in Fig. 8-14, this variation takes place only in a horizontal direction; it is not dependent on the longitudinal coordinate f which is shown to be directed from top to bottom in this diagram. As a consequence the whole process corresponds to pure cross flow as shown in Fig. 8-1 or 8-9, but with the following difference. If it is assumed, as with all the multipass recuperators dealt with so far, that the outside gas is at the same temperature overall within each horizontal cross-section (see Fig. 8-14), each of the two gases will behave in such a way that it not only flows in a direction perpendicular to that of the other gas but flows as if it is completely intermixed in each cross-section perpendicular to its own direction of flow. This case can be calculated precisely by using exactly the same differential equations as set out in Sec. 8-2 and by finding their solution.

However the following more simple method is preferred. It is based on the assumption, which can be justified by performing the calculation exactly, that the temperature of each of the two gases varies in such a way that the other gas is at the same overall temperature at all positions in the heat exchanger; this constant temperature is the same as the average value θ_M or θ'_M of the actual gas temperatures in question. Thus consideration will first be given to the exchange of heat between the outside gas at variable temperature θ and an inside gas at a constant temperature θ'_M. Then the reverse case where heat is exchanged between the inside gas at variable temperature θ' with an outside gas at constant temperature θ_M will be considered. In this way, using Eq. (5-27), it is possible to obtain Eq. (9-27) for the average temperature difference $\Delta\theta_M$ between the two gases, which can also be assumed to be equal to $\theta_M - \theta'_M$,[6]

$$\Delta\theta_M = \frac{\theta_1 - \theta_2}{\ln\dfrac{\theta_1 - \theta'_M}{\theta_2 - \theta'_M}} = \frac{\theta'_2 - \theta'_1}{\ln\dfrac{\theta_M - \theta'_1}{\theta_M - \theta'_2}} = \theta_M - \theta'_M \qquad (9\text{-}27)$$

Furthermore, if the abbreviation (9-28) is introduced

$$\chi = \frac{\theta_1 - \theta'_M}{\theta_2 - \theta'_M} \qquad (9\text{-}28)$$

Eq. (9-29) will follow from the first and second expressions in Eq. (9-27)

$$\Delta\theta_M = \frac{\theta_1 - \theta_2}{\ln\chi} \qquad (9\text{-}29)$$

---

[5] Author's process, first published in the first edition of this book.
[6] Equation (9-27) can be derived exactly from the differential equations.

Using Eq. (9-2) together with Eq. (9-24), Eq. (9-30) for ϕ results

$$\phi = \frac{\Delta\theta_M}{(\Delta\theta_M)_G} = \frac{\ln\dfrac{1-\psi}{1-\dfrac{C'}{C}\psi}}{\left(1-\dfrac{C}{C'}\right)\ln\chi} \tag{9-30}$$

Now ϕ is solely dependent upon ψ and C'/C (see Figs. 9-2, 9-3 and 9-4) and as a consequence the appearance of the dimensionless quantity χ complicates Eq. (9-30). In order to find a connection between ψ, C'/C and χ, $\theta_1 - \theta_1'$ must be calculated in the following manner. Using Eqs. (9-22) and (9-28)

$$\ln\frac{\theta_M - \theta_1'}{\theta_M - \theta_2'} = \ln\frac{\theta_M - \theta_1'}{(\theta_M - \theta_1') - (\theta_2' - \theta_1')} = \frac{C}{C'}\ln\chi$$

results from the second and third expressions in Eq. (9-27) and from this Eq. (9-31) is obtained

$$\theta_M - \theta_1' = \frac{\theta_2' - \theta_1'}{1 - \chi^{-C/C'}} \tag{9-31}$$

In addition Eq. (9-32) follows from Eqs. (9-28) and (9-22)

$$\theta_1 - \theta_M' = \frac{C'}{C}(\theta_2' - \theta_1')\frac{\chi}{\chi - 1} \tag{9-32}$$

Finally, Eq. (9-29) can be written in the form of (9-33) if we take into consideration Eqs. (9-22) and (9-27)

$$\theta_M - \theta_M' = \frac{C'}{C}\cdot\frac{\theta_2' - \theta_1'}{\ln\chi} \tag{9-33}$$

By adding together Eqs. (9-31) and (9-32) and then subtracting Eq. (9-33), the required expression for $\theta_1 - \theta_1'$ is obtained. Using Eq. (9-3), there results Eq. (9-34)

$$\frac{1}{\psi} = \frac{\theta_1 - \theta_1'}{\theta_2' - \theta_1} = \frac{\chi^{C/C'}}{\chi^{C/C'} - 1} + \frac{C'}{C}\left(\frac{\chi}{\chi - 1} - \frac{1}{\ln\chi}\right) \tag{9-34}$$

The required connection between ϕ and ψ and C'/C is yielded by Eqs. (9-30) and (9-34). Indeed the parameter χ cannot be eliminated from these equations in such a way that an equation in ϕ is obtained. However if an arbitrary value is assumed for χ, given C'/C from Eqs. (9-30) and (9-34), two values of ϕ and ψ are obtained which can be fitted together. Thus the appearance of the parameter χ scarcely constitutes a disadvantage in the calculations for a diagram of the type shown in Figs. 9-2, 9-3 and 9-4. However, in an individual case an approximately correct value for ϕ and ψ can quickly be obtained since, according to Eq. (9-28), an approximate value for χ can be determined by estimating θ_M'.

Equations (9-30) and (9-34) can be simplified into the form of Eqs. (9-35) and (9-36) for the case where $C = C'$, again using

$$\lim_{C \to C'} \left(\frac{C'}{C} - 1 \right) = 0$$

$$\phi = \frac{\psi}{(1-\psi)\ln\chi} \qquad \text{(where } C = C') \qquad (9\text{-}35)$$

$$\frac{1}{\psi} = 2\frac{\chi}{\chi-1} - \frac{1}{\ln\chi} \quad \text{(where } C = C') \qquad (9\text{-}36)$$

In this way all the equations are now found which enable one to calculate the correction factor ϕ, using Eq. (9-2), for the limiting case of an infinite number of passages.

Table 9-1, calculated for the case where $C = C'$ and using Eq. (9-26) as well as Eqs. (9-35) and (9-36), demonstrates the fact that, as mentioned above, the values of ϕ for heat exchangers with two passages and with an infinite number of passages differ only relatively slightly.

Within the limits of the accuracy of the calculation, the values of ϕ are the same for when $\psi = 0.3$ and $\psi = 0.4$. With larger values of ψ, ϕ is only slightly smaller for an infinite number of passages than it is for two passages. More or less the same conclusion emerges when the calculation is carried out with other relationships between C and C', even in the extreme cases where $C'/C = 4$ or $C'/C = \frac{1}{4}$.

It is remarkable that ϕ is independent of whether the gas flowing internally, as shown in Fig. 9-7, is flowing through the first passage or even through the second passage in a direction opposite to that of the outside gas. In the second case the gas flowing internally enters from the top, like the gas flowing externally, and, after reversal at the bottom end of the counterflow heat exchanger, flows out again, at the top. In both cases the temperature distribution is shown in Figs. 9-8 and 9-9 where $\theta_1 = 100°$; $\theta_1' = 0°$; $\psi = 0.55$; $\phi = 0.661$ (last line in Table 9-1) and, in addition, where $C = C' = 300 \text{ W/K}$ and $k = 100 \text{ W/(m}^2\text{K)}$.

In the first case, before it leaves the recuperator, the gas flowing internally will again be cooled somewhat after passing through a smooth maximum of

Table 9-1 Correction factor $\phi = \Delta\theta_M/(\Delta\theta_M)_G$ as a function of $\psi = (\theta_2' - \theta_1')/(\theta_1 - \theta_1')$ where $C = C'$, for heat exchangers with two internal passages, with an infinite number of internal passages and for pure parallel flow

ψ	ϕ in heat exchangers		
	With two passages	With an infinite number of passages	Pure parallel flow
0.3	0.969	0.970	0.935
0.4	0.921	0.922	0.828
0.5	0.802	0.795	0.000
0.55	0.661	0.626	

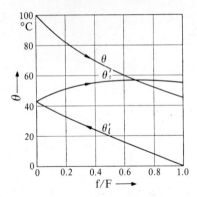

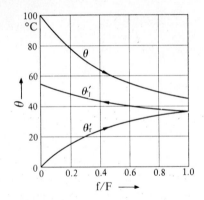

Figure 9-8 Recuperator with two internal passages $\psi = 0.55$; $\phi = 0.661$. 1. Passage in counterflow. 2. Passage in parallel flow.

Figure 9-9 Recuperator with two internal passages $\psi = 0.55$; $\phi = 0.661$. 1. Passage in parallel flow. 2. Passage in counterflow.

temperature; this is caused by the reversal of the temperature difference relative to the gas flowing externally.

The last column of Table 9-1 makes possible a comparison with pure parallel flow. Here the ratio of the average temperature difference in parallel flow and counterflow is introduced as ϕ. It can be recognized from this that two and more passage recuperators are somewhat superior to pure parallel flow devices. The necessary heating surface area is inversely proportional to ϕ.

9-2 EXCHANGE OF HEAT BETWEEN THREE MATERIALS

It is not unusual for a material to give up heat or "cold" simultaneously to two or more other materials. Thus, for example, the heat contained in the exhaust gases from a boiler can be transferred to the feed water and simultaneously employed for preheating combustion air.

This type of multiple exchange of heat can be carried out in a single recuperator in various ways. The tube assembly arrangement shown in Fig. 9-10 is very clear. A counterflow heat exchanger used in low temperature technology is chosen by way of example; this is used to cool, to a very low temperature, a gas mixture which is to be separated.

For example, hydrogen flows through one section of the tubes, methane flows through a second section and, in certain cases, carbon dioxide flows through a third section of the tubes. The gas mixture to be cooled flows in the outer area of the tubes, surrounded by a common tube casing. The tubes themselves are either straight and arranged in parallel or are wound helically in order to achieve cross-counterflow (see Sec. 8-3).

Usually the substances, which are simultaneously being heated or cooled, undergo approximately the same temperature variations. Here the temperature distribution is not very different from the situation where separate heat exchangers are connected in parallel and are used for the exchange of heat between these materials. It follows that calculations for this type of heat exchanger can be performed sufficiently accurately using methods already developed.

If, on the other hand, the entry and exit temperatures differ substantially from one another or the specific thermal capacities of the gases is dependent upon the temperature in a different way, the exchange of heat for each of these gases would be influenced by the simultaneous exchange between the other gases. In this case the temperature distribution differs greatly from that of heat exchangers connected in parallel and can only be determined by calculations particular to the circumstances under consideration.

Calculation of the Temperature Distribution with Specific Thermal Capacities which are Independent of Temperature

Morley [M207] has calculated the exchange of heat between one material and two others in the case where the specific thermal capacities are independent of temperature. In a slightly altered form, the method he proposed will be discussed below.

As well as the specific thermal capacities, the heat transfer coefficients and the heating surfaces areas per unit length of the recuperator are also assumed to be constant. As the gas at temperature θ cools down, the two other gases flowing in the opposite direction at temperatures θ' and θ'', should be warmed (see Fig. 9-11). Let the thermal capacity rates of the gases be C, C' and C''.

Let the total heating surface area of the recuperator be F; the surface F' is available for the exchange of heat with the gas at temperature θ', the surface $F'' = F - F'$ is available for the exchange of heat with the gas at temperature θ''. Let the corresponding heat transfer coefficients be k' and k''. Let the position in the

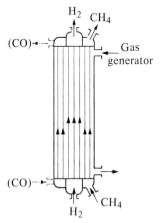

Figure 9-10 Recuperator for the simultaneous transfer of heat from a gas mixture to hydrogen and methane.

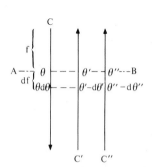

Figure 9-11 Heat exchange between three materials.

recuperator (cross-section A–B in Fig. 9-11) be determined by the total heating surface area f between this position and the entry position of the originally warmer gas (Fig. 9-10, top).

Equation (9-37) represents the rate at which heat is transferred in an infinitely small section of the recuperator with the heating surface area df; note that $d\theta$, $d\theta'$, and $d\theta''$ are negative with a positive df.

$$d\dot{q} = -C \, d\theta = -(C' \, d\theta' + C'' \, d\theta'') \tag{9-37}$$

In addition Eqs. (9-38) and (9-39) can be applied to the exchange of heat with two gases which are to be heated:

$$d\dot{q}' = -C' \, d\theta' = k' \frac{F'}{F} df (\theta - \theta') \tag{9-38}$$

$$d\dot{q}'' = -C'' \, d\theta'' = k'' \frac{F''}{F} df (\theta - \theta'') \tag{9-39}$$

Thus three simultaneous differential Eqs. (9-40), (9-41) and (9-42) are obtained for the distribution of the temperatures θ, θ' and θ'' with respect to f:

$$C \frac{d\theta}{df} = C' \frac{d\theta'}{df} + C'' \frac{d\theta''}{df} \tag{9-40}$$

$$\frac{d\theta'}{df} = -K'(\theta - \theta') \tag{9-41}$$

$$\frac{d\theta''}{df} = -K''(\theta - \theta'') \tag{9-42}$$

in which the abbreviations

$$\frac{k'F'}{C'F} = K' \quad \text{and} \quad \frac{k''F''}{C''F} = K'' \tag{9-43}$$

are used. These differential equations have solutions of the form:

$$\left. \begin{aligned} \theta &= A + B \exp(\beta f) + D \exp(\gamma f) \\ \theta' &= A + B' \exp(\beta f) + D' \exp(\gamma f) \\ \theta'' &= A + B'' \exp(\beta f) + D'' \exp(\gamma f) \end{aligned} \right\} \tag{9-44}$$

If these expressions for θ, θ' and θ'' are inserted into the differential Eqs. (9-40), (9-41) and (9-42) and note is taken of the fact that the coefficients of both $e^{\beta f}$ and $e^{\gamma f}$ must be equal to one another on both sides, the following relationships result between the constants:

$$\left. \begin{aligned} CB &= C'B' + C''B'', & CD &= C'D' + C''D'', \\ B\beta &= -K'(B - B'), & D'\gamma &= -K'(D - D'), \\ B''\beta &= -K''(B - B''), & D''\gamma &= -K''(D - D''). \end{aligned} \right\} \tag{9-45}$$

Equation (9-46) follows from the two equations bottom left

$$B' = \frac{K'}{K'-\beta}B; \quad B'' = \frac{K''}{K''-\beta}B \qquad (9\text{-}46)$$

and by insertion into the top left equation we get

$$\beta^2 - \left[\left(1 - \frac{C'}{C}\right)K' + \left(1 - \frac{C''}{C}\right)K''\right]\beta = \left(\frac{C'+C''}{C} - 1\right)K'K'' \qquad (9\text{-}47)$$

Now γ satisfies the same quadratic equation; thus β and γ correspond to the roots of this Eq. (9-47). We thus obtain

$$\beta = \frac{1}{2}\left[\left(1 - \frac{C'}{C}\right)K' + \left(1 - \frac{C''}{C}\right)K''\right]$$
$$+ \sqrt{\left(\frac{C'+C''}{C} - 1\right)K'K'' + \frac{1}{4}\left[\left(1 - \frac{C'}{C}\right)K' + \left(1 - \frac{C''}{C}\right)K''\right]^2} \qquad (9\text{-}48)$$

$$\gamma = \frac{1}{2}\left[\left(1 - \frac{C'}{C}\right)K' + \left(1 - \frac{C''}{C}\right)K''\right]$$
$$- \sqrt{\left(\frac{C'+C''}{C} - 1\right)K'K'' + \frac{1}{4}\left[\left(1 - \frac{C'}{C}\right)K' + \left(1 - \frac{C''}{C}\right)K''\right]^2} \qquad (9\text{-}49)$$

Utilizing this and taking into consideration Eq. (9-46) the solutions to the differential equations take the following form

$$\theta = A + B\exp(\beta f) + D\exp(\gamma f) \qquad (9\text{-}50)$$

$$\theta' = A + \frac{K'}{K'-\beta}B\exp(\beta f) + \frac{K'}{K'-\gamma}D\exp(\gamma f) \qquad (9\text{-}51)$$

$$\theta'' = A + \frac{K''}{K''-\beta}B\exp(\beta f) + \frac{K''}{K''-\gamma}D\exp(\gamma f) \qquad (9\text{-}52)$$

In these equations the constants A, B and D remain to be determined and can be found from the boundary conditions, for example from the temperatures of the three gases at one end of the heat exchanger or from the three entry temperatures given previously.

For example, if, when $f = 0$, the entry temperature $\theta = \theta_1$ and the exit temperatures $\theta' = \theta'_2$ and $\theta'' = \theta''_2$ are known, the following expressions are obtained from Eqs. (9-50), (9-51) and (9-52) by setting $f = 0$:

$$B = \frac{(\gamma - K')(\theta_1 - \theta'_2) - (\gamma - K'')(\theta_1 - \theta''_2)}{\beta\left(\dfrac{\gamma - K'}{\beta - K'} - \dfrac{\gamma - K''}{\beta - K''}\right)} \qquad (9\text{-}53)$$

$$D = \frac{(\beta - K')(\theta_1 - \theta_2') - (\beta - K'')(\theta_1 - \theta_2'')}{\gamma\left(\dfrac{\beta - K'}{\gamma - K'} - \dfrac{\beta - K''}{\gamma - K''}\right)} \qquad (9\text{-}54)$$

If, on the other hand, the entry temperatures of all the gases are given, that is θ_1 at $f = 0$ and θ_1' and θ_1'' at $f = F$, the following expressions are obtained:

$$B = \frac{\dfrac{\gamma - K'}{\gamma - K'[1 - \exp{(\gamma F)}]}(\theta_1 - \theta_1') - \dfrac{\gamma - K''}{\gamma - K''[1 - \exp{(\gamma F)}]}(\theta_1 - \theta_1'')}{\dfrac{\gamma - K'}{\beta - K'} \cdot \dfrac{\beta - K'[1 - \exp{(\beta F)}]}{\gamma - K'[1 - \exp{(\gamma F)}]} - \dfrac{\gamma - K''}{\beta - K''} \cdot \dfrac{\beta - K''[1 - \exp{(\beta F)}]}{\gamma - K''[1 - \exp{(\gamma F)}]}} \qquad (9\text{-}55)$$

$$D = \frac{\dfrac{\beta - K'}{\beta - K'[1 - \exp{(\beta F)}]}(\theta_1 - \theta_1') - \dfrac{\beta - K''}{\beta - K''[1 - \exp{(\beta F)}]}(\theta_1 - \theta_1'')}{\dfrac{\beta - K'}{\gamma - K'} \cdot \dfrac{\gamma - K'[1 - \exp{(\gamma F)}]}{\beta - K'[1 - \exp{(\beta F)}]} - \dfrac{\beta - K''}{\gamma - K''} \cdot \dfrac{\gamma - K''[1 - \exp{(\gamma F)}]}{\beta - K''[1 - \exp{(\beta F)}]}} \qquad (9\text{-}56)$$

In both cases A is obtained from Eq. (9-50) by setting $f = 0$:

$$A = \theta_1 - (B + D) \qquad (9\text{-}57)$$

If the values of A, B and D have been determined using these equations, the temperature distribution of the three gases can be calculated completely using the solutions (9-50), (9-51) and (9-52) utilizing at the same time Eqs. (9-43), (9-48) and (9-49). However the reverse problem of determining F', F'' and thus $F = F' + F''$ from the entry and exit temperatures given can only be solved by iterative methods.

Method of Approximation for Calculating the Temperature Distribution with Variable Specific Heats

If the thermal capacities C, C' and C'' vary with temperature or k' and k'' and thus K' and K'' from Eq. (9-43) are not constant, it is possible, as long as these variations are only slight, to still use the equations above by inserting suitable average values for the quantities involved. When the variations are large it is often possible to consider the thermal capacities to be constant, at least section by section. It is then useful to consider or to estimate in some way that the temperature differences at one end of the recuperator are given and then to carry through the calculation sequentially for each section using Eqs. (9-50) to (9-54).

It is possible to consider the variations in the thermal capacities more precisely by a step-by-step process in which the temperatures in Eqs. (9-37), (9-38) and (9-39) are replaced by the corresponding enthalpies. Such a process was described briefly in the first edition of this book, pp. 260 to 261, which are reproduced as Appendix 2 in this book.

TEN

THE INCREASE IN ENTROPY IN HEAT EXCHANGERS

10-1 INCREASE IN ENTROPY AND ENERGY EXPENDITURE DUE TO THE IRREVERSIBILITY OF HEAT TRANSFER

In the thermodynamic sense *the processes in a heat exchanger are not reversible*. This is because at each position in the exchanger heat can only be transferred from the gas at the higher temperature to the gas at the lower temperature and a heat transfer in the opposite direction is not physically possible. It follows from the second law of thermodynamics that a constant increase in entropy must take place in the heat exchanger, as long as heat is not lost to the surroundings. This will be determined in the following way.

When a quantity of heat dq is transferred from a body at the absolute temperature T to a body at a lower temperature T', the entropy of the first body decreases by dq/T and that of the second body increases by dq/T'.[1] The increase in

---

[1] It is assumed here that during the exchange of heat, each of these bodies experiences only a reversible change of state because otherwise the definition of entropy $ds = dq/T$, which is only valid for reversible changes in state could not be applied. It could be supposed, for example, that the temperature drop occurs either in a very thin layer at the boundary surface of both bodies (for example the boundary layer that occurs with flow in tubes) or in a very thin partition wall of low conductivity; in such cases the entropy of this layer or wall can be neglected in the entropy equation.

In a heat exchanger which is in a state of equilibrium, a drop in temperature also occurs in walls that are thick and walls with low conductivity. For the following reason this does not come into the calculation of the entropy. The temperature and the thermodynamic condition of the material of construction remain unchanged at each position within the walls. Thus the entropy at each position in the walls also remains unchanged. Although in such cases the walls make far reaching contributions to the irreversibility, the increase in entropy to be calculated accordingly occurs in the gases alone which flow through the heat exchanger.

entropy in such a heat exchanger is therefore equal to $dq(1/T' - 1/T)$. This result can be applied to each individual position in a heat exchanger if T and T' are understood to be the absolute temperatures of the two gases. Now $dq = d\dot{q}$ is the rate at which heat is transferred through the heating surface area df at the position being considered. By integrating the entropy increase over all positions in the exchanger, Eq. (10-1) is obtained where $d\dot{q} = -\dot{m}\,dh = \dot{m}\,dh'$ is the rate at which the total entropy increases in the heat exchanger

$$\Delta S = \dot{m}' \int_{\theta_1}^{\theta_2} \frac{dh'}{T'} - \dot{m} \int_{\theta_2}^{\theta_1} \frac{dh}{T} \tag{10-1}$$

Equation (10-2) follows from this once it is noted that $m\,dh = C\,dT$ and $m'\,dh' = C'\,dT'$ where the values of C and C' are constant

$$\Delta S = C' \ln \frac{T'_2}{T'_1} - C \ln \frac{T_1}{T_2} \tag{10-2}$$

and where T_1 and T'_1 denote the absolute entry temperatures and T_2 and T'_2 denote the absolute exit temperatures of the gases. In contrast, if the specific thermal capacities of the gases, and thus C and C', are dependent on the temperature, ΔS can be determined most easily by obtaining the entropy values s_1 and s'_1 for the entry temperatures and the entropy values s_2 and s'_2 for the exit temperatures from T, s or h, s diagrams for the relevant materials or by determining them from a corresponding state equation. The rate of increase in total entropy is then calculated using

$$\Delta S = \dot{m}'(s'_2 - s'_1) - \dot{m}(s_1 - s_2) \tag{10-3}$$

The establishment that an increase in entropy takes place in a heat exchanger and can be calculated is only of theoretical interest. According to the second law of thermodynamics each entropy increase ΔS_{irr} caused by the irreversibility of a process signifies a loss which manifests itself in practice as an entry loss or an increased use of heat and work.

It is relatively easy to overlook this in processes which use energy and where the materials taking part—apart from a possible internal reciprocal exchange of heat—transfer heat in the outward direction only to bodies at the temperature of the surroundings (for example cooling water). By way of example one can think of the compression of air or the separation of a gas mixture at low temperatures. Before cooling both processes require compression of the air or the gases which are to be separated. The heat of compression which arises here is led away into the cooling water. However, in the long run, the "cold" loss which takes place in compression or separation equipment implies an input of heat from the surroundings so that in fact heat can only be exchanged with the surroundings in the outward direction. Increases in entropy in these processes are due to the incompleteness of the transfer of "cold" in the counterflow heat exchangers and also to the pressure and "cold" losses as well as to the throttling that occurs in the valves etc. This increase in entropy can only be compensated if there is an increase in the expenditure of energy, that is by compressing the air or gas at a correspondingly higher pressure.

A simple relationship between the increase in entropy ΔS_{irr} caused by the irreversibility and the smallest increase $\Delta A'$ in the expenditure of energy A' necessary to achieve this compensation can be derived from the second law of thermodynamics for these conditions where heat is only exchanged with the surroundings in the outward direction. This relationship, which will be established more thoroughly later on, takes the form:

$$\Delta A' = T_u \cdot \Delta S_{irr} \tag{10-4}$$

where T_u denotes the absolute temperature of the surroundings. Using this equation it is possible to specify the increase in the work of compression caused by the irreversibility for each part of the process, as well as for the total process, provided that the corresponding increases in entropy ΔS_{irr} have been determined.

If we apply Eq. (10-4) to heat exchangers we can say: "The calculation of the increase in entropy allows us to make a decision as to how much the imperfection of the heat exchanger influences the total process involved in the exchange of heat. In particular, in many cases such a calculation can determine how large an increase in energy expenditure is brought about by the exchange of heat". These considerations are closely connected with the idea of exergy but this will not be discussed further here.

It follows from previous discussions that it is possible to suppose that in a perfect heat exchanger, as defined in Sec. 5-11 (counterflow where $kF = \infty$), there is no increase in entropy, and thus also no correspondingly detrimental effect on the total process takes place. Moreover a completely reversible transfer of heat is only possible, even in a perfect heat exchanger, if the temperature difference $\theta - \theta' = 0$ is present at all positions. However this assumes that the thermal capacities C and C' of both gases are equal to one another. However, if C and C' are different $\theta - \theta'$ can, in general only be infinitely small at a particular position, even in a perfect heat exchanger, as has already been explained. Thus, even in a perfect heat exchanger, the final temperature differences cause an irreversibility at all other positions and thus a corresponding increase in entropy ΔS_{irr}.

From this could be drawn the incorrect conclusion that the smallest increase in entropy occurs when $C = C'$ even in an actual counterflow heat exchanger with a final average temperature difference $\Delta \theta_M$, that is when the temperature differences $\theta - \theta'$ are the same size at all positions. However, as the following consideration shows, this becomes less and less true the lower the absolute temperature. Indeed, when the entropy increases by $dS = dq/T' - dq/T = dq\dfrac{T - T'}{TT'}$ for the transfer of a small quantity of heat dq, when dq and $\theta - \theta' = T - T'$ remain the same, dS is greater the lower the temperatures T and T'. Thus, if the increase in entropy is to be kept small in a counterflow heat exchanger, the temperature differences in the colder parts of the heat exchanger must be smaller than the temperature differences in the warmer parts.

The Expenditure in Energy Required to Counteract an Irreversible Increase in Entropy (Based on Eq. (10-4))

In the T, S diagram in Fig. 10-1 the irreversible increase in entropy ΔS_{irr} takes place from 1 to 2 at a temperature below that of the surroundings T_u. In order that this increase in entropy should be caused only by irreversibility, it will be assumed that during the process neither the heat nor work from the surroundings will be added to or removed from the working medium which can consist of several substances. Such an increase in entropy can be counteracted by the following irreversible processes involving the working medium:

1. Reversible adiabatic compression from 2 to 2′.
2. Reversible isothermic compression from 2′ to 1′ with the emission of a quantity of heat Q to the surroundings at temperature T_u.
3. Reversible adiabatic expansion from 1′ to 1.

If the irreversible process from 1 to 2 is included, a cyclic process is formed. The first law gives Eq. (10-5) for this cyclic process:

$$-Q = \Delta H + \Delta A = 0 + \Delta A = -\Delta A' \qquad (10\text{-}5)$$

where $\Delta A'$ denotes the work added to the medium. The change ΔH of the enthalpy of the working medium is zero because at the end of each cyclic process the initial state and thus the original value of H will again be attained.

One basic assumption is that heat will only be given off to the surroundings. This heat is equal to $-Q$ as extracted from the cyclic process. The entropy increase ΔS_{irr} is counteracted directly by this emission of heat so that $S'_1 - S'_2 = S_1 = S_2 = -\Delta S_{irr}$. As the emission of heat should be reversible $S'_1 - S'_2 = -\Delta S_{irr} = -Q/T_u$ or

$$Q = T_u \cdot \Delta S_{irr} \qquad (10\text{-}6)$$

However, as shown above, according to the first law

$$Q = \Delta A' \text{ also}$$

and following from Eq. (10-4)

$$\Delta A' = T_u \cdot \Delta S_{irr}$$

yields the *least expenditure of energy* necessary to *cancel* the irreversible *increase in entropy*.

An objection could be raised to all this namely that, in general, the irreversible

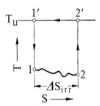

Figure 10-1 Irreversible increase in entropy and reversible return movement.

increase in entropy does not occur in the isolation which is an assumption built into Fig. 10-1. Rather the irreversible increases in entropy are in fact superimposed, to varying extents, over all parts of the process. However, Eq. (10-4) is also applicable to similar, more general cases, as can be shown by comparing an irreversible cyclic process with a reversible cyclic process both of which achieve the same results. This is shown in Fr. Bošnjakovič's book "Technische Thermodynamik", section one [B208] 1935, pp. 89 and 90 (and also the fifth edition, 1967, pp. 107 to 109). Here it is shown that in processes which take place above the *temperature of the surroundings* the irreversibilities cause a *loss of work* which can be calculated using Eq. (10-4).

The derivation of Eq. (10-4) given here is particularly suitable for the processes which take place in a heat exchanger if the heat exchanger is considered to be separate from its surroundings. The irreversible change in state considered does not contradict the fact that in an open-circuit system heat is transferred internally and the pressure drops also contribute to the increase in entropy. The calculation of the increase in entropy is shown above.

However if it is required to apply Eq. (10-4) to the exchange of heat between the open-circuit system and its surroundings, it is possible to include in the open-circuit system those substances in the surroundings from which the heat originates and to apply the ideas presented above to the enlarged system or, more simply, to proceed from the premise that Eq. (10-4) is generally valid. One reaches the conclusion by either of these two possible approaches.

If a quantity of heat Q is taken from the surroundings at temperature T_u and added to the open-circuit system at temperature T, the entropy will be increased by

$$\Delta S = Q\left(\frac{1}{T} - \frac{1}{T_u}\right)$$

The additional work necessary for the total process to take place is, using Eq. (10-4),

$$\Delta A' = Q\frac{T_u - T}{T}$$

This is exactly the same work that is required to enable the quantity of heat Q to be given back to the surroundings in an irreversible Carnot cyclic process.

In the work of P. Grassmann [G208], the connection is discussed between the type of increase in entropy described here and the loss in the technical capacity for doing the work which is known, today, as exergy.

10-2 LITERATURE FOR PART THREE: RECUPERATORS

See Foreword, Sec. 1–5.

A

201. Altenkirch, E.: Graphische Ermittlung von Heiz- und Kühlflächen bei ungleichmäßiger Wärmeaufnahmefähigkeit der Wärmeträger. Z. ges. Kälte-Ind. (1914) 189–193.

B

201. Bahnke, G. D.; Howard, C. P.: The Effect of Longitudinal Heat Conduction on Periodic Flow Heat Exchanger Performance. ASME Paper No. 63-AHGT-16.
202. Bammert, K.; Kläukens, H.; Mukherjee, S. K.: Auslegung und Konstruktion von Wärmeaustauschern für geschlossene Gasturbinenanlagen. Brennstoff-Wärme-Kraft 22 (1970) 275–279.
203. Becker, J.: Ausführungsbeispiele für Wärmeaustauscher in Chemieanlagen. Verf. Tech. 3 (1969) 8, 335–340.
204. Bender, E.: Zum regeltechnischen Verhalten von Kreuzstrom-Wärmeaustauschern. Chem. Ing. Tech. 45 (1973) 350–356.
204a. Bentwich, M.: Multistream Countercurrent Heat Exchangers. Trans. ASME, J. Heat Transfer (1973) 458–463.
205. Billet, R.: Verdampfertechnik. Hochschultaschenbücher No. 85. Bibliographisches Institut Mannheim: 1965.
206. Boehm, J.: Zur Beurteilung der Wärmedurchgangszahlen bei veränderlichem Durchsatz und Heizflächenverschmutzung von Wärmeaustauschern. Gesundh. Ing. 72 (1951) 291–294.
207. Böhm, H.: Versuche zur Ermittlung der konvektiven Wärmeübergangszahlen an gemauerten engen Kanälen. Arch. Eisenhüttenw. 10 (1933) 423–431.
208. Bošnjaković, Fr.: Techn. Thermodynamik, Part. 1 Dresden and Leipzig: Steinkopff 1935 5th ed. Dresden 1967.
209. Bošnjaković, F.; Vilicić, M.; Slipcević, B.: Einheitliche Berechnung von Rekuperatoren. VDI-Forschungsheft No. 432.
210. Bowman, R.·A.: Mean Temperature Difference Correction in Multipass Exchangers. Ind. Eng. Chem. 28 (1936), 541–544; also Bowman, R. A.; Mueller, A. C.; Nagle, W. M. Mean Temperature Difference in Design. Trans. Am. Soc. Mech. Engrs., 62 (1940) 283–294 (Report in Feuerungstechnik, 31 (1943) 159).
211. Browder, T. J.: Shell-Side Heat-Transfer Coefficient. Chem. Eng. 67 (1960) 21, 206 and 207.
212. Burgmüller, P.; Roduner, H.: Der Rekuperator für einen schnellen Leistungsreaktor von 1000 MWe mit direktem Gasturbinenkreislauf. Brennstoff-Wärme-Kraft 21 (1969) 10, 527–531.
213. Buskies, U.: Wärmeaustauscher und Verdampfer (Achema 1970). Chem. Ing. Tech. 42 (1970) 23, 1411–1415.

C

201. Cammerer, I. S.: Die Berechnung des praktischen Wärmeschutzes der Baustoffe aus ihrer Wichte. Heizung und Lüftung (1943) 75–81. I. S. Cammerer, Der Wassergehalt organischer Dämmstoffe in Abhängigkeit von der Luftfeuchtigkeit. Z. ges. Kälteindustrie 51 (1944) 88–91.
202. Clayton, D. G.: New Concepts for Heat Exchanger Performance. (Wirkungsgrad, Gütegrad). Wärme- und Stoffübertragung 7 (1974) 107–112.
203. Collins, S. C.; Keyes, F. G.: Gegenströmer der Tieftemperaturtechnik. J. Physik Chem. 43 (1939) 5.
204. Condamin, R.: Échangeurs compacts et miniaturisés (Températures inférieures à 200°C). Journées de la transmission de la chaleur. Paris 1964, Lecture No. 6.21.
205. Cox, B.: Jallouk, P. A.; Methods for Evaluating the Performances of Compact Heat Transfer Surfaces. (Wärmeübertragung, Energieaufwand, Kosten). Trans. ASME, J. Heat Transfer (1973) 464–469.
206. Cox, J. E.; Kumar, C. A.: Sizing of Heat Exchangers with Non-Uniform Coefficients. Wärme- und Stoffübertragung 6 (1973) 1 and 2.

D

201. Davis, K. F.: Ross Heater and Mfg. Co. Bull. 350 (1931) 72.
202. Dia, T.: Experimentelle Untersuchungen zur Dimensionierung der Zu- und Abflußhauben von Wärmeaustauschern. Diss. Aachen 1958.
203. Dibbern, D.: Reif- und Schneebildung beim Abkühlen von Gas-Dampf-Gemischen in Gegenstrom-Wärmeaustauschern. Abh. d. Deutsch. Kältetechnischen Vereins 1963.

204. Dinglinger, G.: Die Wärmeübertragung im Kratzkühler. Diss. d. TH Karlsruhe (1963). Kältetechn. 16 (1964) 6, 170.
205. Doetsch, H.: Die Wärmeübertragung von Kühlrippen an strömende Luft. Abhandl. Aerod. Inst. Techn. Hochschule Aachen, No. 14, pp. 3–23. Berlin: Springer 1934.
206. Dolz, H.: Anwendungsbereiche und Entwicklungstendenzen von Wärmeübertragern der Kältetechnik. Luft- und Kältetechn. (1968) 6, 274–279.
207. Dummet, G. A.: Plattenwärmeaustauscher. Dechema-Monogr. No. 26 (1956) 168–198.
208. Dupuy, M. R.: Échangeurs rationnels récents. Journées de la transmission de la chaleur, Paris 1964, Lecture No. 6.20.

E

201. Ediss, B. G.: Graphic Aids for Heat Exchanger Computation. Chem. and Process Engineering 43 (1962) 8, 384–388.
202. Eckert, E. R. G.: Einführung in den Wärme- und Stoffaustausch. 1st ed. 1949, pp. 16 and 17; 3rd ed. 1966, p. 27; see [3].

F

201. Fekete, K.: Die Berechnung der Wärmeübergangszahl in Luftkühlern mit Berücksichtigung der Kondensation. Heizung-Lüftung-Haustech. 18 (1967) 3, 95–98.
202. Fischer, E.: Berechnung des Wärmedurchganges bei Gleich- und Gegenstrom für temperaturabhängige Wärmeübergangszahlen. Chem. Ing. Tech. 39 (1967) 7, 438–440.
203. Fischer, H.: Einfluß der Spalte zwischen Umlenkblechen und Rohren auf den Wärmeübergang bei Wärmeaustauschern. Chem. Ing. Tech. 40 (1968) 11, 525–528.
204. Fischer, K. F.: Mean Temperature Difference Correction in Multipass Exchangers. Ind. Eng. Chem. 30 (1938) 377.
205. Flaxbart, E. W.; Schirmer, D. E.: Economic Considerations in Shell and Tube Heat Exchanger Selection. Chem. Eng. Progr. 57 (1961) 98–105.
206. Fricke, L. H.; Morris, H. J.; Otto, R. E.; Williams, Th. J.: Process Dynamics and Analogue-Computer Simulation of Shell and Tube Heat Exchangers. Chem. Eng. Progr. Symp. Ser. 56 (1960) 31, 80–85; Advances in Computational and Mathematical Techniques in Chemical Engineering.
207. Fritzsche, A. F.: Allgemeine Bewertung von Rohrbündel-Wärmeübertragern der Längsstrom- und der Querstrombauart. VDI-Forschungsheft 450 (1955) 5–18.
208. Fukur, S.; Sakamoto, M.: Some Experimental Results on Heat Transfer Characteristic of Air Cooled Heat Exchangers for Air Conditioning Devices. Bull. JSME 11 (1968) 44, 303–311.
209. Funck, K.: Die Bedeutung von Wärmeübertragern für Verdichteranlagen. Allg. Wärmetech. 10 (1961) 6, 105–109.

G

201. Gardener, H. S.; Siller, J.: Shell Side Coefficients of Heat Transfer in a Baffled Heat Exchanger. Trans. Am. Soc. Mech. Engrs. 69 (1947) 687.
202. Gia, V. V.: Récupérateurs et régénérateurs de chaleur. Bucarest and Paris 1970.
203. Glaser, H.: Bewertung von Wärmeaustauschsystemen mit Hilfe einer Leistungszahl. Angew. Chemie B, vol. 20 (1948) 129–133.
204. Glaser, H.: Der Gütegrad von Wärmeaustauschern. Chemie-Ingenieur-Technik (Jahrg. 1949) 95–99.
205. Graßmann, P.: Über den Wirkungsgrad von Wärmeaustauschern. Ann. d. Physik, 5th Series, 42 (1942) 203–210.
206. Grassmann, P.: Zur Berechnung der Ein- und Austrittstemperaturen von Wärmeaustauschern. Z. VDI-Beihefte Verfahrenstechnik 1943, No. 3, 87–90.
207. Graßmann, P.: Bewertung von Wärmeübergang und Druckverlust in Wärmeaustauschapparaten. Angew. Chemie B 20 (1948) 289–292.

208. Grassmann, P.: Zur allgemeinen Definition des Wirkungsgrades. Chem. Ing. Tech. 22 (1950) 77–96.
209. Gregorig, R.: Energieverluste der Wärmeaustauscher. Chem. Ing. Tech. 37 (1965) 108–116, 524–527, 956–962.
210. Gregorig, R.: Über ein Teilproblem der Optimierung beim Dimensionieren eines Wärmeaustauschers. Verfahrenstechnik 1 (1967) 19–22.
211. Gregorig, R.; Alvensleben, B.: Zur elementaren Optimierung von Rohrbündel-Wärmeaustauschern. Verfahrenstechnik 2 (1968) 2, 63–71.
212. Gregorig, R.: Wärmeaustausch und Wärmeaustauscher. 2nd ed. Aarau and Frankfurt: Sauerländer 1973.
213. Gumz, W.: Vorschlag zur Bewertung von Luftvorwärmern. Arch. Wärmewirtschaft 11 (1930) 195 and 196.

H

201. Hackeschmidt, M.; Vogelsang, E.: Zur Berechnung von Verdunstungskühlern unter Berücksichtigung der Strömungsverhältnisse im Bereich der Stoff- und Wärmeübertragung. Luft- und Kältetech. 4 (1968) 3, 114–118.
202. Haddad, M. T.: Régime dynamique des échangeurs de chaleur. Traitement analytique-simulation-calcul iteratif. 2 parts. Revue génerale de thermique 3 (1964) 34, 1251–1269; 35, 1431–1440.
203. Hahnemann, H. W.: Approximate Calculation of Thermal Ratios in Heat Exchangers, Including Heat Conduction in Direction of Flow. National Gas Turbine Establishment Memorandum No. M 36, 1948.
204. Hausen, H.: Über die Berechnung von Luftverflüssigungsanlagen auf Grund neuer Messungen des Thomson-Joule-Effektes. Z. Ges. Kälte-Ind. 32 (1925) 93–98 and 114–122 (in particular in the appendix p. 121).
205. Hausen, H.: Materialtrennung durch Destillation und Rektifikation, in Eucken, A.; Jakob, M.: Der Chemie-Ingenieur. Leipzig. Vol. I, Part 3, 1933, pp. 70–169, in particular A. 117 and 118.
206. Hausen, H.: Die Wirkung des Austausches von Rektifikationsböden. Z. ang. Math. Mech. 17 (1937) 25–37.
207. Hausen, H.: Graphisches Verfahren zur Berechnung der Wirkung von Rektifikationsböden. Z. ges. Kälte-Ind. 44 (1937) 59–65 (in particular pp. 64 and 65).
208. Hausen, H.: Gestaltung und Wirkung der Wärmeaustauscher für strömende Stoffe. Z. VDI, Beiheft "Verfahrenstechnik" (1940) No. 1, 1–6.
209. Hausen, H.: Wärmeübertragung durch Rippenrohre. Z. VDI-Beiheft "Verfahrenstechnik" (1940) No. 2, 55–57.
210. Hausen, H.: Ein allgemeiner Ausdruck für den Wärmedurchgang durch ebene, zylindrische und kugelförmige Wände. Archiv gesamte Wärmetechnik 2 (1951) 123 and 124.
211. Hatfield, M. R.; Ford, C. E.: Development of "Karbate" Materials and their Applications. Trans. Am. Inst. chem. Engrs. 42 (1946) 121.
212. Hilz, R.: Verschiedene Arten des Ausfrierens einer Komponente aus binären strömenden Gasgemischen. Z. ges. Kälteind. 47 (1940) 34, 74 and 88.
213. Hochgesand, G.: Wärmeaustauscher und Verdampfer, Apparate für die Rektifikation. Chem. Ing. Tech. 38 (1966) 7, 761 and 762.
214. Hofmann, E.: Über die Berechnung von Kühlern für Gas-Dampf-Gemische. Z. ges. Kälteind. 49 (1942) 70 and 79.
215. Hofmann, E.: Wärmedurchgangszahlen von Rippenrohren bei erzwungener Strömung. Z. ges. Kälte-Ind. 51 (1944) 84–88.
216. Huber, A.: Ein zusammengesetzter Wärmeaustauscher (Theorie). Österreichisches Ingenieur-Archiv 16 (1961) 174–178.

J

201. Jaroschek, K.: Einfluß der Wärmedurchgangszahl und der Wärmewirtschaft auf die Heizflächenbemessung von Wärmeaustauschern. Z. VDI. Beiheft "Verfahrenstechnik" (1943) No. 2, 52–58.

202. Jaroschek, K.: Bedeutung der Heizflächenbemessung von Wärmeaustauschern, insbesondere in Kraftwerksbetrieben. Z. VDI 87 (1943) 210–213.

203. Johannsen, L. O.; Holschuh, A.: Stahlwinderhitzer für Hochöfen. Stahl und Eisen 57 (1937) 1142.

204. Jung, H.: Die Beanspruchung der Rohrböden von Wärmeaustauschern. Chem. Ing. Tech. 42 (1970) 7, 515–520.

K

201. Kämmerer, C.: Temperaturverlauf und Heizflächenbestimmung bei Gegenstrom-Wärmeaustauschern. Arch. Wärmewirtsch. 22 (1941) 153–156.

202. Kamman, D. T.; Koppel, L. B.: Dynamics of a Flow forced Heat Exchanger (experimental). Ind. Eng. Chem. Fundamentals 5 (1966) 2, 208–211.

203. Kays, W. M.: The Basic Heat Transfer and Flow Friction Characteristics of Six Compact High-Performance Heat Transfer Surfaces. Trans. ASME Ser. A 82 (1960) 1, 27–34. (Bases for calculation obtained from measurement. The arrangements consist of plates with superimposed fins.)

204. Kays, W. M.; London, A. L.: Compact Heat Exchangers. 2nd ed. New York: McGraw-Hill, 1964.

205. Kirschbaum, E.: Wirkung von Rektifizierböden und zweckmäßige Flüssigkeitsführung. Forschung Ing. Wes. 5 (1934) 245.

206. Kirschbaum, E.: Beanspruchungen infolge Wärmedehnung in Wärmeaustauschapparaten. Z. VDI-Beihefte Verfahrenstechnik (1940) 6, 167–170.

207. Kirschbaum, E.: Wärmedurchgang durch Rohre mit Schutzschichten. Z. VDI 86 (1942) 337 and 338.

208. Kirschbaum, E.; Wachendorff, W.: Verkrustung der Heizflächen in Verdampfapparaten. Verfahrenstechnik (1942) 3, 61–71.

209. Kirschbaum, E.: Wärmeübertragung und Druckverlust in Wärmeaustauschern aus Porzellan. Z. VDI Beihefte "Verfahrenstechnik" (1944) 1, 6–12.

210. Kirschbaum, E.: Wärmeübertragung und Druckverlust in Wärmeaustauschern aus Porzellan. Z. VDI-Beihefte "Verfahrenstechnik" (1944) 1, 6–12.

211. Kirschbaum, E.: Destillier- und Rektifiziertechnik. 4th ed. Berlin, Heidelberg, New York: Springer 1969, pp. 185 and 456.

212. Kistner, H.: Bestimmung der Wärmeübergangszahlen und Druckverluste bei doppelt versetzter und nicht versetzter Rostpackung. Arch. Eisenhüttenw. 3 (1929/30) 751–768.

213. Klempt, W.: Kontaktöfen und Kontaktapparate in der chemischen Industrie. Z. VDI-Beihefte "Verfahrenstechnik" (1939) 122–127.

214. Klingen, B.; Eisen, W.: Verbesserung der Rohrform für Konvektionswärmeaustauscher in Gußausführung. Brennstoff-Wärme-Kraft 17 (1965) 9, 449–452.

215. Kraußold, H.: Wärmeaustauscher, ein Überblick über die Entwicklung der letzten Jahre. Verfahrenstechnik 2 (1968) 203–209.

216. Kraußold, H.: Wärmeaustauscher. Lueger, Lexikon der Verfahrenstechnik. Stuttgart: 1970, pp. 562–566.

217. Kühl, H.: Probleme des Kreuzstrom-Wärmeaustauschers. Berlin, Göttingen, Heidelberg: Springer 1959.

218. Kühne, H.: Vorschläge zur genauen Festlegung und Prüfung der Leistungsgarantien von Kreislaufkühlern für Turbogeneratoren. Elektrotechn. Z. 50 (1929) 1543, Section 2.

219. Kühne, H.: Beitrag zur Frage der Aufstellung von Leistungsregeln für Wärmeaustauscher. Z. VDI-Beiheft "Verfahrenstechnik" (1943) 2, 37–46, Section 9.

220. Kühne, H.: Wirkungsgrad und Wirtschaftlichkeit von Wärmeaustauschern. Z. VDI-Beiheft "Verfahrenstechnik" (1944) 2, 47–53.

221. Kühne, H.: Schaubilder zur Ermittlung der Temperaturen von Kreuzstromwärmeaustauschern. Haustechnische Rundschau 49 (1944) 17/18, 161–164.

222. Kühne, H.: Winke für die Bestellung und Leistungsprüfung von Wärmeaustauschern. Halle: C. Marhold 1945 (28 pages); 2nd ed. 1948.

223. Kühne, H.: Die Grundlagen der Berechnung von Oberflächen-Wärmeaustauschern. Göttingen 1949, Table 32, p. 192.

224. Kühne, H.: Die Bewertung von Wärmeaustauschflächen mittels einer energetischen Wärmeübergangsgleichung. Chemiker-Ztg. chem. Apparaturen 85 (1961) 20, 778–784.
225. Kühne, H.: Über die Bewertung von Wärmeaustauschern. Chem. Apparatur 87 (1963) 12, 441–452.
226. Kühne, H.: Wärmeaustauscher—Vorschläge zur Typisierung und Normung. Chem. Ing. Tech. 36 (1964) 972.

L

201. Landolt-Börnstein, Zahlenwerte und Funktionen. Vol. IV 4 b. Berlin, Heidelberg, New York: Springer 1972, pp. 417–433 and 454–481 (Heat and cold protection).
202. Lienerth, A. J.: Auslegung von Gas/Gas-Röhrenwärmeaustauschern bei vorgeschriebenen Druckverlusten. Verfahrenstechnik 1 (1967) 6, 261–267.
203. Linde, R.: Die konstruktive Ausbildung von Anlagen zur Gaszerlegung und ihre Anpassung an die theoretischen Forderungen. Z. ges. Kälte-Ind. 41 (1934) 161–183. (Low temperature technology counterflow heat exchanger.)
204. Linde, H.: Über das Ausfrieren von Dämpfen aus Gas-Dampf-Gemischen bei atmosphärischem Druck. Z. angew. Physik 2 (1950) 49–59.
205. Linke, W.: Untersuchungen über Rohrbündelwärmeübertrager. Chem. Ing. Tech. 27 (1955) 142–148.
206. Linke, W.; Dia, T.; Skupinski, E.: Die Durchflußeistung von Wärmeübertragern mit künstlicher Turbulenz. Allgem. Wärmetech. 11 (1961) 19–25.
207. Linke, W.; Dia, T.: Die Bemessung der Zu- und Abflußhauben von Wärmeaustauschern. Kältetechnik 15 (1963) 85–91.
208. Lueger.: Lexikon der Verfahrenstechnik, Stuttgart 1970. Stichwort Wärmeaustauscher, pp. 562–566; Stichwort Reaktionsapparate, pp. 388–391.

M

201. Matulla, H.; Orlicek, A. F.: Bestimmung der Wärmeübergangskoeffizienten in einem Doppelrohrwärmeaustauscher durch Frequenzganganalyse. Chem. Ing. Tech. 43 (1971) 20, 1127–1130.
202. McKillop, A. A.; Dunkley, W. L.: Heat Transfer Plate Heat Exchangers. Ind. Eng. Chem. 52 (1960) 9, 740–744.
203. Meißner, W.: Über die Vorgänge in den Gegenstromapparaten der Gasverflüssiger. Z. techn. Physik 7 (1926) 235.
204. Meißner, W.; Immler, R.: Über die Temperaturabhängigkeit der Wärmeleitfähigkeit einiger Baumaterialien zwischen −15 und +30°C. Wärme- und Kältetechnik (1937) 10, 1.
205. Meißner, W.; Immler, R.: Einfluß des Wassergehaltes auf die Wärmeleitfähigkeit von Isolierstoffen. Wärme- und Kältetechnik 40 (1938) 9, 129; see also Wärme- und Kältetechnik 39 (1937) 10, 1.
206. Mollenhauer, J.: Wärmeübergang und Druckverlust in Plattenwärmeaustauschern. Diss. Berlin 1967. VDI-Z. 111 (1969) 8, 533.
207. Morley, T. B.: Exchange of Heat between Three Fluids. The Engineer 155 (1933) 134 (short extract appears in "Forschung" 4 (1933) 153).
208. Moussez, C.; Morin, R.: Étude sur un échangeur-bouilleur. Journées de la transmission de la chaleur, Paris 1964. Lecture No. 6.33.

N

201. Nagel, O.: Zusammenhänge zwischen Wärmeübergang und Phasenänderung im Umlaufverdampfer. Chem. Ing. Tech. 35 (1963) 3, 179–185.
202. Nagle, W. M.: Mean Temperature Differences in Multipass Heat Exchangers. Ind. Eng. Chem. 25 (1933) 604–609.
203. Nesselmann, K.: Der Einfluß der Wärmeverluste auf Doppelrohrwärmeaustauscher. Z. ges. Kälteind. 35 (1928) 62 or Wiss. Veröffentlichungen a. d. Siemens-Konzern 6 (1928) 174.
204. Neußel, E.: Gasströmung und Wärmeaufnahme bei Rippenrohr-Vorwärmern. (Definition des Wirkungsgrades). Arch. Wärmewirtsch. 13 (1932) 266–270.

205. Norton, C. L.: Pebble Heaters (Recuperators with moving pebbles). Chem. Eng. 53 (1946) 116.
206. Nußelt, W.: Der Wärmeübergang im Kreuzstrom. Z. VDI 55 (1911) 2021–2024.
207. Nußelt, W.: Eine neue Formel für den Wärmedurchgang im Kreuzstrom. Techn. Mech. u. Therm. 1 (1930) 417–422 (in particular equations nos. (17) and (18)).

P

201. Palmor, Z.; Dayan, J.; Avriel, M.: Evaluating the Performance of Shell and Tube Heat-Exchangers. Israel J. Tech. 11 (1973) 273.
202. Parisot, J.: Graphit als Werkstoff für Wärmeaustauscher. Technische Rundschau No. 54, Dec. 1956.
203. Peters, D. L.: Heat Exchanger Design with Transfer Coefficients Varying with Temperature of Length of Flow Path. Wärme- und Stoffübertragung 3 (1970) 220–226.
204. Peters, D. L.: Cost-Optimized Design of Heat Exchangers and Condensers, particularly Air-Cooled Fin-Tube Units. Chem. Engng. Group; Council for Scientific and Industrial Research. Pretoria 1970.
205. Poßner, L.: Die Gestaltung und Berechnung von Rauchgasvorwärmern. Berlin: Springer 1929.

R

201. Rabes, M.: Theorie der Luftverflüssigung. Z. ges. Kälteind. 37 (1930) 7–12, 26–29 and 48–54 (in particular p. 8).
202. Raisch, E.; Weyh, W.: Die Wärmeleitfähigkeit von Isolierstoffen bei tiefen Temperaturen. Zeitschrift f. ges. Kälteindustrie 39 (1932) 123.
203. Raisch, E.: Untersuchungen der Wärmeleitfähigkeit von Vollwänden in Abhängigkeit von Temperatur und Feuchtigkeit. Z. VDI 80 (1936) 1551.
204. Rische, E. A.: Untersuchungen über das Ausfrieren von Dämpfen aus Gas-Dampf-Gemischen. Diss. Hannover 1957. Chem. Ing. Tech. 29 (1957) 603–614.
205. Robin, M. G.: Quelques aspects de la technique des échangeurs de chaleur à métaux liquides utilisés dans les installations nucléaires. Journées de la transmission de la chaleur. Paris 1964, Lecture No. 6.23.
206. Roetzel, W.: Berücksichtigung veränderlicher Wärmeübergangskoeffizienten und Wärmekapazitäten bei der Bemessung von Wärmeaustauschern. Wärme- and Stoffübertragung 2 (1969) 163–170.
207. Roetzel, W.: Mittlere Temperaturdifferenz bei Kreuzstrom in einem Rohrbündel-Wärmeaustauscher. Brennstoff-Wärme-Kraft 21 (1969) 246–250.
208. Roetzel, W.: Thermal Design of Non-Isothermal Condensers. Wärme- und Stoffübertragung 6 (1973) 228–234.
209. Roetzel, W.: Thermal Design of Condensers; the Limiting Case of Zero Mass Resistance. Wärme- und Stoffübertragung 7 (1974) 60–64.
210. Roetzel, W.: Heat Exchanger Design with Variable Transfer Coefficients for Crossflow and Mixed Flow Arrangements. Int. J. Heat Mass Transfer 17 (1974) 1037–1049.

S

201. Schack, A.: Der physikalische und wirtschaftliche Zusammenhang von Wärmeübertragung und Druckverlust. Arch. Eisenhüttenw. 2 (1928/29) 613–624.
202. Schack, A.: Die Gas- und Luftvorwärmung durch Stahlrekuperatoren in der Großindustrie Z. kompfr. und flüssige Gase 36 (1941) 101.
203. Schack, A.: Échangeurs de chaleur métalliques pour hautes températures. Journées de la transmission de la chaleur. Paris 1964. Lecture No. 7.01.
204. Schack, A.: Der industrielle Wärmeübergang. 7th ed. 1969, p. 283.
205. Schack, K.: Zur Berechnung von Druckverlust und Wärmeübertragung in Wärmeaustauschern mit Umlenkblechen. Gas Wärme Int. 20 (1971) 237–270. Schack, K.: Entwicklung und heutiger Stand des Rekuperatorenbaus. Gas Wärme Int. 21 (1972) 345–347.

206. Schedwill, H.: Thermische Auslegung von Kreuzstromwärmeaustauschern. VDI-Z. 110 (1968) 28, 1245.
207. Schmidt, E.: Die Wärmeübertragung durch Rippen. Z. VDI 70 (1926) 885–889 and 947–951.
208. Schmidt, E.: Wärmeschutz durch Aluminiumfolie. Z. VDI 71 (1927) 1395.
209. Schmidt, E.; Hindenburg, W.: Versuche über die Wärmeabgabe von Rippenrohren. Arch. Wärmewirtschaft 12 (1931) 327–333.
210. Schmidt, E.; Helweg, E.: Temperaturverteilung in den Blöcken im Stoßofen. Forsch. Ing. Wes. 4 (1933) 238–248.
211. Schmidt, E.: The Design of Contra-Flow Heat Exchangers. Proc. Inst. Mech. Engrs. 159 (1948) 351–356.
212. Schmidt, Th. E.: Vergleichszahlen zur Bewertung von Wärmeaustauschern. Kältetechnik 1 (1949) 81–86.
213. Schmidt, Th. E.: Wärmeaustauscher. Brennstoff-Wärme-Kraft 21 (1969) 4, 224–227 (Annual literature review).
214. Schmidt, Th. E.: Zur kostengünstigen Bemessung von Wärmeübertragern mittels thermischer Kenngrößen und Progressionsfunktionen. VDI-Forschungsheft 549 (1972) 39–44.
215. Schneller, J.: Berechnung von Dreikomponenten-Wärmeaustauschsystemen. Chem. Ing. Techn. 42 (1970) 20, 1245–1251.
216. Schukin, V. K.: Verallgemeinerung von Versuchsdaten über die Wärmeabgabe von Schlangenvorwärmern. Teploenergetika (russ.) (1969) 2, 50–52.
217. Schulenburg, F.: Wahl der Bezugslänge zur Darstellung von Wärmeübergang und Druckverlust in Wärmeaustauschern. Chem. Ing. Tech. 37 (1965) 8, 799–810.
218. Söhngen, R.: Graphitwärmeaustauscher. Chem. Ing. Tech. 23 (1951) 4, 81–85.
219. Sonnenschein, H.: Kontinuierlicher Hochtemperatur-Wärmeaustausch mittels bewegter Speicherteilchen. Chem. Ing. Tech. 43 (1971) 5, 240–245.
220. Spalding, D. B.: Neue Wege zur Vorausberechnung der Wirksamkeit von Wärmeaustauschern. Chem. Ing. Tech. 46 (1974) 969–975 (Computer-Models).
221. Stary, F.: Berechnung von Gleich- und Gegenstrom-Wärmeübertragern bei temperaturabhängigen Stoffgrößen. Österr. Ing.-Archiv 16 (1962) 211–230.
222. Steinbach, A.: Die Korrosion in der Kältetechnik und ihre Bekämpfung. Gesundheitsingenieur 67 (1944) 67–72.
223. Stephan, K.: Wärmeaustauscher (Review). Brennstoff-Wärme-Kraft 17 (1965) 221–214.
224. Stephan, K.: Wärmeaustauscher (Survey with 152 bibliographical data). Brennstoff-Wärme-Kraft 18 (1966) 191–194.
225. Stephan, K.: Wärmeleitung und Strömungswiderstand von Spezialrohren für Wärmeaustauscher. Mannesmann-Forschungsber. (1966) No. 363.

T

201. Thiessen, W.: Schweißtechnische Konstruktion von Apparaten unter besonderer Berücktigung der Ausführung von Wärmeaustauschern. Chem. Ing. Tech. 42 (1970) 11, 751–756.
202. Traustel, S.: Über die Übertragungseinheiten und ein "Vierfelderdiagramm" zur Berechnung von Wärmeübertragern. Wärme- und Stoffübertragung 3 (1970) 153–155.
203. Trefny, E.: Wärmeaustausch bei beliebiger Stromart (mittleres Temperaturgefälle, Stufenrechnung). Chem. Ing. Tech. 37 (1965) 8, 835–842.
204. Trommelen, A. M.: Heat Transfer in a Scraped-Surface Heat Exchanger. Trans. Instn. chem. Engrs. 45 (1967) 5, T176–T178.
205. Trumpler, P. R.; Dodge, B. F.: The Design of Ribbon-Packed Exchangers for Low Temperature Air Separation Plants. Trans. Am. Inst. chem. Engrs. 43 (1947) 75–84, also Chem. Eng. Progress 43 (1947) 75.

U

201. Ullmann, Fr.: Enzyklopädie der technischen Chemie, 2nd ed., Vol. X (1932), p. 86; DIN 1062–1068.

202. Underwood, A. J. V.: J. Inst. Petroleum Tech. 20 (1934) p. 145.
203. Upmalis, A.: Wärmeleistung von Heizkörpern aus Drahtspiralrohren. Wärme 75 (1969) 1, 15–18.
204. Urbach, D.: Auslegung von Wärmeaustauschern in Klimaanlagen unter dem Gesichtspunkt guter Regelbarkeit. Heizung-Lüftung-Haustechnik 20 (1969) 167–172.

V

201. VDI-Richtlinien für Leistungsversuche an Wärmeaustauschern. Düsseldorf 1968.
202. VDI-Wärmeatlas, 2nd ed. 1974, Physical properties section.
203. Vollbrecht, H.; Oberstedt, H. W.: Röhrenwärmeaustauscher mit Verdrängerkörpern. Dechema-Monogr. No. 33 (1959) 217–228 and Chem. Ing. Tech. 33 (1961) 19–22.
204. Vollbrecht, H.: Betriebliche Vor- und Nachteile der Glattrohr-Wärmeaustauscher. Chemie-Anlagen und Verfahren (1971) 1, 35–38.
205. Vogelbusch, W.: Verkrustung der Wärmeaustauschflächen in Verdampfern, Vorwärmern und Kondensatoren. Z. VDI. Beihefte "Verfahrenstechnik" (1943) 3, 73.

W

201. Walger, O.: Zur Berechnung von Röhrenwärmeaustauschern. Z. VDI Beihefte "Verfahrenstechnik" (1941) 1, 11–13.
202. Weigand, W. A.: Optimal Control of a Plug-Flow Heat Exchanger with Control Produced by Wall Flux or Wall Temperature. Ind. Eng. Chem. Fundamentals 9 (1970) 4, 641–651 (theoretical).
203. Wenning, H.: Optimierung von Rohrbündel-Wärmeaustauschern mit elektronischen Datenverarbeitungsanlagen. Chem. Ing. Tech. 39 (1967) 9/10, 614–621.
204. Wenzel, H.: Vorausbestimmung der Betriebszeit von Wärmeaustauschern bei Verlegung durch Wasser- und Kohlendioxideis. Kältetech. 20 (1968) 6, 187–191.
205. Whistler, A. M.: (Nachteilige Wirkung eines in der Längsrichtung angeordneten Umlenkbleches bei einem zweigängigen Wärmeaustauscher). Trans. Am. Soc. Mech. Engrs. 69 (1947) 683.
206. Whitt, F. R.: Heat-Transfer Coefficients in Chemical Plant Heat Exchangers. Brit. chem. Eng. 6 (1961) 6, 398–401.

FOUR

REGENERATORS

ELEVEN

REVIEW OF REGENERATOR THEORY

11-1 OPERATION AND CONSTRUCTION OF REGENERATORS

As has been mentioned several times regenerators are understood to be reversible heat exchangers through which the gases flow alternately and in which the heat in transit is temporarily stored in a packing material of high thermal capacity. A continuous operation requires at least two regenerators so that one gas can be cooled and the other heated simultaneously. An exception to this is the case of regenerators with a rotating heat storing mass, of which there need be only one.

Regenerators are clearly distinguished from recuperators by the fact that the operations which take place in regenerators are dependent not only on position but also on time. Account must be taken of this time dependence in precise heat transfer calculations in regenerator theory. This time dependence is relevant also to each element in a rotating heat-storing mass.

Regenerator operation is illustrated by the pair of regenerators shown in Fig. 11-1. Each of the two vessels, which are usually cylindrical, contains a heat-storing mass which is permeable to gas, for example porous, or which has ducts running through it. It is assumed that in practice, the regenerators will always be operated in counterflow. First the gas to be cooled flows into the right-hand regenerator from top to bottom and the gas to be heated flows into the left-hand regenerator from bottom to top. After reversal the opposite process takes place; the originally colder gas passes upwards through the right-hand regenerator and the originally warmer gas passes downwards through the left-hand regenerator. Thus, by regular reversals, the heat-storing mass in each regenerator is washed alternately by hotter and cooler gas, so that its temperature at each position fluctuates periodically. By this means the heat-storing mass is able to absorb the heat to be transferred during

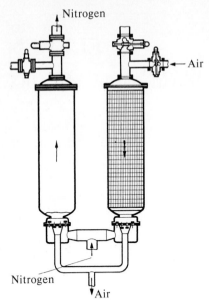

Figure 11-1 Regenerators.

the "hot period" and to return it to the cooler gas during the "cold period". The result is that the originally hotter gas emerges from the regenerator at a lower temperature and the originally colder gas emerges at a higher temperature.

Nature has given us a particular *regenerator prototype* in the nose and also in the trachea and the bronchial tubes of humans and mammals. In winter the spent respiratory air gives up heat to the walls of these airways, not only preventing excessive cooling of the nose and the bronchial tubes but also, when breathing in, preheating the fresh air by the absorption of heat from these walls. This regenerative exchange of heat between the fresh air and the spent respiratory air is thus due to a certain heat storage capacity in these walls. This phenomenon is of course very considerably masked by the continuous supply of heat from the blood flowing through the nose and the other respiratory organs. Nature has solved the problem of regenerator reversal in quite an ideal manner without using valves and the like, by connecting it to the periodic process of respiration.

Industrial Regenerators for High Temperatures

Industrial regenerators are of particularly noteworthy construction in applications for very high or very low temperatures. Even today the *hot blast stoves of blast furnaces* are most prominent by reason of their size and their capacity to exchange heat. With a height of up to 50 m and a diameter of up to about 11 m, two or more hot blast stoves working together can heat 500,000 m^3/h of air, to a temperature of about 1000 to 1300°C. This hot air is called "blast" in the iron and steel industry.

As long as hot blast stoves are built of bricks, as is the usual case, they always function as regenerators. They are reversed every half to one hour. Figure 11-2

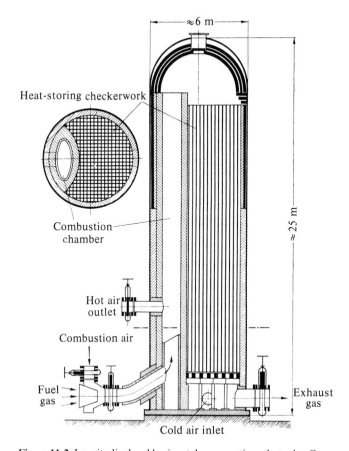

Figure 11-2 Longitudinal and horizontal cross-sections through a Cowper stove.

shows a longitudinal section through a brick hot blast stove. Formerly hot blast stoves usually contained a combustion chamber built of fireproof brick located next to the chequerwork-like heat-storing mass, as shown on the left in Fig. 11-2. Nowadays the combustion chamber is almost always separate. In this chamber the heat which is to be exchanged is produced by burning blast furnace gas or another fuel. The burnt gases enter at the top of the hot blast stove and on their way down heat the heat storing mass and in so doing, are themselves cooled. After the reversal, conversely, the air to be heated flows upwards through the chequerwork. The typical shape of the cross-section of the heat-storing mass is described in Sec. 2-13.

   Another important example of regenerators working at high temperatures is given by *the chambers of a Siemens–Martin Regenerator* in which fuel gas and combustion air are preheated by the heat of the waste combustion gases. The heat-storing mass arranged in these chambers is shaped significantly like the heat-storing mass in the hot blast stove. However, while hot blast stoves are reversed at regular time intervals, as are almost all other regenerators, in the case of the Siemens–Martin furnace, the regenerators are reversed more rapidly towards the end of the period of operation.

The fact that regenerators rather than recuperators are usually used at very high temperatures is principally due to the physical properties of the building materials. Although considerable progress has been made in the development of heat proof steels, (see [S306]), even nowadays fireproof bricks withstand very high temperatures better than do metals. For this reason heat exchangers which work at very high temperatures are generally built of bricks* (see Part Three, Sec. 6-5). In recuperators however, partition walls made of brick are restricted by their limited ability to withstand the pressure difference which exists between the hotter and colder gas and which is approximately 1 bar in hot blast stoves. Such walls have only a slight resistance to pressure, particularly at very high temperatures. Moreover a leakage of small quantities of gas at weak positions in the wall cannot be entirely avoided. On the other hand these difficulties do not arise in regenerators as the heat storing mass is not really exposed to any pressure differences. Of course a certain mixing of carryover gas particles during the reversing process cannot be avoided in regenerators but in most cases this is not very significant.

Regenerators at Medium and Low Temperatures

Metal heat-storing masses are often preferred to brick ones at moderately high temperatures. At low temperatures heat-storing masses were initially only made of metal. Such heat-storing masses were constructed of thin sheet metal, as has already been discussed in Sec. 2-13. The Ljungström air preheater illustrated in cross-section in Fig. 11-3 has a heat-storing mass constructed of flat and corrugated sheet metal.

In the Ljungström air preheater, the heat-storing mass rotates slowly around a perpendicular or horizontal shaft. If the shaft is perpendicular, as shown in Fig. 11-3, the two gases also flow in a perpendicular direction in cross-flow or parallel-flow through a divided cross-section of the heat-storing mass. The process is discontinuous for each very narrow perpendicular section of the heat-storing mass because at any given time it abruptly passes from the cold to the hot gas stream and vice versa. As the process continues each of these sections is subjected to exactly the same temperature variations as the heat-storing mass in a regenerator which is reversed in the normal manner. In both cases the gases enter each section of the heat-storing mass at a constant temperature in both the hot and the cold periods. It follows from this that the following theory which is developed for standard regenerators can also be applied unchanged to the Ljungström air preheaters.

In low temperature technology, which used only recuperators up to about 1930, regenerators have been successfully substituted for pressures up to about 5 bars and occasionally also up to 10 bars. One of the most important reasons for

---

* Editor's note: The words "*ceramic materials*" might be used instead of "*bricks*".

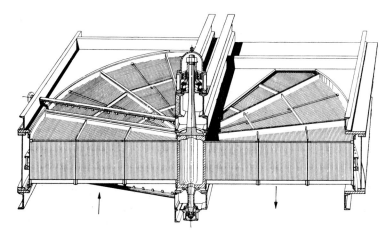

Figure 11-3 Ljungström air preheater.

their introduction is the not inconsiderable pressure differences between the compressed and expanded gases taking part in the exchange of heat. Recuperators are largely restricted to a tubular shape because of stressing due to pressure differences but a metallic regenerator offers almost unlimited possibilities for a suitable geometry of the heat transferring surfaces because the heat storing mass is really only subjected to small pressure differences. The shape of a metal heat storing mass suggested by M. Fränkl [F301] was discussed in Sec. 2-13. Nowadays flint, basalt chippings or the like are usually used as heat-storing masses and this reduces the initial costs. Figure 11-1 which has already been discussed, shows a pair of regenerators used in low temperature technology in which, for example, nitrogen gives up "cold" to air. The gas currents are reversed at intervals of two to three minutes. The large surface area available for heat transfer and the favourable distribution of gas flow in the narrow intersecting ducts or between the small stones produces an exceptional transfer of heat even at a low fluid velocity and, thus, a small pressure loss (see Sec. 2-13). A further important advantage of these regenerators lies in the fact that, as has already been mentioned, the compressed gas, for example air, does not have to be previously purified by the removal of water vapour and carbon dioxide. Rather these components are deposited either in the liquid or solid state, on the heat storing mass and in the following period after evaporation or sublimation, are again taken away by the expanded gases (see Secs. 19-1 to 19-4).

If the regenerators used in low temperature technology are compared with the brick hot blast stoves attached to blast furnaces, the following picture emerges. In hot blast stoves the thickness of the bricks fluctuates between 30 and 200 mm. On the other hand the packing in the regenerators in low temperature technology originally consisted of metal sheets of less than 1 mm thickness and even today the average diameter of the bricks is usually less than 10 mm. The result of this is that a storage volume of $1 \, m^3$ in a normal hot blast stove can only accommodate a heating surface area of from about 20 to $30 \, m^2$; on the other hand a low temperature regenerator can house a heating surface area of from about 400 to $2000 \, m^2$, that is about 20 to 50 times larger. Furthermore, at low temperatures the heat transfer coefficients are substantially higher because of the very small dimensions of the individual ducts although greater fluid velocities are hardly ever used. All this means that in low temperature regenerators with a height of only 4 m a heat exchange efficiency of 98 to 99 per cent can be obtained. However, hot blast stoves have an efficiency of only about 80 to 90 per cent. Thus in low temperature technology an average temperature difference between the two gases of only 2 to 3° is sufficient for the exchange of heat while at high temperatures, a temperature difference of up to 100 times greater than this is necessary.

11-2 DEVELOPMENT OF REGENERATOR THEORY

As the temperature in regenerators varies periodically as well as spatially, an exact calculation is considerably more involved for regenerators than it is for recuperators. Indeed, many of the relationships developed for recuperators can also be used for regenerators such as the heat balance equations, together with the concept of average temperature difference and efficiency and the difference between parallel flow, counterflow and cross flow. However, closer considerations reveal a basic difference in the temperature distribution along the length of the heat exchanger relative to recuperators. In the case of counterflow these differences are only predominantly apparent close to the ends of the regenerator. In the case of parallel flow the variations are much more apparent.

There are really two main directions which can be followed in the mathematical treatment of the processes in regenerators. One direction, which was principally first suggested by representatives of the iron and steel industry, is based on an attempt to transfer the method of calculation developed for recuperators to regenerators with as little alteration as possible and simply to find a suitable modified expression for the heat transmission coefficient. The basis for this is found in the investigation of the temperature distribution in a cross-section of a regenerator brick which serves as a heat storing mass, particularly at high temperatures. Apart from old publications [H315], this treatment appears in works by Heiligenstaedt [H314], Rummel [R304, 305, 306], and Schack [S307] as well as by Hausen [H307, 308], and Stuke [S319].

In these approaches, apart from the last two, it is more or less assumed to be obvious that the temperatures vary in the longitudinal direction of a regenerator as they do in the longitudinal direction of a recuperator. In contrast, the second main line of theory seeks to investigate the problem more deeply by considering the exact variations in this distribution as well as its relationship with the periodic temperature differences. The spatial temperature differences within a cross-section of the heat-storing mass are, on the contrary, generally relegated to the background or completely disregarded. In regenerators built of brick this really amounts to disregarding the internal values of the brick temperature and considering only their spatial mean value in any cross-section of the heat-storing mass. On the other hand, when the heat-storing mass consists of thin metal sheet, no noticeable temperature differences appear within a cross-section of the heat-storing mass. Investigations falling into the second group were originally, to a large extent, suited to the requirements of low temperature technology and have been mainly inspired by this.

Anzelius [A304] and Nusselt [N303] have shown how, employing this assumption, the heat-storing mass is heated or cooled by the continuous flow of only one gas. The author [H303] was the first to develop an exact theory for the equilibrium of regenerators which are usually reversed at regular intervals although it is true that the temperature difference within a cross-section of the heat storing mass was ignored. He thus formulated the conditions for reversal and represented the processes taking place in regenerators as temperature oscillations. Using the same reversal condition Nusselt [N304] has described the equilibrium by means of an integral equation and has given as the solution an infinite series of integrals. But since both types of calculation are very laborious and time consuming in their application the author [H304, 305] has developed methods of approximation, which serve the same purpose but use far simpler and quicker methods of calculation. Step-by-step processes which are similar to one another have been set out by Saunder [S302] and Allen [A302], and later by Lambertson [L301] and Willmott [W303 to 307]. These have enabled for the first time, calculations to be performed for regenerators which have a periodically variable mass flow. In addition to the heat pole method which is discussed in Secs. 15-1 to 15-3, the Illife method [I301] and also the new method of Nahavandi and Weinstein [N301] together with that of Sandner [S301] can be considered as numerical methods for the solution of the relevant integral equation.

Works by Schmeidler [S311], Ackermann [A301] and Lowen [L306] combine the two directions indicated, as also do newer publications by Modest and Tien [M304]. A combination of the theories developed by the author yields a process [H308] in which calculations for regenerators can be carried out relatively simply and with sufficient accuracy for practical purposes.

The influence of deposits of condensing components on the temperature distribution in regenerators can be calculated by means of a step-by-step method developed by the author [H306].

11-3 SUMMARY OF THE PROCESSES IN REGENERATORS

Before describing the regenerator theories which have been developed, a summary will be made of the physical processes which take place in regenerators in a state of equilibrium.

Heat Balance Equations

One first basis for finding the temperature distribution in a regenerator emerges by considering the heat balance equations. For the state of equilibrium the same heat balance equations are applicable as were derived for recuperators, Sec. 5-5.

However, with regenerators these equations must be related to the chronological average value of the temperatures. The time T, between two consecutive reversals during which the originally warmer gas is flowing through the regenerator will be called the "warm period"* and the length of time T', during which the originally colder gas is flowing through the regenerator will be called the "cold period". In hot blast stove operations the terms "gas period" and "air period" are used in place of these. Furthermore, θ and θ' will be the temperatures of the originally warmer and originally colder gas in a particular cross-section of a regenerator at a particular time. The chronological average values $\bar{\theta}$ and $\bar{\theta}'$ are assumed to be formed from these time varying gas temperatures during a warm or cold period; this is shown, for example, in the cross-section indicated by $A-B$ in Fig. 11-4. The originally warmer gas enters at the top of the regenerator at a constant temperature θ_1; the originally colder gas also enters at a constant temperature θ'_1; in parallel flow this entry is at the top of the regenerator also but in counterflow it is at the bottom. The chronologically variable exit temperatures will be denoted by θ_2 and θ'_2 while their time mean values will correspondingly be denoted by $\bar{\theta}_2$ and $\bar{\theta}'_2$.

In addition C and C' will be the thermal capacity flow rates of the two gases flowing through the regenerator.[1] For simplicity C and C' may be considered to be independent of temperatures. CT and $C'T'$ then represent the thermal capacities of the gases which flow through the regenerator during a warm and cold period respectively. The heat balance equations now follow from the consideration that in an arbitrary section of the regenerator and in a state of equilibrium, during the warm period, the hot gas gives up as much heat as the colder gas takes up during the cold period. Thus Eqs. (11-1) are derived, for example, for the section of the regenerator lying above the AB cross-section in Fig. 11-4:

$$
\begin{aligned}
CT(\theta_1 - \bar{\theta}) &= C'T'(\bar{\theta}' - \theta'_1) && \text{for parallel flow} \\
CT(\theta_1 - \bar{\theta}) &= C'T'(\bar{\theta}'_2 - \bar{\theta}') && \text{for counterflow}
\end{aligned}
\tag{11-1}
$$

These relationships are the same as the heat balance Eqs. (5-29) in Sec. 5-5, if θ and θ' are replaced by $\bar{\theta}$ and $\bar{\theta}'$ and C and C' are replaced by CT and $C'T'$. The following Eq. (11-2) is obtained from Eq. (11-1) using Eq. (5-30):

$$
\begin{aligned}
\bar{\theta} - \bar{\theta}' &= \theta_1 - \theta'_1 - \frac{C'T' + CT}{C'T'}(\theta_1 - \bar{\theta}) && \text{for parallel flow} \\
\bar{\theta} - \bar{\theta}' &= \theta_1 - \bar{\theta}'_2 - \frac{C'T' - CT}{C'T'}(\theta_1 - \bar{\theta}) && \text{for counterflow}
\end{aligned}
\tag{11-2}
$$

Equations (11-1) and (11-2) are of great significance in the following way. If the chronological average values of the temperatures of both gases are given at a particular section of the regenerator, for example, $\bar{\theta}_1$ and $\bar{\theta}_2$ at the top end of a counterflow regenerator (see Fig. 11-4), then according to Eq. (11-1) the average temperature $\bar{\theta}'$ of the colder gas is clearly shown to be dependent on the average

---

* Editor's note: The "warm period" is sometimes called the "hot period".
[1] For information regarding the notation C and C' see Footnote 3 on p. 300 and * on page 145.

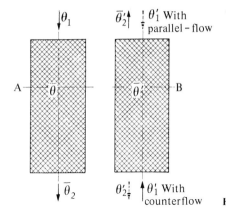

Figure 11-4 Temperatures of the gases in regenerators.

temperature $\bar\theta$ of the warmer gas and vice versa. Thus, using Eq. (11-2) it is also possible to determine how the chronological average value of $\bar\theta - \bar\theta'$ of the temperature difference between the two gases is dependent on θ or θ'. If, particularly in counterflow, CT and $C'T'$ are equal then it will be seen from Eq. (11-2) that $\bar\theta - \bar\theta'$ has the same value in all parts of the regenerator.

Since the heat balance equations are essentially the same for regenerators and recuperators, it thus seems useful to be able to calculate the average temperature difference $\Delta\theta_M$ in the whole regenerator in the same way as for a recuperator simply by replacing θ_2 and θ'_2 in the equations developed previously (Eq. (5-56) and Sec. 5-12) by the time mean values $\bar\theta_2$ and $\bar\theta'_2$ of the exit temperatures of the gases.

Furthermore, as has been shown previously for recuperators, Eqs. (11-1) and (11-2) can also be used to obtain the curves of temperatures $\bar\theta$ and $\bar\theta'$ as a function of the quantity of heat q which is transferred in regenerator sections of a varying length during a warm or cold period. On the basis of this type of consideration exactly the same temperature distribution is obtained in regenerators as is found in recuperators and thus Figs. 5-11 to 5-14, which apply to recuperators, can also be used for regenerators.

However, if the temperature distribution is considered to be a function of the longitudinal coordinate in a regenerator, basic differences arise between recuperators and regenerators, as has already been mentioned. These differences are closely associated with the chronological changes in the temperatures of the gases and the heat storing mass.

The Variations with Time of the Temperature in a Regenerator Brick

In all that follows it is considered that the heat transfer coefficients which determine the rate of heat transfer between the heat-storing mass and the gases flowing through the regenerator will be assumed to be known, and that they can be determined using the results of measurements obtained from numerous experiments as has already been discussed in Sec. 2-13 in Part Two.

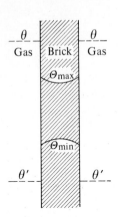

Figure 11-5 Fluctuation of the temperature in the brick cross-section.

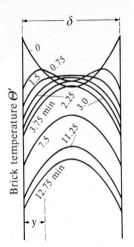

Figure 11-6 Spatial temperature distribution in the brick at different times with chronologically varying gas temperature θ'.

In order to follow the changes in temperature in one of the regenerator bricks, to simplify matters, the brick will be considered to be a flat plate the surface of which is parallel to the direction of gas flow. In addition it will be assumed that the system is in periodic equilibrium and that the thermal capacities CT and $C'T'$ of both the gases, which are related to the period lengths T and T', are equal to one another. A cross-section inside the brick is imagined to be in a position perpendicular to the direction of flow of the gases (Fig. 11-5). The instantaneous temperature of the brick at an arbitrary position in this cross-section will be denoted by Θ.[2] As the surface of the brick will be washed alternately by warm and cold gas, the brick temperature Θ in equilibrium will fluctuate periodically between two limits Θ_{max} and Θ_{min} (Fig. 11-5). These fluctuations are larger in the region of the brick surface than they are inside the brick.

Figure 11-6 shows the cooling of a 40 mm thick brick during the cold period.[3] Here it is assumed that the brick cross-section under consideration is located in a relatively long regenerator and is not too near one of the regenerator ends. The length of the cold period as well as that of the warm period which follows it is assumed to be 12.75 minutes.

At the beginning of the cold period the temperature curve is represented by an open parabola or by a parabola-like curve which curves upwards. This curve is deformed by the initially very rapid cooling of the outer layers of the brick. After three minutes or so, it takes the form of a parabola which is open towards the bottom but which has the same shape as the original parabola. From now on the

---

[2] The use of the letter Θ for the temperature of the heat-storing mass as opposed to the gas temperatures θ and θ' seems to the author to be easier on the eye than to denote the difference through the use of indices. Moreover this also simplifies the writing of the equations.

[3] This temperature distribution has been obtained graphically using the difference equations of Binderschmidt, see [B304, 305] and [S312, 313].

curve, without changing its shape, is displaced towards the bottom at constant velocity. In the following warm period the same temperature changes take place but in the opposite direction.

If the temperature distribution in Fig. 11-6 is displayed as a function of time with the time coordinate for the warm period set in the opposite direction to that for the cold period but on the same axis, Fig. 11-7 will be obtained. The dashed lines represent the gas temperatures θ and θ' and the solid closed lines represent the surface temperatures Θ_0 and Θ'_0 of the brick during the warm and cold period. At positions further inside the brick the temperatures Θ and Θ' also run along closed lines although the distance between these lines is less than for Θ_0 and Θ'_0. Directly after reversal all the temperature-time distributions are initially curved but they soon become linear in form. If the *spatial average value* Θ_m or Θ'_m *of the brick temperature* were formed at every instant across the entire brick thickness, the remarkable result would be obtained that the time variation of this average value would be reproduced by a *single straight line* in Fig. 11-7. This straight line would run in the opposite directions in the two periods.

If the thermal capacities of the gases flowing through the regenerator during the warm and cold period are different, the linear time variation of the average brick temperatures Θ_m and Θ'_m changes into two weakly curved lines which come together almost exactly (see Sec. 12-6, Fig. 12-14).

The temperature variations are fundamentally different at and near the ends of the regenerator; this is mainly a result of the two gases having constant entry temperatures. Thus, for example, at the warm end of the regenerator, there is a temperature variation as shown in Fig. 11-8. The times are again shown to be in opposite directions on the same axis for the warm and cold periods. The topmost

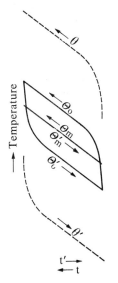

Figure 11-7 Temperature–time curve in a regenerator cross-section.

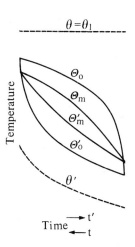

Figure 11-8 Chronological temperature variations at the warm end of the regenerator.

line $\theta = \theta_1$ reproduces the chronologically constant entry temperature of the warmer gas. The two top solid lines characterize the variation of the surface temperature Θ_0 and the average temperature Θ_m of the brick during the warm period. θ', Θ'_0 and Θ'_m are the corresponding temperatures in the cold period. The temperature $\theta' = \theta'_2$ of the colder gas, which leaves the regenerator here, varies less than it does in the more internal parts of the regenerator. More than anything else, however, it is remarkable that the curves for Θ_m and Θ'_m are not co-incident as they are in Fig. 11-7. The curvature and the average distance between these curves becomes smaller as one moves further from the ends of the regenerator. In relatively short regenerators the differences between Θ_m and Θ'_m do not completely disappear in the middle of the regenerator.

Temperature Distribution in the Longitudinal Direction of the Regenerator

The differences mentioned above between the variations with time of the temperature in the inner parts of the regenerator and the variations close to the end of the regenerator have a direct effect on the spatial temperature distribution in the longitudinal direction of the regenerator. In order to illustrate this as simply as possible it must again be assumed that the thermal capacities CT and $C'T'$ of the gases flowing through the regenerator during the warm and cold periods are not only equal to one another but are also independent of temperature. All other quantities such as the thickness of the brick, the temperature conductivity of the bricks, the heat transfer coefficients, etc., should also have the same value at all positions in the regenerator. Now if the regenerator is of sufficient length so that, in a large, inner region, the brick temperature varies chronologically in a way such as is shown in Figs. 11-6 and 11-7, the variation of all the gas and brick temperatures in the longitudinal direction of the regenerator will be linear in this region. The linear variations with time of the average brick temperatures Θ_m and Θ'_m in a cross-section is thus, to a certain degree, mirrored by a linear spatial temperature distribution in the longitudinal direction. On the other hand, the variations with time of Θ_m and Θ'_m which are curved and are further apart, occur close to the ends of the regenerator (see Fig. 11-8), and the temperature distribution is also curved in the longitudinal direction of the regenerator. This can be seen in the curves shown in Fig. 11-10, bottom, which will shortly be discussed in greater detail and in Figs. 13-15, 18-1 and 18-2, 18-6.

Fundamental Oscillations and Harmonic Frequencies in Regenerators

The physical significance of the difference in the temperature behaviour of a long regenerator between its central region and at its ends can be understood if the periodic temperature changes, in the state of equilibrium, are regarded to be a forced temperature oscillation, which can then be divided, like the oscillations of a chord,* into a fundamental oscillation and its harmonics (see Sec. 13-6).

---

* Editor's note: It may be of interest to readers to know that Professor Hausen played the violin for many years.

On the basis of assumptions specified above, i.e., when $CT = C'T'$, when the heat transfer coefficients are constant, etc., the fundamental oscillation is represented, in both the sense of distance and time by a linear variation of temperature: by the linear variation of all the temperatures in the longitudinal direction of the regenerator and by the linear variation with time of the average brick temperature; see Fig. 11-10, top.

In order to understand the significance of the harmonic vibrations, it is useful to assume a heat storing mass with high thermal conductivity, so that the spatial temperature differences within a regenerator cross-section of the heat-storing mass can be ignored. Thus, with a harmonic vibration, it must be imagined that the regenerator has the almost same temperature throughout and that in each period the gas enters the regenerator at a periodically increasing temperature. Thus the entry temperature has the form of periodic vibrations which increase in amplitude. This is shown in Fig. 11-9 where ξ, on a reduced scale which will be discussed later, represents the distance of the regenerator cross-section being considered from the position where the gas enters. $\Lambda - \xi$ denotes the distance from the position where the gas leaves. The temperature oscillations of the gas are shown as dashed lines and those of the heat-storing mass are shown as solid lines.

The number of oscillations per warm or cold period distinguish one harmonic from another; the number can lie between about 1 and ∞. The temperature oscillations of the gas result in a corresponding vibration in the temperature of the heat-storing mass. However the vibrations are weakened by this and thus fade away from the interior of the regenerator (Fig. 11-9, left). In the inner parts of a long regenerator the gas has almost the same temperature as the heat-storing mass. However as the gas reaches the other end of the regenerator its temperature variation alters yet again in a way that is characteristic of the harmonic vibration being considered. This is explained by the fact the effect is felt of the disturbance in the temperature distribution of the heat storing mass caused by the temperature

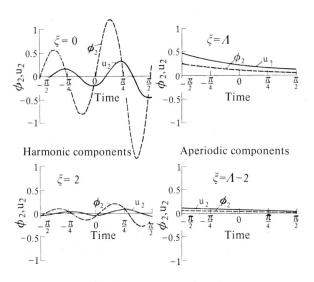

Harmonic components

Aperiodic components

Figure 11-9 The harmonic oscillations of the temperature in a regenerator: Eigen functions u_2 and ϕ_2 for $\chi = 2$.

oscillations which have taken place during the previous period. The gas, which is now flowing by, tries to eliminate this disturbance. The temperature swing which is thus suffered by the gas is chronologically aperiodic (Fig. 11-9, right); the non-linearity is greatest at the exit position of the gas and diminishes from here in the direction of the inside of the regenerator. Each harmonic vibration thus consists of two components which, as has been mentioned, have their highest values at the ends of the regenerator and which fade away exponentially towards the inside of the regenerator.

If the harmonic vibrations are added to the fundamental oscillations in a suitable manner, it is possible to represent the constant entry temperature of the gas flowing through the regenerator during the period under consideration together with the variation with time of the temperatures at end of the regenerator, as is shown fundamentally in Fig. 11-8. On the other hand, however, the

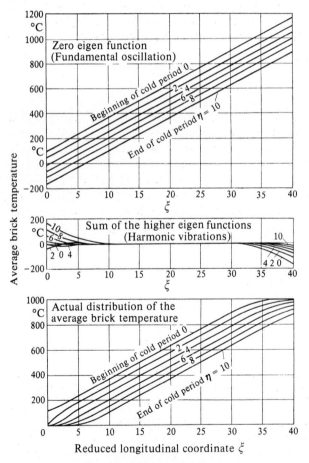

Figure 11-10 Construction of the temperature distribution in the longitudinal direction of a regenerator using the fundamental oscillation and harmonic vibrations.

fundamental oscillation alone could not reproduce a chronologically constant temperature because this represents a gas temperature which varies linearly with time (see Fig. 11-9, top and Fig. 13-12). By way of example, Fig. 11-10 shows how the superposition of the harmonic vibrations influences the temperature distribution in the *longitudinal direction* of the regenerator.

For the sake of clarity only the variation of the average brick temperature Θ'_m during the cold period is shown. The temperature distribution which results solely from the fundamental oscillation is shown at the top. The individual lines correspond to specific times during the cold period so that time is shown on a reduced scale η [see Eqs. (13-24) and (13-25)]. The linear variation of temperature in the longitudinal direction of the regenerator corresponds to the chronological linear decrease of the average brick temperature. The centre section of the figure illustrates the sum of all the harmonic vibrations, each multiplied by a suitable factor. Corresponding to the properties of harmonic vibrations which have already been described, this sum has its greatest values at the ends of the regenerator while it is negligible in a large central region. The bottom section of the figure illustrates the true temperature curve in the regenerator which results when the harmonic vibrations are added to the fundamental oscillation.

The example shown in Fig. 11-10 corresponds somewhat to the situation which arises in low temperature technology. However, in contrast, in the iron and steel industry the harmonic vibrations often reach so far into the interior of the regenerator that even in the centre of the regenerator the variations with time of Θ_m and Θ'_m as shown in Figs. 11-7 and 11-8 do not disappear completely.

The fundamental oscillation and the harmonic vibrations solutions of the differential equations for heat transfer in regenerators, which will be dealt with in Sec. 12-1 and Sec. 13-3, can be considered mathematically. These solutions, which are satisfied by a boundary condition given by the reversal, will be called eigen functions which are characterized by an eigen value χ which can take any real value in the complete range between 0 and ∞. The zero eigen function, which is denoted by $\chi = 0$, represents the fundamental oscillation, while the higher eigen functions for $\chi \geqslant 1$ represent the harmonic vibrations.

11-4 THE HEAT TRANSMISSION COEFFICIENT FOR REGENERATORS

The Influence on Heat Transfer of the Curved Temperature Distribution at the Ends of the Regenerator

The chronological and spatial temperature distribution discussed in the previous paragraphs has the following effect on heat transfer at the different positions in the regenerator. Figure 11-8 shows that the curves for the average temperatures Θ_m and Θ'_m of the heat-storing mass do not coincide at the ends of the regenerator but form a type of hysteresis loop. Such hysteresis loops occur at all the positions in the regenerator at which the harmonic vibrations have considerable importance. The

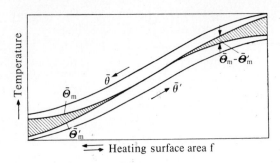

Figure 11-11 Chronological average value of the temperatures in a regenerator.

average height $\bar{\Theta}_m - \bar{\Theta}'_m$[4] of this hysteresis cycle is at its greatest at the ends of the regenerator and decreases towards the interior of the regenerator with a magnitude similar to that of the harmonic vibrations themselves. See Fig. 11-11.

This hysteresis cycle influences heat transfer very considerably. The quantity of heat being transferred at each position in the regenerator during the warm period is, given constant heat transfer coefficients, proportional to the chronological average value $\bar{\theta} - \bar{\Theta}_m$ of the difference $\theta - \Theta_m$ between the gas temperature and the average brick temperature (see Sec. 13-1). Correspondingly the chronological average value $\bar{\Theta}'_m - \bar{\theta}'$ determines the rate of heat transfer during the cold period. Heat transfer during a full cycle is thus determined by $(\bar{\theta} - \bar{\Theta}_m) + (\bar{\Theta}'_m - \bar{\theta}')$. If this value is first calculated for the fundamental oscillation and it is noticed that $\bar{\Theta}_m = \bar{\Theta}'_m$ in this case, see Fig. 11-7, it follows that

$$(\bar{\theta}_m - \bar{\Theta}_m) + (\bar{\Theta}'_m - \bar{\theta}') = \bar{\theta} - \bar{\theta}'$$

Thus the total average temperature difference between the two gases is utilized for heat transfer in the case of the fundamental oscillation. On the other hand, in the case of harmonic vibrations, the difference between $(\bar{\theta} - \bar{\Theta}_m) + (\bar{\Theta}'_m - \bar{\theta}')$ on the one hand and $\bar{\theta} - \bar{\theta}'$ on the other is equal to the average height $\bar{\Theta}_m - \bar{\Theta}'_m$ of the hysteresis cycle. Thus the transfer of heat is correspondingly reduced at such positions in the regenerator. The *influence of the harmonic vibrations* and thus also that of the temperature distribution at the ends of the regenerator, which curve deviates from the fundamental vibration, consists throughout of *a deterioration in the exchange of heat*. However the total effect of this deterioration is often only very slight; nevertheless this cannot be neglected in precise calculations.

The Heat Transmission Coefficient

Rummel [R304] suggested a heat transmission coefficient k for the state of equilibrium in regenerators and this coefficient is defined as follows. Again T denotes the duration of the warm period and T' denotes the duration of the cold period. The duration $T + T'$ of successive warm and cold periods is called the cycle

---

[4] The bar values denote the chronological average values in the warm or cold period. The indices m or M, on the other hand, characterize the spatial averaging, m being over a cross-section and M being over the longitudinal direction of the regenerator.

time.* According to Rummel the heat transmission coefficient k is obtained by dividing the quantity of heat Q_{Per} which is transferred in *one* regenerator during a full cycle by the heating surface area F of the regenerator, by the average temperature difference between the two gases and by the duration $T + T'$ of a complete cycle. Thus if $\Delta\theta_M = (\bar{\theta} - \bar{\theta}')_M$ is the chronological and spatial average value of the temperature difference between the two gases, the quantity of heat transferred in *one* regenerator during a complete cycle amounts to

$$Q_{Per} = kF(T + T')\Delta\theta_M \qquad (11\text{-}3)$$

This quantity of heat is thus taken up by the total heat-storing mass of a regenerator during the warm period and is released again during the cold period. It is thus assumed (see page 299), that the average temperature difference $\Delta\theta_M$ is calculated using the entry temperatures θ_1 and θ'_1 and the chronological average values $\bar{\theta}_2$ and $\bar{\theta}'_2$ of the exit temperatures of the gases as for a recuperator employing Eq. (5-56) and, where the temperature dependence of C and C' is to be taken into account within the calculation, this average temperature difference can be computed using the method of approximation described in Sec. 5-12.

Calculation of the Heat Transmission Coefficient

A sufficiently accurate and relatively simple process for calculating the heat transmission coefficient k, given constant physical properties and constant heat transfer coefficients, has been developed by the author [H308] on the basis of regenerator theories which will be discussed in Sec. 12-5 and Sec. 13-6. This process consists of two main steps in which the influence of the fundamental oscillation and the influence of the harmonic vibrations are considered separately. This process, which makes use of equations and diagrams, will be discussed now so that it can be used without having to work through its underlying theories.

The heat transmission coefficient which results from the fundamental oscillation alone will be denoted by k_0. α and α' will again be the heat-transfer coefficients in the warm and cold periods. In addition δ will be the thickness of a plane wall shaped brick or the diameter of a cylindrical or spherical packing brick; λ_s is the thermal conductivity of the brick. Equation (11-4) is developed for k_0 from the theory and uses a function which is yet to be discussed, namely Φ.

$$\frac{1}{k_0} = (T + T')\left[\frac{1}{\alpha T} + \frac{1}{\alpha' T'} + \left(\frac{1}{T} + \frac{1}{T'}\right)\frac{\delta}{\lambda_s}\Phi\right] \qquad (11\text{-}4)$$

The function Φ reproduces the influence of the very rapid temperature changes which are suffered by the gas and the brick surface immediately after reversal, (see Figs. 11-6 and 11-7). The value of Φ is dependent on the dimensionless group $\frac{\delta^2}{2a}\left(\frac{1}{T} + \frac{1}{T'}\right)$, where $a = \lambda_s/\rho c$ denotes the thermal diffusivity of the material of which the heat-storing mass is constructed, ρ is its density and c is its specific heat.

---

* Editor's note: Hausen uses the expression "full period".

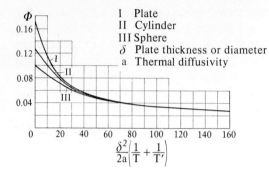

I Plate
II Cylinder
III Sphere
δ Plate thickness or diameter
a Thermal diffusivity

$$\frac{\delta^2}{2a}\left(\frac{1}{T}+\frac{1}{T'}\right)$$

Figure 11-12 Auxiliary function Φ for calculating the heat transmission coefficient.

After calculating this dimensionless group which reduces to the form δ^2/aT if the duration of both periods is the same ($T = T'$), the value of Φ can be read off from Fig. 11-12. In this Figure curve I represents plate-shaped bricks, curve II represents cylindrical bricks and curve III represents spherical bricks, all with a "diameter" δ.

If the heat-storing mass has a shape which deviates from these three basic types it is easy to estimate which to use as the closest approximation. An average plate thickness δ or an average diameter δ is determined, together with the volume V and the surface area F of the heat-storing mass from which an "equivalent" plate thickness $\delta_{gl} = \delta/2 + V/F$ can be calculated. This expression is obtained from the consideration that the volume is a determining factor for heat storage while the surface area of the heat-storing mass is a determining factor for heat transfer and

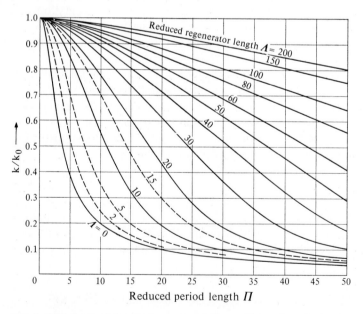

Figure 11-13 Ratio of the true heat transmission coefficient k to the heat transmission coefficient k_0 based on the fundamental oscillation.

thus δ_{gl} not only depends on the actual diameter but also on the ratio V/F. Using this equivalent plate thickness $\delta = \delta_{gl}$ the abscissa in Fig. 11-12 can then be determined, Φ can be read off from curve I and thus $\delta = \delta_{gl}$ and Φ can be used in Eq. (11-4). This process is based on the fact that when δ_{gl} is used curves II and III coincide very closely with curve I (see Sec. 12-7 and Fig. 12-13).

As has already been mentioned the harmonic vibrations or higher eigen functions reduce heat transfer. The true heat transfer coefficient is thus somewhat smaller than the heat transmission coefficient developed in Eq. 11-4. The ratio k/k_0 depends on two dimensionless groups

$$\Lambda = 4\frac{k_0(T+T')F}{CT+C'T'} = 2\frac{k_0(T+T')F}{C_{Per}}$$

and

$$\Pi = 2\frac{k_0(T+T')F}{C_s}$$

(11-5)

Here $C_{Per} = \frac{1}{2}(CT+C'T')$ is the average value of the thermal capacities of the quantities of gas flowing during the hot and cold periods, C_s is the thermal capacity and F the heating surface area of the heat-storing mass. If k_0, C, C' and F/C_s are assumed to be given, Λ is essentially proportional to the heating surface area and thus to the length of the regenerator while Π on the other hand is proportional to the period lengths T and T'. Thus Λ is called the "reduced regenerator length" and Π is called the "reduced period length".[5] After determining the values of Λ and Π from Eq. (11-5), k/k_0 is determined from Fig. 11-13. Finally if the value of k_0, calculated from Eq. (11-4), is multiplied by k/k_0, the true heat transmission coefficient will be found.

Explanations and Transformations

The calculation process described is only strictly valid when α, α', C, C', λ_s, ρ and c are not dependent on temperature and $CT = C'T'$. Equation (11-5) for the dimensionless groups Λ and Π are so selected, however, that as a rule the process can also be used, providing a good approximation, for the case when $CT \neq C'T'$ (see Sec. 13-8).

By calculating Π for plates, cylinders and spheres using the volume V, density ρ and specific heat c of the heat-storing mass, using $F/C_s = F/V\rho c$, Eq. (11-5) can be converted to

$$\Pi = 4m\frac{k_0(T+T')}{\rho c\delta} \quad \begin{cases} \text{plates:} & m = 1 \\ \text{cylinders:} & m = 2 \\ \text{spheres:} & m = 3 \end{cases}$$

(11-6)

Thus Eq. (11-5) has the advantage of being uniformly valid for each arbitrarily shaped heat storing mass.

---

[5] The meaning of this nomenclature, which perhaps seems somewhat artificial, will be fully understood in the third chapter of this section on regenerators (pp. 353 and 354).

The fact that the values of Φ and k/k_0 must be read off from diagrams with the aid of the dimensionless groups, will perhaps be considered to be a disadvantage of the process described. Although precise equations are known for Φ (see Secs. 12-6 and 12-7), these are too complicated for practical purposes. However, in almost all cases in practice, Φ can be calculated with sufficient accuracy using the following approximate equations:

$$
\left.
\begin{array}{llll}
\text{Plates:} & \Phi = \dfrac{1}{6} - 0.00556\dfrac{\delta^2}{2a}\left(\dfrac{1}{T} + \dfrac{1}{T'}\right) & \text{for} & \dfrac{\delta^2}{2a}\left(\dfrac{1}{T} + \dfrac{1}{T'}\right) \leqslant 10 \\[3mm]
\text{Cylinders:} & \Phi = \dfrac{1}{8} - 0.00261\dfrac{\delta^2}{2a}\left(\dfrac{1}{T} + \dfrac{1}{T'}\right) & \text{for} & \dfrac{\delta^2}{2a}\left(\dfrac{1}{T} + \dfrac{1}{T'}\right) \leqslant 15 \\[3mm]
\text{Spheres:} & \Phi = \dfrac{1}{10} - 0.00143\dfrac{\delta^2}{2a}\left(\dfrac{1}{T} + \dfrac{1}{T'}\right) & \text{for} & \dfrac{\delta^2}{2a}\left(\dfrac{1}{T} + \dfrac{1}{T'}\right) \leqslant 20
\end{array}
\right\} \quad (11\text{-}7)
$$

and for all larger values of $\dfrac{\delta^2}{2a}\left(\dfrac{1}{T} + \dfrac{1}{T'}\right)$, can be calculated using the relationship:

$$
\Phi = \frac{0.357}{\sqrt{\varepsilon + \dfrac{\delta^2}{2a}\left(\dfrac{1}{T} + \dfrac{1}{T'}\right)}} \qquad \text{with} \quad
\begin{cases}
\varepsilon = 0.3 \text{ for plates} \\
\varepsilon = 1.1 \text{ for cylinders} \\
\varepsilon = 3.0 \text{ for spheres}
\end{cases}
\qquad (11\text{-}8)
$$

The values of k/k_0 illustrated in Fig. 11-13 were calculated electronically in 1971 and 1972 using the processes described in Secs. 16-2, 16-4, 16-5 and 14-4 and are as accurate as possible.[6] An exact closed equation for k/k_0 is not known at the present time, apart from the limiting cases of $\Lambda = \infty$ and $\Lambda = 0$ for which Eq. (11-9) is valid (see Sec. 13-5):

$$
\begin{aligned}
\frac{k}{k_0} &= 1 & &\text{with } \Lambda = \infty \\[2mm]
\frac{k}{k_0} &= \frac{2}{\Pi}\tanh\frac{\Pi}{2} & &\text{with } \Lambda = 0
\end{aligned}
\qquad (11\text{-}9)
$$

However the following empirical approximation can be used for k/k_0:

$$
\frac{k}{k_0} = 1 - \frac{1}{\Lambda}\left(0.8\Pi - 3\tanh(0.2\Pi)\right) \qquad \text{for} \quad \frac{\Pi}{\Lambda} = \frac{C_{Per}}{C_s} < 0.5 \qquad (11\text{-}10)
$$

By not making too great a demand on accuracy, *the calculation of the heat transmission coefficient can be considerably simplified* if it is noted that in many areas of application the dimensionless groups at any given time only vary within certain limits. Thus, for example, in the iron and steel industry $\dfrac{\delta^2}{2a}\left(\dfrac{1}{T} + \dfrac{1}{T'}\right)$ lies in the

---

[6] Willmott (Sec. 16-4) and Schellman using the methods described in Sec. 16-5, personally put their values at my disposal. The values of Lambertson (Sec. 16-2) and Sandner (Sec. 14-4) have been taken by myself from Sandner's dissertation [S301].

region of 4, Λ lies between 5 and 20 and the relationship $\Pi/\Lambda = C_{Per}/C_s$ [see Eq. (11-5)] lies between 0.2 and 0.3. Within these limits it is possible to calculate the true heat transmission coefficient k using the equation:

$$\frac{1}{k} = \left[1 + \frac{F}{400}\left(\frac{\alpha}{C} + \frac{\alpha'}{C'}\right)\right](T + T')\left[\frac{1}{\alpha T} + \frac{1}{\alpha' T'} + \left(\frac{1}{T} + \frac{1}{T'}\right)\frac{\delta}{7\lambda_B}\right] \quad (11\text{-}11)$$

In the given area this equation is valid to an accuracy of about 2 per cent. It would be even simpler to calculate if an average value of $k/k_0 = 0.93$ were used, but at the boundaries of this area of application this gives an error of about 5 per cent.

The following short proof of this equation is reproduced from the first edition of this book.

Λ can be set approximately equal to the arithmetic mean of $\dfrac{\bar{\alpha}F}{C}$ and $\dfrac{\bar{\alpha}'F}{C'}$ (see Sec. 13-1) because when $T = T'$, $\bar{\alpha} = \bar{\alpha}'$ and $C = C'$, it follows from Eqs. (11-5), (11-6) and (13-28) that $\Lambda = \dfrac{\bar{\alpha}F}{C} = \dfrac{\bar{\alpha}'F}{C'}$. In addition if, for $\dfrac{\bar{\alpha}}{\alpha}$ and $\dfrac{\bar{\alpha}'}{\alpha'}$, we choose the value 0.833 which results from Eq. (13-3) for $\alpha = \alpha' = 20$, $\delta = 0.07$, $\lambda_B = 1$ and $\Phi = \frac{1}{7}$, we will obtain the given equation for Λ.

Dependence of the Heat Transmission Coefficient on the Length of the Period and the Thickness of the Brick

Figures 11-14 and 11-15 show how, using the calculation processes described above, the heat transmission coefficient k is dependent on the length of the period and the thickness of the plane wall-shaped bricks. Here it is assumed that $CT = C'T'$, $\lambda_s = 1.163$ W/(m K), $\rho c = 2070$ kJ/(m^2 K) and $\alpha = \alpha' = 23.26$ W/(m^2 K). In Fig. 11-14 the thickness of the brick is assumed to be $\delta = 0.03$ m $= 30$ mm and, for various values of the reduced regenerator length Λ, k is shown as a function of the length of period $T = T'$. The smaller the value of Λ, i.e., the shorter the regenerator,

W/m^{2}K

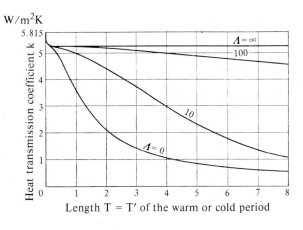

Figure 11-14 Heat transmission coefficient k for various values of the reduced regenerator length Λ, dependent on the period length $T = T'$. Brick thickness $\delta = 30$ mm.

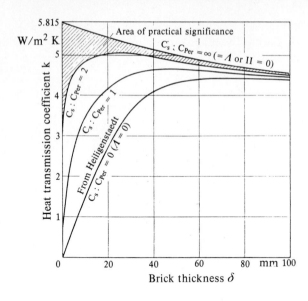

Figure 11-15 Heat transmission coefficient k as a function of the brick thickness δ for the period length $T = T' = 1$ h.

the more rapidly k decreases with increasing period length. The topmost curve for $\Lambda = \infty$ represents the heat transmission coefficient k_0 which corresponds solely to the fundamental oscillation. The decrease in k_0 with a very short period length, which occurs very rapidly initially, is caused by the fact that the function Φ in Fig. 11-12 increases in this region from 0 to 1/6. Physically this is connected with the fact that the rapid temperature changes at the beginning of each period play an increasingly smaller part with greater period length. This is because during these initial temperature changes, in which the thermal flow affects only the outer parts of the brick and does not penetrate the bricks very deeply, the conditions for the transfer of heat are more favourable than later on in the process.

Figure 11-15 shows the heat transmission coefficient k dependent on the *brick thickness* δ for a constant period length of $T = T' = 1$ h. The curves drawn relate to constant values of the ratio C_s/C_{Per}, where, as previously, C_s denotes the thermal capacity of the total heat storing mass of the regenerator and C_{Per} denotes the average value of the thermal capacities of the quantities of gas flowing through the regenerator during a warm and a cold period. The topmost curve deals with the case $C_s/C_{Per} = \infty$. Since, using Eq. (11-5), $C_s/C_{Per} = \Lambda/\Pi$, either $\Lambda = \infty$ or $\Pi = 0$. This topmost curve thus refers to either $\alpha F = \infty$ or to infinitely short period lengths, usually to long regenerators.

The bottommost case corresponds to the limiting case of an infinitely short regenerator; this case has also been handled by Heiligenstaedt (see Sec. 12-2). Under normal operating conditions C_s/C_{Per} must always be greater than 2, for otherwise regenerator efficiency would be too low. Thus the shaded surface between the curves for $C_s/C_{Per} = 2$ and $C_s/C_{Per} = \infty$ represents the area which is of practical significance.

The fact that the curves in Fig. 11-15, apart from the curve for $C_s/C_{Per} = \infty$, first rise and then fall away again and finally tend towards the limiting value

$k = 4.4$, is explained as follows. The decrease, which also occurs in the topmost curve is due to the fact that the inner parts of the bricks play an increasingly smaller part in the exchange of heat with increasing thickness of brick; see Figs. 12-4 and 12-6 later on in this book.

The rise of the curves, particularly when the brick thickness is small, is explained by the hysteresis cycle of the average brick temperatures Θ_m and Θ'_m (see Fig. 11-8). As has already been explained above, the ratio of the average height of this hysteresis cycle to the average temperature difference between the two gases indicates the deterioration in heat transfer compared with the fundamental oscillation. However, given a constant period length, on average this ratio is greater the thinner the brick and the smaller the ratio C_s/C_{Per}. Conversely therefore, the quantity of heat transferred must be greater with an increasing δ and increasing C_s/C_{Per}.

The Calculation of Two or More Regenerators Working Together. Comparison with a Recuperator

If the heat transmission coefficient k is known a regenerator can be largely calculated like a recuperator using the relationships developed in Sec. 5-5 ff. However there is a difference between the calculations of both types of heat exchanger as in both cases the heat transmission coefficient is defined in a different way. In recuperators the heat is transferred through two heating surface areas, for example through the inner and outer surfaces of tubes; in regenerators, using Rummel's definition, heat transfer only takes place through *one* heating surface area, that is through the heating surface are of *one* regenerator. In regenerators the rate of heat transfer representing the performance of a recuperator is only achieved through the operation of two or more regenerators working together. The total thermal efficiency of these regenerators must be taken into account when a heat transmission coefficient is required that will correspond to that for a recuperator.

The number of regenerators taking part will be n. Usually $n = 2$ (see Fig. 11-1); however occasionally, and this was more usual in days past than it is today, the operation with $n = 3$ or $n = 4$ regenerators working in cyclic permutation can be undertaken, for example in the hot blast stoves of blast furnaces. When there are three regenerators working together, this is achieved by having the warm period twice as long as the cold period. Thus there are two regenerators through which the originally warm gas flows simultaneously and one regenerator through which the originally colder gas flows.

The quantity of heat transferred in n regenerators in the time $T + T'$ is obtained by multiplying the value of Q_{Per} obtained from Eq. (11-3) by n. In a precise calculation it must be considered that the time requirement T_{ges} for a full period is somewhat longer that $T + T'$ because the amount of time required for the reversal itself is not included in either T or T'.

By way of comparison, a recuperator will also be considered, the total heating surface area of which, for example the sum of the inner and outer heating surface area of tubes, will be the same as the total heating surface area of n regenerators, namely nF. However, in the case of recuperators, it is usual to relate the heat transmission coefficient to only one heating surface area so in this sense calculations will be made using a recuperator heating surface area of $\frac{n}{2}F$. In the recuperator used in this comparison however the same quantity of heat $n \cdot Q_{Per}$ will be transferred in time T_{ges} as in the n regenerators. Thus Eq. (11-12) results for the recuperator and corresponds to Eq. (5-20) with the heat transmission coefficient k_{rek} and the heating surface area $\frac{n}{2}F$

$$n \cdot Q_{Per} = k_{rek} \cdot \frac{n}{2} F T_{ges} \Delta \theta_M \tag{11-12}$$

Comparison with Eq. (11-3) gives

$$k_{rek} = 2k \cdot \frac{T+T}{T_{ges}} (=k_{reg}) \tag{11-13}$$

or, if the difference between T and T' and T_{ges} is neglected

$$k_{rek} = 2k \, (=k_{reg}) \tag{11-14}$$

The following results from this comparison: *n regenerators with a heating surface area F are equivalent to a recuperator with an average heating surface area nF/2 when the heat transmission coefficient of the recuperator is given by:*

$$k_{rek} = 2k \frac{T+T'}{T_{ges}} (=k_{reg}) \tag{11-15}$$

or approximately by $k_{rek} = 2k$. Here k is the heat transmission coefficient for regenerators defined by Eq. (11-3). However k_{rek} can also be conceived as a heat transmission coefficient for regenerators which is defined as the heat transmission coefficient of recuperators. In this sense k_{rek} can also be written k_{reg}; this is shown in brackets in Eqs. (11-13), (11-14) and (11-15).

Accordingly, in order to define a heat transmission coefficient for regenerators so that it is equivalent to that for recuperators, it must be twice as large as that in Rummel's definition. Then, given n regenerators, it should not be related to the surface area nF but only to the surface area $\frac{n}{2}F$. This could be followed through quite logically. However Rummel's definition is simpler and requires no special considerations.

TWELVE

CALCULATION OF THE TEMPERATURE DISTRIBUTION AND THE TRANSFER OF HEAT IN COUNTERFLOW REGENERATORS FROM THE CHRONOLOGICAL TEMPERATURE CHANGES IN A BRICK CROSS-SECTION

12-1 ESTABLISHMENT OF THE DIFFERENTIAL EQUATIONS AND THE BOUNDARY CONDITIONS

In this chapter will be discussed the methods of calculation which originate from investigations into the temperature distribution in the cross-section of a regenerator brick and which, as a final result, aim to find an equation for determining the heat transmission coefficient. To this end the differential equations and the boundary conditions which have to be fulfilled must first be discussed.

The Differential Equations

The heat-storing mass will be presumed to be constructed of flat plates, for example, bricks with an overall thickness δ, which are arranged adjacent to each other at a uniform distance apart and between which the gas flows at a constant speed and direction. The thermal conductivity of the heat-storing mass can be very low, for example as in fire-proof bricks. Thermal conductivity in the direction of flow will be neglected as this does not have any significance.[1]

A position in a brick will be considered which is sited in a cross-section vertical to the direction of flow and which is at a distance y from the surface area is shown

---

[1] If necessary this influence can be taken into consideration separately in the calculation; see Sec. 7-3.

on the left of Fig. 11-6. If the temperature at this position is Θ the usual equation for heat conduction (12-1) is applicable to within this brick cross-section

$$\frac{\partial \Theta}{\partial t} = a \frac{\partial^2 \Theta}{\partial y^2} \qquad (12\text{-}1)$$

where t denotes the time and $a = \lambda_s/\rho c$ denotes the thermal diffusivity of the building material.

An additional differential equation results from the thermal balance for an infinitely short section of the total regenerator at the position of the brick cross-section being considered; the size of this section is determined by the surface area df of the corresponding heat-storing mass. Its position in the regenerator is defined by the surface area f of that heat-storing mass which lies between the position where the gas enters the regenerator and the cross-section under consideration (see Fig. 12-1). The gas flows in the direction of the positive f axis; thus in Fig. 12-1 it flows from top to bottom. C is the thermal capacity of the quantity of gas flowing through the regenerator in unit time. This gas enters the very short section of the regenerator being considered at position f and at temperature θ and leaves it at position $f + df$ and at temperature $\theta + (\partial\theta/\partial f)_t\, df$. In the warm period, during which the gas is cooled on its way through the regenerator, $(\partial\theta/\partial f)_t$ is negative. Thus, in the section of the regenerator under consideration the gas gives up, in time dt, the quantity of heat

$$dQ = -C\, dt \left(\frac{\partial\theta}{\partial f} \right)_t df$$

This quantity of heat[2] will be transferred to the heat-storing mass. If, again, α is the heat transfer coefficient and Θ_0 is the surface area temperature of the heat-storing mass so that a temperature difference $\theta - \Theta_0$ exists between the gas and the surface, the quantity of heat being transferred through the surface df will be

$$dQ = \alpha\, df\, (\theta - \Theta_0)\, dt \qquad (12\text{-}2)$$

Comparing both expressions for dQ we obtain:

$$\left(\frac{\partial\theta}{\partial f} \right)_t = \frac{\alpha}{C}(\Theta_0 - \theta) \qquad (12\text{-}3)$$

The quantity of heat dQ heats the heat-storing mass and thus increases the heat

---

[2] Strictly speaking the differential quotient $\partial\theta/\partial f$ in the expression for dQ should not refer to $t = $ const. because $d\theta$ represents the temperature change in a specific particle of gas when flowing past df; the time taken for the gas to flow past df is very short. Furthermore, in a precise calculation one must account, in addition, for a lower heat consumption such that, the quantity of gas in this regenerator section which is warmer at time $t + dt$ by $(\partial\theta/\partial t)\, dt$ than at time t. If these two influences are considered (see Anzelius [A304] and H. Hausen [H303]), $t - t_0$ must be brought into the subsequent differential equations in place of t, where t_0 denotes the time needed by the gas to move from the entry position to the position in question. However this time can be neglected as a rule. Willmott and Hinchcliffe [W311] have examined these processes during a reversal but without considering any pressure balance.

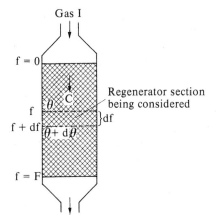

Gas I

f = 0

f

f + df

f = F

Regenerator section
being considered

df

Figure 12-1 Regenerator and a section of a regenerator
with a heating surface area *df*.

accumulated there. It follows that:

$$dQ = dC_s \left(\frac{\partial \Theta_m}{\partial t} \right)_f dt$$

where dC_s denotes the thermal capacity of the heat-storing mass in the very short section of the regenerator under consideration and, as in Sec. 11-3, Θ_m denotes the spatial average value of the temperature of the heat-storing mass in the considered cross-section at time t. Comparison of the last two expressions obtained for dQ yields

$$\left(\frac{\partial \Theta_m}{\partial t} \right)_f = \frac{\alpha \, df}{dC_s} (\theta - \Theta_0) \qquad (12\text{-}4)$$

Restricting our considerations to a plate-shaped heat-storing element the volume associated with df is equal to $\delta \cdot df / 2$ so that

$$dC_s = \frac{\rho c \, \delta}{2} \cdot df$$

Equation (12-4) then takes the following form:

$$\left(\frac{\partial \Theta_m}{\partial t} \right)_f = \frac{2\alpha}{\rho c \, \delta} (\theta - \Theta_0) \qquad \text{(plate-shaped heat-storing mass)} \qquad (12\text{-}5)$$

Equations (12-1), (12-3) and (12-4) are then the required differential equations which determine the total temperature distribution in the regenerator with two boundary conditions which have yet to be discussed. Equations (12-3) and (12-4) were originally derived for the warm period only but they are equally applicable to the cold period provided the corresponding values α' and C' replace α and C and the idea that during the cold period f is measured positively in the direction of flow, that is from the cold end of the regenerator, is taken into consideration.

The Boundary Conditions

In the state of equilibrium an important boundary condition arises from the consideration that the temperature at each position in the heat-storing mass must have the same value at the end of the cold period as it has at the beginning of the warm period, and vice versa. This boundary condition is known as the *reversal condition*. The equations largely developed in this chapter and to a certain extent in the next chapter, primarily fulfil only this reversal consideration.

A more precise and complete method of calculation of the processes within regenerators, as will be developed later, must also satisfy a second boundary condition which states that the gas enters the regenerator at a positive, usually constant, temperature. This condition is called the *entry condition*. An additional condition (12-6) for the surface area of the heat-storing mass can also be added.

$$\alpha(\theta - \Theta_0) = -\lambda_s \left(\frac{\partial \Theta}{\partial y}\right)_{y=0} \tag{12-6}$$

This condition states that the quantity of heat which will be transferred from the gas to the surface as a consequence of the temperature difference $\theta - \Theta_0$ will be passed into the heat-storing mass via a corresponding drop in temperature. However this condition is also satisfactorily fulfilled by Eq. (12-4) or (12-5). Equation (12-4) or (12-5) states that the quantity of heat being transferred by virtue of the temperature drop $\theta - \Theta_0$ will be added to the store of heat in the heat-storing mass; the amount of heat stored is determined by Θ_m. However Eq. (12-6) is sometimes quite important.*

12-2 THE METHOD OF CALCULATION OF HEILIGENSTAEDT

The original method of calculation of Heiligenstaedt [H313, 314] and Rummel [R304, 305, 306] which at one time were quite significant, will be discussed only in outline since the type of calculation which has been already outlined in Secs. 11-3 and 11-4 gives more precise results which are more generally applicable; the justification for this will be thoroughly substantiated in the following paragraphs. Schack's methods of calculation [S305, 307] have only a limited validity but nevertheless they give very usable values for the normal conditions of the iron and steel industry.

Heiligenstaedt's method rests on the following considerations. As has already been mentioned in Sec. 11-3 and in the establishment of the boundary conditions in Sec. 11-4, as a rule the two gases enter, alternately, at the opposite ends of the regenerator at a chronologically constant temperature (see Fig. 11-8). On the other hand however, at other positions in the regenerator the temperatures of the gases are dependent on time (see Fig. 11-7). In order to simplify matters Heiligenstaedt

---

* Editor's note: Equation (12-6) must be included if the unsteady state conduction of heat is explicitly represented by Eq. (12-1) [W307].

makes the assumption that *the two gas temperatures θ and θ' are also chronologically constant at every regenerator cross-section.*

In order to calculate the distribution of the temperature Θ in a brick cross-section, using this assumption, Heiligenstaedt uses the following particular solution of the differential Eq. (12-1)

$$\Theta = A + B \exp\left(-\beta^2 at\right) \cdot \cos \beta\left(y - \frac{\delta}{2}\right) \tag{12-7}$$

where t denotes time and A, B and β denote free parameters. This solution embodies the consideration that the temperature distribution must always be symmetrical around the centre of the brick $y = \frac{\delta}{2}$. If this relationship for Θ is inserted into the boundary condition (12-6) it will be recognized that the assumption $\theta = \text{const.}$ will only be satisfied if it is assumed that $A = \theta$ and β satisfies the relationship

$$\left(\beta\frac{\delta}{2}\right) \cdot \tan\left(\beta\frac{\delta}{2}\right) = \frac{\alpha}{\lambda_s}\frac{\delta}{2} \tag{12-8}$$

Of the infinite number of roots of this equation only the first is used and this is denoted by β_0. Correspondingly if B_0 is written for B, Eq. (12-7) converts to

$$\Theta = \theta + B_0 \exp\left(-\beta_0^2 at\right) \cdot \cos \beta_0\left(y - \frac{\delta}{2}\right) \tag{12-9}$$

This equation applies to the warm period. In a similar manner Eq. (12-10) is obtained for the cold period

$$\Theta' = \theta' + B_0' \exp\left(-\beta_0'^2 at'\right) \cdot \cos \beta_0'\left(y - \frac{\delta}{2}\right) \tag{12-10}$$

where β_0' is the first root of (12-8) if α is replaced by α' in this equation. Heiligenstaedt gives tables from which β_0 and β_0' can be quickly determined. In Eqs. (12-9) and (12-10) the time t or t' is calculated from the beginning of each warm and cold period respectively.

In order to deal with the state of equilibrium, Heiligenstaedt basically considered an infinite number of successive warm and cold periods. Thus he started from a brick which was isothermal in its cross-section and then calculated the constants B_0 or B_0' for each new period from the temperature distribution at the end of the preceding period (see $\Theta_{\max}$ or $\Theta_{\min}$ in Fig. 11-5). The requirement that must be fulfilled here is that the temperature distribution in the cross-section under consideration does not change during a reversal and that Θ should be equal to Θ' everywhere at the moment of reversal. However this is not completely satisfied when the simple Eqs. (12-9) and (12-10) are used. Instead Heiligenstaedt calculated new values of B_0 or B_0' so that at any given time, on average at least, the closest possible agreement between Θ and Θ', in the sense of the method of least squares will be achieved.

The many complicated relationships obtained by Heiligenstaedt in this manner for calculating B_0 and B_0' for the state of equilibrium will not be replicated here. Equations which are equally precise but which take a simpler form can be obtained by a shorter method if the reversal condition is only applied to the average brick temperatures Θ_m and Θ_m' at the beginning and end of the period. In this case it is only necessary simply for the average brick temperature to remain unchanged at the moment of reversal.

Equation (12-11) is obtained from Eq. (12-9) for the average brick temperature at an arbitrary time t in the warm period

$$\Theta_m = \frac{1}{\delta}\int_0^\delta \Theta \, dy = \theta + B_0 \exp\left(-\beta_0^2 at\right) \cdot \frac{\sin\beta_0\frac{\delta}{2}}{\beta_0\frac{\delta}{2}} \qquad (12\text{-}11)$$

A corresponding relationship for Θ_m' follows from Eq. (12-10) for the cold period. Θ_m and Θ_m' can now be calculated for the beginning and end of each period from the expressions so obtained. By comparing the corresponding values at the change over from the warm to the cold period and from the cold to the warm period two equations are finally obtained which determine B_0 and B_0'.

Knowing B_0 and B_0' the temperature distribution in the brick cross-section under consideration can be calculated for the state of equilibrium using Eqs. (12-9) and (12-10). Figure 12-2 illustrates such a calculation for the cold period. The temperature distribution in the brick is shown at each position in time by an arc of a cosine curve. If we move from the temperature Θ_{max}, which occurs at the end of the warm period, to that at the beginning of the cold period, then according to Heiligenstaedt's calculations the cos-arc, which is open upwards initially, suddenly reverses into a strongly curved arc which faces downwards but which lies at the same average height. The ordinates of this reversed cos-arc which are measured from θ' then decrease chronologically and exponentially during the cold period as indicated by Eq. (12-10), and as can be seen in Fig. 12-2 from the curves drawn for successive time intervals.

Heiligenstaedt has derived a relationship for the heat exchange coefficient

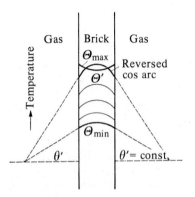

Figure 12-2 Distribution of the brick temperature in the air period according to the Heiligenstaedt theory.

$\varepsilon = k(T + T')$ from the temperatures he calculated with the aid of an equation which was developed in basically the same way as was Eq. (12-34), which will be presented in Sec. 12-6. The relationship obtained by Heiligenstaedt in this way is again very complicated. On the other hand, employing Eqs. (12-34) and (12-9) and using the values of B_0 and B'_0, obtained in the manner just described, the considerably simpler Eq. (12-12) is obtained for the heat transmission coefficient:

$$k = \frac{\rho c \, \delta}{2(T + T')} \cdot \frac{(1 - \psi)(1 - \psi')}{1 - \psi\psi'} \tag{12-12}$$

using the abbreviation

$$\psi = \exp(-\beta_0^2 aT); \qquad \psi' = \exp(-\beta_0'^2 aT') \tag{12-13}$$

Heiligenstaedt's assumption that the gas temperature does not vary with time in each period is not, in general, true as will be seen later in Secs. 12-5 and 12-7. Nevertheless the fundamental significance of Heiligenstaedt's work remains important. This is because he demonstrated for the first time, that the processes can be represented theoretically if but approximately, and that a heat exchange coefficient, and thus also a heat transmission coefficient can be obtained.

12-3 THE METHOD OF RUMMEL

Heiligenstaedt himself, in 1931, pointed out in his book [H313] p. 169, that the assumption of the chronologically constant gas temperatures was introduced only as a means of simplification, and that it does not hold good in practice. Rummel [R304] suggested a more empirical method in order to get away from this assumption and to find an expression for the heat transmission coefficient which is simpler than that of Heiligenstaedt. He looked for an expression for the heat transmission coefficient k which has a form similar to that of the simple heat transmission equation for plane walls (2-6). He thus arrived at the equation

$$\frac{1}{k} = (T + T')\left(\frac{1}{\alpha T} + \frac{1}{\alpha' T'} + \frac{2}{\zeta \eta \rho c \, \delta}\right) \tag{12-14}$$

where η and ζ denote functions which have yet to be discussed.

η is the utilization factor of the brick, that is, the ratio of the quantity of heat actually stored in one section of the regenerator to the quantity of heat which could be stored with the same variation of surface temperature if the thermal conductivity λ_s of the brick was infinitely large (Fig. 12-3, left). Rummel calculated η from the approximation

$$\frac{1}{\eta} = 1 + \frac{\delta^2}{4a(T + T')} \quad \text{with } a = \frac{\lambda_s}{\rho c} \tag{12-15}$$

The proportionality factor ζ is determined solely from the variation of surface temperature which can be imagined to be similar to that shown in Fig. 12-3, right. The chronological average values $\bar{\Theta}_0$ and $\bar{\Theta}'_0$ of this temperature in both periods

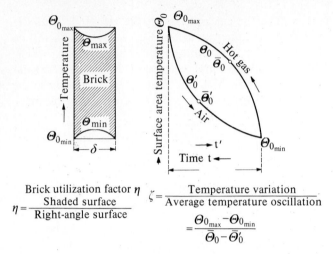

Brick utilization factor η

$\eta = \dfrac{\text{Shaded surface}}{\text{Right-angle surface}}$

$\zeta = \dfrac{\text{Temperature variation}}{\text{Average temperature oscillation}}$

$= \dfrac{\Theta_{0\max} - \Theta_{0\min}}{\bar{\Theta}_0 - \bar{\Theta}'_0}$

Figure 12-3 The Rummel method for hot blast stove calculations.

vary considerably. Rummel placed the largest temperature ratio $\bar{\Theta}_{0\max} - \bar{\Theta}_{0\min}$ as the numerator and the difference $\bar{\Theta}_0 - \bar{\Theta}'_{0\min}$ between the chronological average values as the denominator in the expression for ζ, so that

$$\zeta = \frac{\Theta_{0\max} - \Theta_{0\min}}{\bar{\Theta}_0 - \bar{\Theta}'_0} \qquad (12\text{-}16)$$

Rummel determined ζ by experiment; he found that ζ varies only very slightly for hot blast stoves and usually lies between 2 and 3.5.

If ζ and η are known, the calculation of k using Eq. (12-14) is very simple. However this equation cannot be completely satisfactory from a theoretical point of view because it contains a proportionality factor which has to be determined experimentally while η can only be determined imprecisely.

It is remarkable that the Eq. (11-4) developed by the author for k_0, that is, for the heat transmission coefficient derived from the fundamental oscillation, has a similar form to Rummel's Eq. (12-14).

12-4 SCHACK'S PROCESS

As has already been mentioned, Schack [S307] published a method of approximation, the results of which are in very close agreement with the indirect measurements taken in a blast furnace stove. Like Heiligenstaedt, and to a certain extent like Rummel, Schack proceeded from considerations of the temperature distribution in a brick cross-section. Schack formed the empirical expressions (12-17) and (12-18) for the chronological variation of the average brick temperature Θ, the surface area temperature Θ_0 and the gas temperature θ in such a brick cross-section (see Fig. 11-7);

$$\Theta_0 - \Theta_m = C_1 + C_2 \exp\left(n\frac{t}{T}\right) \tag{12-17}$$

and

$$\theta = C_3 + C_4 \exp\left(m\frac{t}{T}\right) \tag{12-18}$$

In these expressions C_1, C_2, C_3, C_4, n and m are constants. Schack determined the values of C_1, C_2, C_3 and C_4 by using a relationship developed by Gröber for the brick utilization factor η (see Sec. 12-3 and [S303]). He took into account on the one hand the restriction that the brick temperatures Θ_0 and Θ_m cannot change suddenly from one discrete state to another at a reversal and on the other an approximation of the chronological fluctuations $\Delta\theta$ and $\Delta\theta'$ in the gas temperatures during the warm and cold periods. In addition, by comparing with experimental results, Schack found $n = -8$ and $m = 0.1$. By inserting the expression thus obtained for $\Theta_0 - \Theta_m$ and θ into Eq. (12-4) a differential equation arises which has only Θ_m as an unknown and which can easily be integrated. Using a further intermediate calculation Schack arrived at a rather complicated equation for the heat exchange coefficient $K = k(T + T')$ which can be understood more easily diagrammatically.

Schack's process yields the heat exchange coefficient K, and thus also the heat transmission coefficient k, for a particular position in the regenerator.

If an average heat transmission coefficient is formed from this for the total regenerator and this value of k is then compared with the values given by the exact process described in Sec. 11-4, the following impression emerges. Schack's process generally proves quite usable for the brick thicknesses and period lengths normally found in the iron and steel industry and is in close agreement with Heiligenstaedt's theories, as well as those of the author, in the limiting case of very short regenerators. However, it is, on the contrary, less reliable in the case of very large reduced regenerator lengths Λ (see Eq. (13-28)) and heat-storing masses with very thin walls, i.e., in conditions which exist, above all, in low temperature technology.

12-5 FUNDAMENTAL OSCILLATION OF A REGENERATOR WITH PLATE-SHAPED PACKING AND WITH THE SAME THERMAL CAPACITIES PER PERIOD FOR BOTH GASES [H308] $(CT = C'T')$

A regenerator theory will now be developed, which is practically free of all assumptions[3] and thus replicates, very precisely, the actual behaviour of a regenerator. The most important fundamental concepts incorporated into this theory have already been discussed in Sec. 11-3. They consist, above all, of perceiving the processes, which take place in a regenerator at a state of equilibrium,

---

[3] A small, practical but insignificant discrepancy occurs only in the calculation of the harmonic vibrations, such that a heat transfer coefficient $\bar{\alpha}$ related to the average brick temperature Θ_m will be introduced (see Sec. 13-1).

as forced temperature oscillations and separating these into a fundamental oscillation and harmonic vibrations. In order to obtain the mathematical relationships which replicate the fundamental oscillation and the harmonic vibrations we require those solutions of the differential Eqs. (12-1), (12-3) and (12-4) which satisfy the reversal conditions (see Sec. 12-1). There is an infinite number of solutions which are known as the "eigen functions", following one of the normal methods of notation used in the theory of partial differential equations. The functions are characterized by an "eigen value" χ, which can take one of a range of values of all whole real numbers between 0 and ∞. As has already been mentioned at the end of Sec. 11-3, the fundamental oscillation corresponds to the eigen functions for $\chi = 0$ while the higher eigen functions replicate the harmonic vibrations.

First of all, the zero eigen functions will be derived for the case where the thermal capacities of both the gases flowing through the regenerator are independent of temperature and, relative to the length of the warm or cold period, are equal to each other. If, as above, C and C' are the thermal capacities of the gases per unit time and T and T' are the lengths of the warm and cold periods respectively, this will mean that $CT = C'T'$. For the time being, the bricks will again be considered to be plates with an overall thickness δ.

Derivation of the Zero Eigen Function

In most of the early works concerning the calculation of brick regenerators it was assumed, as a matter of course, that, in the state of equilibrium, the temperature distribution in the longitudinal direction in a regenerator was the same as that in a recuperator. In making this comparison the temperature distribution in a regenerator is perceived as existing at a specific point in time or at some average position in time within a period. It is always possible to find, among the solutions mentioned to the differential equations, one which represents this most simple temperature curve in the longitudinal direction, and it can be shown that this solution is exactly the zero eigen function ($\chi = 0$) which corresponds to the fundamental oscillation.[4] Thus the zero eigen function can also be defined as being the same temperature distribution in the longitudinal direction as arises in a recuperator.

In a recuperator a linear temperature distribution occurs when $C = C'$ and α and α' are constant. Thus the zero eigen function for $CT = C'T'$ must represent, at each instant of time, a linear temperature distribution in the longitudinal direction of the regenerator. Having chosen the heating surface area f which extends to the cross-section under consideration as the longitudinal coordinate of the regenerator, the expression (12-19) is formed for the gas temperature θ and the brick temperature Θ for constant time t:

$$\left(\frac{\partial \theta}{\partial f}\right)_t = \left(\frac{\partial \Theta}{\partial f}\right)_t = \left(\frac{\partial \Theta_m}{\partial f}\right)_t = \text{const} \tag{12-19}$$

---

[4] This is shown in Secs. 13-6 and 13-8.

If in addition to α and C, λ_s, ρ, c and δ are assumed to be constant, Eq. (12-3) shows, primarily, that $\Theta_0 - \theta = \text{const}$. Accordingly the temperature difference between the brick surface and the gas at a given time t has the same value at all positions in the regenerator. It follows that the same amount of heat must be transferred at all positions over a given time interval with the result that the temperatures vary uniformly and at the same rate throughout the regenerator. Thus the straight lines which represent the temperature distribution in the longitudinal direction of the regenerator must be displaced parallel to each other. Now the constants in Eq. (12-19), and thus also $\Theta_0 - \theta$ are not dependent on time. Equation (12-20) is thus obtained from Eq. (12-4):

$$\left(\frac{\partial \Theta_m}{\partial t}\right)_f = \text{const} \tag{12-20}$$

A result of the linear variation in the longitudinal direction is that the *average temperature* Θ_m in each brick cross-section *varies linearly chronologically*. This rule,[5] which was regarded previously only as a very close approximation, is shown here to be an *exact mathematical relationship for the zero eigen function* when $CT = C'T'$.

It follows from the reversal condition that the average brick temperatures Θ_m and Θ'_m in the warm and cold periods at the moment of reversal must be equal to one another since, at the end of the cold period, the average brick temperature must have the same value as at the beginning of the warm period, and vice versa. It follows from this, using Eq. (12-20) that, for each regenerator cross-section, the curve of Θ_m and Θ'_m is represented by a *single shared, straight line* if, as in Fig. 11-7, the times t and t' in the warm and cold periods, are drawn as the abscissa in the opposite directions and are scaled in such a way that T and T' appear as straight lines of equal size. Θ_m and Θ'_m pass through this straight line in alternately opposite directions, as has already been shown in Fig. 11-7. Furthermore, if the times t and t' of both periods, which correspond to the same abscissa values in Fig. 11-7, are known as the times "which correspond to each other", Eq. (12-21) is applicable at such times:

$$\Theta_m = \Theta'_m \tag{12-21}$$

This equation also expresses an important feature of the zero eigen function.

According to these findings the average brick temperatures Θ_m and Θ'_m are of fundamental significance in later considerations. To a certain extent their curves form a simply constructed framework on which the rather complicated curve of the individual brick temperatures Θ and Θ' and the gas temperatures θ and θ' can be constructed.

Figures 11-6 and 11-7 give a rough idea of the basic relationships of the gas temperatures and the brick temperatures within the framework of the zero eigen function. In order to be able to calculate this variation in temperature under very different conditions, we need the following solution of the differential Eq. (12-1):

$$\Theta = A + \left(\frac{\partial \Theta_m}{\partial t}\right)_f \cdot t - \frac{1}{2a}\left(\frac{\partial \Theta_m}{\partial t}\right)_f y(\delta - y) + \sum_{n=1}^{\infty} \beta_n \cdot \exp\left(-\beta_n^2 a t\right) \cdot \cos \beta_n \left(y - \frac{\delta}{2}\right)$$

$$\tag{12-22}$$

[5] See H. Hausen [H307].

where A, B_n and β_n denote, above all, arbitrarily selected constants and $\left(\dfrac{\partial \Theta_m}{\partial t}\right)_f$ is constant following on from Eq. (12-20). The letter n, which first only appears as an index, is a positive, real, whole number which can have the values 0, 1, 2, 3, etc. The first term after A corresponds to the requirement that Θ_m should vary linearly with time. The second term represents the parabolic temperature distribution as it tends to develop towards the end of the period (see Fig. 11-6). Finally, the summation of the cos-terms replicates the deviations from the parabolic curve which usually occur directly after a reversal. The way in which these terms are dependent on y takes into account that the solution must be symmetrical about $y = \delta/2$,[6] where y, as shown in Fig. 11-6 is the distance of the position in the brick being considered, from one of its two surfaces. In order to maintain the chronologically linear variation of Θ_m, in spite of the appearance of the cos-terms, the value of β_n is determined in such a way that the average value of each cos-term, formed over the total thickness of the brick δ, disappears. Thus

$$\frac{1}{\delta} \int_0^\delta \cos \beta_n \left(y - \frac{\delta}{2}\right) \cdot dy = 0$$

is formed and Eq. (12-23) is obtained:

$$\beta_n = \frac{2n\pi}{\delta} \tag{12-23}$$

The average value of the parabolic term disappears when $\delta^2/6$ is subtracted from $y(\delta - y)$. Furthermore, as the sign of β_n can be chosen arbitrarily, Eq. (12-24) can be written in place of Eq. (12-22), taking into account Eqs. (12-5) and (12-23):

$$\Theta = \Theta_m - \frac{\alpha}{\lambda_s \delta}(\theta - \Theta_0)\left[y(\delta - y) - \frac{\delta^2}{6}\right]$$
$$+ \sum_{n=1}^{\infty} B_n \exp\left[-\left(\frac{2n\pi}{\delta}\right)^2 at\right]\cos\left(2n\pi\frac{y}{\delta}\right) \tag{12-24}$$

If this equation refers to the warm period, the following Eq. (12-25) correspondingly refers to the cold period:

$$\Theta' = \Theta'_m - \frac{\alpha'}{\lambda_s \delta}(\theta' - \Theta'_0)\left[y(\delta - y) - \frac{\delta^2}{6}\right]$$
$$+ \sum_{n=1}^{\infty} B'_n \exp\left[-\left(\frac{2n\pi}{\delta}\right)^2 at'\right]\cos\left(2n\pi\frac{y}{\delta}\right) \tag{12-25}$$

[6] The two terms with $\left(\dfrac{\partial \Theta_m}{\partial t}\right)_f$ in Eq. (12-22) form a particular integral f in Eqs. (12-1) for the special case

$$\left(\frac{\partial \Theta}{\partial t}\right)_f = \left(\frac{\partial \Theta_m}{\partial t}\right)_f = \text{const.}$$

Mathematically it is remarkable that these two terms can also be understood as the zero term of the cos-series as can easily be shown by taking the limit to $\beta_n = 0$ with B_n becoming infinitely large. These two terms are anticipated in Eq. (12-22) as this is the most simple way of making use of the properties mentioned of the average brick temperatures.

This last equation can be rearranged somewhat if we take into account the fact that, at equilibrium, the same quantity of heat q_{Per} will be transferred per unit surface area in the cold period, as will be transferred in the warm period. This yields Eq. (12-26):

$$q_{Per} = \alpha'(\Theta_0' - \theta')T' = \alpha(\theta - \Theta_0)T \qquad (12\text{-}26)$$

Thus Eq. (12-25) takes on the form

$$\Theta' = \Theta_m' + \frac{\alpha}{\lambda_s \delta}(\theta - \Theta_0)\frac{T}{T'}\left[y(\delta - y) - \frac{\delta^2}{6}\right]$$

$$+ \sum_{n=1}^{\infty} B_n' \exp\left[-\left(\frac{2n\pi}{\delta}\right)^2 at'\right]\cos\left(2n\pi\frac{y}{\delta}\right) \qquad (12\text{-}27)$$

The origin for time t or t' must always be reestablished in each period as being the moment of reversal.

The values of B_n and B_n' are obtained from the *reversal condition*, according to which, at all positions in the cross-section the brick temperature should have the same value at the beginning of the warm period as at the end of the cold period and vice versa. Thus, if we determine the temperature Θ at the beginning of the warm period ($t = 0$) using Eq. (12-24) and compare it with the temperature Θ' at the end of the cold period ($t' = T'$) using Eq. (12-27), Eq. (12-28) is obtained upon taking Eq. (12-21) into consideration:

$$\sum_{n=1}^{\infty}\left\{B_n - B_n'\exp\left[-\left(\frac{2n\pi}{\delta}\right)^2 aT'\right]\right\}\cos\left(2n\pi\frac{y}{\delta}\right)$$

$$= \frac{\alpha}{\lambda_s \delta}(\theta - \Theta_0)\left(1 + \frac{T}{T'}\right)\left[y(\delta - y) - \frac{\delta^2}{6}\right] \qquad (12\text{-}28)$$

On the left-hand side of this equation is a Fourier series with coefficients which still have to be determined

$$B_n - B_n'\exp\left[-\left(\frac{2n\pi}{\delta}\right)^2 aT'\right]$$

If the right-hand side of Eq. (12-28) is set equal to $f(y)$, using the well known equation for determining the Fourier coefficients we obtain

$$B_n - B_n'\exp\left[-\left(\frac{2n\pi}{\delta}\right)^2 aT'\right] = \frac{2}{\delta}\int_0^\delta f(y)\cos\left(2n\pi\frac{y}{\delta}\right)\cdot dy$$

and after integration

$$B_n - B_n'\exp\left[-\left(\frac{2n\pi}{\delta}\right)^2 aT'\right] = -\frac{1}{(n\pi)^2}\frac{\alpha\,\delta}{\lambda_s}(\theta - \Theta_0)\frac{T+T'}{T'} \qquad (12\text{-}29)$$

A similar examination of the end of the warm period and the beginning of the cold period yields:

$$B_n' - B_n\exp\left[-\left(\frac{2n\pi}{\delta}\right)^2 aT\right] = +\frac{1}{(n\pi)^2}\frac{\alpha\,\delta}{\lambda_s}(\theta - \Theta_0)\frac{T+T'}{T'} \qquad (12\text{-}30)$$

If Eqs. (12-29) and (12-30) are solved for B_n and B'_n, Eq. (12-31) is obtained, by inserting the expression resulting for B_n in Eq. (12-24), as the *most general form of the zero eigen function for* $CT = C'T'$;

$$
\begin{aligned}
\Theta = \Theta_m &- \frac{\alpha\delta}{\lambda_s}(\theta - \Theta_0)\left[\frac{y}{\delta}\left(1 - \frac{y}{\delta}\right) - \frac{1}{6}\right] \\
&+ \frac{T+T'}{T'}\sum_{n=1}^{\infty}\frac{1}{(n\pi)^2}\cdot\frac{1 - \exp\left[-\left(\frac{2n\pi}{\delta}\right)^2 aT'\right]}{1 - \exp\left[-\left(\frac{2n\pi}{\delta}\right)^2 a(T+T')\right]}\cdot\exp\left[-\left(\frac{2n\pi}{\delta}\right)^2 at\right] \\
&\times \cos\left(2n\pi\frac{y}{\delta}\right)\Big]
\end{aligned}
$$

(12-31)

Here Θ_m is given by Eq. (12-32) taking into account Eq. (12-5)

$$
\Theta_m = (\Theta_m)_a + \left(\frac{\partial\Theta_m}{\partial t}\right)_f t = (\Theta_m)_a + \frac{2\alpha}{\rho c\,\delta}(\theta - \Theta_0)t
$$

(12-32)

where $(\Theta_m)_a$ denotes the value of Θ_m at the beginning of the period. Using Eqs. (12-31) and (12-32) the total variation in the temperature distribution can be calculated for the period under consideration and, upon making all necessary adjustments, that for the succeeding period can be computed also. The gas temperature Θ is obtained by calculating the surface temperature Θ_0 from Eq. (12-31) with $y = 0$ and then adding the time invariant value of $\theta - \Theta_0$.

Results of Calculation

Figure 12-4 shows the temperature distribution in an 80 mm thick brick calculated using Eqs. (12-31) and (12-32) for $CT = C'T'$ where the period lengths are assumed to be $T = T' = 1$ h. In addition $\lambda_s = 1.163$ W/(m K), $\rho = 2000$ kg/m³, $c = 1.047$ kJ/(kg K) are assumed. The curves on the left-hand side of Fig. 12-4 replicate the time variations of temperature in a specific brick cross-section, at the brick surface $(y = 0)$, at 15 and 25 mm from the surface and at the centre of the brick itself $(y = \delta/2 = 40$ mm).

It will be seen that the temperature distribution at all positions within the brick is bent directly after reversal, as in Fig. 11-7, and gradually changes into a chronologically linear variation. The completely straight curve of the average brick temperatures Θ_m and Θ'_m is shown as a dot-and-dash line. The right-hand side shows, as does Fig. 11-6, the temperature distribution within the brick at various times t. By way of comparison, Fig. 12-5 contains the experimental results obtained by Schumacher [S317] in an experimental hot blast stove; these relate to the same data as shown in Fig. 12-4. The agreement between calculation and experiment indicates that very accurate measurements were made.

The picture of the temperature distribution in the brick changes when thicker

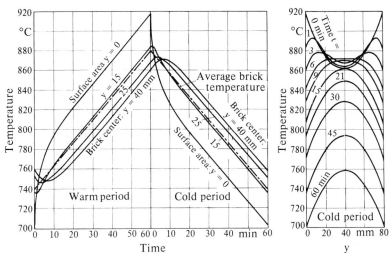

Figure 12-4 Calculated temperature distribution in an 80 mm thick regenerator brick.

bricks or shorter period lengths are considered. The thicker the brick, the longer it takes for the inverting of the parabola shaped, or parabola-like, curves from being open upwards to being open downwards; this inversion is shown in Figs. 11-6 and 12-4 (right). In the case of very thick bricks, this can take so much time that even at the end of the period the complete parabolic shape is not fully realized. The deviations from this shape are greater the shorter the period. Figure 12-6 illustrates the result of a calculation for a 200 mm thick brick with conditions otherwise the same as those in Fig. 12-4 and where the period length is again assumed to be 1 hour. It will be observed that the centre of the brick plays a relatively small part in the fluctuations in temperature and thus in heat transfer. Beyond a certain brick thickness the temperature fluctuations in the centre of the brick cease altogether,

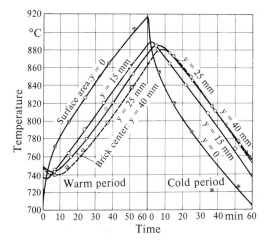

Figure 12-5 Time variation of temperature in a checker brick using Schumacher's measurements.

assuming an unchanged period length. A further increase in the thickness of the brick thus has no further influence on the exchange of heat and thus the heat transmission coefficient must also be independent of the brick thickness.

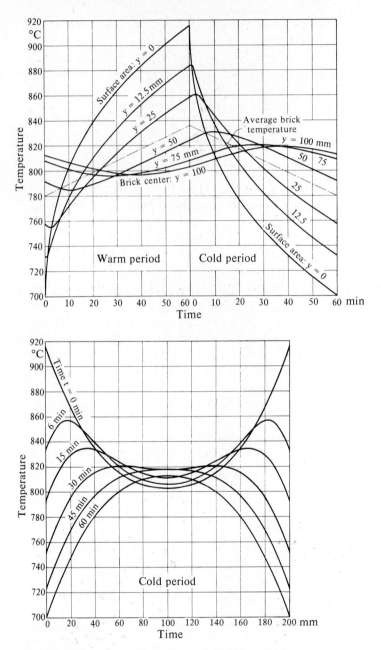

Figure 12-6 Temperature distribution in a brick 200 mm thick.

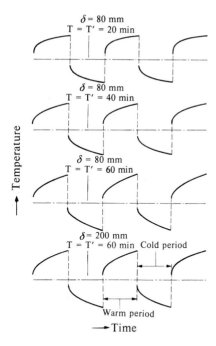

Figure 12-7 Calculated variation of gas temperature as a function of time.

Figure 12-7 illustrates, for this very case, the calculation of the variation of the gas temperature with time. The three top curves apply to the 80 mm thick brick with a period length of $T = T' = 20$, 40 and 60 minutes; the bottom curve applies to the 200 mm thick brick with a period length of $T = T' = 60$ min. Here it can be seen that, within each period, the temperatures initially change very rapidly but then tend more and more towards a variation which is linear.

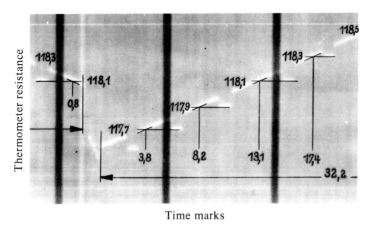

Time marks

Figure 12-8 Chronologically linear variation in temperature in a regenerator with a metal hea'-storing mass using Glaser's measurements. Diagonal light lines: resistance of the thermometer. Horizontal light lines: standard resistances. Vertical dark lines: time marks.

In regenerators with a heat-storing mass made of sheet metal, δ/λ_s is, as a rule, so small that it can be assumed that $\delta/\lambda_s = 0$ without making any noticeable error. It follows from Eq. (12-31) that $\Theta = \Theta_m$. Since Θ_0 also equals Θ_m in this case, then using the zero eigen function, not only does the temperature of the heat-storing mass vary linearly with time but since $\theta - \Theta_0$ is constant, so also the gas temperature follows a completely linear variation with time.

These conclusions are confirmed by Fig. 12-8 which shows the chronological variation of the gas temperature in the centre of such a regenerator obtained from Glaser's measurements [G301]. At the test position in the regenerator the conditions relevant to the zero eigen function were practically completely satisfied. For the ordinate is used the resistance of the resistance thermometer which follows the variations in gas temperature. As the changes in resistance in the region in question are proportional to the temperature changes, a chronological linear variation in temperature is clearly displayed by these measurements.

12-6 GENERAL RELATIONSHIPS FOR THE HEAT TRANSMISSION COEFFICIENT AND THE APPLICATION OF THE ZERO EIGEN FUNCTION [H308]

The aim of calculations of the spatial and chronological variation in temperature in regenerators is usually to find an expression for the heat transmission coefficient k. Knowledge of this coefficient makes it possible, employing Eq. (11-3), to calculate the most important data for the construction and operation of regenerators in a manner as simple as for recuperators. Using this method, for example, the quantity of heat Q_{Per} transferred in each period can be determined, as can the two unknown exit temperatures $\bar{\theta}_2$ and $\bar{\theta}'_2$

To begin with, two generally valid relationships for the heat transmission coefficient will be derived. This will be followed by the application of these relationships to the zero eigen function as was developed in previous paragraphs.

According to Eq. (12-2), in the warm period the quantity of heat transferred via an element df of the heating surface area in time dt amounts to

$$dQ = \alpha \, df \, dt (\theta - \Theta_0)$$

The total quantity of heat transferred in the warm period is obtained using:

$$Q_{Per} = \alpha F T (\bar{\theta} - \bar{\Theta}_0)_M$$

if the heat transfer coefficient α is assumed to be constant over the whole regenerator and the chronological and spatial average value of the temperature difference $\theta - \Theta_0$ is denoted by $(\bar{\theta} - \bar{\Theta}_0)_M$. Comparing this relationship for Q_{Per} with Eq. (11-3) we get

$$k = \frac{\alpha T}{T + T'} \cdot \frac{(\bar{\theta} - \bar{\Theta}_0)_M}{\Delta \theta_M} \qquad \text{(average value for the whole regenerator)} \qquad (12\text{-}33)$$

From this can be found the desired general equation for k. Employing this general equation, the calculation of k leads back to the calculation of the average temperature differences $(\bar{\theta}-\bar{\Theta}_0)_M$ and $\Delta\theta_M$ which can be determined from the solutions of the differential equations.

The heat transmission coefficient calculated using Eq. (12-33) represents an average value for the whole regenerator. However the heat transmission coefficient k_ξ at a specific position ξ in the regenerator can also be calculated using Eq. (12-33) by inserting only the chronological average values $\bar{\theta}-\bar{\Theta}_0$ and $\bar{\theta}-\bar{\theta}'$ which relate to the cross-section under consideration, instead of $(\bar{\theta}-\bar{\Theta}_0)_M$ and $\Delta\theta_M$. In this way Eq. (12-34) is obtained:

$$k_\xi = \frac{\alpha T}{T+T'}\frac{\bar{\theta}-\bar{\Theta}_0}{\bar{\theta}-\bar{\theta}'} \qquad \text{(for a specific cross-section of the regenerator)} \quad (12\text{-}34)$$

The heat transmission coefficient k determined by Eq. (12-34) varies, in general, along the longitudinal direction of the regenerator. even when α has the same overall value. This is because the ratio of the temperature differences in Eq. (12-34) differs from place to place as a result of the influence, mentioned in Sec. 11-4, of the hysteresis loop, particularly towards the ends of the regenerator. However, this variation does not need to be taken into consideration at this stage provided one is only dealing with the fundamental oscillation of a regenerator. In this case the heat transmission coefficient can always be calculated using the simple Eq. (12-34) and the average heat transmission coefficient k can be immediately obtained for the fundamental oscillation, since Eqs. (12-33) and (12-34) are then the same.

The Heat Transmission Coefficient k_0 from the Zero Eigen Function

Once the temperature distribution has been obtained using Eqs. (12-31) and (12-32), the heat transmission coefficient k_0, which corresponds to the zero eigen function, can be calculated with the help of Eq. (12-34).

First we calculate the chronological average value of the surface area temperature Θ_0 in the warm period by putting $y = 0$ into Eq. (12-31) and integrating with respect to time t from 0 to T. We thus obtain

$$\bar{\Theta}_0 - \bar{\Theta}_m = \frac{\alpha\delta}{\lambda_s}(\bar{\theta}-\bar{\Theta}_0)\Phi \qquad (12\text{-}35)$$

where $\bar{\Theta}_m$ is the time mean value of Θ_m, and where Φ is given by:

$$\Phi = \frac{1}{6} - \frac{\delta^2}{4a}\left(\frac{1}{T}+\frac{1}{T'}\right)$$

$$\times \sum_{n=1}^{\infty}\frac{1}{(n\pi)^4}\cdot\frac{\left\{1-\exp\left[-\left(\frac{2n\pi}{\delta}\right)^2 aT\right]\right\}\left\{1-\exp\left[-\left(\frac{2n\pi}{\delta}\right)^2 aT'\right]\right\}}{1-\exp\left[-\left(\frac{2n\pi}{\delta}\right)^2 a(T+T')\right]} \qquad (12\text{-}36)$$

Equation (12-37) arises in a corresponding way for the cold period, utilizing Eq. (12-26)

$$\bar{\Theta}'_m - \bar{\Theta}'_0 = \frac{\alpha\delta}{\lambda_s}(\theta - \Theta_0)\frac{T}{T'}\Phi \tag{12-37}$$

Since $\theta - \Theta_0 = \text{const}$, $\Theta'_0 = \theta' = \text{const}$ and $\bar{\Theta}_m = \bar{\Theta}'_m$ (corresponding to Eq. (12-21)), the chronological average value $\bar{\theta} - \bar{\theta}'$ of the difference between the gas temperatures is determined by

$$\bar{\theta} - \bar{\theta}' = (\theta - \Theta_0) + (\bar{\Theta}_0 - \bar{\Theta}_m) + (\bar{\Theta}'_m - \bar{\Theta}'_0) + (\Theta'_0 - \theta') \tag{12-38}$$

By inserting Eq. (12-35) and (12-37) into Eq. (12-38) and then utilizing Eq. (12-26), we obtain

$$\bar{\theta} - \bar{\theta}' = (\theta - \Theta_0)\left[1 + \frac{\alpha T}{\alpha' T'} + \left(1 + \frac{T}{T'}\right)\frac{\alpha\,\delta}{\lambda_s}\Phi\right] \tag{12-39}$$

With this, because $\theta - \Theta_0 = \text{const}$ and thus also $\bar{\theta} - \bar{\theta}' = \text{const}$, using Eq. (12-34) the following relationship results for the heat transmission coefficient k_0:

$$\frac{1}{k_0} = (T + T')\left[\frac{1}{\alpha T} + \frac{1}{\alpha' T'} + \left(\frac{1}{T} + \frac{1}{T'}\right)\frac{\delta}{\lambda_s}\Phi\right] \tag{12-40}$$

This equation, which has been mentioned already in Sec. 11-4 as Eq. (11-4), relates to the zero eigen function and is valid irrespective of the thickness of the brick or the length of the warm or cold period. Although it is derived on the assumption that $CT = C'T'$, as will be shown subsequently it provides a very close approximation for other cases.

Further Equations for the Function Φ

The factor Φ in the last term of Eq. (12-40), consists primarily according to Eq. (12-36), of the constant $1/6$ originating from the parabolic temperature curve. On the other hand the series expresses the influence of the very rapid temperature changes which take place immediately after a reversal, see for example, Figs. 11-7, 12-4 or 12-6. The calculation of k_0 using Eq. (12-40) does not depend upon the complexity of this function. However Eq. (12-36) can be replaced by the following approximating relationships which are very precise:

$$\text{for} \quad \frac{\delta^2}{2a}\left(\frac{1}{T} + \frac{1}{T'}\right) \leqslant 10: \quad \Phi = \frac{1}{6} - \frac{1}{180}\frac{\delta^2}{2a}\left(\frac{1}{T} + \frac{1}{T'}\right) \tag{12-41}$$

$$\text{for} \quad \frac{\delta^2}{2a}\left(\frac{1}{T} + \frac{1}{T'}\right) \geqslant 10: \quad \Phi = \frac{0.357}{\sqrt{0.3 + \frac{\delta^2}{2a}\left(\frac{1}{T} + \frac{1}{T'}\right)}} \tag{12-42}$$

It is even simpler to read Φ off from Fig. 12-9 where Φ is displayed as a function of

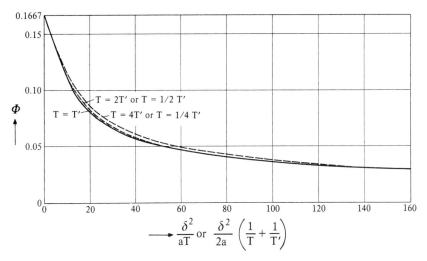

Φ

$\dfrac{\delta^2}{aT}$ or $\dfrac{\delta^2}{2a}\left(\dfrac{1}{T}+\dfrac{1}{T'}\right)$

Figure 12-9 Auxiliary function Φ for the calculation of the heat transmission coefficient.

$\dfrac{\delta^2}{aT}$ (with $T = T'$), or more generally as a function $\dfrac{\delta^2}{2a}\left(\dfrac{1}{T}+\dfrac{1}{T'}\right)$ using Eq. (12-36). The solid curve is valid for $T = T'$, while the dashed curves refer to the cases $T = 2T'$ or $T = \frac{1}{2}T'$ and to $T = 4T'$ or $T = \frac{1}{4}T'$. The dashed lines show that even when $T = 4T'$ or $T = \frac{1}{4}T'$ the deviations from the corresponding values for $T = T'$ are very small. Thus, for almost all practical cases, Φ can be read off from the solid curve (see Fig. 11-12 also) with sufficient accuracy or can be calculated using the corresponding Eqs. (12-41) and (12-42). All these considerations are based on the assumption that $CT = C'T'$.

Some more, very precise equations for Φ will be given for the sake of completion and as a basis for further theoretical research. Equation (12-36) has the disadvantage that the series in that equation converges very slowly. It is therefore expedient to rearrange it somewhat by adding and subtracting $\dfrac{1}{(n\pi)^4}$ to each term

in the series under the summation sign. Further, since $\displaystyle\sum_{n=1}^{\infty}\dfrac{1}{(n\pi)^4}=\dfrac{1}{90}$, Eq. (12-43) can be used in place of Eq. (12-36):

$$\Phi = \frac{1}{6} - \frac{1}{90}\frac{\delta^2}{4a}\left(\frac{1}{T}+\frac{1}{T'}\right) + \frac{1}{\pi^4}\frac{\delta^2}{4a}\left(\frac{1}{T}+\frac{1}{T'}\right)$$

$$\times \sum_{n=1}^{\infty}\frac{1}{n^4}\cdot\frac{\exp\left[-\left(\dfrac{2n\pi}{\delta}\right)^2 aT\right]+\exp\left[-\left(\dfrac{2n\pi}{\delta}\right)^2 aT'\right]-2\exp\left[-\left(\dfrac{2n\pi}{\delta}\right)^2 a(T+T')\right]}{1-\exp\left[-\left(\dfrac{2n\pi}{\delta}\right)^2 a(T+T')\right]}$$

$$(12\text{-}43)$$

The sum of this series converges so quickly that in most practical cases the sum need not be considered at all (for $\dfrac{\delta^2}{2a}\left(\dfrac{1}{T}+\dfrac{1}{T'}\right)$ up to about 10 or only its first term need be taken into consideration. If the sum is totally neglected, the approximation (12-41) is obtained with its theoretical basis thereby provided.

For $T = T'$ Eqns. (12-36) and (12-43) take the form

$$\Phi = \frac{1}{6} - \sum_{n=1}^{\infty} \frac{1}{(n\pi)^2} \cdot \frac{\tanh\left(\left(\dfrac{n\pi}{\delta}\right)^2 2aT\right)}{\left(\dfrac{n\pi}{\delta}\right)^2 2aT} \tag{12-44}$$

and

$$\Phi = \frac{1}{6} - \frac{1}{90}\cdot\frac{\delta^2}{2aT} + \frac{1}{\pi^4}\frac{\delta^2}{2aT}\sum_{n=1}^{\infty}\frac{1}{n^4}\left[1 - \tanh\left(\left(\frac{n\pi}{\delta}\right)^2 2aT\right)\right] \tag{12-45}$$

Here also the series in Eq. (12-45) converges considerably more quickly than does the series in Eq. (12-44).

The Limiting Cases of Very Thin and Very Thick Bricks

For very thin bricks the value of $\dfrac{\delta^2}{aT}$ or $\dfrac{\delta^2}{2a}\left(\dfrac{1}{T}+\dfrac{1}{T'}\right)$ comes close to the value 0. It thus follows from the equations given for Φ and from Fig. 12-9 that Φ is equal to 1/6. An outcome of this, is that Eq. (12-40) assumed a somewhat simpler form. Within the framework of the zero eigen function this simpler relationship is more precise the thinner the bricks and is at its most exact when the heat-storing mass is made of thin sheets. Both Eq. (12-42) and an exact consideration of the limit [H308] show that with increasing brick thickness, Eq. (11-36) tends towards the expression:

$$\Phi = \frac{0.357}{\delta}\sqrt{\frac{2a}{\dfrac{1}{T}+\dfrac{1}{T'}}} \qquad \text{(for very thick bricks)} \tag{12-46}$$

Together with this equation, (12-40) takes the form

$$\frac{1}{k_0} = (T+T')\left[\frac{1}{\alpha T} + \frac{1}{\alpha' T'} + \sqrt{\frac{1}{T}+\frac{1}{T'}}\cdot\frac{0.505}{\sqrt{\lambda_s c\rho}}\right] \qquad \text{(for very thick bricks)} \tag{12-47}$$

The brick thickness δ no longer appears in this equation. The earlier requirement that k_0 should be independent of the brick thickness in the case of very thick bricks is thus fulfilled. The asymptotic representation (12-46) is so good that Eq. (12-47) can be usually used, with sufficient accuracy, for $\dfrac{\delta}{2a}\left(\dfrac{1}{T}+\dfrac{1}{T'}\right) \geqslant 10$.

Comparison of the Heat Transmission Equation for Regenerators with the Equation for Recuperators

If we consider two regenerators working together with the same warm or cold period length $T = T'$ and with the total period length $T_{ges} = T + T'$ then within the framework of the zero eigen function, using Eqs. (11-15) and (12-40) we obtain the relationship (12-48) for an equivalent heat transmission coefficient, defined as for recuperators and relative to both regenerators:

$$\frac{1}{(k_0)_{reg}} = \frac{1}{2k_0} = \frac{1}{\alpha} + \frac{1}{\alpha'} + \frac{\delta_{reg}}{\lambda_s} 2\Phi \text{ (with } T = T') \tag{12-48}$$

If we compare this relationship with Eq. (2-6) or (5-2) which applies to recuperators

$$\frac{1}{k} = \frac{1}{\alpha} + \frac{1}{\alpha'} + \frac{\delta_{rek}}{\lambda_s}$$

it will be recognized that in a pair of regenerators with the brick thickness δ_{reg} given the values of α, α' and λ_s, an exchange of heat will be achieved which is as good as that in a recuperator with brick thickness $\delta_{rek} = 2\delta_{reg}\Phi \leqslant \dfrac{\delta_{reg}}{3}$, since $\Phi \leqslant 1/6$. It follows that with respect to the same transfer of heat, a brick partition wall in a recuperator can be at most 1/3 as thick as the plate-shaped packing bricks in a regenerator which are made of the same material. In this respect the regenerator is superior to the recuperator as long as only the zero eigen function is considered. The basis for this consideration is as follows:

For the sake of simplicity the brick thickness δ is assumed to be the same in both cases and relatively small so that, according to Fig. 12-9, Φ can be equal, or close, to 1/6. While, in recuperators, the heat must flow right through the brick, in regenerators it only penetrates both sides, on average, by about 1/4 of the brick thickness and then flows back again in the opposite direction. Thus, in regenerators, the heat has to cover, at most, about 1/2 the thickness of the brick (see Fig. 12-10).

In addition to this the flow of heat decreases towards the interior of the brick

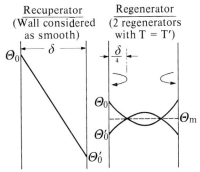

$\Theta_0 - \Theta_0$ is three times smaller in the case of regenerators than in the case of recuperators

Figure 12-10 Comparison of the temperature distribution in a brick in a recuperator and a regenerator.

and thus the main burden of the transport of heat is displaced more towards the brick surface. Beyond that, the drop in temperature which is thus created is minimized by the single intersection of the temperature curves Θ and Θ' shown on the right in Fig. 12-10, such that, according to Eq. (12-21) at corresponding times in both periods Θ_m should be equal to Θ'_m. This accounts for the fact that a specific quantity of heat can be transferred in regenerator bricks of a given thickness δ with a drop in temperature $\Theta_0 - \Theta'_0$ which is at least three times smaller than that in the bricks of a recuperator. However this only applies so long as the material of which the heat-storing mass and the solid walls of the recuperator are made, have the same thermal conductivity λ_s. The partition walls in recuperators are made of metal more often than are the heat storing masses in regenerators* and such metal has a considerably greater thermal conductivity than bricks.

Influence of Dust Deposits

The influence of dust deposits which was first examined by Schack can easily be calculated employing a small extension of the theories which have been so far developed. At both sides of a brick of thickness δ there is a layer of dust with a thickness δ_0 and with a thermal conduction coefficient λ_0. If the dust layer thickness δ_0 is so small that the thermal capacity of the dust layer in relation to the thermal capacity of the brick can be neglected, all the heat given off by the gas to the brick must penetrate the dust layer. Thus the heat obtained by the brick surface from the gas has not only to overcome the $1/\alpha$ which is proportional to the resistance to heat transfer but must also overcome the resistance of the dust layer which is proportional to δ_0/λ_0. Therefore in the equations used hitherto the following replacements must be made:

$$\frac{1}{\alpha} \quad \text{by} \quad \frac{1}{\alpha} + \frac{\delta_0}{\lambda_0}$$

and

$$\frac{1}{\alpha'} \quad \text{by} \quad \frac{1}{\alpha'} + \frac{\delta_0}{\lambda_0}$$

In this way Eq. (12-40) for the heat transmission coefficient k_0 takes the form:

$$\frac{1}{k_0} = (T + T') \left[\frac{1}{\alpha T} + \frac{1}{\alpha' T'} + \left(\frac{1}{T} + \frac{1}{T'} \right) \left(\frac{\delta}{\lambda_s} \Phi + \frac{\delta_0}{\lambda_0} \right) \right] \qquad (12\text{-}49)$$

The additional term δ_0/λ_0 thus expresses the influence of the dust deposit when compared with Eq. (12-40). It should be observed moreover that the heat transfer coefficients α and α' can also have different values for heat transfer to a dust layer than for transfer to a dust-free brick surface.

---

* Editor's note: This is certainly true of high temperature regenerators but many industrial rotary regenerators have packings made of metal plates nowadays.

12-7 FUNDAMENTAL OSCILLATION OF A REGENERATOR WITH CYLINDRICAL OR SPHERICAL PACKING WHEN $CT = C'T'$

In order to clear up the question of how the shape of the heat-storing mass influences heat transfer, B. Stuke [S319] considered the two cases of a heat-storing mass constructed of cylindrical or spherical bricks with a diameter δ. He also assumed an equal thermal capacity $CT = C'T'$ of the quantities of gas flowing in the warm or cold periods. The method of calculation is basically the same as that developed in Sec. 12-5 for plate-shaped bricks; the difference in the resultant relationships consists of the fact that two of the differential equations are now rather different.

Calculation of the Chronological Variation of Temperature in a Cylinder or Sphere

Equations (12-1a) and (12-1b) replace Eq. (12-1)[7] for the conduction of heat in the brick:

(a) Differential equation with cylindrical symmetry

$$\frac{\partial \Theta}{\partial t} = \frac{a}{r}\frac{\partial}{\partial r}\left(r\frac{\partial \Theta}{\partial r}\right) \qquad [12\text{-}1a]$$

(b) with spherical symmetry

$$\frac{\partial \Theta}{\partial t} = \frac{a}{r^2}\frac{\partial}{\partial r}\left(r^2\frac{\partial \Theta}{\partial r}\right) \qquad [12\text{-}1b]$$

where r is the distance of the position in the brick being considered from the centre point of the circle or sphere.

Equation (12-3) remains unchanged. Equation (12-5) takes on a general form

$$\left(\frac{\partial \Theta_m}{\partial t}\right)_t = m\cdot\frac{2\alpha}{\rho c\,\delta}(\theta - \Theta_0) \qquad \begin{array}{ll} \text{Plate} & m = 1 \\ \text{Cylinder} & m = 2 \\ \text{Sphere} & m = 3 \end{array} \qquad [12\text{-}5a]$$

In the two cases handled by Stuke $\left(\dfrac{\delta \Theta_m}{\delta t}\right)_f$ was set const for $CT = C'T'$ within the framework of the zero eigen function.

Considering first Eq. (12-5a), then Eqs. (12-1a) and (12-1b) with cylinder or

[7] Apart from the additional terms a or b, wherever possible the following equations will have the same notation as the corresponding equations in Sec. 12-5.

sphere radius R, have the solutions:

$$\text{Cylinder} \quad \Theta = \Theta_m + \frac{\alpha}{2\lambda_s R}(\theta - \Theta_0)\left(r^2 - \frac{R^2}{2}\right) + \sum_{n=1}^{\infty} A_n \exp(-\omega_n^2 at)J_0(\omega_n r)$$

[12-22a]

$$\text{Sphere} \quad \Theta = \Theta_m + \frac{\alpha}{2\lambda_s R}(\theta - \Theta_0)\left(r^2 - \frac{3}{5}R^2\right) + \sum_{n=1}^{\infty} \frac{B_n}{r} \exp(-\beta_n^2 at)\sin(\beta_n r)$$

[12-22b]

In Eq. (12-22a) J_0 is the zero order Bessel function of the first kind. A_n, ω_n, B_n and β_n are constants which will be determined below. It is remarkable that, as in the case of plate-shaped bricks, the term following Θ_m in Eqs. (12-22a) and (12-22b) represent a parabolic temperature curve. The constant in this term is chosen so that it adds nothing to Θ_m. The defining Eqs. (12-23a) and (12-23b) for the eigen values ω_n and β_n are obtained from the requirement that each of the summation terms should add nothing to Θ_m,

$$\text{Cylinder} \quad J_1(\omega_n R) = 0 \qquad \text{[12-23a]}$$

$$\text{Sphere} \quad \tan(\beta_n R) = \beta_n R \qquad \text{[12-23b]}$$

where J_1 denotes the Bessel function of the first order.

If Eqs. (12-22a) and (12-22b) are related to the warm period and the corresponding relationships are also written down for the cold period, the following relationships are obtained from the reversal condition:

$$A_n = -\frac{2\alpha}{\lambda_s R} \cdot \frac{\theta - \Theta_0}{\omega_n^2 J_0(\omega_n R)} \cdot \frac{T+T'}{T'} \cdot \frac{1 - \exp(-\omega_n^2 aT')}{1 - \exp[-\omega_n^2 a(T+T')]}$$

[12-29a]

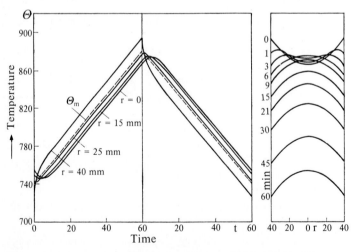

Figure 12-11 Temperature distribution with a cylindrical heat-storing mass. Cylinder diameter = 80 mm.

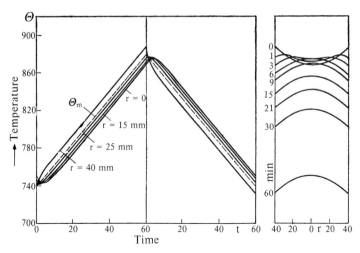

Figure 12-12 Temperature distribution with a spherical heat-storing mass. Sphere diameter = 80 mm.

$$B_n = -\frac{2\alpha}{\lambda_s} \cdot \frac{\theta - \Theta_0}{\beta_n^2 \sin(\beta_n R)} \cdot \frac{T + T'}{T'} \cdot \frac{1 - \exp(-\beta_n^2 a T')}{1 - \exp[-\beta_n^2 a(T + T')]}$$ [12-29b]

The chronological variation of temperature in a cylindrical or spherical brick is clearly defined by Eqs. (12-22a, b), (12-23a, b) and (12-29a, b).

Figure 12-11 illustrates the calculated temperature distribution in a cylindrical cross-section and Fig. 12-12 illustrates the calculated temperature distribution in a spherical cross-section. The diameters of the cylinder and the sphere are assumed to be the same as the thickness of the plate-shaped bricks in Fig. 12-4, that is $\delta = 80$ mm. In all other respects the same assumptions embodied in Fig. 12-4 are used here.

For the most part, the same temperature distribution appears in all three cases. At each moment the temperature differences between the surface and a position inside the brick are smaller in the case of cylinders, and particularly in the case of spheres, than they are in the case of plate-shaped bricks. Accordingly the inner parts of the cylinder, and particularly of the sphere, play a much greater role in the temperature changes than do the inner parts of the plate-shaped bricks. This is explained by the fact that, given equal diameter or thickness δ a smaller volume of heat-storing mass is associated with unit surface area in the case of cylinders than in the case of plates, and in the case of spheres an even smaller volume is associated with unit surface area than in the case of cylinders. Thus a smaller heat-storing mass must also be heated or cooled.

Calculation of the Heat Transmission Coefficient k_0

As in Sec. 12-6, the Eq. (12-40) for the heat transmission coefficient k_0 takes the form

$$\frac{1}{k_0} = (T + T') \left[\frac{1}{\alpha T} + \frac{1}{\alpha' T'} + \left(\frac{1}{T} + \frac{1}{T'} \right) \frac{\delta}{\lambda_s} \Phi \right] \qquad [12\text{-}40]$$

although the function Φ takes on a different form for cylinders and spheres than it has for smooth plates. Equations (12-36a) and (12-36b) relate to Φ for the case where $\delta = 2R$

for cylinders:

$$\Phi = \frac{1}{8} - \frac{\delta^2}{4a} \left(\frac{1}{T} + \frac{1}{T'} \right) \sum_{n=1}^{\infty} \frac{[1 - \exp(-\omega_n^2 aT)][1 - \exp(-\omega_n^2 aT')]}{\left(\omega_n^2 \frac{\delta}{2} \right)^4 \{1 - \exp[-\omega_n^2 a(T + T')]\}} \qquad [12\text{-}36a]$$

for spheres:

$$\Phi = \frac{1}{10} - \frac{\delta^2}{4a} \left(\frac{1}{T} + \frac{1}{T'} \right) \sum_{n=1}^{\infty} \frac{[1 - \exp(-\beta_n^2 aT)][1 - \exp(-\beta_n^2 aT')]}{\left(\beta_n \frac{\delta}{2} \right)^4 \{1 - \exp[-\beta_n^2 a(T + T')]\}} \qquad [12\text{-}36b]$$

With these expressions for Φ the same transformations and boundary considerations can be applied as were described in Sec. 11-4 for plate-shaped bricks. Thus the approximations (11-7) and (11-8) result, which were given in Sec. 11-4. The curve of the function Φ for plates, cylinders and spheres, as a function of $\dfrac{\delta^2}{2a} \left(\dfrac{1}{T} + \dfrac{1}{T'} \right)$ are shown in Fig. 11-12. The curves are only really valid for $T = T'$. However the influence of large deviations from the case where $T = T'$ can practically always be neglected as this influence is no greater for cylindrical and spherical bricks than it is for the plate-shaped bricks as shown in Fig. 12-9.

It will be seen in Fig. 12-13 that the three curves for Φ are very close together if the equivalent plate thickness $\delta_{gl} = R + V/F$ is introduced for δ in Eqs. (12-40), (12-40a) and (12-40b) in place of the diameter; R denotes the radius, V the volume and F the surface of a cylindrical or spherical brick or of a large number of bricks. Thus it can be expected that, using this equivalent plate thickness, the value of Φ can be read from Fig. 12-13 yielding sufficient accuracy, even for an arbitrarily shaped heat storing mass. This was discussed in greater detail in Sec. 11-4.

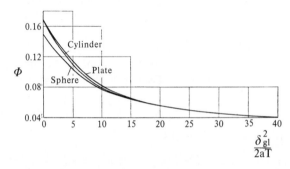

Figure 12-13 Auxiliary function Φ, as a function of the equivalent plate thickness

$$\delta'_{gl} = R + V/F = \delta/2 + V/F.$$

12-8 FUNDAMENTAL OSCILLATION OF A REGENERATOR WITH UNEQUAL THERMAL CAPACITIES PER PERIOD FOR BOTH GASES ($CT \neq C'T'$)

It has already been mentioned several times that the relationships given so far for k_0, in particular Eqs. (12-40) and (12-49) can be used without giving rise to too great an error if $CT \neq C'T'$. In order to demonstrate this and to make an estimate of the possible errors, in Sec. 65 in the first edition (in German) of this book, the author has shown how the zero eigen function can be calculated exactly for the general case $CT \neq C'T'$. Here, as previously, C, C' and α and α' were considered to be independent of temperature.

As the same case, with a few omissions, will be considered somewhat more simply in Sec. 13-5 only the basic ideas behind the exact method of calculation will be indicated here. For further details of the derivation reference should be made to Appendix 3.

The theory will be developed from the fact that, in the case of the zero eigen function, the temperatures in the longitudinal direction of the regenerator, curve in the same way as in a recuperator operating under the same conditions. It follows from Eq. (5-43) that in a recuperator, when $C \neq C'$ the wall temperatures and the gas temperatures are determined by equations of the type

$$\Theta = D + A \cdot \exp(bf) \tag{12-50}$$

where A, b and D are constants and f again denotes the heating surface area which serves as the longitudinal coordinate. The requirement that the zero eigen function of the regenerator should be dependent on f in the way determined by Eq. (12-50) will be satisfied in the following way. The more general solution (12-51) of the differential Eq. (12-1) will be used in place of Eq. (12-22), namely

$$\Theta = D + \sum_{n=0}^{\infty} B_n \exp(-\beta_n^2 at) \cos \beta_n \left(y - \frac{\delta}{2} \right) \tag{12-51}$$

where D and β_n are constants and the coeffient B_n is a function of f. Equation (12-51) is transformed into the form of Eq. (12-50) if the relationship

$$B_n = \text{const} \cdot \exp(bf)$$

is used.

The values of β_n can be determined in such a way that Eq. (12-51) also satisfies the differential Eqs. (12-3) and (12-4).

If Eq. (12-51) is considered to be valid for the warm period and a corresponding equation is formed for the cold period, then, employing the reversal condition, the values of B_n and B_n' can be finally obtained for the regenerator cross-section under consideration. By this means very complicated expressions arise for the brick temperatures Θ and Θ' and thus also, using Eq. (12-4), for the gas temperatures θ and θ'. In this way the total chronological variation of temperature in a brick cross-section can be calculated. A similarly complicated equation can be derived for the heat transmission coefficient k_0 corresponding to the one for the zero eigen function when $CT = C'T'$.

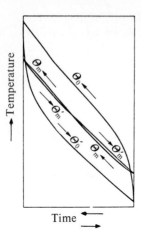

Figure 12-14 Chronological temperature variations in a regenerator brick with $CT = 2C'T'$.

Figure 12-14 shows the result of an exact calculation for $C = 2C'$ and $T = T'$, thus where $CT = 2C'T'$. In all other respects the same data is used as that employed in Fig. 12-4. The variation of the surface temperatures Θ_0 and Θ'_0, as also the variation of the brick temperatures Θ_m and Θ'_m, is displayed as a function of time and, again, as earlier (see Fig. 11-7), the time in the cold period and the time in the warm period are drawn on the same axis but in opposite directions to one another. There is a contrast between the temperature variation with time, in Fig. 12-14 which is curved downwards less strongly, and the corresponding curve in Fig. 12-4. The values of Θ_m and Θ'_m are not superimposed in Fig. 12-14 but form a very narrow hysteresis loop where the curves cross one another. However, the difference between Θ_m and Θ'_m is so small that this difference had to be magnified fivefold in the drawing in order to make it clearly recognizable. Moreover, the temperature difference $\Theta_m - \Theta'_m$, although finite in value, is both positive and negative so that the time average is very small. The very small hysteresis loop in the chronological temperature variation mentioned above disappears when the simplified theory, mentioned in Sec. 13-5, using the zero eigen function is applied; see Fig. 13-9.

Thus, it follows from Figs. 12-14 and 12-4 that the differences in the temperature curves between $CT = 2C'T'$ and $CT = C'T'$ are only very small. In particular it is possible to show, by calculation, that the chronological average values of the temperature differences $\theta - \Theta_0$, $\Theta_0 - \Theta_m$, etc., in Figs. 12-4 and 12-14 are almost exactly the same. Since these temperature differences are a determining factor in the rate of heat transfer, the heat transmission coefficient k_0 in the case shown in Fig. 12-14, is not noticeably different from the value calculated from the earlier Eqs. (11-4) or (12-40) developed for $CT = C'T'$.

THIRTEEN

PRECISE CALCULATION OF THE COMPLETE TEMPERATURE DISTRIBUTION BETWEEN THE ENTRANCE AND EXIT OF A REGENERATOR WITH A HEAT-STORING MASS OF VERY HIGH THERMAL CONDUCTIVITY

13-1 DEVELOPMENT OF THE CASES HANDLED SO FAR IN THE CASE OF A HEAT-STORING MASS OF VERY HIGH THERMAL CONDUCTIVITY

The relationships developed in the previous chapter for the zero eigen function make it possible to calculate the fundamental oscillation of a regenerator operating in counterflow. These relationships give an insight into the temperature changes which occur in the central part of long regenerators. It has been mentioned already in Sec. 11-3, that the behaviour at the ends of the regenerator deviates substantially from this temperature curve as a result of the influence of the higher eigen functions (the harmonics) (Fig. 11-8). The precise calculation of these phenomena will now be set out. The only deviation from a very exact treatment will occur at the point where a heat transfer coefficient related to the average brick temperature is introduced in order to simplify the calculation quite considerably. However, this will have hardly any effect on the result of the calculation.*

Heat Transfer Coefficient $\bar{\alpha}$ Related to the Average Brick Temperature Θ_m

The spatial temperature variations within a brick cross-section, and thus also the difference $\theta - \Theta_m$ between the temperature of the gas and the average brick

---

* Editor's note: This view was held until quite recently but it is clear now the situation is rather more complicated than was first thought. See Fig. 13-3.

temperature were calculated precisely in the treatment above for the zero eigen function. It follows that the chronological average value $\bar{\theta} - \bar{\Theta}_m$ of $\theta - \Theta_m$ during a period with $CT = C'T'$ can easily be determined. This is because the chronological average value of $\Theta_0 - \Theta_m$ is determined by Eq. (12-35) for the case of the zero eigen function. Moreover $\theta - \Theta_0$ is constant. Thus if $\theta - \Theta_0 = \bar{\theta} - \bar{\Theta}_0$ is added to both sides of this equation, Eq. (13-1) is obtained:

$$\bar{\theta} - \bar{\Theta}_m = \left(\frac{1}{\alpha} + \frac{\delta}{\lambda_s}\Phi\right) \cdot \alpha(\theta - \Theta_0) \tag{13-1}$$

Now, $q_{Per} = \alpha(\theta - \Theta_0)T$ represents the quantity of heat transferred per unit surface area during a period. Thus, taking a previous equation into consideration, we can also write

$$q_{Per} = \bar{\alpha}(\bar{\theta} - \bar{\Theta}_m)T \tag{13-2}$$

from which we obtain

$$\frac{1}{\bar{\alpha}} = \frac{1}{\alpha} + \frac{\delta}{\lambda_s}\Phi \tag{13-3}$$

Thus, according to Eq. (13-2) $\bar{\alpha}$ can be considered as a heat transfer coefficient relative to the average brick temperature Θ_m.

Equations (13-2) and (13-3) are applicable precisely in the case of the time average for the zero eigen function. However, in the following considerations, for the sake of simplicity, some assumptions will be made, which have yet to be established, that in general, for example even the sum of the higher eigen functions are included and, moreover, at each point of time, calculations can be made using the heat transfer coefficient related to Θ_m. It follows from Eq. (13-2), that Eq. (13-4) may be written for a quantity of heat transferred per unit area in the short time dt;

$$dq = \bar{\alpha}(\theta - \Theta_m)\,dt \tag{13-4}$$

If $\bar{\alpha}$ has been determined using Eq. (13-3), the quantity of heat transferred can be obtained solely from the gas temperature θ and the average brick temperature Θ_m, without needing to know the surface temperature Θ_0. Calculations can be made, within certain limitations as though, within a given cross-section, the heat-storing mass had the same temperature $\Theta = \Theta_m$ at all positions at each moment and, correspondingly, as though the temperature difference $\theta - \Theta_m$ determined the rate of heat transfer at the surface.

The general introduction of $\bar{\alpha}$ thus considerably facilitates the theory as all cases can be reduced to the limiting case where $\Theta = \Theta_m = \Theta_0$. It follows therefore that it is only necessary to examine this case precisely. In practice, this limiting case is approached more closely the thinner are the elements of the heat-storing mass and the larger is their thermal conductivity perpendicular to the direction of flow of the gas.

However, the spatial and chronological distribution of the average temperatures Θ_m and Θ'_m of the heat-storing mass are obtained precisely, even with thick bricks, by methods of calculation which use $\bar{\alpha}$, as will be shown in what follows. However the rapid, chronological temperature variations in the gas and the surface of the heat-storing mass, which occur after reversal, as shown in Figs. 12-4, 12-6 and 12-7, will be accurately determined utilizing $\bar{\alpha}$ only if time averages are considered. If these chronological variations must also be determined accurately, further considerations may be required, such as will be described in Sec. 13-2.

Heat Transmission Coefficient k_0, Expressed Using $\bar{\alpha}$ and $\bar{\alpha}'$

By introducing the heat transfer coefficients $\bar{\alpha}$ or $\bar{\alpha}'$ which relate to Θ_m or Θ'_m, as in Eq. (13-3), the heat transfer Eq. (12-40) which was developed for the zero eigen function Sec. 12-6, can be readily converted into the simple form:

$$\frac{1}{k_0} = (T + T')\left[\frac{1}{\bar{\alpha}T} + \frac{1}{\bar{\alpha}'T'}\right] \tag{13-5}$$

Errors in Using the Heat Transfer Coefficient $\bar{\alpha}$ when Eq. (13-3) is Employed

It will be shown now that the errors which arise when the average brick temperatures Θ_m and Θ'_m and the quantity of heat which is transferred, are calculated using Eqs. (13-3) and (13-4), contribute very little to the final result of the calculation. Using the value of Φ in Eq. (13-3) the influence of the very rapid temperature changes which occur immediately after a reversal, is evenly distributed over the whole period and, using Eq. (13-5), this has no effect upon the calculation of the heat transmission coefficient. Leaving this aside, considerable errors basically occur for each of the higher eigen functions taken individually because the variation in temperature represented by these functions is different from that involving only the zero eigen function. Thus it is not to be expected that, where these higher eigen functions are involved, Eq. (13-4) will provide a completely satisfactory approximation. However, as far as the accuracy of the final result is concerned the only decisive factor is how much these errors accumulate within the sum of the high eigen functions, which is formed in a suitable manner; this will be discussed later (see Fig. 11-10, centre). However, as the following reasoning demonstrates, this error is always small.

The largest deviations from the chronological linear variation of the average brick temperature appear at the ends of the regenerator at the positions of entry of the gases (see Fig. 11-8); this linear variation is characterized, according to Sec. 12-5, by the zero eigen function in the case of counterflow. As the entry temperature of the gas $\theta = \theta_1$ is assumed to be constant with time, Θ_m must vary exponentially as will be seen in Eq. (13-20); this equation is presented later. In this case, by way of comparison, the quantity of heat transferred will be calculated precisely but will then be calculated using $\bar{\alpha}$. For plate-shaped bricks the Eqs. (12-8), (12-9) and (12-11), developed by Heiligenstaedt refer to the accurate determination of Θ_0, Θ_m and the quantity of heat $dq = \alpha(\theta - \Theta_0)dt$ which is transferred at each moment. If, for $\theta = \theta_1$, Eq. (12-9) is solved with $y = 0$ using $\theta_1 - \Theta_0$ and Eq. (12-11) is solved using $\theta_1 - \Theta_m$, then

$$\frac{\theta_1 - \Theta_m}{\theta_1 - \Theta_0} = \frac{2\alpha}{\beta_0^2 \lambda_s \delta}$$

is obtained from a consideration of Eq. (12-8).

Thus Eq. (13-6) emerges from Eq. (12-8)

$$dq = \alpha(\theta_1 - \Theta_0)dt = \beta_0^2 \lambda_s \frac{\delta}{2}(\theta_1 - \Theta_m)dt \qquad (13\text{-}6)$$

The approximate calculation using $\bar{\alpha}$, on the other hand, gives:

$$d\bar{q} = \bar{\alpha}(\theta_1 - \Theta_m)dt = \frac{\theta_1 - \Theta_m}{\dfrac{1}{\alpha} + \dfrac{\delta}{\lambda_s}\Phi}dt \qquad (13\text{-}7)$$

As in the case of the equations of Heiligenstaedt, the rapid temperature changes in the brick's surface after reversal will also be neglected in this second type of calculation. Here Φ will be equal to 1/6 because in Eq. (12-36) the summation terms within the expression for Φ disappear. Thus Eq. (13-8) is obtained for the ratio of the approximate value $d\bar{q}$ to the value of dq:

$$\frac{d\bar{q}}{dq} = \frac{1}{\left(\beta_0 \dfrac{\delta}{2}\right)^2} \cdot \frac{\alpha}{\lambda_s}\frac{\delta}{2} \cdot \frac{1}{1 + \dfrac{1}{3}\dfrac{\alpha}{\lambda_s}\dfrac{\delta}{2}} \qquad (13\text{-}8)$$

It follows from this equation that the required ratio can be calculated as a function of $\dfrac{\alpha}{\lambda_s}\dfrac{\delta}{2}$ by assuming an arbitrary value for $\beta_0 \dfrac{\delta}{2}$. In this way $\dfrac{\alpha}{\lambda_s}\dfrac{\delta}{2}$ can be determined by using Eq. (12-8). Figure 13-1 shows the result of such a calculation. $\dfrac{\alpha}{\lambda_s}\dfrac{\delta}{2}^*$ is drawn as the abscissa; the ordinate consists of the percentage error which

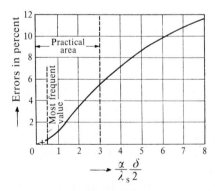

Figure 13-1 The largest errors in the calculation of the quantity of heat transferred when $\theta = \text{const}$ if $\bar{\alpha}$ is used together with Eq. (13-3).

---

* Editor's note: The quantity is a form of the Biot modulus.

measures by how much the quantity of heat transferred at each moment, calculated using Eq. (13-4), is too large. The values of $\dfrac{\alpha}{\lambda_s}\dfrac{\delta}{2}$ which appear in practice in the iron and steel industry lie around an average of about 0.5 [e.g. $\alpha = 20\ \mathrm{W/(m^2\ K)}$, $\delta = 0.05\ \mathrm{m}$ and $\lambda_s = 1\ \mathrm{W/(m\ K)}$], but particularly unfavourable cases occur when the value is approximately 3.0 (e.g. $\alpha = 60$; $\delta = 0.1$; $\lambda_s = 1$). Thus it follows from Fig. 13-1 that the errors amount only to about 0.4 per cent and, but exceptionally, can be as large as 5 or 6 per cent.

In the same way that the use of $\bar{\alpha}$ yields too large a quantity of heat transferred, so also in the same way, it generates too rapid a variation in the average brick temperature. However, errors of this order of magnitude are only found at the ends of the regenerator during the entry of the gas; during the exit of the gas they are substantially smaller. Moreover, the errors rapidly die away towards the interior of the regenerator as do the higher eigen functions, (see Sec. 11-3). Thus. the influence that these errors have on the total transfer of heat usually lies well below 1 per cent. When $\Phi < 1/6$ fewer errors can be expected than when $\Phi = 1/6$, because, according to Eq. (12-35), the temperature difference $\Theta_0 - \Theta_m$ is, on average smaller and thus an inaccuracy in its calculation has a correspondingly weaker effect.

The small size of the errors which are to be expected can be seen by comparing the temperature distribution within a brick cross-section in the two cases just mentioned, that is for the exponential chronological variation of Θ_m and the linear variation of Θ_m. Figure 13-2 displays the temperature distribution which is developed once the rapid temperature changes, which follow a regenerator reversal, have faded away. Curve a shows the spatial temperature distribution in a brick for the exponential time variation of Θ_m using the particularly unfavourable assumption that $\alpha\delta/2\lambda_s = 3.088$. Curve b reproduces the parabolic temperature curve with the linear time variation of Θ_m, given the same value of $\alpha\delta/2\lambda_s$. For almost all practical cases, very slight differences can be expected here also.

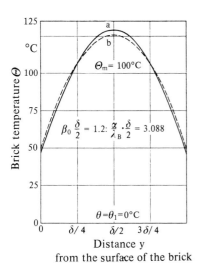

Figure 13-2 Temperature distribution in the heat-storing mass. a. with chronologically exponential variation of Θ_m; b. with chronologically linear variation of Θ_m.

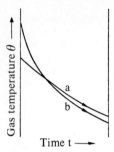

Figure 13-3 Chronological variation of the gas temperature in the cold period (schematic). a. calculated with $\bar{\alpha}$; b. calculated with α.

13-2 CALCULATION OF THE GAS TEMPERATURE θ FROM THE CHRONOLOGICAL VARIATION OF THE AVERAGE BRICK TEMPERATURE Θ_m

It has already been shown that the chronological variation of Θ_m can be calculated accurately at each position in the regenerator by using the heat transfer coefficient $\bar{\alpha}$ related to the average brick temperature $\Theta = \Theta_m$. Methods of calculation based on this will be presented in Secs. 13-6, 15-1, 15-2, 16-4, 16-5 and elsewhere. Once Θ_m is known, the chronological variation of the gas temperature can also be determined by using $\bar{\alpha}$ in Eqs. (13-20) and (13-21) set out below. The result as a function of time, corresponds to curve a in Fig. 13-3. The chronological variation of θ will thus only be represented by a time average. This curve does not express the rapid temperature variation in the gas directly following a reversal; this variation is shown, for example, in Fig. 12-7 and as curve b in Fig. 13-3. This particular variation of the gas temperature with time can be calculated in the following way from a known variation of Θ_m; this is achieved by employing the true heat transfer coefficient α instead of $\bar{\alpha}$ [H311].

The heat-storing mass is considered as consisting of flat plates of thickness δ, density ρ and specific heat c. For such flat plates, Eq. (13-9) is obtained, using Eq. (12-5), for the difference between the gas temperature θ and the surface temperature Θ_0 of the heat storing mass;

$$\theta - \Theta_0 = \frac{\rho c \,\delta}{2\alpha} \frac{d\Theta_m}{dt} \tag{13-9}$$

This difference can then be calculated from the known chronological variation of Θ_m.

In order to determine θ itself, it is first necessary to determine the chronological variation of Θ_0. The zero eigen function provides an important reference point for this. With a linear variation of Θ_m Eq. (13-10) results from Eq. (12-31), for $y = 0$, at the brick surface

$$\Theta_0 = \Theta_m + \frac{\alpha \,\delta}{\lambda_s} (\theta - \Theta_0) \left(\frac{1}{6} - \psi\right) \tag{13-10}$$

with the time function;

$$\psi = \frac{T+T'}{T} \sum_{n=1}^{\infty} \frac{1}{(n\pi)^2} \frac{1-\exp\left[-\left(\frac{2n\pi}{\delta}\right)^2 aT'\right]}{1-\exp\left[-\left(\frac{2n\pi}{\delta}\right)^2 a(T+T')\right]} \exp\left[-\left(\frac{2n\pi}{\delta}\right)^2 at\right] \qquad (13\text{-}11)$$

In this expression π denotes the "circle ratio", while λ_s and a denote the thermal conductivity and the thermal diffusivity of the material of the heat-storing mass respectively. Θ_0 and thus θ can be calculated at each point in time using these two equations for a linear variation of Θ_m, since $\theta - \Theta_0$ is already known from the Eq. (13-9). The function ψ represented by the series in (13-11) reproduces, after multiplication by $\frac{\alpha \delta}{\lambda_s}(\theta - \Theta_0)$, the rapid chronological changes in Θ_0, which have already been mentioned, at the beginning of the period under consideration.

Equations (13-10) and (13-11) can provide a very good approximation, for a non-linear chronological variation of Θ_m provided the deviation from the linear case is not too severe. This is based, above all, on the fact that the function ψ from Eq. (13-11), multiplied by an arbitrary constant, represents a particular solution of the differential Eq. (12-1). This solution is independent of the variation of Θ_m. However, with a non-linear curve for Θ_m, ψ can no longer be simply multiplied by $\frac{\alpha\delta}{\lambda_s}(\theta - \Theta_0)$ because, according to Eq. (13-9), $\theta - \Theta_0$ is now a function of time. A constant ε which is basically arbitrary should be chosen in place of this as a new factor.

The expression $\frac{\alpha \delta}{6\lambda_s}(\theta - \Theta_0)$ which appears immediately before ψ in Eq. (13-10) has as its origin the parabolic temperature profile mentioned previously in a brick cross-section as it manifests itself towards the end of a sufficiently long period (see Fig. 11-6 or 12-4). This temperature distribution does not remain purely parabolic if the time variation of Θ_m is no longer linear. It can be seen from Fig. 13-2 that what deformation does occur is so slight that it can be neglected in most practical cases. Thus the expression $\frac{\alpha \delta}{6\lambda_s}(\theta - \Theta_0)$ in Eq. (13-10) can always be retained and yet provide a sufficiently good approximation, even with a time variation of Θ which is non-linear. However, the variation of $\theta - \Theta_0$ with time from Eq. (13-9) must be taken into consideration. It follows from Eq. (13-10) that Eq. (13-12) can be obtained using the factor ε, which has already been mentioned, for an arbitrary time variation of Θ_m:

$$\Theta_0 = \Theta_m + \frac{\alpha \delta}{6\lambda_s}(\theta - \Theta_0) - \varepsilon\psi \qquad (13\text{-}12)$$

This equation relates to the cold period. The corresponding Eq. (13-13) for the warm period can be written in the form

$$\Theta_0' = \Theta_m' + \frac{\alpha'\,\delta}{6\lambda_s}(\theta' - \Theta_0') - \varepsilon'\psi' \tag{13-13}$$

where ε' again denotes a constant and ψ' is obtained from Eq. (13-11) by replacing T by T' and t by t'.

The values of the constants ε and ε' are obtained by imposing the boundary condition that the spatial temperature distribution in a brick cross-section must be the same at the beginning of a period as it is at the end of the previous period. It will be seen from Fig. 13-2 that this requirement can be satisfied by not only setting $\Theta_m = \Theta_m'$ at the moment of reversal but also by setting $\Theta_0 = \Theta_0'$; thus the surface temperatures are also made equal to one another. If the beginning of a period is denoted by the suffix A and the end of a period is denoted by the suffix E then by imposing this reversal condition, Eq. (13-14) is obtained for the beginning of the cold period:

$$\frac{\alpha\,\delta}{6\lambda_s}(\theta - \Theta_0)_A - \varepsilon \cdot \psi_A = \frac{\alpha'\,\delta}{6\lambda_s}(\theta' - \Theta_0')_E - \varepsilon'\psi_E' \tag{13-14}$$

and Eq. (13-15) is obtained for the beginning of the warm period:

$$\frac{\alpha\,\delta}{6\lambda_s}(\theta - \Theta_0)_E - \varepsilon\psi_E = \frac{\alpha'\,\delta}{6\lambda_s}(\theta' - \Theta_0')_A - \varepsilon'\psi_A' \tag{13-15}$$

Equations (13-16) are presented as a set of abbreviations:

$$\left. \begin{array}{ll} \dfrac{\alpha\,\delta}{6\lambda_s}(\theta - \Theta_0)_A = A, & \dfrac{\alpha\,\delta}{6\lambda_s}(\theta - \Theta_0)_E = E \\[2ex] \dfrac{\alpha'\,\delta}{6\lambda_s}(\theta' - \Theta_0')_A = A', & \dfrac{\alpha'\,\delta}{6\lambda_s}(\theta' - \Theta_0')_E = E' \end{array} \right\} \tag{13-16}$$

Equations (13-17) and (13-18) then result as the solution of Eqs. (13-13) and (13-14).

$$\varepsilon = \frac{(A - E')\psi_A' + (A' - E)\psi_E'}{\psi_A \cdot \psi_{A'} - \psi_E \cdot \psi_{E'}} \tag{13-17}$$

and

$$\varepsilon' = \frac{(A' - E)\psi_A + (A - E')\psi_E}{\psi_A \cdot \psi_{A'} - \psi_E \cdot \psi_{E'}} \tag{13-18}$$

Once the constants ε and ε' have been determined the chronological variation of the gas temperature θ for an arbitrary variation of Θ_m with time can be calculated using Eqs. (13-11), (13-12), (13-13) and (13-9). Even the rapid changes which take place at the beginning of each period can almost always be reproduced sufficiently accurately.

If the heat-storing mass is constructed of cylindrical or spherical elements this calculation can be carried out in a similar manner; reference should be made to Eqs. (12-5a) and (12-22a) or (12-22b).

13-3 THE DIFFERENTIAL EQUATIONS IN DIMENSIONLESS FORM

If it is not required to consider the temperature differences within an individual element of the heat-storing mass but to use only Eqs. (13-3) and (13-4), the differential equations derived in Sec. 12-1 can be considerably simplified in the following way. The unsteady state heat conduction Eq. (12-1) can be dropped. Further the quantity of heat $dq = \alpha(\theta - \Theta_0)\,dt$ transferred per unit surface in time dt can be expressed using Eq. (13-4), that is $\alpha(\theta - \Theta_0)$ is replaced by $\bar{\alpha}(\theta - \Theta_m)$. Thus the differential Eqs. (12-3) and (12-4) take the form

$$\left(\frac{\partial\theta}{\partial f}\right)_t = \frac{\bar{\alpha}}{C}(\Theta_m - \theta) \tag{13-19}$$

$$\left(\frac{\partial\Theta_m}{\partial t}\right)_f = \frac{2\bar{\alpha}}{\rho c\,\delta}(\theta - \Theta_m) \tag{13-20}$$

or, for an arbitrarily shaped heat-storing mass

$$\left(\frac{\partial\Theta_m}{\partial t}\right)_f = \frac{\bar{\alpha}\,df}{dC_s}(\theta - \Theta_m) \tag{13-21}$$

These equations represent a considerable simplification of the mathematical problem since Θ_m no longer appears as a function of the brick temperature Θ which itself is a function of y. In order to simplify notation the suffix m will be omitted from Θ_m from now on. This can be done if it is always borne in mind that Θ only represents the true temperature in the case of a heat-storing mass constructed of thin sheets and that in general it represents the spatial average value of the temperature in the cross-section under consideration.

   If the heat transfer coefficient and the physical properties are not dependent on temperature, Eqs. (13-19) and (13-20) can be converted into an even simpler form. For this purpose two dimensionless variables ξ and η are introduced based on the following definitions:

$$d\xi = \frac{\bar{\alpha}}{C}df \quad \text{and} \quad d\eta = \frac{2\bar{\alpha}}{\rho c\,\delta}dt \tag{13-22}$$

or more generally

$$d\eta = \frac{\bar{\alpha}\,df}{dC_s}dt = \frac{\bar{\alpha}F}{C_s}dt \tag{13-23}$$

If $\bar{\alpha}$, C and $\rho c\,\delta$ are constant, Eq. (13-22) can also be written in the form:[1]

$$\xi = \frac{\bar{\alpha}}{C}f \quad \text{and} \quad \eta = \frac{2\bar{\alpha}}{\rho c\,\delta}t \tag{13-24}$$

---

[1] It will be noted that ξ is defined here somewhat differently than in the case of recuperators where k replaces $\bar{\alpha}$.

or, more generally, as

$$\eta = \frac{\bar{\alpha}\, df}{dC_s}\, t = \frac{\bar{\alpha} F}{C_s}\, t \tag{13-25}$$

where df and dC_s denote the surface area and thermal capacity of a very small part of the heat-storing mass. As ξ utilizes the heating surface area f used as longitudinal coordinate and η contains the time t, ξ will be called the "reduced longitudinal coordinate" and η will be called the "reduced time". Using Eq. (13-22) or (13-24) and Θ in place of Θ_m, the differential Eqs. (13-19) and (13-20) take the form:

$$\left(\frac{\partial \theta}{\partial \xi}\right)_\eta = \Theta - \theta \tag{13-26}$$

and

$$\left(\frac{\partial \Theta}{\partial \eta}\right)_\xi = \theta - \Theta \tag{13-27}$$

Thus, as a result of this transformation, θ and Θ are associated only with the independent invariables ξ and η instead of the seven quantities $\bar{\alpha}$, C, ρ, c, δ, f and t.

If the total heating surface area F and the period length T, for example in the warm period, are substituted into the reduced quantities then it follows from Eq. (13-24), that Eqs. (13-28) and (13-29) can be obtained for the case where $\bar{\alpha} = $ const, $C = $ const and $\rho c\, \delta = $ const:

$$\text{Reduced regenerator length} \qquad \Lambda = \frac{\bar{\alpha} F}{C} \tag{13-28}$$

$$\text{Reduced period length} \qquad \Pi = \frac{2\bar{\alpha}T}{\rho c\, \delta} = \frac{\bar{\alpha} F}{C_s}\, T \tag{13-29}$$

Where $\bar{\alpha}$, C and $\rho c\, \delta$ vary, Λ and Π may be defined by corresponding integrals. In the subsequent period, for example the cold period, the reduced regenerator length and the reduced period length usually have other values Λ' and Π' because, as a rule, $\bar{\alpha}'$, C' and T' are different from $\bar{\alpha}$, C and T.

The differential equations can be easily transformed further so that a differential equation with only one unknown, for example Θ, is obtained. Indeed, if Eq. (13-27) is partially differentiated with respect to ξ and the derivative $\left(\dfrac{\partial \theta}{\partial \xi}\right)_\eta$ in Eqs. (13-26) and (13-27) is replaced by $-\left(\dfrac{\partial \Theta}{\partial \eta}\right)_\xi$, the following partial differential equation in Θ alone emerges:

$$\frac{\partial^2 \Theta}{\partial \xi\, \partial \eta} + \left(\frac{\partial \Theta}{\partial \xi}\right)_\eta + \left(\frac{\partial \Theta}{\partial \eta}\right)_\xi = 0 \tag{13-30}$$

Similar equations can also be derived for θ or $\theta - \Theta$. Equation (13-31) in $\theta - \Theta$ is obtained if Eq. (13-26) is partially differentiated with respect to η, Eq. (13-27) is partially differentiated with respect to ξ and then the equations are subtracted one from another:

$$\frac{\partial^2(\theta-\Theta)}{\partial\xi\,\partial\eta} + \frac{\partial(\theta-\Theta)}{\partial\xi} + \frac{\partial(\theta-\Theta)}{\partial\eta} = 0 \qquad (13\text{-}31)$$

In the paragraphs which follow the exact solutions of the differential equations which have been derived will be explained. Moreover, in order to facilitate an analysis, methods of approximation will be discussed in Sec. 15-1 ff, which with a sufficient number of steps, allow a high degree of accuracy to be achieved. Moreover, such methods have the advantage that they can also be used for varying physical properties and heat transfer coefficients.

13-4 SOLUTION FOR THE FIRST WARMING OR COOLING OF THE HEAT-STORING MASS

At the start, that is when time $\eta = 0$, the temperature Θ_1 is assumed to be constant at all positions. From this moment onwards the gas enters at $\xi = 0$ with the chronologically constant temperature θ_1 and flows through the regenerator in the direction of the positive ξ axis. At a later time η the temperature Θ of the heat-storing mass, and the temperature θ of the gas, is a function of ξ, rather as shown in Fig. 13-4. This temperature distribution is calculated when Θ_1 and θ_1 are given.

This problem was first solved by A. Anzelius [A304] and, independently, by W. Nusselt [N303].[2] The basic ideas behind the method of the Anzelius solution will be explained but in a form somewhat altered from the original paper.

If the function

$$\Delta = \exp(\xi+\eta)(\theta-\Theta) \qquad (13\text{-}32)$$

is first considered to be the unknown variable, a relatively simple closed solution can be found. Indeed, if Eq. (13-32) is partially differentiated once with respect to

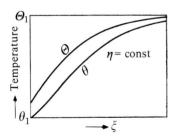

Figure 13-4 Schematic representation of the temperature distribution in the heat-storing mass at a specific time η during the first cooling.

[2] Nusselt's solution is really more general than that of Anzelius as it is also applicable for an arbitrary initial temperature distribution; see Sec. 14-1.

ξ and then again with respect to η, the result will be

$$\frac{\partial^2 \Delta}{\partial \xi \, \partial \eta} = \exp(\xi + \eta) \left[\frac{\partial^2 (\theta - \Theta)}{\partial \xi \, \partial \eta} + \frac{\partial (\theta - \Theta)}{\partial \xi} + \frac{\partial (\theta - \Theta)}{\partial \eta} \right] + \exp(\xi + \eta)(\theta - \Theta)$$

It will be seen from Eq. (13-31) that the bracketed expression on the right-hand side disappears, and by then taking into consideration Eq. (13-32), that Δ satisfies the differential equation

$$\frac{\partial^2 \Delta}{\partial \xi \, \partial \eta} = \Delta \qquad (13\text{-}33)$$

If the expression[3]

$$\Delta = a_0 + a_1 \xi \eta + a_2 (\xi \eta)^2 + a_3 (\xi \eta)^3 + \cdots \qquad a_n (\xi \eta)^n + \cdots$$

is now inserted into (13-33) and the coefficients compared, the following relationships:

$$a_1 = \frac{a_0}{1^2}, \quad a_2 = \frac{a_1}{2^2}, \ldots \qquad a_n = \frac{a_{n-1}}{n^2}, \ldots$$

are obtained, from which it will be seen that:

$$a_1 = \frac{a_0}{(1!)^2}, \quad a_2 = \frac{a_0}{(2!)^2}, \ldots \qquad a_n = \frac{a_0}{(n!)^2} \cdots$$

Thus we obtain

$$\Delta = a_0 \sum_0^\infty \frac{(\xi \eta)^n}{(n!)^2} \qquad (13\text{-}34)$$

The basically arbitrary constant a_0 can be determined from this so that when $\xi = 0$ and $\eta = 0$, θ should be equal to θ_1 and Θ should be equal to Θ_1. It follows from Eqs. (13-34) and (13-32) that $a_0 = \theta_1 - \Theta_1$. The Bessel function of the first type and zero order is defined by the equation

$$J_0(z) = \sum_0^\infty \frac{(-1)^n \left(\dfrac{z}{2}\right)^{2n}}{(n!)^2}$$

Thus Eq. (13-34) can also be written in the form[4]

$$\Delta = (\theta_1 - \Theta_1) \cdot J_0(2i \sqrt{\xi \eta}) \qquad (13\text{-}35)$$

The use of Eq. (13-32) enables Eq. (13-36) to be obtained finally as the solution of

---

[3] In addition to using this expression, a solution can also be found by using the Riemann Integration method; see, for example, Frank, P.; von Mises, R.: Die Differential- und Integralgleichungen der Mechanik und Physik, 1st Part. Braunschweig: Vieweg 1925, p. 605.

[4] Nowadays the forms $I_0(x)$, $I_1(x)$ are often used instead of $J_0(ix)$, $i^{-1} J_1(ix)$, etc., giving expression to the idea that, in spite of the imaginary form, one is dealing with a real function of x. However the mathematical relationships are not as clear.

the differential Eq. (13-21)

$$\theta - \Theta = (\theta_1 - \Theta_1) \cdot \exp[-(\xi + \eta)] \cdot J_0(2i \sqrt{\xi\eta}) \qquad (13\text{-}36)$$

θ and Θ can be obtained from this solution of Eqs. (13-26) and (13-27) by integration. The boundary conditions

$$\Theta = \Theta_1 \quad \text{when} \quad \eta = 0 \quad \text{and} \quad \theta = \theta_1 \quad \text{at} \quad \xi = 0$$

can be satisfied if the integration is undertaken in the following way:

$$\theta = \theta_1 - \int_0^\xi (\theta - \Theta) \, d\xi$$

$$\Theta = \Theta_1 + \int_0^\eta (\theta - \Theta) \, d\eta$$

By inserting $\theta - \Theta$ from Eq. (13-36) we obtain:

$$\theta = \theta_1 - (\theta_1 - \Theta_1) \int_0^\xi \exp[-(\xi + \eta)] J_0(2i \sqrt{\xi\eta}) \, d\xi \qquad (13\text{-}37)$$

$$\Theta = \Theta_1 + (\theta_1 - \Theta_1) \int_0^\eta \exp[-(\xi + \eta)] J_0(2i \sqrt{\xi\eta}) \, d\eta \qquad (13\text{-}38)$$

These are the relationships which have been developed by Anzelius. However, if it is required to calculate Θ_1 at a time η, for various values of ξ, another equation is to be preferred to Eq. (13-38) which is obtained by subtracting Eq. (13-36) from Eq. (13-37). As the integral in Eq. (13-37) (with the plus signs) transforms, by partial integration, into

$$-\exp[-(\xi + \eta)] \cdot J_0(2i \sqrt{\xi\eta}) + \exp(-\eta) + \int_0^\xi \exp[-(\xi + \eta)] \frac{\partial J_0(2i \sqrt{\xi\eta})}{\partial \xi} \, d\xi$$

(13-39) is reached by taking into consideration the relationship $\dfrac{\partial J_0(x)}{\partial x} = -J_1(x)$

$$\Theta = \theta_1 - (\theta_1 - \Theta_1) \exp(-\eta) + (\theta_1 - \Theta_1) \int_0^\xi \exp[-(\xi + \eta)] \sqrt{\frac{\eta}{\xi}} i J_1(2i \sqrt{\xi\eta}) \, d\xi$$

$$(13\text{-}39)$$

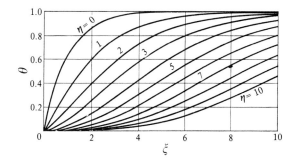

Figure 13-5 Temperature of the gas during the first cooling of the heat-storing mass.

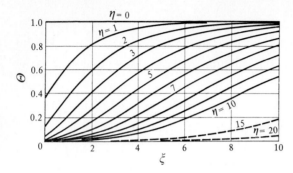

Figure 13-6 Temperature of the heat-storing mass during the first cooling.

The integrals which appear in these solutions can, in general, only be evaluated approximately, for example by using Simpson's rule.

Figures 13-5 and 13-6 illustrate the temperature distribution of the gas and the heat-storing mass given by Eqs. (13-37), (13-38) or (13-39) for the case where the heat-storing mass has an initial, overall temperature of $\Theta_1 = 1$; the gas enters at $\xi = 0$ at a temperature of $\theta_1 = 0$. θ and Θ are drawn as ordinates and, in both cases, ξ is drawn as the abscissa. The individual curves represent the temperature distribution at specific times η. The heat-storing mass is increasingly cooled by the gas, so that as time progresses the gas itself can pick up less and less heat. Schumann [S318] has displayed graphically the temperature distributions rather more precisely than are shown in Figs. 13-5 and 13-6.

Finally, it will be shown that Eqs. (13-37) and (13-38) or (13-39), can also represent the temperature distribution in a recuperator operating in pure cross flow, by an appropriate alteration in the meaning of the variable, (see Sec. 8-2). This is based on the formal equivalence of the differential Eqs. (8-4) and (8-5) derived for cross flow in Sec. 8-2 and the differential Eqs. (13-26) and (13-27) presently being discussed.

Anzelius's relationships are only precisely applicable to a heat-storing mass of high conductivity. However, thanks to the introduction of $\bar{\alpha}$ via Eq. (13-3), Anzelius's results, and thus also Schumann's curves, can be applied to the case where there are large temperature differences within a brick cross-section. However the values of Φ are not as valid as they are for a strictly periodic state of equilibrium for which the function Φ has been developed. In order to make clear the difference in the chronological temperature variation between a heat-storing mass of good conductivity and a heat-storing mass of poor conductivity, it will be assumed that the surface heat transfer coefficient α has the same value in both cases. In the first case $\bar{\alpha} = \alpha$; however in the second case $\bar{\alpha} = \frac{2}{3}\alpha$. The chronological variation of the gas temperature at the position $\xi = af/C = 6$ will be determined. The curve for the heat storing mass of good conductivity can be directly obtained from Fig. 13-5 by reading off θ when $\xi = 6$ for various times η, as is shown in Fig. 13-7 for $\bar{\alpha} = \alpha$.

However, in the second case it will be seen from Eq. (13-24) that $\xi = \bar{\alpha}f/C$ and a smaller value of ξ must be selected, in the ratio $\bar{\alpha}/\alpha = 2/3$, even though the same regenerator cross-section is being dealt with. Thus the gas temperatures θ for the

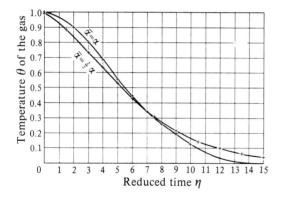

Figure 13-7 Chronological temperature variation of the gas at the position $\xi = 6$ with a constant initial temperature $\Theta_1 = 1$ of the heat-storing mass. Heat-storing mass of very good conductivity: $\bar{\alpha} = \alpha$. Heat-storing mass of poor conductivity: $\bar{\alpha} = 2/3\alpha$.

various values of η must be read off from Fig. 13-5 for $\xi = (2/3)6 = 4$. However, in this second case, it will be observed from Eq. (13-24) that not only is ξ smaller in the ratio $\bar{\alpha}/\alpha$ but so also is η smaller; thus ξ and all values of η must subsequently again be multiplied by 3/2 in order to be able to display this curve on the same scale as that of the first curve. This is how the second curve, which is shown in Fig. 13-7 and which applies to a heat-storing mass of poor conductivity, is obtained.

The fact that the values of Φ, obtained from Eqs. (11-7) or (11-8), are less applicable to the present case than to the state of equilibrium, is based on the following observations. In the state of equilibrium, the parabolic curve which originally opens upwards, changes into a parabolic curve which is facing downwards immediately after a reversal. See Fig. 11-6. However during the first cooling of the heat-storing mass, isothermal temperature conditions prevail initially within the brick cross-section. Thus, from this constant temperature, a transition takes place in which the curve opens downwards. The initially rapid temperature changes represented by the second term in Eq. (12-36), are thus only half as big. Thus, in the case at present being considered, a suitable value of Φ can be obtained if this equation is evaluated using only half the value of the second term.

Handley and Heggs [H301] have calculated more precisely this chronological variation of temperature for various heat-storing masses of high thermal conductivity, using a step-by-step process, which also precisely determines the temperature differences within a brick cross-section. They also obtained a visible flattening of the curve, as is shown in Fig. 13-7, for decreasing thermal conductivity of the material of the heat-storing mass. They demonstrate, in particular, that this phenomenon makes difficult the evaluation of tests where the chronological variation of the gas temperature at the exit to the heat-storing mass is recorded and used to find the size of the heat transfer coefficient. Such an evaluation basically consists of comparing the observed temperature variation with Schumann's (or Anzelius's) curves and selecting those curves which come closest to those observed. It will be recognized from these considerations that only $\bar{\alpha}$ can be obtained from such a comparison and this value must be modified by using Eq. (13-3) if the true heat transfer coefficient is required. However the correct determination of Φ can cause further difficulties.

13-5 THE MOST SIMPLE METHOD OF CALCULATION FOR THE STATE OF EQUILIBRIUM IN COUNTERFLOW

(In particular the infinitely short regenerator and the zero eigen function)

The methods of Evaluation of Tipler and Traustel

Tipler [T301] and Traustel [T302] have developed extremely simplified theories for a rapid evaluation of heat transfer in regenerators, but these will only be mentioned briefly here. Tipler assumed a linear distribution of the gas temperature in the longitudinal direction of the regenerator whereby the rise of the corresponding straight lines decreases with time within each period. Tipler simply calculates the quantity of heat transferred from the temperatures of the gases and the heat-storing mass averaged over the total length of the regenerator.

Traustel [T302], making the assumption that $CT = C'T'$, has suggested that the same linear temperature distribution as occurs in a recuperator should be assumed and thus the performance of a regenerator should be calculated in a similar manner as that of a recuperator.

Traustel proposed that the variation in the temperature of the heat-storing mass should be determined by dividing the quantity of heat given up by the gas to the heat-storing mass during a period by the thermal capacity of the heat-storing mass.

The deployment of this type of method of calculation could hardly be simpler than, but certainly not as precise as, the calculation of k employing Eq. (11-4) using average values of Φ and k/k_0, for example, $\Phi = 1/7$ and $k/k_0 = 0.93$ (see p. 311).

The Simplest Solutions to the Differential Equations[5]

A unique way of simplifying the solutions of the differential equations mentioned below, consists of either considering the heat-storing mass to be an infinitely good conductor or calculating the heat transfer coefficient $\bar{\alpha}$ which relates to the average brick temperature Θ_m using Eq. (13-3). By the introduction of $\bar{\alpha}$ the differential equations are reduced to their most simple form, namely (13-26) and (13-27). The advantages associated with these simplified differential equations can be recognized very clearly when they are applied to those cases where a precise theory has already been developed, as presented in the preceding paragraphs. Thus, before going any further, it is desirable first to deal with the infinitely short regenerator and with the zero eigen function for the arbitrary ratio $CT : C'T'$. In both cases, in spite of the considerable amount of simplification involved, it will be shown that, as long as details of the temperature variations within the heat-storing mass are not required, results to a very high degree of accuracy can be achieved.

---

[5] A method of calculation devised by the author and published initially in the first edition of this book.

Infinitely Short Regenerator

As can be seen from Fig. 11-15, the infinitely short regenerator is not only of interest as a limiting case theoretically but also because Heiligenstaedt developed, for the first time, an exact regenerator theory (Sec. 12-2) for this limiting case. A regenerator is regarded as being infinitely short when its reduced regenerator length Λ is infinitely small. According to Eqs. (13-28) and (13-29), given finite thermal capacities C and C' of the quantities of gas flowing through the regenerator in unit time and given finite reduced period lengths Π and Π', F and C_s will be infinitely small. Because in such a regenerator, only an infinitely small quantity of heat can be transferred in finite time, even though there are big changes in the temperature of the heat-storing mass, the temperatures θ and θ' of the gases flowing through the regenerator can also only undergo infinitely small changes. Thus:

$$\theta = \theta_1 = \text{const and } \theta' = \theta'_1 = \text{const}$$

must apply where θ_1 and θ'_1 denote the entry temperatures of the gas which are assumed to be constant with time. Noting these equations, Eq. (13-40) is obtained as a solution of the differential Eq. (13-27) where D and D' are constants.

$$\left. \begin{array}{ll} \Theta = \theta_1 + D \cdot \exp(-\eta) & \text{(warm period)} \\ \Theta' = \theta'_1 + D' \cdot \exp(-\eta') & \text{(cold period)} \end{array} \right\} \tag{13-40}$$

It will be further assumed that $\Pi = \Pi'$ and thus that the warm period extends from $\eta = 0$ to $\eta = \Pi$ and the cold period extends from $\eta' = 0$ to $\eta' = \Pi$. It is recalled that the average brick temperature Θ or Θ' cannot change at the moment of reversal; this leads to the reversal conditions

$$\Theta(\eta = 0) = \Theta'(\eta' = \Pi)$$
$$\Theta(\eta = \Pi) = \Theta'(\eta' = 0)$$

Taking these reversal conditions into account, the constants in Eq. (13-40) take the form:

$$D = -D' = -(\theta_1 - \theta'_1)\frac{1 - \exp(-\Pi)}{1 - \exp(-2\Pi)} = -\frac{\theta_1 - \theta'_1}{1 + \exp(-\Pi)} \tag{13-41}$$

It follows from Eqs. (13-40) and (13-41) that

$$\Delta\Theta = (\theta_1 - \theta'_1)\frac{1 - \exp(-\Pi)}{1 + \exp(-\Pi)} = (\theta_1 - \theta'_1)\tanh\frac{\Pi}{2}$$

where $\Delta\Theta = \Theta(\Pi) - \Theta(0)$, that is the temperature variation of the heat-storing mass during the warm period. Furthermore with plate-shaped bricks or sheets, since the thermal capacity of the heat-storing mass C_s is equal to $\dfrac{\rho c \delta}{2} F$, the quantity of heat transferred during a period is given by

$$Q_{Per} = \Delta\Theta \cdot \frac{\rho c \delta}{2} \cdot F = \frac{\rho c \delta}{2} F(\theta_1 - \theta'_1)\tanh\frac{\Pi}{2}$$

If this is compared with Eq. (11-3) and consideration is taken of the fact that, in an infinitely short regenerator the average temperature difference $\Delta\theta_M$ is equal to the difference $\theta_1 - \theta'_1$ of the entry temperatures Eq. (13-42) is obtained as the *heat transmission coefficient of the infinitely short regenerator*

$$k = \frac{\rho c \, \delta}{2(T + T')} \tanh \frac{\Pi}{2} \qquad (13\text{-}42)$$

In fact, this equation is simpler than the relationships (12-12) and (12-13), given in Sec. 12-2. Further this equation can be closely compared with the equation for k derived by Heiligenstaedt [H314]. Nevertheless Eq. (13-42), within the limits of accuracy of the Heiligenstaedt-equation, is almost as accurate as that equation so that, for example, no difference between the values given by both these equations can be detected up to $\delta = 50$ mm; see the bottom curve in Fig. 11-15. Using Eq. (13-42) the ratio k/k_0 can also easily be calculated for the infinitely short regenerator. It follows that the heat transmission coefficient k_0, which alone corresponds to the zero eigen function and thus to the fundamental oscillation of the regenerator, is given by

$$k_0 = \frac{\rho c \, \delta}{2(T + T')} \cdot \frac{\Pi}{2}$$

This is obtained from Eq. (11-6) with $m = 1$ or from Eq. (13-106). Equation (13-43) is obtained in this way, using Eq. (13-42);

$$\lim_{\Lambda = 0} \frac{k}{k_0} = \frac{2}{\Pi} \tanh \frac{\Pi}{2} \qquad (13\text{-}43)$$

This equation, which has been already set out as Eq. (11-9) is represented by the bottom curve in Fig. 11-13.

The Most Simple Form of the Zero Eigen Function with $CT \neq C'T'$ and $CT = C'T'$

The zero eigen function for $CT \neq C'T'$, that is for unequal thermal capacity of the quantities of gas flowing through the regenerator during the warm and cold periods, is derived in Appendix 3, using a precise calculation of the spatial temperature differences in each brick cross-section. Here only the basic principles of this derivation are indicated in Sec. 12-8. The relationships obtained are so complex that they can only be used with difficulty in practical calculations. On the other hand, relatively simple equations result if the zero eigen function is considered in the context of the differential Eqs. (13-26) and (13-27) or (13-30).

We start, as previously, with the assumption that with the zero eigen function, at a given moment the temperatures are distributed in the longitudinal direction of a regenerator, in the same way as they are distributed in a recuperator. If C and C' are constant but CT is different from $C'T'$, the temperature Θ distribution in the heat-storing mass in the longitudinal direction of the regenerator will be represented by an equation of the form (see Eq. (12-50)), as discussed in Sec. 12-8:

$$\Theta = D + A \cdot \exp(b\xi)$$

where D and b denote constants; A is a function of the reduced time η. ξ is the longitudinal coordinate determined by Eqs. (13-22) or (13-24). By inserting these expressions for Θ into the differential Eq. (13-30) we obtain, with a new constant B:

$$A = B \cdot \exp\left(-\frac{b}{1+b}\eta\right)$$

The solution of the differential Eq. (13-30) is thus

$$\Theta = D + B \exp\left(b\xi - \frac{b}{1+b}\eta\right) \tag{13-44}$$

This *solution* should refer to the cold period since in what follows the cold period will, for the most part, be the focus of our attention.[6] Accordingly, for the warm period we have

$$\Theta' = D' + B' \exp\left(b'\xi' - \frac{b'}{1+b'}\eta'\right) \tag{13-45}$$

The values of b and b', as also the relationships between D and D' and between B and B' will be determined via the *reversal requirement*. A specific position in the regenerator will be considered which is at a distance ξ from the entry position of the cold gas. This same position will be treated as being at a distance ξ' from the entry position of the warm gas in the warm period. Moreover ξ and ξ' measure the distances on different scales, namely in the ratio of Λ and Λ' which are generally not equal (see Eq. (13-28)). Thus, *at a specific position in the regenerator*

$$\frac{\xi}{\Lambda} = \frac{\Lambda' - \xi'}{\Lambda'}$$

or

$$\xi' = \Lambda' - \frac{\Lambda'}{\Lambda}\xi \tag{13-46}$$

The time scale is set in rather a different way from that set out in Secs. 12-2 to 12-8, in particular the reduced times η and η' are set equal to zero in the middle of the cold or warm period. Thus in the cold period η extends from $-\Pi/2$ to $+\Pi/2$ and in the warm period η' extends from $-\Pi'/2$ to $+\Pi'/2$. It follows from Eq. (13-44) that Eq. (13-47) applies to the beginning of the cold period

$$\Theta = \Theta_a = D + B \exp\left(b\xi + \frac{b}{1+b}\cdot\frac{\Pi}{2}\right) \tag{13-47}$$

---

[6] Contrary to the system used so far, all values of Θ, b, etc., without a single ' (superfix) will represent the cold period and all values of Θ', b', etc., with a superfix will represent the warm period.

Similarly from Eqs. (13-45) and (13-46), it follows that Eq. (13-48) applies to the end of the warm period:

$$\Theta' = \Theta'_c = D' + B' \exp\left[b'\left(\Lambda' - \frac{\Lambda'}{\Lambda}\xi \right) - \frac{b'}{1+b'} \cdot \frac{\Pi'}{2} \right] \qquad (13\text{-}48)$$

According to the reversal requirement Θ_a and Θ'_b are equal to one another at every position ξ. This is represented by Eqs. (13-47) and (13-48) if

$$D = D', \quad B = B' \exp(b'\Lambda')$$

$$b = -\frac{\Lambda'}{\Lambda}b', \quad \frac{b}{1+b}\Pi = -\frac{b'}{1+b'}\Pi'$$

Equations (13-44) and (13-45) take on the form of Eqs. (13-49) and (13-50) upon utilizing the values of D' and B' from the first two of the equations above. The expressions for b and b' which result from the solution of the last two of these equations are incorporated in an appropriate way.

$$\Theta = D + B \exp\left[\left(\frac{\Lambda'}{\Lambda} - \frac{\Pi'}{\Pi} \right)\left(\frac{\Pi}{\Pi+\Pi'}\xi - \frac{\Lambda}{\Lambda+\Lambda'}\eta \right) \right] \qquad (13\text{-}49)$$

$$\Theta' = D + B \exp\left[\left(\frac{\Lambda'}{\Lambda} - \frac{\Pi'}{\Pi} \right)\left(\frac{\Pi}{\Pi+\Pi'}\xi + \frac{\Lambda}{\Lambda+\Lambda'} \cdot \frac{\Pi}{\Pi'}\eta' \right) \right] \qquad (13\text{-}50)$$

In order to make things clearer, ξ is introduced into Eq. (13-50) in place of ξ', using Eq. (13-46). Equation (13-51) follows from Eqs. (13-49) and (13-27) for the gas temperature θ in the cold period

$$\theta = D + B\frac{\Lambda}{\Lambda+\Lambda'} \cdot \frac{\Pi+\Pi'}{\Pi} \exp\left[\left(\frac{\Lambda'}{\Lambda} - \frac{\Pi'}{\Pi} \right)\left(\frac{\Pi}{\Pi+\Pi'}\xi - \frac{\Lambda}{\Lambda+\Lambda'}\eta \right) \right] \qquad (13\text{-}51)$$

Correspondingly, Eq. (13-52) is obtained for the warm period

$$\theta' = D + B\frac{\Lambda'}{\Lambda+\Lambda'} \cdot \frac{\Pi+\Pi'}{\Pi'} \exp\left[\left(\frac{\Lambda'}{\Lambda} - \frac{\Pi'}{\Pi} \right)\left(\frac{\Pi}{\Pi+\Pi'}\xi + \frac{\Lambda}{\Lambda+\Lambda'} \cdot \frac{\Pi}{\Pi'}\eta' \right) \right] \qquad (13\text{-}52)$$

Equations (13-49) to (13-52) represent the required *zero eigen function*.

The temperature distribution calculated using these equations shows some remarkable properties which can be seen from Fig. 13-8. The curve of the average temperature Θ of the heat-storing mass obtained from Eq. (13-49) for various times η of the cold period, as a function of ξ is illustrated on the left-hand side of this figure.

The temperature curves relating to the heat-storing mass for the successive times η shown in the figure, are separated by the fixed distance $\Delta\xi$. A further mathematical relationship can be recognized from the right-hand side of the figure which shows, for any specific position ξ, the chronological variation of Θ and Θ' as also that of the gas temperatures θ and θ'. The times η and η', displayed in opposite directions and scaled in the ratio $\Pi':\Pi$, are drawn as the abscissa. Θ and Θ' pass through one and the same curve but in the opposite directions. Thus their

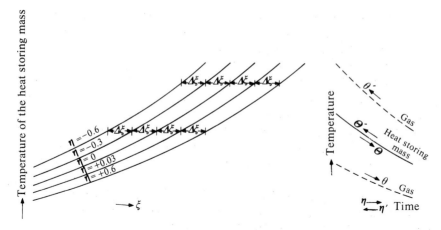

Figure 13-8 Temperature distribution in the heat-storing mass using the simplified zero eigen function with $CT > C'T'$.

values on the same vertical axis, coincide at corresponding times in both periods as shown in the figure. The very slim hysteresis loop which is present in Fig. 12-14 completely disappears here.

The mathematical interrelationships mentioned can also be directly obtained from Eqs. (13-49) and (13-50). It will be seen from Eq. (13-49) that Θ remains unchanged when time is increased by $\Delta\eta$ if ξ is increased by $\Delta\xi = \dfrac{\Pi+\Pi'}{\Pi} \cdot \dfrac{\Lambda}{\Lambda+\Lambda'} \cdot \Delta\eta$.

Thus Θ remains constant if an increase in ξ is accompanied by time being changed by $\Delta\eta$. In order to demonstrate the disappearance of the hysteresis loop, it can be seen quite clearly from Fig. 13-8 that $\eta' = -\dfrac{\Pi'}{\Pi}\eta$ for all corresponding times in each period. If this relationship is inserted into Eq. (13-50), Θ is equal to Θ' for all corresponding times as will be observed from Eq. (13-49). This is in agreement with Eq. (12-21) which was derived only for the case where $CT = C'T'$; Θ and Θ' now denote the average brick temperatures.

The heat transmission coefficient k_0 for the zero eigen function can now be calculated using Eqs. (13-49) to (13-52). Setting $\eta' = -\dfrac{\Pi'}{\Pi}\eta$ within Eqs. (13-49), (13-51) and (13-52) the following expression arises for the same corresponding times in both periods:

$$\frac{\theta-\Theta}{\theta-\theta'} = \frac{\dfrac{\Lambda}{\Pi} - \dfrac{\Lambda+\Lambda'}{\Pi+\Pi'}}{\dfrac{\Lambda}{\Pi} - \dfrac{\Lambda'}{\Pi'}}$$

As the expression on the right-hand side is a constant, the relationships for the corresponding chronological average values of $\bar{\theta}-\bar{\Theta}$ and $\bar{\theta}-\bar{\theta'}$ must take the same form. Accordingly, when Λ and Π are placed on the right-hand side of the last equation, it follows from Eqs. (13-28) and (13-29) that:

$$\frac{\theta-\Theta}{\theta-\theta'} = \frac{\bar{\theta}-\bar{\Theta}}{\bar{\theta}-\bar{\theta'}} = \frac{\tilde{\alpha}'T'}{\tilde{\alpha}T+\tilde{\alpha}'T'} \tag{13-53}$$

Further if it is considered that the quantity of heat $q_{Per} = \alpha(\bar{\theta}-\bar{\Theta}_0)T$ transferred per unit area during the cold period can also be calculated using Eq. (13-2), then Eq. (13-54) follows upon setting $\Theta = \Theta_m$

$$\alpha(\bar{\theta}-\bar{\Theta}_0) = \tilde{\alpha}(\bar{\theta}-\bar{\Theta}) \tag{13-54}$$

Using Eq. (12-34) we obtain the following relationship for k_0;

$$k_0 = \frac{\tilde{\alpha}T}{T+T'} \cdot \frac{\bar{\theta}-\bar{\Theta}}{\bar{\theta}-\bar{\theta'}}$$

Equation (13-55) can be finally derived, with the help of Eq. (13-53), as the expression for the *heat transmission coefficient based on the zero eigen function*:

$$\frac{1}{k_0} = (T+T')\left(\frac{1}{\tilde{\alpha}T} + \frac{1}{\tilde{\alpha}'T'}\right) \tag{13-55}$$

This equation is in complete agreement with Eq. (13-5) and thus also with Eq. (12-40). However this result is particularly important because, in contrast to Eqs. (12-5) and (12-40), relationship (13-55) is derived for a completely arbitrary ratio of CT to $C'T'$. In this way it can again be pointed out that Eqs. (12-40) and (11-4) are also approximately valid for the case where $CT \neq C'T'$.

Using Eq. (13-55) Eqs. (13-49) to (13-52), which relate to the temperature distribution, can now be developed into another form. If Λ, Λ', Π, Π' as well as ξ, η and η' are introduced into Eqs. (13-24), (13-28) and (13-29), the following relationships emerge.

$$\Theta = D + B \exp\left[\left(\frac{1}{C'T'} - \frac{1}{CT}\right)(T+T')k_0 f - \frac{CT-C'T'}{\tilde{\alpha}C'+\tilde{\alpha}'C} \cdot \frac{2\tilde{\alpha}\tilde{\alpha}'}{\rho c \delta} \frac{t}{T}\right] \tag{13-56}$$

$$\Theta' = D + B \exp\left[\left(\frac{1}{C'T'} - \frac{1}{CT}\right)(T+T')k_0 f + \frac{CT-C'T'}{\tilde{\alpha}C'+\tilde{\alpha}'C} \cdot \frac{2\tilde{\alpha}\tilde{\alpha}'}{\rho c \delta} \frac{t'}{T'}\right] \tag{13-57}$$

$$\theta = D + B \frac{C'}{T} \cdot \frac{\tilde{\alpha}T+\tilde{\alpha}'T'}{\tilde{\alpha}C'+\tilde{\alpha}'C} \exp\left[\left(\frac{1}{C'T'} - \frac{1}{CT}\right)(T+T')k_0 f - \frac{CT-C'T'}{\tilde{\alpha}C'+\tilde{\alpha}'C} \cdot \frac{2\tilde{\alpha}\tilde{\alpha}'}{\rho c \delta} \frac{t}{T}\right] \tag{13-58}$$

$$\theta' = D + B \frac{C}{T'} \cdot \frac{\tilde{\alpha}T+\tilde{\alpha}'T'}{\tilde{\alpha}C'+\tilde{\alpha}'C} \exp\left[\left(\frac{1}{C'T'} - \frac{1}{CT}\right)(T+T')k_0 f + \frac{CT-C'T'}{\tilde{\alpha}C'+\tilde{\alpha}'C} \cdot \frac{2\tilde{\alpha}\tilde{\alpha}'}{\rho c \delta} \cdot \frac{t'}{T'}\right] \tag{13-59}$$

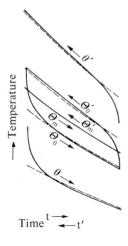

Figure 13-9 Chronological temperature curve with $CT = 2C'T'$ using a precise method of calculation (solid lines) and using the simplified zero eigen function (dashed lines).

Figure 13-9 shows how far these relationships[7] are in agreement with the more precise method of calculation mentioned in Sec. 12-8; this approach also took into account the temperature changes which take place within a brick cross-section. As in Fig. 12-14 this diagram shows, for the zero eigen function, the chronological variation of the brick temperatures $\Theta = \Theta_m$ and Θ_0 and the gas temperature θ in the cold period, together with the corresponding variations in the warm period. This figure relates to the case where $CT = 2C'T'$.[8]

The solid curves for θ and θ' are calculated using the precise method, the dashed lines are computed using Eqs. (13-58) and (13-59). The dashed lines for the surface temperatures Θ_0 and Θ'_0 were determined on the basis of Eq. (13-4). It was assumed that the relevant relationship, namely Eq. (13-54)

$$\alpha(\theta - \Theta_0) = \bar{\alpha}(\theta - \Theta_m)$$

is applicable for each point of time. If this same equation is applied to the cold period, the value of Θ_0 can be calculated from the values of θ and Θ_m. Only a single line is drawn for Θ_m and Θ'_m since the differences between these are very slight, even when calculated using the precise method. Indeed it is not possible to illustrate these differences on the scale being used. The dashed lines for θ, θ', Θ_0 and Θ'_0 do not directly express the rapid temperature changes which occur at the beginning of the period. However they do take them into account insofar as they have the same average value as the solid lines. Moreover, after the rapid temperature changes have died away, these two groups form basically the same curve.

---

[7] Nusselt [N303] arrived, essentially, at the same equations but without the chronological term within the exponential functions. This was because he assumed an infinitely short period length and thus neglected the influence of time from the outset.

[8] It should be remembered that although previously the average brick temperature was denoted by Θ_m, from Sec. 13-3 onwards it has been denoted by Θ for the sake of simplicity.

However, the initial rapid temperature changes in the surface of the heat-storing mass and in the gas can also be accurately calculated. The chronological variation of the average temperature $\Theta = \Theta_m$ of the heat-storing mass is first determined using the Eqs. (13-56) and 13-57) above; the process outlined in Sec. 13-2 is then applied.

Special Case of $CT = C'T'$

When $CT = C'T'$ the exponential terms in Eqs. (13-56) to (13-59) become infinitely small. The same behaviour applies also to Eqs. (13-28) to (13-52); it follows from Eqs. (13-28) and (13-29) that when $CT = C'T'$, Λ'/Λ will also be equal to Π'/Π, and thus $\dfrac{\Lambda}{\Lambda+\Lambda'} = \dfrac{\Pi}{\Pi+\Pi'}$. The exponential function in Eq. (13-49) is therefore developed as a power series which is truncated after the second term. The constants D and B are then allowed to become infinitely large. Equation (13-49) thus takes the form:

$$\Theta = D - B(\xi - \eta) \tag{13-60}$$

where D and B now denote new, arbitrary constants which are infinitely large. A relationship (13-61) for θ is obtained by inserting Eq. (13-60) into Eq. (13-27):

$$\theta = D - B(\xi - \eta - 1) \tag{13-61}$$

Corresponding equations are obtained for Θ' and θ'. Taking ξ and η from Eq. (13-24), Eqs. (13-60) and (13-61) can also be brought into the form

$$\Theta = D - B\left(\frac{\bar{\alpha}}{C}f - \frac{2\bar{\alpha}}{\rho c\,\delta}t\right) \tag{13-62}$$

and

$$\theta = D - B\left(\frac{\bar{\alpha}}{C}f - \frac{2\bar{\alpha}}{\rho c\,\delta}t - 1\right) \tag{13-63}$$

Furthermore, when the time $t = 0$ is set in the middle of the period Eq. (13-64) takes the form below. This corresponds to Eq. (13-59),

$$\theta' = D - B\left(\frac{\bar{\alpha}}{C}f + \frac{2\bar{\alpha}}{\rho c\,\delta}\cdot\frac{C'}{C}\cdot t + \frac{\bar{\alpha}}{\bar{\alpha}'}\cdot\frac{C'}{C}\right) \tag{13-64}$$

The average brick temperature Θ varies spatially and chronologically, according to Eqs. (13-60) and (13-62), in a linear manner when $CT = C'T'$, a variation which corresponds to the zero eigen function. This is in agreement with the findings in Sec. 12-5. However, deviating somewhat from the calculations set out in Sec. 12-5, it follows from Eqs. (13-61) or (13-63) that the gas temperature θ exhibits a completely linear variation, both chronologically and spatially. This is because the $\bar{\alpha}$ used in Eq. (13-63) does not represent directly the rapid temperature changes which follow a reversal, rather the chronological average of the effect of these changes.

Equations (13-60) and (13-63) play an important role in the still missing proof that the relationships developed hitherto for the zero eigen function really do represent the zero eigen function. This evidence will be furnished in the following paragraphs which relate to Eqs. (13.60) to (13-63). However, the fact that the equations developed for $CT \neq C'T'$ also correspond to the zero eigen function, follows from the fact that all these equations transform into Eqs. (13-60) to (13-63) if $CT = C'T'$ and if $\bar{\alpha}$ is introduced, on the basis of Eq. (13-3).

13-6 THE FUNDAMENTAL OSCILLATION AND THE HARMONIC VIBRATIONS OF A COUNTERFLOW REGENERATOR [H304] $CT = C'T'$*

It has already been shown in Sec. 11-3 that it is only possible to completely calculate the temperature distribution for a regenerator's state of equilibrium and to include the effect of what happens at the ends of the regenerator if account is taken of the higher eigen functions (harmonic vibrations) as well as the zero eigen function (fundamental oscillation). Therefore the equations which relate to all the eigen functions will be derived below. In order to avoid relationships which are too complex, we will limit ourselves to the case where the warm and cold periods are of equal length ($T = T'$), where the quantities of gas flowing through in each period per unit time have the same thermal capacities $C = C'$ and also to the case where the heat transfer coefficients $\bar{\alpha}$ and $\bar{\alpha}'$ in the two periods are equal to one another. In this case $CT = C'T'$ and, according to Eqs. (13-28) and (13-29), $\Lambda = \Lambda'$ and $\Pi = \Pi'$ as well. Furthermore as in the case of the differential Eqs. (13-26), (13-27), etc. we want to assume that the thermal conductivity of the heat-storing mass in the direction of flow is so small that it can be neglected.

In this complete treatment of the state of equilibrium of a regenerator, it will be useful to point out that, as a consequence of the introduction of $\bar{\alpha}$ from Eq. (13-3), it is only necessary to pay attention to the average temperature Θ of the heat-storing mass. It is no longer required to consider the spatial temperature differences of the heat-storing mass in a direction perpendicular to that of the flow. Only by this means is it possible to find equations for the higher eigen functions which can easily be recognized in some way or other.

Transformation of the Reversal Requirement in the Case of Counterflow

The required eigen functions are solutions of the differential Eqs. (13-26) and (13-27) which satisfy the reversal condition. As has already been mentioned in Sec. 12-1, the reversal requirement signifies that, in each regenerator cross-section, the temperatures Θ and Θ' of the heat-storing mass must have the same value at the end of the cold period as it has at the beginning of the warm period and vice versa.

---

* From H. Hausen [H303].

By assuming that $\Lambda = \Lambda'$ and $\Pi = \Pi'$, in the case of counterflow, this reversal requirement can be brought into a form that can be handled mathematically.

The positive longitudinal coordinate ξ or ξ' can, as previously, be computed in the direction of flow starting at the gas entrance, so that, in counterflow, $\xi = 0$ and $\xi' = 0$ denote the two different regenerator ends. As has been indicated previously the middle of the warm or cold periods respectively is chosen as the zero point in time η or η'. Thus one period lasts from $\eta = -\Pi/2$ to $\eta = +\Pi/2$ and the other period extends from $\eta' = -\Pi'/2$ to $\eta' = +\Pi'/2$. Furthermore, in contrast to what has been used previously, the zero on the temperature scale will be at the middle point in between the entry temperatures θ_1 and θ'_1 of the two gases.

Since $\Lambda = \Lambda'$ and $\Pi = \Pi'$, at equilibrium although the temperature distribution in the warm and cold periods must be a function of ξ and η for one period and a function of ξ' and η' for the other, it must be the same for both periods, if the sign $(+$ or $-)$ is ignored. θ and θ' together with Θ and Θ' must have opposite signs because, given the zero point of the coordinate system, $\theta_1 = -\theta'_1$. Hence, if we consider two positions ξ and ξ', which lie symmetrically about the centre of the regenerator, Eq. (13-65) applies to those temperatures of the heat-storing mass which prevail at the ends of the two periods:

$$\Theta(\xi, \eta = +\Pi/2) = -\Theta'(\xi', \eta' = +\Pi'/2) \tag{13-65}$$

Position ξ' is identical to position $\Lambda - \xi$. Furthermore, as the temperature of the heat-storing mass does not change at the reversal, the temperature $\Theta'(\xi', \eta' = +\Pi'/2)$ at the end of the period under consideration must be equal to the temperature $\Theta(\Lambda - \xi, \eta = -\Pi/2)$ at the beginning of the following period. It thus follows from the previous equation that Eq. (13-66) must also be valid:

$$\Theta(\xi, +\Pi/2) = -\Theta(\Lambda - \xi, -\Pi/2) \tag{13-66}$$

In this form the reversal condition specifies that the temperature distributions of the heat-storing mass at the beginning and the end of the same priod, which are shown as a function of ξ, lie symmetrically about the centre of the regenerator and the chosen zero point of the temperature scale. (See Fig. 13-10).

Derivation of the Eigen Functions

Particular solutions of the differential equations are sought which satisfy the reversal condition. If, by way of experiment, $\theta - \Theta$ is taken as the product of two

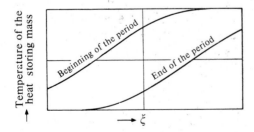

Figure 13-10 Radial symmetry of the temperature distribution in the heat-storing mass at the beginning and end of a period for the case where $\Lambda = \Lambda'$ and $\Pi = \Pi'$.

functions* the first of which is a function only of ξ, and the second is a function only of η, it can easily be shown that the particular integrals, which satisfy the differential Eq. (13-31) take the form

$$\theta - \Theta = A \cdot \exp\left(-\frac{2\xi}{1+n} - \frac{2\eta}{1-n} \right) \tag{13-67}$$

where **A** denotes a constant of integration and n denotes an arbitrary real or complex number. Equations (13-68) and (13-69) follow by integration if this expression is inserted into the differential Eqs. (13-26) and (13-27):

$$\theta = +\frac{1+n}{2} A \cdot \exp\left(-\frac{2\xi}{1+n} - \frac{2\eta}{1-n} \right) \tag{13-68}$$

and

$$\Theta = -\frac{1-n}{2} A \cdot \exp\left(-\frac{2\xi}{1+n} - \frac{2\eta}{1-n} \right) \tag{13-69}$$

Since n can assume any arbitrary value there are an infinite number of particular solutions of the form (13-68) and (13-69). However none of these solutions (apart from $n = \infty$) can alone satisfy the reversal condition (13-66). It is only possible to find solutions which satisfy the reversal requirement by using the expedient of combining two such particular solutions. The expression can thus be formed:

$$\Theta = A_1 \exp\left(-\frac{2\xi}{1+n_1} - \frac{2\eta}{1-n_1} \right) + A_2 \cdot \exp\left(-\frac{2\xi}{1+n_2} - \frac{2\eta}{1-n_2} \right) \tag{13-70}$$

which corresponds to Eq. (13-69). This expression satisfies the reversal condition (13-66) if Eqs. (13-71), (13-72) and (13-73) are formed using $i = \sqrt{-1}$ and an arbitrary whole and real number $\varkappa$, which thus takes the values 0, 1, 2, 3, etc., up to ∞:

$$n_1 = -1 + 2\sqrt{1 + i\frac{\Pi}{\varkappa\pi}} \tag{13-71}$$

$$n_2 = -1 - 2\sqrt{1 + i\frac{\Pi}{\varkappa\pi}} \tag{13-72}$$

and

$$A_2 = -(-1)^{\varkappa} \cdot A_1 \tag{13-73}$$

Complex solutions to the differential equations are obtained in this way. In order to obtain real solutions, an expression must be added to Eq. (13-70) in which $\varkappa$ is replaced by $-\varkappa$. A_1 can also be treated in a similar way in the two expressions; two possibilities exist for dealing with this but these will not be discussed further here. Using apparently complicated intermediate calculations[9] the following real solutions or "*eigen functions*", which will be designated by $u_{\varkappa}$ and $v_{\varkappa}$, are finally

---

* Editor's note: Method of separation of variables.
[9] See the detailed derivation in [H303].

obtained for the temperature Θ of the heat-storing mass

$$
\begin{aligned}
u_{\varkappa} = \sqrt{\frac{d_{\varkappa}}{Q_{\varkappa}}} \Bigg\{ & \exp\left(\frac{\eta}{2a_{\varkappa}} - c_{\varkappa}\xi\right) \sin\left[(1+a_{\varkappa})\frac{\varkappa\pi}{\Pi}\eta - d_{\varkappa}\xi - \arctan\frac{Q_{\varkappa}}{P_{\varkappa}}\right] \\
& - (-1)^{\varkappa} \cdot \exp\left[-\frac{\eta}{2a_{\varkappa}} - c_{\varkappa}(\Lambda - \xi)\right] \\
& \times \sin\left[(1-a_{\varkappa})\frac{\varkappa\pi}{\Pi}\eta - d_{\varkappa}(\Lambda - \xi) - \arctan\frac{Q_{\varkappa}}{P_{\varkappa}}\right]\Bigg\}
\end{aligned}
\tag{13-74}
$$

and

$$
\begin{aligned}
v_{\varkappa} = \sqrt{\frac{d_{\varkappa}}{Q_{\varkappa}}} \Bigg\{ & \exp\left(\frac{\eta}{2a_{\varkappa}} - c_{\varkappa}\xi\right) \cos\left[(1+a_{\varkappa})\frac{\varkappa\pi}{\Pi}\eta - d_{\varkappa}\cdot\xi - \arctan\frac{Q_{\varkappa}}{P_{\varkappa}}\right] \\
& - (-1)^{\varkappa} \cdot \exp\left[-\frac{\eta}{2a_{\varkappa}} - c_{\varkappa}(\Lambda - \xi)\right] \\
& \times \cos\left[(1-a_{\varkappa})\frac{\varkappa\pi}{\Pi}\eta - d_{\varkappa}(\Lambda - \xi) - \arctan\frac{Q_{\varkappa}}{P_{\varkappa}}\right]\Bigg\}
\end{aligned}
\tag{13-75}
$$

In these equations $a_{\varkappa}, b_{\varkappa}$, etc., are abbreviations for the following expressions:

$$
a_{\varkappa} = \sqrt{\frac{1}{2}\left[\sqrt{1 + \left(\frac{\Pi}{\varkappa\pi}\right)^2} + 1\right]} \qquad b_{\varkappa} = \sqrt{\frac{1}{2}\left[\sqrt{1 + \left(\frac{\Pi}{\varkappa\pi}\right)^2} - 1\right]}
$$

$$
c_{\varkappa} = \frac{a_{\varkappa}}{\sqrt{1 + \left(\frac{\Pi}{\varkappa\pi}\right)^2}} \qquad d_{\varkappa} = \frac{b_{\varkappa}}{\sqrt{1 + \left(\frac{\Pi}{\varkappa\pi}\right)^2}}
\tag{13-76}
$$

$$
P_{\varkappa} = 1 + \frac{1}{2a_{\varkappa}} \qquad Q_{\varkappa} = (1+a_{\varkappa})\frac{\varkappa\pi}{\Pi}
$$

The following corresponding *eigen functions* emerge from Eq. (13-27) for the *gas temperature*

$$
\begin{aligned}
\phi_{\varkappa} = & \exp\left(\frac{\eta}{2a_{\varkappa}} - c_{\varkappa}\xi\right) \cdot \sin\left[(1+a_{\varkappa})\frac{\varkappa\pi}{\Pi}\eta - d_{\varkappa}\xi\right] \\
& - (-1)^{\varkappa}\frac{b_{\varkappa}}{1+a_{\varkappa}} \cdot \exp\left[-\frac{\eta}{2a_{\varkappa}} - c_{\varkappa}(\Lambda - \xi)\right] \\
& \times \sin\left[(1-a_{\varkappa})\frac{\varkappa\pi}{\Pi}\eta - d_{\varkappa}(\Lambda - \xi) - \arctan\frac{1}{b_{\varkappa}}\right]
\end{aligned}
\tag{13-77}
$$

and

$$
\begin{aligned}
\psi_{\varkappa} = & \exp\left(\frac{\eta}{2a_{\varkappa}} - c_{\varkappa}\xi\right) \cdot \cos\left[(1+a_{\varkappa})\frac{\varkappa\pi}{\Pi}\eta - d_{\varkappa}\xi\right] \\
& - (-1)^{\varkappa}\frac{b_{\varkappa}}{1+a_{\varkappa}} \cdot \exp\left[-\frac{\eta}{2a_{\varkappa}} - c_{\varkappa}(\Lambda - \xi)\right] \\
& \times \cos\left[(1-a_{\varkappa})\frac{\varkappa\pi}{\Pi}\eta - d_{\varkappa}(\Lambda - \xi) - \arctan\frac{1}{b_{\varkappa}}\right]
\end{aligned}
\tag{13-78}
$$

As the eigen value $\varkappa$ can assume all positive real values between 0 and ∞, an infinite number of eigen functions of the form (13-74) and (13-75) or (13-77) and (13-78) are obtained.

Zero eigen functions ($\varkappa = 0$). For the limiting case where $\varkappa = 0$ it can be shown that v_0 and ψ_0 disappear but u_0 and ϕ_0 assume the following simple form:

$$u_0 = \eta - \xi + \frac{\Lambda}{2} \tag{13-79}$$

$$\phi_0 = \eta - \xi + 1 + \frac{\Lambda}{2} \tag{13-80}$$

These eigen functions reproduce the spatial and chronological linear temperature distribution, which has been mentioned many times, as it appears in the interior of a long regenerator for the case where $CT = C'T'$. Apart from the constants, Eqs. (13-79) and (13-80) are essentially identical to Eqs. (13-60) and (13-61) which were derived earlier. Equations (13-79) and (13-80) provide the proof which was hitherto lacking, that all the relationships for Θ and θ which were developed in Secs. 12-5, 12-7 and in the second part of Sec. 13-5 are none other than the zero eigen functions.

The eigen functions for $\varkappa > 0$. The properties of the eigen functions for $\varkappa > 0$ can best be examined by considering the cases where the coefficients $a_\varkappa$, $b_\varkappa$, $c_\varkappa$, etc., tend towards the following limiting values for increasing values of $\varkappa$:

$$\left.\begin{array}{ll} \lim_{\varkappa=\infty} a_\varkappa = \lim_{\varkappa=\infty} c_\varkappa = 1, & \lim_{\varkappa=\infty} (1 - a_\varkappa) = -\frac{1}{8}\left(\frac{\Pi}{\varkappa\pi}\right)^2 \\[2mm] \lim_{\varkappa=\infty} b_\varkappa = \lim_{\varkappa=\infty} d_\varkappa = \frac{\Pi}{2\varkappa\pi}, & \\[2mm] \lim_{\varkappa=\infty} P_\varkappa = 1.5, & \lim_{\varkappa=\infty} Q_\varkappa = \frac{2\varkappa\pi}{\Pi} \end{array}\right\} \tag{13-81}$$

The approximation to these limiting values occurs the more rapidly, the smaller the value of Π. Thus, for example, where $\Pi = \pi = 3.1416$:

$$a_1 = 1.0987, \qquad a_5 = 1.0049 \qquad \left(\text{instead of } \lim_{\varkappa=\infty} a_\varkappa = 1\right)$$

while, on the other hand, when $\Pi = 5\pi$:

$$a_1 = 1.7460, \qquad a_5 = 1.0987$$

Consequently, as the limiting values obtained above are already quite good approximations, even with small values of $\varkappa$, a picture of the eigen functions can be obtained, which is basically correct, if these limiting values are inserted into Eqs. (13-74), (13-75), (13-77) and (13-78). The following relationships are found in this way for *large values of* $\varkappa$:

$$u_\varkappa = \frac{\Pi}{2\varkappa\pi}\left\{\exp\left(\frac{\eta}{2}-\xi\right)\cdot\sin\left(\frac{2\varkappa\pi}{\Pi}\eta-\frac{\Pi}{2\varkappa\pi}\xi-\frac{\pi}{2}\right)\right.$$
$$\left.-(-1)^\varkappa\cdot\exp\left[-\frac{\eta}{2}-(\Lambda-\xi)\right]\cdot\sin\left[-\frac{\Pi}{8\varkappa\pi}\eta-\frac{\Pi}{2\varkappa\pi}(\Lambda-\xi)-\frac{\pi}{2}\right]\right\} \quad (13\text{-}82)$$

$$v_\varkappa = \frac{\Pi}{2\varkappa\pi}\left\{\exp\left(\frac{\eta}{2}-\xi\right)\cdot\cos\left(\frac{2\varkappa\pi}{\Pi}\eta-\frac{\Pi}{2\varkappa\pi}\xi-\frac{\pi}{2}\right)\right.$$
$$\left.-(-1)^\varkappa\cdot\exp\left[-\frac{\eta}{2}-(\Lambda-\xi)\right]\cdot\cos\left[-\frac{\Pi}{8\varkappa\pi}\eta-\frac{\Pi}{2\varkappa\pi}(\Lambda-\xi)-\frac{\pi}{2}\right]\right\} \quad (13\text{-}83)$$

$$\phi_\varkappa = \exp\left(\frac{\eta}{2}-\xi\right)\cdot\sin\left(\frac{2\varkappa\pi}{\Pi}\eta-\frac{\Pi}{2\varkappa\pi}\xi\right)$$
$$-(-1)^\varkappa\frac{\Pi}{4\varkappa\pi}\cdot\exp\left[-\frac{\eta}{2}-(\Lambda-\xi)\right]\cdot\sin\left[-\frac{\Pi}{8\varkappa\pi}\eta-\frac{\Pi}{2\varkappa\pi}(\Lambda-\xi)-\frac{\pi}{2}\right] \quad (13\text{-}84)$$

$$\psi_\varkappa = \exp\left(\frac{\eta}{2}-\xi\right)\cdot\cos\left(\frac{2\varkappa\pi}{\Pi}\eta-\frac{\Pi}{2\varkappa\pi}\xi\right)$$
$$-(-1)^\varkappa\frac{\Pi}{4\varkappa\pi}\cdot\exp\left[-\frac{\eta}{2}-(\Lambda-\xi)\right]\cdot\cos\left[-\frac{\Pi}{8\varkappa\pi}\eta-\frac{\Pi}{2\varkappa\pi}(\Lambda-\xi)-\frac{\pi}{2}\right] \quad (13\text{-}85)$$

For a particular position in the regenerator $\xi = $ const, the first terms in these equations assume the form

$$\text{const}\cdot\exp\left(\frac{\eta}{2}\right)\cdot\frac{\sin}{\cos}\left(\frac{2\varkappa\pi}{\Pi}\eta\right)$$

if the slight phase displacement which occurs with increasing ξ is neglected. As the argument $\dfrac{2\varkappa\pi}{\Pi}\eta$ changes by $\varkappa\cdot 2\pi$ during the period Π, the first terms represent oscillations with $\varkappa$ cycles during a regenerator period. The amplitudes of the oscillations thus increase with increasing η in proportion to $\exp\left(\dfrac{\eta}{2}\right)$. On the other hand for the second terms, the sin and cos functions are almost independent given large values of $\varkappa$, η and ξ. Thus, given constant ξ, the second term decreases aperiodically, essentially as a function of $\exp\left(-\dfrac{\eta}{2}\right)$.

The dependence on ξ of both terms is remarkable. The first term has its largest value at $\xi = 0$, that is, at the end of the regenerator in which the gas enters during the period being considered. The first term rapidly reduces with increasing distance ξ from this position. At $\xi = 6.9$ it only amounts to 1 per cent of its value when $\xi = 0$, provided $\varkappa$ is large. Given small values of $\varkappa$ the decrease is slower.

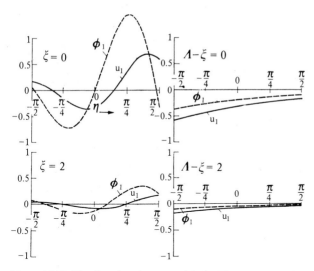

Figure 13-11 The eigen functions u_1 and ϕ_1 for $\Pi = \pi$.

The second aperiodic term, on the other hand, is only dependent upon the distance $\Lambda - \xi$ from the opposite end of the regenerator. At this end, where the second gas enters, the second term has its highest value. On the same principle, this second term, like the first term, becomes smaller with increasing distance from the entry position of the gas. In long regenerators the eigen functions for $\varkappa > 0$ only have significant values in the region of the ends of the regenerator, while they amount to almost zero in the middle of the long central region. The reason for this is that, in the middle of the regenerator, the temperature distribution can easily be solely represented by the zero eigen function. The two terms of the higher eigen functions only have significant values in the central parts of relatively short regenerators and are added here to the zero eigen function. Such cases occur, for example, in the iron and steel industry where Λ has a value between about 10 and 20 but they do not occur in low temperature technology where Λ almost always has a value of more than 100.

Figures 13-11 and 13-12 show the eigen functions for $\varkappa = 1$ when $\Pi = \pi = 3.1416$ computed using the precise Eqs. (13-74), (13-75), (13-77) and (13-78). Figure 11-9 shows the eigen functions $u_\varkappa$ and $v_\varkappa$ for $\varkappa = 2$, a case which has already been discussed in Sec. 11-3. In these illustrations the oscillation terms for constant values of ξ are shown on the left while the aperiodic terms for constant values of $\Lambda - \xi$ are shown as a function of time η on the right. The dashed lines represent the temperature of the gas, the solid lines represent the temperature of the heat-storing mass. In the case of the oscillation terms, the temperature of the heat-storing mass lags behind the temperature of the gas by a phase displacement, which increases for higher values of $\varkappa$, up to a limiting value of $\pi/2$ (90°).

The physical significance of these higher eigen functions is that, as has been mentioned many times, they replicate the harmonic vibrations of the regenerator. This has already been explained in Sec. 11-3. It follows from this, and from these

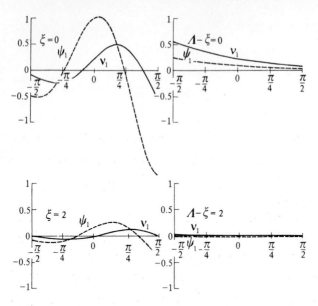

Figure 13-12 The eigen functions v_1 and ψ_1 for $\Pi = \pi$.

latest considerations, that the first terms in these eigen functions represent oscillations around a zero temperature position. On the other hand the second terms express the balancing of deviations from this zero temperature position which carry over the oscillations which took place in the previous period in the region of the other end of the regenerator. The first terms are therefore known, for short, as "oscillation terms" while the second terms are known as the "balancing terms".

13-7 CONSTRUCTION OF THE TOTAL TEMPERATURE DISTRIBUTION IN THE REGENERATOR FROM THE FUNDAMENTAL OSCILLATION AND THE HARMONIC VIBRATIONS, WITH CONSTANT GAS ENTRY TEMPERATURES

Once the eigen functions and thus the fundamental oscillation and the harmonic vibrations have been calculated, it is now possible to determine the total temperature distribution for the state of equilibrium of a counterflow regenerator. To this end these vibrations must be combined together in a suitable way such that the requirement is fulfilled that the entrance gas temperature during the period under consideration is constant. Mathematically such a combination is possible because the differential equations are linear. Thus the solutions representing the eigen functions, of which each linear combination is composed, are each a solution of the differential equations. On the basis of initially arbitrary constants $\alpha_\varkappa$ and $\beta_\varkappa$ the gas temperature θ and the temperature Θ of the heat-storing mass can be developed as the following series of the eigen functions:

$$\theta = \alpha_0\phi_0 + \alpha_1\phi_1 + \alpha_2\phi_2 + \alpha_3\phi_3 + \cdots \qquad \alpha_\varkappa\phi_\varkappa + \cdots$$
$$+ \beta_1\psi_1 + \beta_2\psi_2 + \beta_3\psi_3 + \cdots \qquad \beta_\varkappa\psi_\varkappa + \cdots \tag{13-86}$$

$$\Theta = \alpha_0 u_0 + \alpha_1 u_1 + \alpha_2 u_2 + \alpha_3 u_3 + \cdots \qquad \alpha_\varkappa u_\varkappa + \cdots$$
$$+ \beta_1 v_1 + \beta_2 v_2 + \beta_3 v_3 + \cdots \qquad \beta_\varkappa v_\varkappa + \cdots \tag{13-87}$$

These series meet the reversal requirement because each individual term does so. All that is thus required is to determine the constants $\alpha_1, \alpha_2, \ldots \beta_1, \beta_2, \ldots$, etc., in such a way that the series (13-86) yields the constant value $\theta = \theta_1$ at position $\xi = 0$. This problem, however, is made difficult by the fact that the usual procedures available for determining such coefficients as occur in series of harmonic functions, particularly in the Fourier series, cannot be used directly in this case.

This difficulty arises from the fact that the eigen functions $\phi_\varkappa$ and $\psi_\varkappa$ are not orthogonal on the region $-\Pi/2 < \eta < +\Pi/2$. That is, for $\iota \neq \varkappa$, the so-called inner products

$$\int_{-\Pi/2}^{+\Pi/2} \phi_\iota\phi_\varkappa \, d\eta, \qquad \int_{-\Pi/2}^{+\Pi/2} \phi_\iota\phi_\varkappa \, d\eta, \qquad \int_{-\Pi/2}^{+\Pi/2} \psi_\iota\psi_\varkappa \, d\eta$$

do not disappear. Although a method is known whereby each arbitrary function system can be transformed into a system of orthogonal functions, this method is so complicated when it is applied that it is preferable to use the following procedure which yields only partial orthogonality. It is necessary to assume that Λ is so large that when $\xi = 0$ the balancing terms of the eigen functions can be neglected. In this case, it follows from Eqs. (13-84) and (13-85) that the eigen functions $\phi_\varkappa$ and $\psi_\varkappa$ assume the following form when $\xi = 0$ for very large values of $\varkappa$:

$$\phi_\varkappa = \exp\left(\frac{\eta}{2}\right) \cdot \sin\left(\frac{2\varkappa\pi}{\Pi}\eta\right), \quad \psi_\varkappa = \exp\left(\frac{\eta}{2}\right) \cdot \cos\left(\frac{2\varkappa\pi}{\Pi}\eta\right)$$

As the functions $\sin\left(\dfrac{2\varkappa\pi}{\Pi}\eta\right)$ and $\cos\left(\dfrac{2\varkappa\pi}{\Pi}\eta\right)$ are orthogonal in the region $-\dfrac{\Pi}{2}$ to $+\dfrac{\Pi}{2}$, the functions $\Phi_\varkappa = \exp\left(-\dfrac{\eta}{2}\right)\phi_\varkappa$ and $\Psi_\varkappa = \exp\left(-\dfrac{\eta}{2}\right)\psi_\varkappa$ come closer and closer to becoming orthogonal for increasing $\varkappa$. This can be called "asymptotic orthogonality".

If Eq. (13-86) is divided by $\exp\left(\dfrac{\eta}{2}\right)$ the series shown in (13-88) for $\xi = 0$ is obtained:

$$\exp\left(-\frac{\eta}{2}\right)\theta_1 = \alpha_0\Phi_0 + \alpha_1\Phi_1 + \alpha_2\Phi_2 + \alpha_3\Phi_3 + \cdots + \alpha_\varkappa\Phi_\varkappa + \cdots$$
$$+ \beta_1\Psi_1 + \beta_2\Psi_2 + \beta_3\Psi_3 + \cdots + \beta_\varkappa\Psi_\varkappa + \cdots \tag{13-88}$$

Now $\Phi_\varkappa = \exp\left(-\dfrac{\eta}{2}\right)\phi_\varkappa$ and $\Psi_\varkappa = \exp\left(-\dfrac{\eta}{2}\right)\psi_\varkappa$ denote the values of these functions at position $\xi = 0$. If the right-hand side of this equation is set equal to S it is required to determine the coefficients $\alpha_\varkappa$ and $\beta_\varkappa$ in such a way that $\exp\left(-\dfrac{\eta}{2}\right)\theta_1$, with $\theta_1 = \text{const}$, is as close to S as possible in the sense of the method of least squares. Thus, if $\varepsilon = \exp\left(-\dfrac{\eta}{2}\right)\theta_1 - S$, it is required that $M = \displaystyle\int_{-\Pi/2}^{+\Pi/2} \varepsilon^2 \, d\eta$ should be a minimum. Note that, $\dfrac{\partial M}{\partial \alpha_\varkappa}$ must be equal to $\displaystyle\int 2\varepsilon \dfrac{\partial \varepsilon}{\partial \alpha_\varkappa} \, d\eta = 0$ and correspondingly $\dfrac{\partial M}{\partial \beta_\varkappa} = \displaystyle\int 2\varepsilon \dfrac{\partial \varepsilon}{\partial \beta_\varkappa} \, d\eta = 0$. Equations (13-89) and (13-90) follow by inserting the expression for ε determined utilizing Eq. (13-88), into the last two integrals:

$$\alpha_0 \int \Phi_0 \Phi_\varkappa \, d\eta + \alpha_1 \int \Phi_1 \Phi_\varkappa \, d\eta + \cdots \alpha_\varkappa \int \Phi_\varkappa^2 \, d\eta + \cdots \alpha_\iota \int \Phi_\iota \Phi_\varkappa \, d\eta + \cdots$$

$$+ \beta_1 \int \Psi_1 \Phi_\varkappa \, d\eta + \cdots \beta_\varkappa \int \Psi_\varkappa \Phi_\varkappa \, d\eta + \cdots \beta_\iota \int \Psi_\iota \Phi_\varkappa \, d\eta + \cdots = \int \exp\left(-\dfrac{\eta}{2}\right)\theta_1 \Phi_\varkappa \, d\eta$$

$$\tag{13-89}$$

and

$$\alpha_0 \int \Phi_0 \Psi_\varkappa \, d\eta + \alpha_1 \int \Phi_1 \Psi_\varkappa \, d\eta + \cdots \alpha_\varkappa \int \Phi_\varkappa \Psi_\varkappa \, d\eta + \cdots \alpha_\iota \int \Phi_\iota \Psi_\varkappa \, d\eta + \cdots$$

$$+ \beta_1 \int \Psi_1 \Psi_\varkappa \, d\eta + \cdots \beta_\varkappa \int \Psi_\varkappa^2 \, d\eta + \cdots \beta_\iota \int \Psi_\iota \Psi_\varkappa \, d\eta + \cdots = \int \exp\left(-\dfrac{\eta}{2}\right)\theta_1 \Psi_\varkappa \, d\eta$$

$$\tag{13-90}$$

Since the inner products exhibit asymptotic orthogonality $\int \Phi_\iota \Phi_\varkappa \, d\eta$, $\int \Psi_\iota \Phi_\varkappa \, d\eta$ and $\int \Psi_\iota \Psi_\varkappa \, d\eta$ (apart from $\int \Phi_\varkappa^2 \, d\eta$) and $\int \Psi_\varkappa^2 \, d\eta$) become smaller as ι and $\varkappa$ increase. As a consequence they can be ignored for sufficiently large ι and $\varkappa$. It is necessary to assume that all the inner products, provided they do not contain Φ_0, for non-equal eigen functions disappear when ι or $\varkappa$ are greater than a certain value K. For $\varkappa = 1, 2, 3$ to $\varkappa = K$, Eqs. (13-89) and (13-90) terminate after α_K and β_K; thus $2K$ equations are obtained in $2K+1$ unknowns. On the other hand, for $\varkappa = 0$ Eq. (13-89) has the basic form of an infinite series, while Eq. (13-90) does not exist for $\varkappa = 0$. But even this infinite series can be changed into a finite series as will be shown below.

Equations (13-91) and (13-92) follow for $\varkappa > K$ based on the assumption about the inner products embodied in Eqs. (13-89) and (13-90). v denotes a positive whole number between 1 and ∞.

$$\alpha_{K+v} = \dfrac{\displaystyle\int \exp\left(-\dfrac{\eta}{2}\right)\theta_1 \Phi_{K+v} \, d\eta - \alpha_0 \int \Phi_0 \Phi_{K+v} \, d\eta}{\displaystyle\int \Phi_{K+v}^2 \, d\eta}$$

$$\tag{13-91}$$

and

$$\beta_{K+v} = \frac{\int \exp\left(-\frac{\eta}{2}\right)\theta_1 \Psi_{K+v}\, d\eta - \alpha_0 \int \Phi_0 \Psi_{K+v}\, d\eta}{\int \Psi_{K+v}^2\, d\eta} \qquad (13\text{-}92)$$

If Eqs. (13-91) and (13-92) are inserted into Eq. (13-89) for $\varkappa = 0$, only the $2K+1$ unknowns $\alpha_0, \alpha_1, \ldots, \alpha_K, \beta_1, \ldots, \beta_K$ will appear in this equation. These coefficients can thus be calculated by solving in a suitable manner, the $2K+1$ linear equations, which have been thus obtained.* The higher coefficients are then obtained by inserting the value obtained for α_0 into Eqs. (13-91) and (13-92). In practice the number of linear equations less than $2K+1$ can be reduced since some of the inner products can be neglected.

For accuracy not to be lost, the larger the value of Π, the larger K must be. Before the development of computers calculations with $K > 2$ or 3 were very tedious. Nowadays K can be considerably higher. In certain circumstances rather than increasing the value of K, it is more expedient to increase accuracy by repeating the calculation in the following way, with a relatively small value of K. The values of $\alpha_{K+1}, \alpha_{K+2}, \ldots, \beta_{K+1}, \beta_{K+2}$, etc., which were obtained in the previous calculation can be inserted into Eqs. (13-89) and (13-90), and approximate allowance made for the inner products of unequal eigen functions previously neglected. $2K+1$ linear equations again result for all values of $\varkappa$ between 0 and K and improved values of α_0, α_1 to α_K and β_1 to β_K are obtained from the solutions of these equations. New values of $\alpha_{K+1}, \alpha_{K+2}, \ldots, \beta_{K+1}$, etc., can again be found using Eqs. (13-91) and (13-92).

Instead of using the method of least squares it is also possible to proceed in the following way. It will be assumed that the series in Eq. (13-86) will be terminated beyond a certain value of $\varkappa$, for example beyond $\varkappa = 4$, so that $2\varkappa + 1 = 9$ coefficients are to be determined, namely α_0 to α_4 and β_1 to β_4. It will then be supposed that the period length T can be divided into $2\varkappa$ equal intervals of time. It is possible to calculate the values ϕ_0 to ϕ_K and ψ_1 to ψ_K for each of the $2\varkappa + 1$ points in time and to set these values into Eq. (13-86) as was done for $\theta = \theta_1$. $2\varkappa + 1$ linear equations are thus obtained. The solution of these linear equations determines suitable values for α_0 to α_K and β_1 to β_K. However, an exact replication of θ_1 is only obtained at the data points in time, although the deviations which are found in between these points become smaller, the more the number of terms of the series which are taken into consideration.

Once the values of $\alpha_\varkappa$ and $\beta_\varkappa$ have been determined it is then possible to calculate the whole temperature distribution in the regenerator, using Eqs. (13-86) and (13-87). Figure 13-13 shows the step-by-step compilation of the eigen functions $\phi_\varkappa$ and $\psi_\varkappa$, multiplied by $\alpha_\varkappa$ and $\beta_\varkappa$, for the gas temperature at position $\xi = 0$.

---

* Editor's note: Hausen mentions the use of determinants whereas Gaussian elimination and back substitution are likely to be used nowadays.

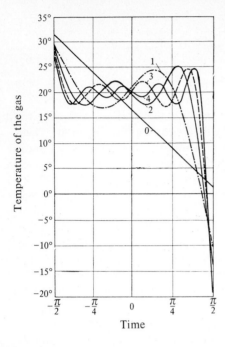

Figure 13-13 Oscillating approximation to the constant entry temperature $\theta = 20°C$ when $\Lambda = 10$ and $\Pi = \pi$.

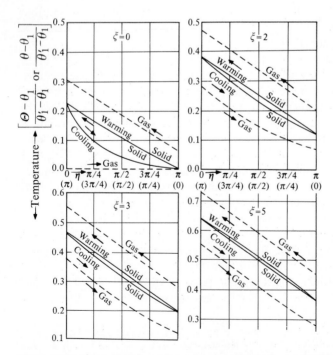

Figure 13-14 Chronological variations of temperature in the regenerator when $\Lambda = 10$ and $\Pi = \pi$.

Here $\Lambda = 10$, $\Pi = \pi = 3.1416$ and $\theta_1 = 20°C$. The reduced time η, which extends from $-\pi/2$ to $+\pi/2$, is drawn as the abscissa and the temperature of the gas is drawn as the ordinate. The straight line marked by 0 corresponds to the expression $\alpha_0\phi_0$; curve 1 corresponds to the expression $\alpha_0\phi_0 + \alpha_1\phi_1 + \beta_1\psi_1$; curve 4 takes into consideration all the eigen functions from $\varkappa = 0$ to $\varkappa = 4$. One can recognize the development of the approximation of the constant entry temperature $\theta_1 = 20°C$; this approximation improves as the number of eigen functions included is increased.

The end result of the calculation of the chronological and spatial temperature distribution for the example chosen, can be seen in Figs. 13-14 and 13-15. Figure 13-14 illustrates the chronological variation of temperature in both periods at positions $\xi = 0, 2, 3$ and 5, where $\xi = 0$ corresponds to the cold end of the regenerator and since $\Lambda = 10$, $\xi = 5$ corresponds to the centre of the regenerator.

The temperature Θ of the heat-storing mass goes through a hysteresis loop which is greatest at the ends of the regenerator and which becomes smaller with increasing ξ until, in the centre of the regenerator, it almost becomes a straight line; this is the same for both periods. The fact that this is not an exact straight line and is not one in most of the regenerator, is because the regenerator, in the example chosen, is relatively short. Thus the deviations from the zero eigen function, even as far as the centre of the regenerator, are relatively large. Figure 13-15 illustrates the spatial distribution of the gas temperature θ and the temperature Θ of the heat-storing mass in the cold period and at various times η, as a function of ξ.

The example considered replicates conditions as they are found in the iron and steel industry. A case which corresponds more to low temperature technology has already been mentioned in Sec. 11-3 and shown in Fig. 11-10. Further examples will be shown in later sections, in particular in Sec. 18-1.

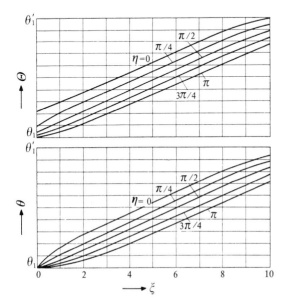

Figure 13-15 Spatial temperature distribution in the regenerator when $\Lambda = 10$ and $\Pi = \pi$.

13-8 EFFICIENCY OF COUNTERFLOW REGENERATORS AND THE TRUE HEAT TRANSMISSION COEFFICIENT

As the efficiency of a heat exchanger is uniquely determined by the temperature distribution at equilibrium, it must be possible to calculate the efficiency of regenerators with the help of the coefficients α_χ and β_χ from the eigen functions. As with recuperators (see Part Three, Sec. 5-11) efficiency is defined by the ratio of the actual quantity of heat exchanged to the quantity of heat which would be exchanged in an infinitely large heat exchanger. In the present case, in which the thermal capacities, per period, of both gases are assumed to be the same, that is $CT = C'T'$, the gas which enters an infinitely large heat exchanger at temperature θ_1 could be completely cooled or heated to the entry temperature θ'_1 of the gas entering at the other end of the regenerator. Then, because $\theta'_1 = -\theta_1$, the quantity of heat being transferred in the regenerator, during a period, would be:

$$Q_{id} = \pm CT(\theta_1 + \theta'_1) = \pm CT \cdot 2\theta_1$$

In an actual regenerator the gas leaves at position $\xi = \Lambda$, at time η, at the temperature θ_2. The average exit temperature during the period under consideration is equal to $\bar{\theta}_2 = \dfrac{1}{\Pi} \displaystyle\int_{-\Pi/2}^{+\Pi/2} \theta_2 \, d\eta$. During this period the gas exchanges the quantity of heat (13-93) with the heat-storing mass:

$$Q_{Per} = \pm CT(\theta_1 - \bar{\theta}_2) \tag{13-93}$$

The efficiency of the regenerator thus works out to be*

$$\eta_{Reg} = \frac{Q_{Per}}{Q_{id}} = \frac{\theta_1 - \bar{\theta}_2}{\theta_1 - \theta'_1} = \frac{\theta_1 - \bar{\theta}_2}{2\theta_1} \tag{13-94}$$

The penultimate expression in this equation is applicable when $CT = C'T'$; the last expression is valid only when $\Lambda = \Lambda'$ also, that is when $\bar{\alpha}/C = \bar{\alpha}'/C'$ (see Eq. (13-28)).

Finally, by integrating θ over the total length of the period, using Eqs. (13-86), $\bar{\theta}_2$ will be obtained. Thus, the efficiency η_{Reg} can be found also employing Eqs. (13-94).

Figure 13-16 illustrates efficiency calculated in this manner.[10] η_{Reg} is shown as a function of the reduced length Λ for various period lengths Π. Here, as above, it is assumed that $\Lambda = \Lambda'$ and $\Pi = \Pi'$. It can be seen that efficiency increases with Λ, but, for a given Λ, it decreases with increasing period length. The topmost curve was calculated using Eq. (13-95).

Given an infinitely short period length $\Pi = 0$ the eigen functions μ_0 and ϕ_0 are alone sufficient to represent precisely the temperature distribution. Indeed, by

---

* Editor's note: η_{Reg} is known as the "thermal ratio" in the English literature.

[10] The calculation has in fact been carried out on a computer by Willmott, Schellmann and Sandner using the methods described in Secs. 16-4, 16-5, 14-4, and 16-2. See Footnote 6 on p. 310.

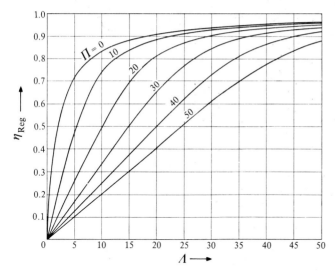

Figure 13-16 Efficiency of the counterflow regenerator when $C = C'$ and $T = T'$.

neglecting all the higher eigen functions,

$$\theta = \alpha_0\left(\eta - \xi + 1 + \frac{\Lambda}{2}\right)$$

is first obtained by using Eqs. (13-80) and (13-86). Since $\Pi = 0$, η is also equal to 0. It is thus possible to satisfy the requirement that $\theta = \theta_1$ when $\eta = 0$ by setting $\alpha_0 = \dfrac{2\theta_1}{2+\Lambda}$. Using this the relationship one obtains is

$$\bar{\theta}_2 = \theta_2 = \theta_1 \frac{2-\Lambda}{2+\Lambda}$$

for the average exit temperature of the gas for $\Pi = 0$. Thus Eq. (13-95) follows from Eq. (13-94):

$$\lim_{\Pi=0} \eta_{Reg} = \frac{\Lambda}{2+\Lambda} \qquad (13\text{-}95)$$

This equation corresponds to the topmost curve in Fig. 13-16.

Equation (13-95) points towards an informative comparison between regenerators and recuperators. If it is assumed that $\bar{\alpha} = \bar{\alpha}'$ and $T = T'$ and note is taken of the fact that, according to Fig. 11-13, $k = k_0$ when $\Pi = 0$, it follows from Eqs. (13-5), (11-14) and (13-28) that $\Lambda = 2k_{reg}F/C$, where $k_{reg} = 2k$; k_{reg} denotes the heat transmission coefficient of at least two regenerators working together. Thus Eq. (13-96) can be written in place of Eq. (13-95):

$$\lim_{\Pi=0} \eta_{Reg} = \frac{k_{reg}F/C}{1+k_{reg}F/C} \qquad (13\text{-}96)$$

This relationship agrees with Eqs. (5-66) and (5-40) in Secs. 5-11 and 5-6 for recuperators when $C = C'$, from which it follows that

$$\eta_{Reg} = \frac{kF/C}{1+kF/C}$$

where k now denotes the heat transmission coefficient of a recuperator. It follows from this that for infinitely short period lengths, regenerators have the same efficiency as recuperators if the dimensionless quantities $k_{reg} \cdot F/C$ and kF/C are equal to one another in both cases.

Relationship Between the Efficiency η_{Reg} and the Heat Transmission Coefficient

The true heat transmission coefficient k of the total regenerator and the spatial heat transmission coefficient $k\xi$ which have already been defined in Sec. 11-4, can be calculated using Eqs. (12-33) and (12-34) employing the eigen functions and their coefficients $\alpha_{\varkappa}$ and $\beta_{\varkappa}$.

A relationship between k and the efficiency η_{Reg} can be obtained in the following way where $CT = C'T'$. The relationship (13-97) is obtained from the observation that $\theta - \theta' = \bar{\theta}_2 - \theta'_1 = $ const. Using Eqs. (11-3), for the quantity of heat transferred during the period under consideration, this equation takes the form:

$$Q_{Per} = k(T+T')F(\bar{\theta}_2 - \theta'_1) = CT(\theta_1 - \bar{\theta}_2) \tag{13-97}$$

Equation (13-98) follows upon taking into consideration Eq. (13-94):

$$k = \frac{C}{F} \frac{T}{T+T'} \frac{\eta_{reg}}{1 - \eta_{Reg}} \tag{13-98}$$

Thus when $CT = C'T'$, the true heat transmission coefficient k can be easily calculated if η_{Reg} is known, using this equation. It is possible to read off the value of η_{Reg} from Fig. 13-16.

In addition, the ratio k/k_0 of the true heat transmission coefficient k, which embodies all the eigen functions, to the heat transmission coefficient k_0, which corresponds only to the zero eigen function, can be calculated from η_{Reg}. In order to illustrate this, similar considerations will be applied as were used above for the *zero eigen functions* alone. Equations (13-60) and (13-61) will be used for this purpose as these are more generally valid than Eqs. (13-79) and (13-80); they also include the cases where Λ and Λ' are different.

If $\bar{\theta}_{2,0}$ is the average exit temperature that will be attained by the gas in the period under consideration when only the zero eigen function is valid, Eq. (13-99), which corresponds to Eq. (13-97), represents the quantity of heat thus transferred in this period:

$$Q_0 = k_0(T+T')F(\bar{\theta}_{2,0} - \theta'_1) = CT(\theta_1 - \bar{\theta}_{2,0}) \tag{13-99}$$

Strictly speaking, θ_1 and θ'_1 must also be regarded as chronological average values.

Based on the fact that the regenerator extends from $\xi = 0$ to $\xi = \Lambda$, Eq. (13-100) is obtained from Eq. (13-61):

$$\theta_1 - \bar{\theta}_{2,0} = B\Lambda \tag{13-100}$$

and correspondingly, for the next period:

$$\theta_1' - \bar{\theta}_{2,0} = -(\theta_1 - \bar{\theta}_{2,0}) = B'\Lambda'$$

Equation (13-101) follows from these two equations:

$$B' = -\frac{\Lambda}{\Lambda'} B \tag{13-101}$$

In addition, it must be noted that

$$\theta - \Theta = B$$

and

$$\theta' - \Theta' = B'$$

result from Eqs. (13-60) and (13-61). Moreover, since $\Theta = \Theta'$ for corresponding times in both periods for the fundamental oscillation, it will be seen from the last two equations that

$$\theta - \theta' = \bar{\theta}_{2,0} - \theta_1' = B - B'$$

or, taking Eq. (13-101) into account,

$$\bar{\theta}_{2,0} - \theta_1' = B\left(1 + \frac{\Lambda}{\Lambda'}\right) \tag{13-102}$$

Inserting Eqs. (13-100) and (13-102) into (13-99) yields

$$k_0(T + T')F\left(\frac{1}{\Lambda} + \frac{1}{\Lambda'}\right) = CT \tag{13-103}$$

Finally, using Eq. (13-98) we obtain

$$\frac{k}{k_0} = \left(\frac{1}{\Lambda} + \frac{1}{\Lambda'}\right)\frac{\eta_{Reg}}{1 - \eta_{Reg}} \tag{13-104}$$

If consideration is again limited to the case where $\Lambda = \Lambda'$, in place of Eq. (13-104) can also be written

$$\frac{k}{k_0} = \frac{2}{\Lambda} \cdot \frac{\eta_{Reg}}{1 - \eta_{Reg}} \tag{13-105}$$

Thus, it follows from this equation that the relationship k/k_0 can be easily calculated from the value η_{Reg} of the efficiency as obtained, for example, from Fig. 13-16. The values of k/k_0 obtained in this way are shown in Fig. 11-13.

In order to be able to use Fig. 11-13 even when $CT \neq C'T'$ the more general relationships (11-5) or even (11-6), which have been presented already in Sec. 11-4, can be used with dimensionless groups Λ and Π in place of Eqs. (13-28) and (13-29). These relationships are obtained in the following way. It is necessary to proceed on the assumption, which will be justified by later calculations, that sufficiently precise values for k/k_0 can be read off from Fig. 11-13, even when $\Lambda' \neq \Lambda$ and $\Pi' \neq \Pi$ so long as only average values which are developed from these dimensionless groups are introduced into Fig. 11-13. It seems appropriate here to form the harmonic mean, that is the arithmetic mean, of the reciprocal values. This has been suggested already for Λ and Λ' in Eq. (13-104) and is suggested for Π and Π' in Eq. (15-21), introduced later. Thus, from Eqs. (13-28) and (13-29) the following average values are obtained

$$\frac{1}{\Lambda_m} = \frac{1}{2}\left(\frac{1}{\Lambda} + \frac{1}{\Lambda'}\right) = \frac{1}{2F}\left(\frac{CT}{\bar{\alpha}T} + \frac{C'T'}{\bar{\alpha}'T'}\right) \approx \frac{CT + C'T'}{4F}\left(\frac{1}{\bar{\alpha}T} + \frac{1}{\bar{\alpha}'T'}\right)$$

and

$$\frac{1}{\Pi_m} = \frac{\gamma c\,\delta}{4}\left(\frac{1}{\bar{\alpha}T} + \frac{1}{\bar{\alpha}'T'}\right)$$

It will be seen from Eq. (13-5) that these expressions transform into

$$\Lambda_m = 4\frac{k_0(T + T')F}{CT + C'T'} \quad \text{and} \quad \Pi_m = 4\frac{k_0(T + T')}{\rho c\,\delta} \tag{13-106}$$

These relationships are in agreement with Eqs. (11-5) and (11-6), which have already been mentioned; the only difference is that in these two last equations Λ and Π are used for the average value with simplification in mind, in place of Λ_m and Π_m, and $\rho c\,\delta/2$ is replaced by the more general expression C_s/F. The precise calculation of k and k/k_0 for the case where $CT \neq C'T'$ will be discussed in Sec. 15-3. In this section the curve for $k\xi$, developed from Eq. (12-34), as a function of the longitudinal direction of the regenerator will be discussed by way of example; see Fig. 15-7.

The Relationship Between the Hysteresis Loop and the Heat Transmission Coefficient of a Regenerator

A relationship which exists between the hysteresis loop and the heat transmission coefficient has already been discussed as a fundamental idea in Sec. 11-4. In what follows, it will be shown that this relationship can be precisely expressed by an equation when $CT = C'T'$.

It will be observed from Eq. (13-4) that the quantity of heat which is transferred through a small surface area df during the length of time T within the period under consideration is equal to:

$$dQ_{Per} = \bar{\alpha}\,df\int_0^T (\theta - \Theta)\,dt = \bar{\alpha}\,df\,T(\bar{\theta} - \bar{\Theta})$$

where $\bar{\theta}$ and $\bar{\Theta}$ again denote the chronological average values. In a similar way

$$dQ_{Per} = \bar{\alpha}'\,df\,T'(\bar{\Theta}' - \bar{\theta}')$$

applies to the second period of duration T'. If both equations are solved using the temperature differences, the following relationship is obtained by addition and by using Eq. (13-5);

$$dQ_{Per} = k_0(T+T')df[(\bar{\theta}-\bar{\theta}')-(\bar{\Theta}-\bar{\Theta}')]$$

In this equation, as can be seen in Fig. 13-14, for example, the difference $\bar{\Theta}-\bar{\Theta}'$ between the chronological average values of the temperatures of the heat-storing mass, is equal to the average height of the hysteresis loop. Equation (13-107) is obtained for the total quantity of heat transferred in the regenerator during time $T+T'$. This equation is obtained by integration over df, taking advantage of the fact that $\bar{\theta}-\bar{\theta}'$ is constant since $CT = C'T'$:

$$Q_{Per} = k_0(T+T')F[(\bar{\theta}-\bar{\theta}')-H] \qquad (13\text{-}107)$$

Here $H = \dfrac{1}{F}\displaystyle\int_0^F (\bar{\Theta}-\bar{\Theta}')\,df$ can be understood as the total average value of the height of all the hysteresis loops. In Fig. 11-11 H is equal to the average height of the shaded sections of the surface area relative to the total length of the regenerator.

Finally, if the expression obtained for Q_{Per} using Eq. (13-107) is set equal to the expression obtained using Eq. (11-3), then Eq. (13-108) is obtained as the final, notable, result:

$$\frac{k}{k_0} = 1 - \frac{H}{\bar{\theta}-\bar{\theta}'} \qquad (13\text{-}108)$$

It can be seen from this that the ratio of the total average H of all heights of the hysteresis loops and the average temperature difference $\bar{\theta}-\bar{\theta}'$ between the two gases, directly determines the value of k/k_0.

13-9 TEMPERATURE DISTRIBUTION IN THE PARALLEL FLOW REGENERATOR [H303, H304]

In parallel flow the precise calculation of the strictly periodic state of equilibrium is considerably simpler than in the case where the gases flow in the opposite direction to one another.

Essentially this is based on the fact that the requirement that the gas entry temperatures should be chronologically constant during the warm and cold period refers to the same end of the regenerator. From the mathematical point of view it does not matter if two different types of gas flow through the regenerator during the two periods, which are, again, considered to be of equal length. Rather, the temperature distribution remains the same if we assume that one and the same gas enters at $\xi = 0$ in square wave fashion with the constant temperatures $\theta = \theta_1$ and $\theta = \theta'_1 = -\theta_1$. Thus, at $\xi = 0$, the boundary condition (13-109) is obtained if one period lasts from $\eta = -\Pi$ to $\eta = 0$ and the second period lasts from $\eta = 0$ to $\eta = \Pi$:

$$\theta = \theta_1 \text{ for } -\Pi < \eta < 0; \quad \theta = -\theta_1 \text{ for } 0 < \eta < \Pi \qquad (13\text{-}109)$$

This boundary condition can be satisfied for the state of equilibrium for the case where $\Lambda = \Lambda'$ and $\Pi = \Pi'$ by the development of eigen functions which are purely periodic relative to η and run through a whole number of cycles in the basic interval 2Π. The solutions (13-68) and (13-69) given earlier become purely periodic in η if we set

$$\frac{2}{1-n} = i\frac{\varkappa\pi}{\Pi} \tag{13-110}$$

where i is $\sqrt{-1}$ and $\varkappa$ is a real whole number. After some intermediate calculations, which will not be reproduced here, the following real eigen functions are obtained for the temperature Θ of the heat-storing mass:

$$u_\varkappa = \sqrt{\frac{\Pi^2}{\Pi^2 + \varkappa^2\pi^2}} \cdot \exp\left(-\frac{\varkappa^2\pi^2}{\Pi^2 + \varkappa^2\pi^2}\xi\right)\sin\left\{\frac{\varkappa\pi}{\Pi}\eta - \frac{\Pi\varkappa\pi}{\Pi^2 + \varkappa^2\pi^2}\xi - \arctan\frac{\varkappa\pi}{\Pi}\right\} \tag{13-111}$$

$$v_\varkappa = \sqrt{\frac{\Pi^2}{\Pi^2 + \varkappa^2\pi^2}} \cdot \exp\left(-\frac{\varkappa^2\pi^2}{\Pi^2 + \varkappa^2\pi^2}\xi\right)\cos\left\{\frac{\varkappa\pi}{\Pi}\eta - \frac{\Pi\varkappa\pi}{\Pi^2 + \varkappa^2\pi^2}\xi - \arctan\frac{\varkappa\pi}{\Pi}\right\} \tag{13-112}$$

and correspondingly, for the gas temperature

$$\phi_\varkappa = \exp\left(-\frac{\varkappa^2\pi^2}{\Pi^2 + \varkappa^2\pi^2}\xi\right)\sin\left\{\frac{\varkappa\pi}{\Pi}\eta - \frac{\Pi\varkappa\pi}{\Pi^2 + \varkappa^2\pi^2}\xi\right\} \tag{13-113}$$

$$\psi_\varkappa = \exp\left(-\frac{\varkappa^2\pi^2}{\Pi^2 + \varkappa^2\pi^2}\xi\right)\cos\left\{\frac{\varkappa\pi}{\Pi}\eta - \frac{\Pi\varkappa\pi}{\Pi^2 + \varkappa^2\pi^2}\xi\right\} \tag{13-114}$$

For $\varkappa = 0$ one has $u_0 = \phi_0 = 0$ and $v_0 = \psi_0 = 1$.

The temperature functions Θ and θ of the heat-storing mass and the gas can again be represented as a series of the eigen functions, as in Eqs. (13-86) and (13-87). Because the eigen functions (13-111) to (13-114) are, contrary to counterflow, truly orthogonal, the coefficients $\alpha_0, \alpha_1, \ldots, \beta_1, \ldots$, etc., can easily be determined using the usual Fourier method[11] in such a way that the boundary condition (13-109) is satisfied. Thus Eqs. (13-115) and (13-116) yield the desired solution:

$$\theta = -\frac{4}{\pi}\theta_1\left[\phi_1 + \frac{1}{3}\phi_3 + \frac{1}{5}\phi_5 + \frac{1}{7}\phi_7 + \cdots\right], \tag{13-115}$$

$$\Theta = -\frac{4}{\pi}\theta_1\left[u_1 + \frac{1}{3}u_3 + \frac{1}{5}u_5 + \frac{1}{7}u_7 + \cdots\right]. \tag{13-116}$$

As in the case of counterflow, Eq. (13-94) gives the *efficiency*. The average exit temperature $\bar\theta_2$ which is to be inserted into these equations is obtained by integrating Eq. (13-115) with respect to η, with $\xi = \Lambda$, from $\eta = 0$ to $\eta = \Pi$ bearing in mind Eq. (13-113).

---

[11] See, for example, Hütte, Bd. 1, 28th edition, 1955, p. 107.

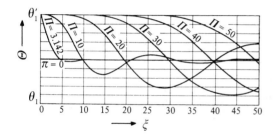

Figure 13-17 Temperature of the heat-storing mass at the beginning of the cold period, for parallel flow at equilibrium.

The following diagrams reproduce the results of various calculations for a parallel flow regenerator at the state of equilibrium when $\Lambda = \Lambda'$ and $\Pi = \Pi'$; they show that the oscillatory nature of the periodic temperature changes are much more apparent in parallel flow than they are in counterflow. Thus, for example, it will be seen in Fig. 13-17 that the temperature distribution in the heat-storing mass at the beginning of the cold period is represented by a temperature oscillation about the mean temperature $\Theta = 0$ which fades away with increasing ξ.

Only with an infinitely short period length ($\Pi = 0$) are the oscillatory variations so small that the temperature distribution coincides fully with the horizontal $\Theta = 0$. If all the curves are reflected against this horizontal line, the temperature distribution at the end of the cold period is obtained for each of the values of Π plotted. In a way that is in contrast to the counterflow regenerator, the temperature distribution in parallel flow is independent of Λ. Thus the temperature distribution when $\Lambda = 20$ is obtained, for example, if the curves in Fig. 13-17 are imagined to be broken off at $\xi = 20$.

Figures 13-18 and 13-19 display the temperature distribution for the case where $\Lambda = 10$ and $\Pi = \pi$ at various times during the cold period. Figure 13-18 shows the temperature Θ of the heat-storing mass, Fig. 13-19 shows the temperature θ of the gas. It is to be noted that the gas, during its passage through the regenerator, first rises above the average temperature $\theta = 0$. It is then cooled a little. During this cooling process, the gas transports the heat, which it has obtained from the heat-storing mass, in the direction of flow. When $\xi = \Lambda$ the gas usually leaves at a temperature which is somewhat higher than $1/2(\theta_1 + \theta_1')$. This is particularly so for the case where $\Lambda = 5$ where the regenerator is thus cut off at

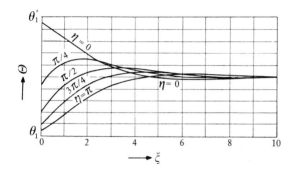

Figure 13-18 Distribution of the temperature Θ of the heat-storing mass during the cold period. Parallel flow $\Pi = \pi$.

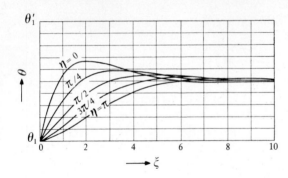

Figure 13-19 Distribution of the gas temperature θ during the cold period. Parallel flow.

$\xi = 5$. It follows from this that, under certain circumstances, the efficiency of the regenerator is greater in the case of parallel flow than that of recuperators which cannot reach more than 50 per cent under these conditions.

Figure 13-20 illustrates the efficiency of the regenerator as a function of Λ for various values of Π. For an infinitely short period length ($\Pi = 0$) the efficiency is as great as that for a recuperator; in this case, as Λ increases it approaches asymptotically the value of 0.5. This efficiency occurs in practice for $\Lambda = 5$. For larger values of Π, the curves take the form of an attenuating oscillation about the value 0.5. Each of these curves has a pronounced maximum value which far outstrips the efficiency of the recuperator. Indeed, when $\Pi = 40$, this amounts to more than 80 per cent. As can be seen from the diagram, the maximum value occurs close to $\Pi = \Lambda$. This condition, it follows from Eq. (11-5), corresponds to the case where the thermal capacity C_{Per} of the quantity of gas flowing through the regenerator during the warm or cold period is equal to the thermal capacity C_s of the total heat-storing mass.

By way of comparison Fig. 13-20 also shows the values of the efficiency for counterflow operation when $\Pi = \Lambda$ and $\Pi = 0$. It will be seen from this, that when $\Pi = \Lambda$ the difference between parallel flow and counterflow is not very great;

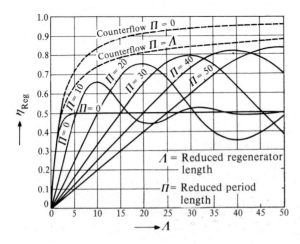

Figure 13-20 Efficiency of the parallel-flow regenerator.

indeed, given Λ, the efficiency of the counterflow regenerator can be perceptibly increased by reducing Π. From purely thermodynamic considerations therefore, counterflow operation is always superior to parallel flow.

This superiority of counterflow operation clearly emanates from the fact that, according to Fig. 13-20, an efficiency of 83 per cent, for example, can only be achieved in parallel flow if the regenerator is at least three times as long ($\Lambda = 40$) as for counterflow ($\Lambda = 12$ approximately for $\Pi = 4$).

13-10 INITIAL WARMING OR COOLING OF THE HEAT-STORING MASS; THE LOWAN THEORIES OF PARALLEL FLOW OPERATION OF A REGENERATOR

Lowan [L306] has developed a method for calculating the temperature distribution in regenerators which takes into account all the details of the temperature differences within the heat-storing mass. *The shape of the heat-storing mass* is assumed by Lowan to be *cylindrical* and not plate-shaped. Thus the gas flows axially to the surface area of the supposedly solid cylinder under conditions where the heat transfer coefficient α or α' is constant. Lowan first of all deals with the *first warming or cooling* of the heat-storing mass which is initially isothermal and then goes on to deal with the subsequent temperature changes which occur in *parallel-flow*, when reversals take place at constant time intervals Lowan obtained as a final result the temperature distribution in a regenerator operated in parallel flow at cyclic equilibrium.

Lowan's underlying reasoning was briefly explained in the first edition of this book. His theory is somewhat complicated. It is made more difficult by the fact that orthogonal functions are not used. His ideas can only be completely appreciated by studying his original work and therefore it will not be reproduced here.

FOURTEEN

METHODS FOR CALCULATING THE STATE OF EQUILIBRIUM BASED ON THE SOLUTION OF AN INTEGRAL EQUATION

Methods of calculation of the performance of a counterflow regenerator at equilibrium with equal thermal capacities of both gases per period ($CT = C'T'$) were presented in Secs. 13-6 and 13-7 employing eigen functions. These allowed the fundamental oscillation and harmonic vibrations to be represented. The temperature distribution at equilibrium can also be determined by means of an integral equation which both Nusselt [N304] and the author [H303] established at the same time but independently of one another.[1]

Some of the work reproduced below is concerned with the solution of this integral equation for the case where the heat-storing mass has a high thermal conductivity in a direction perpendicular to the direction of flow. Among this work are to be found the publications of Nusselt himself as well as that of Nahavandi and Weinstein, and Sandner. Included also among these is the solution, which was suggested briefly by the author, which uses the method of Gaussian integration. Schmeidler and Ackermann have added more generally applicable integration equations which also determine precisely the detailed temperature changes that occur within the brick cross-section. These authors endeavour to find solutions to these equations.

The heat pole method which is also going to be discussed, can be considered, particularly in its improved form, as an approximate solution of the integral equation for the case of a heat-storing mass with high thermal conductivity.

---

[1] The author's solution, like that of Nusselt, consisted of an infinite series of integrals but is more complex and was not published.

14-1 NUSSELT'S METHOD FOR THE STATE OF EQUILIBRIUM OF A REGENERATOR WITH A HEAT-STORING MASS OF HIGH THERMAL CONDUCTIVITY WITH $C = C'$ AND $T = T'$

The solution of the integral equation which was presented by Nusselt [N304] will be set out below in a somewhat modified form which is rather more simple. Here it is again assumed that the heat-storing mass conducts the heat with an infinitely large conductivity in a direction perpendicular to that of the flow of the gas. However the results of this theory can be extended without a significant loss of accuracy to regenerators with a heat-storing mass of lower thermal conductivity by introducing the heat transfer coefficient $\bar{\alpha}$. This follows from Eq. (13-3) and relates to the average brick temperature.

First the reversal is disregarded and it is assumed that the gas enters the regenerator at position $\xi = 0$ starting from time $\eta = 0$, with the constant temperature $\theta = \theta_1$. At time $\eta = 0$, the heat-storing mass has the arbitrary temperature distribution $\Theta = \theta_1 + f(\xi)$, represented as a function of ξ. This $f(\xi)$ denotes the initial overall temperature distribution of the heat-storing mass relative to the entrance gas temperature. Nusselt [N303] first analysed this case exactly. As will be shown in more detail in Sec. 15-3 the differential Eqs. (13-26) and (13-27) in this case have the solution (14-1) and (14-2):

$$\theta = \theta_1 + \int_0^\xi f(\varepsilon) \exp[-(\xi - \varepsilon + \eta)] \cdot J_0(2i\sqrt{(\xi - \varepsilon)\eta})\, d\varepsilon \tag{14-1}$$

$$\Theta = \theta_1 + f(\xi)\exp(-\eta) - \int_0^\xi f(\varepsilon)\exp[-(\xi - \varepsilon + \eta)]\sqrt{\frac{\eta}{\xi - \varepsilon}}$$
$$\times\, iJ_1(2i\sqrt{(\xi - \varepsilon)\eta})\, d\varepsilon \tag{14-2}$$

J_0 and J_1 are Bessel functions of the first type and of zero or first order[2] and ε represents a variable of integration. This can also be demonstrated by subsequent insertion into the differential equations. It can easily be seen that these solutions fulfil the requirement mentioned that $\theta = \theta_1$ when $\xi = 0$ and $\Theta = \theta_1 + f(\xi)$ when $\eta = 0$.[3]

In order to apply Eqs. (14-1) and (14-2) to the *state of equilibrium* of a counterflow regenerator which is reversed at constant time intervals, $\theta_1 + f(\xi)$ will be considered as the temperature distribution of the heat-storing mass at the beginning of the period under consideration ($\eta = 0$). Θ_e is the temperature of the heat-storing mass at the end of this period ($\eta = \Pi$). With an arbitrary zero point on the temperature scale, the reversal requirement (13-66) for $\Lambda = \Lambda'$ and $\Pi = \Pi'$

---

[2] For a new way of writing the Bessel functions with imaginary arguments see the footnote on p. 356 in Sec. 13-4.

[3] The equations given by Nusselt (loc. cit) which he found using the Riemann integration method contained integrals over η as well as integrals over ξ.

can also be written in the form

$$\Theta_e(\xi) = -f(\Lambda - \xi) + \theta_1'$$ (14-3)

If the expression obtained from Eq. (14-2) with $\eta = \Pi$ is inserted into Eq. (14-3) for Θ_e the following integral equation is obtained:

$$f(\xi)\exp(-\Pi) + f(\Lambda - \xi) - \int_0^\xi f(\varepsilon)\exp[-(\xi - \varepsilon + \Pi)]\sqrt{\frac{\Pi}{\xi - \varepsilon}}$$
$$\times iJ_1(2i\sqrt{(\xi - \varepsilon)\Pi})\,d\varepsilon - (\theta_1' - \theta_1) = 0$$ (14-4)

The temperature distribution in the heat-storing mass at the beginning of the period is determined by this integral equation.

In order to solve the integral equation for $f(\xi)$, the abbreviation

$$-\exp[-(\xi - \varepsilon + \Pi)]\sqrt{\frac{\Pi}{\xi - \varepsilon}} \cdot iJ_1(2i\sqrt{(\xi - \varepsilon)\Pi}) = K(\xi - \varepsilon)$$ (14-5)

is first used[4] and then by adding a factor λ, which will later be made equal to 1, the integral equation is brought into the form

$$f(\xi)\exp(-\Pi) + f(\Lambda - \xi) + \lambda \int_0^\xi f(\varepsilon)K(\xi - \varepsilon)\,d\varepsilon = \theta_1' - \theta_1$$ (14-6)

For $f(\xi)$ Nusselt formed the series

$$f(\xi) = f_0(\xi) + \lambda f_1(\xi) + \lambda^2 f_2(\xi) + \lambda^3 f_3(\xi) + \cdots$$ (14-7)

If this infinite series is inserted into (14-6) this integral equation can be satisfied by setting the sum of all the terms with equal powers of λ equal to zero. In this way Eq. (14-8) is obtained:

$$\left.\begin{array}{l}
f_0(\xi)\exp(-\Pi) + f_0(\Lambda - \xi) = \theta_1' - \theta_1 \\
f_1(\xi)\exp(-\Pi) + f_1(\Lambda - \xi) = -\Phi_0(\xi) \\
f_2(\xi)\exp(-\Pi) + f_2(\Lambda - \xi) = -\Phi_1(\xi) \\
f_n(\xi)\exp(-\Pi) + f_n(\Lambda - \xi) = -\Phi_{n-1}(\xi)
\end{array}\right\}$$ (14-8)

in which the further abbreviations

$$\left.\begin{array}{l}
\displaystyle\int_0^\xi f_0(\varepsilon)K(\xi - \varepsilon)\,d\varepsilon = \Phi_0(\xi) \\[2mm]
\displaystyle\int_0^\xi f_1(\varepsilon)K(\xi - \varepsilon)\,d\varepsilon = \Phi_1(\xi) \\[2mm]
\displaystyle\int_0^\xi f_{n-1}(\varepsilon)K(\xi - \varepsilon)\,d\varepsilon = \Phi_{n-1}(\xi)
\end{array}\right\}$$ (14-9)

---

[4] The equation of approximation (15-6) in Sec. 15-2 is recommended for the practical calculation of $K(\xi - \varepsilon)$.

are introduced. Further, if $\Lambda - \varepsilon$ is substituted into Eq. (14-8) for ξ, this equation is modified upon multiplication by $\exp(\Pi)$, into the form of Eq. (14-10):

$$
\left.
\begin{aligned}
f_0(\Lambda - \xi) + f_0(\xi)\exp(\Pi) &= (\theta_1' - \theta_1)\exp(\Pi) \\
f_1(\Lambda - \xi) + f_1(\xi)\exp(\Pi) &= -\Phi_0(\Lambda - \xi)\exp(\Pi) \\
f_2(\Lambda - \xi) + f_2(\xi)\exp(\Pi) &= -\Phi_1(\Lambda - \xi)\exp(\Pi) \\
f_n(\Lambda - \xi) + f_n(\xi)\exp(\Pi) &= -\Phi_{n-1}(\Lambda - \xi)\exp(\Pi)
\end{aligned}
\right\}
\tag{14-10}
$$

Finally $f_0(\Lambda - \xi)$, $f_1(\Lambda - \xi)$, etc., can easily be eliminated from Eqs. (14-8) and (14-10), and (14-11) obtained:

$$
\left.
\begin{aligned}
f_0(\xi) &= \frac{1 - \exp(-\Pi)}{1 - \exp(-2\Pi)}(\theta_1' - \theta_1) \\[2mm]
f_1(\xi) &= \frac{\exp(-\Pi)\Phi_0(\xi) - \Phi_0(\Lambda - \xi)}{1 - \exp(-2\Pi)} \\[2mm]
f_2(\xi) &= \frac{\exp(-\Pi)\Phi_1(\xi) - \Phi_1(\Lambda - \xi)}{1 - \exp(-2\Pi)} \\[2mm]
f_n(\xi) &= \frac{\exp(-\Pi)\Phi_{n-1}(\xi) - \Phi_{n-1}(\Lambda - \xi)}{1 - \exp(-2\Pi)}
\end{aligned}
\right\}
\tag{14-11}
$$

It follows from these equations that $f_0(\xi), f_1(\xi), \ldots$ can be evaluated successively. $\Phi_0(\xi)$ and thus $\Phi_0(\Lambda - \xi)$ can be found using Eq. (14-9) from $f_0(\xi)$ which itself can be found using Eq. (14-11). Similarly $f_1(\xi)$ can be determined from $\Phi_0(\xi)$ and $\Phi_0(\Lambda - \xi)$ using (14-11); from $f_1(\xi)$, $\Phi_1(\xi)$ and thus $\Phi_1(\Lambda - \xi)$ can be determined using Eq. (14-9). In this way, using Eq. (14-7), the solution to the integral equation can be found. If λ is now set equal to 1 Eq. (14-7) becomes

$$
f(\xi) = f_0(\xi) + f_1(\xi) + f_2(\xi) + f_3(\xi) + \cdots
\tag{14-12}
$$

This solution is remarkable from a theoretical point of view, but unfortunately has the disadvantage from the point of view of practical calculations that, apart from $f_0(\xi)$, each term of this series, as shown in Eqs. (14-9) and (14-11), contains integrals which, in general, cannot be determined analytically. Each term can only be evaluated approximately using numerical quadrature for example, employing Simpson's rule or the Gaussian method of numerical integration. Indeed, the heat pole method mentioned in Secs. 15-1 and 15-2 is suitable for the evaluation of these terms; this method can be regarded mathematically as an approximate method for the calculation of integrals of the form $\int_0^\xi f(\varepsilon)K(\xi - \varepsilon)\,d\varepsilon$. It is usually considerably simpler to use the heat pole method to calculate the state of equilibrium directly, as will be described later.

As soon as the function $f(\xi)$ has been determined for the state of equilibrium using Nusselt's method, it is possible to calculate also the efficiency of the regenerator and thus, using Eq. (13-105), k/k_0. The means of doing this will be considered at the end of the discussion of approximate methods (see Sec. 15-3).

14-2 THE METHODS OF NAHAVANDI AND WEINSTEIN FOR SOLVING THE INTEGRAL EQUATIONS

Nahavandi and Weinstein [N301] have proposed a different procedure to solve the integral Eqs. (14-4) and (14-6) which relate to the steady state with $CT = C'T'$. However their method can also be applied to the more general case where $CT \neq C'T'$.* Their approach can be understood, without loss of generality, in terms of the solution to the integral Eqs. (14-4) and (14-6). They set up the following expression (14-13) for the required excess temperature $\Theta - \theta_1 = f(\xi)$ for the start of the cold period:

$$f(\xi) = \sum_{n=0}^{N} a_n \cdot \xi^n \quad \text{and correspondingly} \quad f(\varepsilon) = \sum_{n=0}^{N} a_n \varepsilon^n \quad (14\text{-}13)$$

The authors suggest the following method in order to determine the $N+1$ coefficients a_n of this polynomial. The polynomial (14-13) for $f(\xi)$ and $f(\varepsilon)$ is inserted into the integral equation. In this way an equation is produced, which with the exception of the unknown coefficients a_n contains only integrals, which can be evaluated for different values of ξ from the outset. If $N+1$ values of ξ between $\xi = 0$ and $\xi = \Lambda$ are now chosen arbitrarily and the necessary calculation is carried out for each of these values of ξ, $N+1$ linear equations in the coefficients a_n are obtained.

However, Nahavandi and Weinstein do not say how they calculate the integrals contained in the integral equation for the various values of ξ. The integral (14-14) is first evaluated for specific values of n with the help of methods of the approximation

$$\int_0^\xi \varepsilon^n \cdot K(\xi - \varepsilon) \, d\varepsilon \quad (14\text{-}14)$$

The value of

$$\int_0^\xi f(\varepsilon) \cdot K(\xi - \varepsilon) \, d\varepsilon = \sum_0^N a_n \int_0^\xi \varepsilon^n \cdot K(\xi - \varepsilon) \, d\varepsilon$$

is easily determined by multiplying the values of the integrals (14-14) by a_n and adding them together. This process must be carried out separately for each value of ξ; thus in all $N(N+1)$ integrals of the type (14-14) must be evaluated. Account is taken of the fact that the value $\xi = 0$ is among the chosen values of ξ. For this special case all $N+1$ integrals of the type (14-14) disappear.

In the more general case in particular where $CT \neq C'T'$, the temperature distribution at the start of the cold period is again represented by Eq. (14-13) which is then inserted into Eq. (14-2) applied to this cold period. At the start of the hot

---

* Editor's note: At the request of the author, the text has been corrected at this point. In the German text, reference is made only to the symmetric case whereas the method of Nahavandi and Weinstein can also be applied to the more general case $CT \neq C'T'$ as can be seen from their original paper.

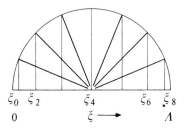

ξ_0 ξ_2 ξ_4 ξ_6 ξ_8

0          $\xi \longrightarrow$ Λ **Figure 14-1** Chebychew distribution of data points.

period however the temperature profile is represented by the expressions

$$\sum_{n=0}^{N} a_n^* (\xi')^n \qquad \text{and} \qquad \sum_{n=0}^{N} a_n^* (\varepsilon')^n$$

This is inserted into Eq. (14-2) for the hot period. The problem becomes then one of evaluating the $N+1$ coefficients a_n and the $N+1$ coefficients a_n^*. By application of the reversal conditions, the necessary $2N+2$ simultaneous linear equations for the $2N+2$ unknown coefficients are obtained.*

An examination of the practicability of the Nahavandi and Weinstein method has been carried out by Willmott and Duggan [W308, 309] in an area up to $\Lambda = 10$ and $\Pi = 10$ for $CT = C'T'$. They suggest that the number of power terms can be reduced without loss of accuracy if the ξ positions are not uniformly distributed but are more numerous around the ends of the regenerator than they are in the central parts of the regenerator. They recommend the Chebyshev distribution which is specified by the equation

$$\xi_i = \frac{\Lambda}{2}\left[1 - \cos\frac{i\pi}{N}\right] \qquad \text{for } i = 0, 1, 2, 3, \text{ etc. up to } N$$

Figure 14-1 shows this distribution for $N = 8$.

A difficulty which, according to Willmott and Duggan, can occur in the method of Nahavandi and Weinstein will be explained in Sec. 15-3. This difficulty is even more apparent in the heat pole method which will be discussed later.

It may be also mentioned at this juncture that Nahavandi and Weinstein have developed the integral equation employing Laplace transforms.

14-3 THE METHOD OF H. HAUSEN FOR THE SOLUTION OF THE DIFFERENTIAL EQUATION
(Not yet published)

In the solution of Nahavandi and Weinstein, which has just been discussed, the number of constants which have to be determined becomes larger as Λ increases.

---

* Editor's note: At the request of the author, the text has been corrected at this point. In the German text, reference is made only to the symmetric case whereas the method of Nahavandi and Weinstein can be applied to the more general case $CT \neq C'T'$ as can be seen from their original paper.

The rapid increase in computing time which is caused by this can be reduced considerably by choosing an expression for the curve $f(\xi)$ for the temperature distribution at the beginning of the cold period at equilibrium. From the outset a choice is made such that $f(\xi)$ is likely to be more suitable for actual regenerator conditions than is a polynomial. As can be seen from Fig. 11-10, bottom, the deviations at the beginning of the period represented in this figure, from the straight-line of the zero eigen function are well reproduced visually by hyperbolic functions. The author's suggestion consists therefore of forming Eq. (14-15) for $CT = C'T'$:

$$f(\xi) = a + \frac{b}{c+\xi} + d[(\xi - e) - \sqrt{f + (\xi - e)^2}] \qquad (14\text{-}15)$$

where a, b, c, d, e, and f are constants. $\dfrac{b}{c+\xi}$ represents, as a hyperbolic function, the deviation from the zero eigen function in the region of the gas entrance and the other expression represents the zero eigen function and the deviations from this in the region of the gas exit. This second expression replicates a hyperbole which on one side has the zero eigen function and on the other side has a horizontal asymptote at height a. This horizontal asymptote corresponds to the isotherm $\theta = \theta'_1$ that is, a is almost the same as $\theta'_1 - \theta_1$. e is the value of ξ at which the asymptotes mentioned separate. Hence, as ξ decreases, the second expression generally takes on the form of the zero eigen function; it thus includes the zero eigen function.

The advantage of this expression lies in the fact that it is independent of the size of Λ and has only six constants. These can be determined in principle by inserting Eq. (14-15), for six different values of ξ, into the integral equation. However this is made difficult by the fact that the six resulting equations are not linear. One might proceed by eliminating approximate values of c, e and f, using similar cases which have already been worked out. The resultant equations mentioned can then be linearized by separately partially differentiating them with respect to these three constants. This is to be preferred since in addition to a, b and d the resultant equations then only contain the variations Δc, Δe and Δf. If the improved values of c, e and f still do not fulfil the original equations with sufficient accuracy the type of calculation indicated must be carried out one or more times again.*†

This method has still to be worked out in detail. Because of the integrals

---

* Editor's note: Given six distinct values of ξ on the interval $0 < \xi < \Lambda$, one might proceed to find all the values of a, b, c, d, e and f by solving all the resultant equations by the well known method of Newton Raphson. On the other hand, by using more values ξ, these values of a, b, c, d, e and f can be found using the method of least squares as proposed by Dr. Schellmann and Dr. Müller.

† Author's note: Thankfully Dr. Schellmann (see [S309]) and Dr. Müller in Leverkusen have carried out calculations using the above expression (14-15) for the region $\Lambda = 5$ to 40 and $\Pi = 5$ to 20. Schellmann has written to the author, informing him of the result that the state of equilibrium could be accurately reproduced using this method but that it was not easy to determine the constants in Eq. (14-15), for which they used the method of least squares.

involved this may require much further consideration. Only then will it be possible to state whether or not it has any advantages over other methods.

14-4 THE METHOD OF CALCULATION OF SANDNER

In a similar way to Nahavandi and Weinstein, H. Sandner [S301] developed equations using the Laplace transform and these are, essentially, the same as Eqs. (14-1) and (14-2). However Sandner transformed these equations again using the Laplace transformation. He chose equidistant positions $\xi = 0$, $\xi = \Delta\varepsilon$, $\xi = 2\Delta\varepsilon$, etc., over the total length of the regenerator at which positions the initial overall temperature of the heat-storing mass relative to the entry temperature of the gas is equal to $f(0)$, $f(\Delta\varepsilon)$, $f(2\Delta\varepsilon)$, etc. He obtained Eqs. (14-16) and (14-17) for the temperature Θ of the heat-storing mass and θ of the gas at position ξ at the later time η;

$$\Theta(\xi,\eta) = \theta_1' - R_0(\xi,\eta)\left[(\theta_1'-\theta_1)-f(0)\right] + \frac{R_1(\xi,\eta)}{\Delta\varepsilon}(f(\Delta\varepsilon)-f(0))$$

$$+ \frac{1}{\Delta\varepsilon}\sum_{n=1}^{N-1} R_1(\xi-n\,\Delta\varepsilon,\eta)\left[f((n+1)\,\Delta\varepsilon)-2f(n\,\Delta\varepsilon)+f((n-1)\,\Delta\varepsilon)\right] \qquad (14\text{-}16)$$

$$\theta(\xi,\eta) = \theta_1' - R_0^*(\xi,\eta)\left[(\theta_1'-\theta_1)-f(0)\right] + R_1^*(\xi,\eta)\,(f(\Delta\varepsilon)-f(0))$$

$$+ \frac{1}{\Delta\varepsilon}\sum_{n=1}^{N-1} R_1^*(\xi-n\,\Delta\varepsilon,\eta)\left[f((n+1)\,\Delta\varepsilon)-2f(n\,\Delta\varepsilon)+f((n-1)\,\Delta\varepsilon)\right] \qquad (14\text{-}17)$$

where R_0, R_1, R_0^* and R_1^* take the following meaning:

$$R_0(\xi,\eta) = \exp\left[-(\xi+\eta)\right]J_0(2i\sqrt{\xi\eta}) + \exp(-\eta)\int_0^\xi \exp(-\xi')J_0(2i\sqrt{\xi'\eta})\,d\xi'$$

$$\qquad (14\text{-}18)$$

$$= 1 - \exp(-\xi)\int_0^\eta \exp(-\eta')J_0(2i\sqrt{\xi\eta'})\,d\eta'^5 \qquad (14\text{-}19)$$

and is the temperature of the heat-storing mass $\Theta(\xi,\eta)$ at position ξ at time η based on the assumption that the heat-storing mass has the constant temperature $\Theta(\xi,0) = \theta_1 + f(\xi) = \theta_1 + 1$ at time $\eta = 0$. Thus R_0 represents one of the curves in Fig. 13-6; $R_1(\xi,\eta) = \int_0^\xi R_0(\xi',\eta)\,d\xi'$ is the integral of R_0 and thus the area under the curve shown in Fig. 13-6; $R_0^*(\xi,\eta)$ and $R_1^*(\xi,\eta)$ are the corresponding expressions for the gas temperature θ; thus R_0^* represents one of the curves in Fig. 13-5 and R_1^* represents the area lying under this curve.

In order to determine the temperature distribution at time η Sandner has endeavoured to evaluate the functions R_0, R_1, R_0^* and R_1^* which are defined by

---

[5] In line with a modern proposal, Sandner writes $I_0(x)$ instead of $J_0(ix)$ and correspondingly $I_1(x)$ instead of $-i\cdot J_1(ix)$; see the footnote on page 356 in Sec. 13-4.

integrals, by using approximate equations which are as accurate as possible over a wide region. These approximate equations are asymptotic expansions whose accuracy improves with increasing ξ and η but which are also very precise, as a rule, even for the smallest values of ξ or η which occur.

First Sandner introduces the following auxiliary functions T_1, T_2, T_3, T_4:

$$T_1 = 1 - \frac{1}{2}[\text{erf}(\sqrt{\eta} - \sqrt{\xi}) + \text{erf}(\sqrt{\eta} + \sqrt{\xi})] \tag{14-20}$$

$$T_2 = \exp[-(\xi + \eta)]J_0(2i\sqrt{\xi\eta}) \tag{14-21}$$

$$T_3 = \frac{1}{\sqrt{\pi\eta}}\exp[-(\xi + \eta)]\cosh(2\sqrt{\xi\eta}) \tag{14-22}$$

$$T_4 = \frac{1}{\sqrt{\pi\xi}}\exp[-(\xi + \eta)]\sinh(2\sqrt{\xi\eta}) \tag{14-23}$$

where erf denotes the error functions and $\text{erf}(x) = \dfrac{2}{\sqrt{\pi}}\displaystyle\int_0^x \exp(-t^2)\,dt$. Using these functions Sandner was able to obtain very good approximating equations, namely:

$$R_0(\xi, \eta) = T_1 + \frac{1}{2}T_2 - \frac{1}{8}T_3 + \frac{1}{8}T_4 \tag{14-24}$$

$$R_0^*(\xi, \eta) = T_1 - \frac{1}{2}T_2 - \frac{1}{8}T_3 + \frac{1}{8}T_4 \tag{14-25}$$

$$R_1(\xi, \eta) = (\xi - \eta)T_1 - \frac{1}{4}T_2 + \left(\eta + \frac{1}{16}\right)T_3 + \left(\xi + \frac{1}{16}\right)T_4 \tag{14-26}$$

$$R_1^*(\xi, \eta) = (\xi - \eta - 1)T_1 + \frac{1}{4}T_2 + \left(\eta + \frac{3}{16}\right)T_3 + \left(\xi - \frac{1}{16}\right)T_4 \tag{14-27}$$

With the aid of these equations it is possible to calculate the temperature distribution at a later time η if an arbitrary temperature distribution $f(\xi)$ at time $\eta = 0$ is given for the positions $\xi = 0$, $\Delta\varepsilon$, $2\Delta\varepsilon$, etc., up to $N\Delta\varepsilon$. However, in order to improve the accuracy, Sandner suggested the performing of the calculation in this manner, for the positions $\xi = 0$, $2\Delta\varepsilon$, $4\Delta\varepsilon$, etc., only. For positions lying between these $\xi = \Delta\varepsilon$, $3\Delta\varepsilon$, etc., on the other hand, he modified the process so that the quantity of heat stored in the regenerator could be computed as accurately as possible between two neighbouring positions $\xi = n\Delta\varepsilon$ and $(n+2)\Delta\varepsilon$ for which the calculations had already been carried out. For this he formed the integral

$$\int_{n\Delta\varepsilon}^{(n+2)\Delta\varepsilon} \Theta\,d\xi'$$

and, using Simpson's rule, set this equal to

$$\frac{\Delta\varepsilon}{2}(\Theta_n + \Theta_{n+2} + 2\Theta_{n+1})$$

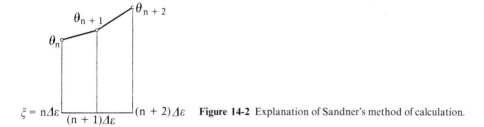

Figure 14-2 Explanation of Sandner's method of calculation.

that is, equal to the area underneath the two straight lines in Fig. 14-2.* The temperature of the heat-storing mass at an intermediate position $\xi = (n+1)\Delta\varepsilon$ then becomes

$$\Theta_{n+1} = \frac{1}{\Delta\varepsilon} \int_{n\Delta\varepsilon}^{(n+2)\Delta\varepsilon} \Theta\, d\xi' - \frac{1}{2}(\Theta_n + \Theta_{n+2})$$

$$= \frac{1}{\Delta\varepsilon}\left[\int_0^{(n+2)\Delta\varepsilon} \Theta\, d\xi' - \int_0^{n\Delta\varepsilon} \Theta\, d\xi' \right] - \frac{1}{2}(\Theta_n + \Theta_{n+2}) \qquad (14\text{-}28)$$

Sandner developed the relationship (14-29) for the integral which occurs in (14-28)

$$\int_0^\xi \Theta\, d\xi'$$

with $\xi = (n+2)\Delta\varepsilon$ or $\xi = n\Delta\varepsilon$

$$\int_0^\xi \Theta\, d\xi' = \int_0^\xi \Theta(\xi',\eta)\, d\xi' = \xi(\theta'_1 - \theta_1) - R_1(\xi,\eta)[(\theta'_1 - \theta_1) - f(0)]$$

$$+ \frac{1}{\Delta\varepsilon} R_2(\xi,\eta)[f(\Delta\varepsilon) - f(0)]$$

$$+ \frac{1}{\Delta\varepsilon} \sum_{n=1}^{N-1} R_2(\xi - n\Delta\varepsilon)\{f[(n+1)\Delta\varepsilon] - 2f(n\Delta\varepsilon) + f[(n-1)\Delta\varepsilon]\}$$

$$(14\text{-}29)$$

where the following is an approximation of $R_2(\xi,\eta) = \int_0^\xi R_1(\xi',\eta)\, d\xi'$

$$R_2(\xi,\eta) = \left[\frac{\xi^2}{2} + \frac{\eta^2}{2} - \xi\eta + \eta\right]T_1 + \frac{1}{8}T_2$$

$$+ \frac{1}{2}\left[\xi\eta - \eta^2 - \frac{3}{2}\eta - \frac{1}{16}\right]T_3$$

$$+ \frac{1}{2}\left[\xi^2 - \xi\eta - \frac{\xi}{2} - \frac{1}{16}\right]T_4. \qquad (14\text{-}30)$$

* Editor's note: Actually, the area beneath the parabola passing through Θ_n, Θ_{n+1} and Θ_{n+2}, see Fig. 14-2.

The calculation of the temperature distribution in the heat-storing mass at time η proceeds by determining the temperatures $\Theta_0, \Theta_2, \Theta_4, \ldots, \Theta_n, \Theta_{n+2}$, etc., using Eq. (14-16) and then determining the temperatures which lie between these by using Eqs. (14-28) to (14-30). These are $\Theta_1, \Theta_3, \ldots, \Theta_{n+1}$.

The state of equilibrium of the regenerator is obtained by first carrying out the calculation for the end of the period $\eta = \Pi$ using the values of $f(0), f(\Delta \varepsilon), f(2\Delta \varepsilon)$ etc., yet to be determined, and then applying the reversal requirement (14-3) at the selected positions of ξ, that is at $\xi = 0$, $\xi = \Delta \varepsilon$, $\xi = 2\Delta \varepsilon$ etc., up to $\xi = \Lambda$. Thus $N + 1$ equations are obtained from which $f(0), f(\Delta \varepsilon), f(2\Delta \varepsilon)$ etc., up to $f(\Lambda)$ can be determined. The solution of these equations is used in the determination of the temperature distribution $f(\xi)$ at the beginning of the period for the state of equilibrium. In the more general case where $\Lambda \neq \Lambda'$ and $\Pi \neq \Pi'$ the state of equilibrium is found by performing the calculation over a larger number of successive periods.

This method by Sandner is considerably more complex than the approximate procedures which will be described later in this book, for example the heat pole method (see Sec. 15-1). However, as Sandner himself points out, the "asymptotic" expansions for T_1 to T_4, and thus for R_0 to R_2 and R_0^* to R_2^*, fit so closely to the actual curve that it is only necessary to carry out calculations for a few basic positions namely at $\xi = 0$, $\xi = \Delta \varepsilon$, $\xi = 2\Delta \varepsilon$, etc. Thus 16 such positions are sufficient up to $\Lambda = 200$ and 32 positions are sufficient up to $\Lambda = 1000$ in order to be able to calculate accurately the efficiency of the regenerator to one part in a thousand.

14-5 CALCULATION OF REGENERATORS USING THE GAUSSIAN INTEGRATION METHOD

It will be shown in what follows that the method of Gaussian quadrature offers a possible means of calculating the temperature Θ of the heat-storing mass using Eqs. (14-2) and, more particularly, enables the state of equilibrium to be determined using the integral Eq. (14-4) or (14-6). Moreover, as will be seen, an additional complication will not be neglected. It is thus to be expected that this method should yield more accurate results [H312].

The Method of Gaussian Quadrature

A function $y(\varepsilon)$ can be integrated over a prescribed region $\Delta \varepsilon$ of the independent variable ε. According to Gauss, knowledge of the value of this function at several prescribed positions is sufficient to enable the value of the integral to be computed with great accuracy. In the following way only two such values of the function y_a and y_b will be used. The positions a and b, at which these values of the function are determined within the region $\Delta \varepsilon$, will be at a distance $0.2113 \Delta \varepsilon$ from the bounds of the interval (see Fig. 14-3). The value of the integral is then found from the simple

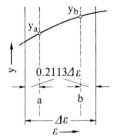

Figure 14-3 Explanation of the method of Gaussian quadrature.

relationship

$$\int_{\varepsilon}^{\varepsilon+\Delta\varepsilon} y(\varepsilon)\,d\varepsilon = \frac{\Delta\varepsilon}{2}(y_a+y_b) \qquad (14\text{-}31)$$

This relationship is exact if the function is parabolic in the region of $\Delta\varepsilon$. It yields a good approximation when the deviations from a parabolic curve are only slight.

Evaluation of the Integrals in Eqs. (14-2), (14-4) or (14-6) Using the Gaussian Method

The initial overall temperature $f(\xi)$ of the heat-storing mass, relative to the entry temperature θ_1 of the colder gas, will be represented by $f(\varepsilon)$ in the integration. In order to evaluate the integrals in Eqs. (14-2), (14-4) or (14-6) the function $f(\varepsilon)\cdot K(\xi-\varepsilon)$ must be integrated and thus $K(\xi-\varepsilon)$ determined for the state of equilibrium by utilizing Eq. (14-5). If, on the other hand, as in Eq. (14-2), the temperature of the heat-storing mass at an arbitrary time η is required, Π should be replaced by η in Eq. (14-5).

In order to use the method of Gaussian quadrature one should imagine a diagram in which $f(\xi)$ is drawn as a function of ε over the total length of the regenerator Λ, and then divided into N sections of equal width $\Delta\varepsilon$ (Fig. 14-4). The Gaussian method is applied separately to each of these sections. Even here the prescribed positions a and b are determined as in Fig. 14-3, in such a way that they are at a distance $0.2113\,\Delta\varepsilon$ from the limits of the section under consideration.

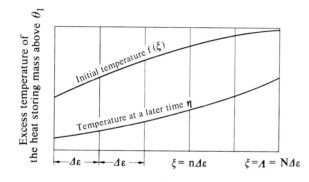

Figure 14-4 Division of the regenerator into N sections of equal width.

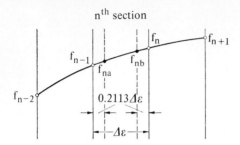

Figure 14-5 Application of the method of Gaussian quadrature to the nth section.

In the nth section the initial temperature at these positions is denoted by f_{na} and f_{nb} and the corresponding values of the function $K(\varepsilon)$ are denoted by K_{na} and K_{nb} (Fig. 14-5). The same approach is applied in a corresponding manner to the other sections. If, as in Eqs. (14-2), (14-4) or (14-6), integration is performed over the interval from $\xi = 0$ to $\xi = n\Delta\varepsilon$, that is over n sections, the corresponding application of Eq. (14-31) yields

$$\int_0^{\xi = n\Delta\varepsilon} f(\varepsilon)K(\xi-\varepsilon)\,d\varepsilon = \frac{\Delta\varepsilon}{2}[(f_{1a}K_{nb}+f_{1b}K_{na})+(f_{2a}K_{n-1,b}+f_{2b}K_{n-1,a})$$

$$+\cdots+(f_{n-1,a}K_{2b}+f_{n-1,b}K_{2a})+(f_{n,a}K_{1b}+f_{n,b}K_{1a})] \qquad (14\text{-}32)$$

Such an equation is valid for each value of the upper limit of the interval of integration namely $\xi = n\,\Delta\varepsilon$ with $n = 1, 2, 3$, etc.; the value of the integral disappears when $n = 0$. Thus, with a given initial temperature distribution, and thus with known values of $f_{1a}, f_{1b}, f_{2a}, f_{2b}, f_{3a}$, etc., the final temperatures $\Theta(\varepsilon)$ at a later time η at the position $\xi = n\,\Delta\varepsilon$ can be calculated using Eq. (14-2).

The determination of the conditions when the state of equilibrium prevails, requires additional considerations. In principle there is here a difficulty, namely that it follows from Eq. (14-32) that only the temperatures at the particular positions within each section appear in the integral, whereas the temperatures $f(\xi)$ and $f(\Lambda-\xi)$ at the limits of the intervals of integration appear in the first two expressions in Eq. (14-6). The total number of these unknown temperatures for the state of equilibrium is about three times larger than the number of defining equations, such as can be formed in the manner represented by Eq. (14-6) together with Eq. (14-32). This difficulty can only be eliminated by expressing the temperatures f_{1a}, f_{1b}, f_{2a}, etc., which lie within the intervals in terms of the temperatures f_0, f_1, f_2, etc., up to f_N at the boundaries of the intervals. This can be done very accurately as will be shown, by utilizing an interpolation equation of the third order.

In the nth section (Fig. 14-5) the temperatures f_{na} and f_{nb} are determined from the temperatures f_{n-2}, f_{n-1}, f_n and f_{n+1} at the limits of the intervals of integration. The constants of an equation of the third order of the type

$$f\left(\frac{\varepsilon}{\Delta\varepsilon}\right) = a+b\frac{\varepsilon}{\Delta\varepsilon}+c\left(\frac{\varepsilon}{\Delta\varepsilon}\right)^2+d\left(\frac{\varepsilon}{\Delta\varepsilon}\right)^3$$

can now be determined so that the temperatures f_{n-2}, f_{n-1}, f_n and f_{n+1} are exactly reproduced (see Fig. 14-5). f_{na} and f_{nb} can then be also determined from this equation. Here it is expedient to position the zero point in the centre of the nth section. In this way is obtained

$$f_{na} = -0.0497f_{n-2} + 0.8544f_{n-1} + 0.2289f_n - 0.0336f_{n+1} \qquad (14\text{-}33)$$

$$f_{nb} = -0.0336f_{n-2} + 0.2289f_{n-1} + 0.8544f_n - 0.0497f_{n+1} \qquad (14\text{-}34)$$

For the first and last sections, which lie at the ends of the regenerator, the process must be modified in the following way. In order to determine f_{1a} and f_{1b} the third order curve is laid through the four temperature points f_0, f_1, f_2 and f_3. Thus, for the first section, is obtained

$$f_{1a} = 0.6557f_0 + 0.5270f_1 - 0.2324f_2 + 0.0497f_3 \qquad (14\text{-}35)$$

$$f_{1b} = 0.0943f_0 + 1.0562f_1 - 0.1842f_2 + 0.0337f_3 \qquad (14\text{-}36)$$

Correspondingly, for the last of the N section is yielded:

$$f_{Na} = 0.0337f_{N-3} - 0.1842f_{N-2} + 1.0562f_{N-1} + 0.0943f_N \qquad (14\text{-}37)$$

$$f_{Nb} = 0.0497f_{N-3} - 0.2324f_{N-2} + 0.5270f_{N-1} + 0.6557f_N \qquad (14\text{-}38)$$

If all these Eqs. (14-33) to (14-38) are inserted into Eq. (14-32) for the integral, $N + 1$ equations are finally obtained from Eq. (14-6) for the unknown temperatures f_0, f_1, f_2 etc. up to f_N at the state of equilibrium.

The relationships which finally result are very complex. However, with the high efficiency of the method of Gaussian quadrature it is to be expected that a considerable degree of accuracy will be achieved with relatively few intervals $\Delta\varepsilon$.

14-6 GENERAL INTEGRAL RELATIONSHIPS DEVELOPED BY SCHMEIDLER AND ACKERMANN. THE THEORY OF LARSEN

In the integral equations which have been dealt with so far, the spatial temperature differences in the cross-sections of the regenerator bricks, as are shown in Figs. 11-6 and 12-4, are only considered summarily through introduction of the heat transfer coefficient $\bar{\alpha}$. This is achieved using Eq. (13-3), which is related to the average brick temperature $\Theta = \Theta_M$. It was shown on p. 349 that in practically all cases of regenerators working in counterflow, a considerable degree of accuracy is always achieved in this way. Nevertheless, when dealing with basic theory and the examination of the results thereof it may be significant that Schmeidler [S311] and Ackermann [A301] were able to precisely express all details of the spatial temperature differences at positions including the ends of the regenerator in extended integral equation form.

The fundamental basics of Schmeidler and Ackermann's mathematical derivations and the form of the integral equation obtained by them was discussed in the first edition of this book. In order to preserve the conciseness required of this second edition, only the considerations under which these developments are applicable will be mentioned here and the type of solutions found will be discussed in brief. This can hardly be a disadvantage as it is not possible to penetrate the very complex interrelationships without studying the original literature.

Integral Equations of Schmeidler for the State of Equilibrium

Schmeidler [S311] has derived two integral equations for the state of equilibrium where $C = C'$, $T = T'$ and $\alpha = \alpha'$. His considerations are valid for an arbitrary cross-section of the heat-storing mass which has the same shape at all positions in the regenerator. Thus his calculations involve two space coordinates within a brick cross-section. The integral equations developed by him are given in such a form that the temperatures of the two gases at the state of equilibrium appear as unknowns. The variation of these temperatures, with space and time, can be determined basically by solving the integral equations. However, Schmeidler only gives an approximate solution for the chronological average values of the gas temperatures and certain proportionality factors must be estimated.

Method of Ackermann

Ackermann's method of calculation [A301] is as general as the integral equations method of Schmeidler, if one ignores the not very important difference that Ackermann treats the heat-storing mass as though it were constructed of smooth plates of equal thickness. Thus Ackermann's procedure computes all the detailed temperature differences in the heat-storing mass. However, an assumption is made at the beginning of each period relating to the temperature distribution so that, as a rule, the state of equilibrium can only be determined by performing the calculations over a larger number of successive periods. His equation contains one integral—which enables the distribution of the temperature Θ of the heat-storing mass to be calculated—and can be considered to be a generalization of Eq. (14-2). Ackermann's solution, which is in the form of a series, is absolutely exact and can thus be evaluated numerically for practical cases with a reasonable degree of accuracy, although it is somewhat laborious. Ackermann has dealt with the example case where $\Lambda = 20$, $\Pi = 4$, $\Lambda' = 10$, $\Pi' = 2$ for the state of equilibrium. In so doing he was able, on the whole, to truncate the series after the fifth term.

Above all Ackermann's method is of significance in its treatment of the state of equilibrium. By employing this method it is possible to test the accuracy which can be achieved with the procedures described in Secs. 12-5 to 13-7 and with the methods of approximation which will be discussed in Sec. 15-1. If necessary, the degree of accuracy can be increased. Before using Ackermann's method it is always useful to first determine the temperature distribution in the heat-storing mass at the beginning or the end of a period using one of the simpler methods mentioned. If one then carries out further calculations using Ackermann's method, starting from this as an initial temperature distribution, a high degree of accuracy, as a rule, will be achieved. Only a few periods need be calculated and a decision can be made as to the errors, probably only very few, in the procedure which has been used.

The Theory of Lowan

In Sec. 13-10 it was shown that Lowan also developed a theory of similar general applicability but this relates only to regenerators operating in parallel flow.

The Theory of Larsen

The theory of Larsen [L302] is noteworthy; this shows not only how the temperature distribution in regenerators can be calculated from an arbitrary initial temperature distribution within the heat-storing mass but can also be calculated in the case where there is an arbitrary variation in the entry temperature of the gas.

CHAPTER
FIFTEEN

CALCULATION OF THE TEMPERATURE DISTRIBUTION IN THE HEAT-STORING MASS OF A REGENERATOR USING THE HEAT POLE METHOD

15-1 SIMPLE HEAT POLE METHOD

At the beginning of Chapter 14 it was shown that the heat pole method can also be regarded as an approximate solution of the integral Eq. (14-4) or (14-6) for the state of equilibrium in a regenerator. This is particularly clear in the case of the refined heat pole methods.

In many practical cases, for which regenerator calculations are being planned, the step-by-step methods dealt with later in Chapter 16 will be preferable. However the heat pole methods also have advantages of their own, as will be seen from the following considerations.

Starting from a previously specified temperature distribution in the longitudinal direction of the regenerator, only the temperature distribution is sought as it occurs at a considerably later time. With the step-by-step method this final temperature curve can only be determined by first calculating the temperature at time steps $\Delta\eta$ over a large number of intermediate positions in time. This type of calculation is often desirable because in this way the total dependence of the temperature upon ξ and η over a wide area is obtained.

On the other hand, by using the heat pole method the temperature distribution can also be obtained directly at an arbitrarily later time, that is without having to calculate intermediate values. However, above all, equilibrium can be determined straight away, without having to carry out calculations, using a step-by-step process, over a large number of successive periods.

In the heat pole method, as in most of the previous considerations, a heat-storing mass must be assumed which has a high thermal conductivity perpendicular

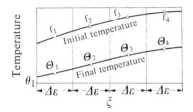

Figure 15-1 Division of the regenerator into sections of equal width $\Delta\varepsilon$ with average temperatures $f_1, f_2 \ldots$ and $\Theta_1, \Theta_2 \ldots$ etc.

to the direction of flow. However, as previously, it is possible subsequently, through the introduction of a heat transmission coefficient $\tilde{\alpha}$ which is related to the average brick temperature Θ_m, to apply the results of such computations to regenerators with a heat-storing mass of low thermal conductivity, and in so doing provide a very good approximation.

The *simple heat pole method* is based on the following considerations: since the differential Eq. (13-30) is linear, its particular solutions can be added together arbitrarily. Thus the final temperature distribution can be found by separating the initial temperature into several parts. The final temperature is determined individually for each of these parts and subsequently the final results are added together. The separation is carried out in such a way that the regenerator length Λ can be thought of as being divided into an arbitrary number N of equal parts and represented in a diagram in which the temperature of the heat-storing mass is drawn as a function of ξ (Fig. 15-1). Thus, in this diagram, there are N sections of width $\Delta\varepsilon$.

The cold period will be considered. Here it is appropriate to carry out calculations using the "excess" temperature relative to the temperature θ_1 of the entry gas. The average values of the initial, excess temperatures in the first, second, third sections, etc., are considered to be $f_1, f_2, f_3 \ldots f_N$.

An individual section of unit height will be termed a heat pole; this section thus corresponds to the initial excess temperature $\Theta - \theta_1 = 1$ and has a width $\Delta\varepsilon$ between the positions $\xi = \varepsilon$ and $\xi = \varepsilon + \Delta\varepsilon$ (Fig. 15-2). By a heat pole is to be understood the accumulation of heat at a narrow, limited position while outside this zone the excess temperature is $\Theta - \theta_1 = 0$.

If the gas which enters at temperature θ_1 now flows through the regenerator with a temperature distribution as illustrated in Fig. 15-2 (in this figure it is assumed that the gas flows from left to right), the section with the heat pole will be progressively cooled and the heat taken up by the gas will be transferred to the

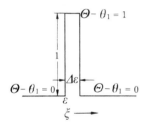

Figure 15-2 Heat pole.

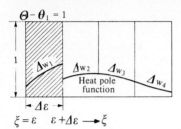

Figure 15-3 Heat pole function (schematic).

adjoining regenerator mass on the right-hand side. After the elapse of a specific time η a temperature distribution develops within the heat-storing mass, rather as it is shown in Fig. 15-3. This curve of the excess temperature, the calculation of which will be described later, is known as the heat pole function Δw at time η. Now Δw is a function of the pole width $\Delta \varepsilon$, of the distance $\xi - \varepsilon$ of the position ξ in question from the heat pole and of the time η. Further Δw_1 to Δw_N are the average values of this function in sections 1 to N.[1] Calculations are performed using only these average values which are determined as accurately as possible but they do not include the variation of the heat pole function within each individual section.

Knowing the average value of the heat pole function one can easily outline the influence of each individual section in Fig. 15-1 upon the final temperature distribution. For example, since the average initial excess temperature f_1 in section 1 can be obtained by multiplying a heat pole located in this section by f_1, the influence of this first section upon the final temperature of the heat-storing mass in the N-sections is:

$$f_1 \cdot \Delta w_1, f_1 \cdot \Delta w_2, f_1 \cdot \Delta w_3 \cdots f_1 \cdot \Delta w_N$$

Similarly the influence of the second section with the average temperature f_2 on the final temperature in the N sections is:

$$0, f_2 \cdot \Delta w_1, f_2 \cdot \Delta w_2, \cdots, f_2 \cdot \Delta w_{N-1}$$

If the influence of the remaining sections is determined in a similar manner, Eq. (15-1) is obtained by adding together these influences to yield the required excess temperature at time η:

$$\left.\begin{array}{lll} \text{in the 1st section} & \Theta_1 - \theta_1 = f_1 \Delta w_1 \\ \text{in the 2nd section} & \Theta_2 - \theta_1 = f_1 \Delta w_2 + f_2 \Delta w_1 \\ \text{in the 3rd section} & \Theta_3 - \theta_1 = f_1 \Delta w_3 + f_2 \Delta w_2 + f_3 \Delta w_1 \\ \text{in the } N\text{th section} & \Theta_N - \theta_1 = f_1 \Delta w_N + f_2 \Delta w_{N-1} + f_3 \Delta w_{N-2} + \cdots f_N \Delta w_1 \end{array}\right\} \quad (15\text{-}1)$$

It thus follows from Eq. (15-1) that the final temperature distribution at time η can be calculated relatively quickly using the heat pole function Δw for each arbitrary initial distribution. Thus the average values Θ_1, Θ_2, etc., of the final temperatures in the individual sections are obtained. Accuracy can be improved arbitrarily by increasing the number of heat poles.

---

[1] Δw (except Δw_1) becomes infinitely small for $\Delta \varepsilon = 0$. The limiting value $\lim (\Delta w / \Delta \varepsilon)$ represents the Green function used in the theory of partial differential equations.

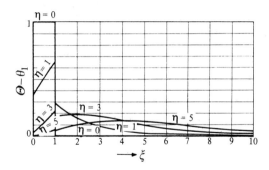

Figure 15-4 Heat pole function Δw for a heat pole of width $\Delta \varepsilon = 1$.

The gas temperature distribution can be calculated in the same way by taking as a basis a corresponding heat pole function for the gas temperature θ. However, it is simpler just to determine Θ by means of the heat pole method and then to determine θ by a graphical method, which has yet to be described, as shown in Fig. 16-3. Alternatively a corresponding numerical step-by-step process, for example Eq. (16-11) can be employed.

Calculation of the Heat Pole Function

The heat pole function can be determined basically from the initial temperature distribution of the heat pole (Fig. 15-2) using one of the step-by-step methods described in Sec. 16-4 or Sec. 16-5. Thus one obtains, for example, the curve of Δw as a function of $\xi - \varepsilon$ for various times η as is shown in Fig. 15-4 for $\Delta \varepsilon = 1$.

It follows from Fig. 13-6 that Δw can be obtained more simply by assuming, appropriately, that $\theta_1 = 0$. Consider a heat pole which exists between $\xi = \varepsilon = 0$ and $\xi = \Delta \varepsilon$. This can be considered as a rectangular constant initial temperature distribution $\Theta - \theta_1 = 1$ beginning at $\xi = 0$, from which is subtracted this same initial temperature but displaced by $\Delta \varepsilon$ to the right. This second temperature distribution thus has the value $\Theta - \theta_1 = 0$ up to $\xi = \Delta \varepsilon$ and from $\xi = \Delta \varepsilon$ onwards has the value $\Theta - \theta_1 = 1$. In this way can be obtained also the heat pole function for a later time η: the temperature distribution which results for η, see Fig. 13-6, and which has been displaced towards the right by $\Delta \varepsilon$ is subtracted from the same temperature curve which has not been displaced towards the right (see Fig. 15-5). The accuracy with which the heat pole function is obtained in this manner is dependent upon how accurately the curves in Fig. 13-6 have been determined. The

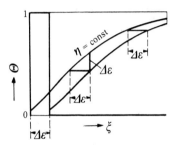

Figure 15-5 Production of the heat pole function by displacing the temperature curve from Fig. 13-4 by $\Delta \varepsilon$ for time η.

very accurate curves obtained by Schumann [S318] are highly recommended for this purpose.

However, the heat pole function can also be determined by calculation only; this is done by taking the integral in Eq. (13-9) which has the upper limit of integration ξ and subtracting from it the same integral with the upper limit of integration $\xi - \Delta\varepsilon$. The average value Δw_N for the invidual sections can thus be finally formed from the curve of the heat pole function which has been so obtained, and from use of Fig. 13-6.[2]

Calculation of the Temperature Distribution at Equilibrium

The heat pole method is particularly suitable for determining the temperature distribution in the heat-storing mass at the beginning of a period for the state of equilibrium. This will be illustrated here for the cold period for the case where $C = C'$ and $T = T'$. The reversal requirements which have to be fulfilled for counterflow are

$$\Theta\left(\xi, \frac{\Pi}{2}\right) - \theta_1 = \theta'_1 - \Theta\left(\Lambda - \xi, -\frac{\Pi}{2}\right)$$

This is somewhat different from Eq. (13-66), in that the zero point of the temperature scale is selected arbitrarily. Equation (15-2) can also be written to represent this if the initial excess temperature of the heat-storing mass relative to θ_1 is denoted by $f(\xi)$ and the excess temperature at the end of the period by $g(\xi)$ (see Eq. (14-3)).

$$g(\xi) = \theta'_1 - \theta_1 - f(\Lambda - \xi) \tag{15-2}$$

Furthermore, if average values are introduced into the individual sections instead of the true temperatures, the reversal requirement takes the form of N equations

$$\left.\begin{aligned} g_1 &= \theta'_1 - \theta_1 - f_N \\ g_2 &= \theta'_1 - \theta_1 - f_{N-1} \\ &\vdots \\ g_N &= \theta'_1 - \theta_1 - f_1 \end{aligned}\right\} \tag{15-3}$$

Equation (15-4) is obtained from Eq. (15-1) if $g_1 = \Theta_1 - \theta_1$ and $g_N = \Theta_N - \theta_1$, etc., Δw_1 etc. should likewise relate to the end of the cold period and these values are placed in the Eq. (15-3) above:

$$\left.\begin{aligned} \Delta w_1 f_1 &\qquad\qquad +f_N &= \theta'_1 - \theta_1 \\ \Delta w_2 f_1 + \Delta w_1 f_2 &\qquad\qquad +f_{N-1} &= \theta'_1 - \theta_1 \\ \Delta w_3 f_1 + \Delta w_2 f_2 + \Delta w_1 f_3 &\qquad +f_{N-2} &= \theta'_1 - \theta_1 \\ &\vdots \\ \Delta w_N f_1 + \Delta w_{N-1} f_2 + \cdots + \Delta w_1 f_N + f_1 &\quad = \theta'_1 - \theta_1 \end{aligned}\right\} \tag{15-4}$$

[2] Further details of this way of determining the heat pole function can be found in [H305] pp. 110 to 112.

The solution of these equations gives the required excess temperatures f_1, f_2, etc., up to f_N for the beginning of the period. The solution begins, appropriately, by inserting f_N, f_{N-1}, etc., from the first half of Eq. (15-4) into the remaining equations, by which means the number of equations is quickly reduced by half.

Obviously, the methods described can also be applied to *parallel flow*. The heat pole method can also be used for calculating the equilibrium temperatures for cases where Λ and Λ' and Π and Π' are different in the warm and cold periods. One first imagines Eq. (15-1) with the initial values f_1, f_2, etc., as yet undetermined, related again to the cold period. The expressions for the final temperatures in the cold period which result from these can then be considered to be the initial temperatures in the warm period. Equation (15-1) is again used here, although correspondingly altered values of the heat pole function must be selected because of the different length of the warm period. The section widths $\Delta \varepsilon$ must also be different with different values of Λ and Λ' in the warm and cold periods since the computations can only be carried through without difficulty if the number of sections remains unchanged. Furthermore, in counterflow the numbering of the sections, which must always be in the direction of flow, reverses with the beginning of a new period. Finally, if the expressions for the final temperatures in the warm period are set equal to the initial temperatures in the cold period, defining equations for the state of equilibrium will result, from which the unknown values of f_1, f_2, etc., up to f_N can be calculated.

The variations in the initial temperature from period to period which occur *before the state of equilibrium is achieved*, can be calculated very easily using the heat pole method. To do this one starts from the temperature distribution in the heat-storing mass at the moment of commencement of operation and applies Eq. (15-1) alternately to the warm and cold period. By carrying through the calculations relating to a sufficiently large number of periods the state of equilibrium will be finally reached.

15-2 REFINEMENT OF THE HEAT POLE METHOD ON THE BASIS OF ITS RELATIONSHIP TO AN INTEGRAL EQUATION[3]

It has already been indicated, and it will now be demonstrated that the calculation of the initial temperature for the state of equilibrium using Eq. (15-4) can be considered as an approximate method for solving an integral equation. This integral equation is obtained in the following way by examining the limiting case of infinitely narrow heat poles.

The heat pole function for a heat pole between $\xi = 0$ and $\xi = \Delta \varepsilon$ (see Fig. 15-5),

---

[3] The method described here was worked out by the author in 1945 or 1946 but only published in the first edition of this book. It was only during the correction of this first edition that I became aware of the work of Iliffe [1301] which is based on similar considerations.

can be calculated from Fig. 13-6 by displacing the curve for the required time η towards the right by the section width $\Delta\varepsilon$ and by subtracting the value obtained from that of the curve which has not been displaced. If the temperature from Fig. 13-6 at time η is denoted by $\Theta^*(\xi)$, the heat pole function at each position $\xi > \Delta\varepsilon$ has the value

$$\Delta w = \Theta^*(\xi) - \Theta^*(\xi - \Delta\varepsilon)$$

On the other hand for $\xi < \Delta\varepsilon$, that is for the first section (see Fig. 15-1) in which the subtraction does not take place, we obtain

$$\Delta w_1 = \Theta^*(\xi)$$

For a heat pole between ε and $\varepsilon + \Delta\varepsilon$ is yielded, correspondingly,

$$\Delta w = \Theta^*(\xi - \varepsilon) - \Theta^*(\xi - \varepsilon - \Delta\varepsilon) \quad \text{for} \quad \xi > \varepsilon + \Delta\varepsilon$$
$$\Delta w_1 = \Theta^*(\xi - \varepsilon) \quad\quad\quad\quad\quad \text{for} \quad \varepsilon \leqslant M < \varepsilon + \Delta\varepsilon$$

If the width $\Delta\varepsilon$ of the heat pole is now made infinitely small, the expression for the heat pole function takes the form:

$$dw = \frac{\partial\Theta^*(\xi - \varepsilon)}{\partial(\xi - \varepsilon)} \cdot d\varepsilon^4 \quad \text{for} \quad \xi > \varepsilon$$

$$\Delta w_1 = \Theta^*(0) \quad\quad\quad\quad \text{for} \quad \xi = \varepsilon$$

As an abbreviation

$$\frac{\partial\Theta^*(\xi - \varepsilon)}{\partial(\xi - \varepsilon)} = K(\xi - \varepsilon)$$

can be written in the last but one equation. This expression can be obtained by partially differentiating Eq. (13-6) with respect to ξ and taking into account Eq. (13-26). Thus Eq. (15-5) results when $\Theta_1 = 1$ and $\theta_1 = 0$ and ξ is replaced by $\xi - \varepsilon$:

$$K(\xi - \varepsilon) = \frac{\partial\Theta^*(\xi - \varepsilon)}{\partial(\xi - \varepsilon)} = -\exp[-(\xi - \varepsilon + \eta)]\sqrt{\frac{\eta}{\xi - \varepsilon}} \cdot iJ_1(2i\sqrt{(\xi - \varepsilon)\eta}) \quad (15\text{-}5)$$

where i is $\sqrt{-1}$ and J_1 denotes the Bessel function of the first order. Equation (15-6) can also be used instead of Eq. (15-5) for larger values of $(\xi - \varepsilon)\eta$ by taking the development of the series

$$\lim_{x = \infty} J_1(ix) = i\frac{\exp(x)}{\sqrt{2\pi x}}\left(1 - \frac{3}{1!\,8x} - \frac{3\cdot 5}{2!\,(8x)^2}\cdots\right)$$

into consideration.[4]

---

[4] Strictly speaking $d(\xi - \varepsilon)$ should be written in this expression in place of $d\varepsilon$, as the latter integration in Eqs. (15-7) and (15-8) must be carried out in the direction of increasing $\xi - \varepsilon$. However, it does conform with general opinion, and with the considerations developed so far, to write $d\varepsilon$. The correct result is achieved using $d\varepsilon$ so as long as integration in Eqs. (15-7) and (15-8) takes place in the direction of increasing ε.

$$K(\xi-\varepsilon) = \exp[-(\sqrt{\xi-\varepsilon}-\sqrt{\eta})^2] \cdot \frac{1}{\sqrt{4\pi}} \sqrt[4]{\frac{\eta}{(\xi-\varepsilon)^3}} \left(1 - \frac{0.188}{\sqrt{(\xi-\varepsilon)\eta}} - \frac{0.0293}{(\xi-\varepsilon)\eta}\right)$$

(15-6)

With $\sqrt{(\xi-\varepsilon)\eta} = 2$ the error in this equation is already smaller than 2 per cent and decreases rapidly with increasing $(\xi-\varepsilon)\eta$.

On the other hand, for $\Theta^*(0)$

$$\Theta^*(0) = \exp(-\eta)$$

is obtained by integrating the differential Eq. (13-27) with $\xi = 0$, again assuming $\theta = \theta_1 = 0$ and $\Theta_1 = 1$. It follows that

$$dw = K(\xi-\varepsilon)d\varepsilon \quad \text{for} \quad \xi > \varepsilon$$
$$\Delta w_1 = \exp(-\eta) \quad \text{for} \quad \xi = \varepsilon$$

If $f(\xi)$ is now the initial excess temperature of the heat-storing mass at position ξ, and $f(\varepsilon)$ the initial excess temperature at position ε the following relationship is obtained, using Eq. (15-1), at the limit of an infinite number of infinitely small heat poles:

$$\Theta(\xi) - \theta_1 = \int_0^\xi f(\varepsilon)\, dw + f(\xi)\Delta w_1$$

or, utilizing the expressions for dw and Δw_1 above:

$$\Theta(\xi) - \theta_1 = f(\xi)\exp(-\eta) + \int_0^\xi f(\varepsilon)K(\xi-\varepsilon)\, d\varepsilon$$

(15-7)

Into this equation $K(\xi-\varepsilon)$ can be inserted, as in Eqs. (15-5) and (15-6).

In addition, considerations will be limited to the state of equilibrium and to the case where $\Lambda = \Lambda'$ and $\Pi = \Pi'$. The following integral Eq. (15-8) results from the application of the reversal requirement (15-2) and from Eq. (15-7); $g(\xi)$ is set equal to $\Theta(\xi) - \theta_1$ and thus eliminated and the value $\eta = \Pi$ is used:

$$f(\xi)e^{-\Pi} + f(\Lambda-\xi) + \int_0^\xi f(\varepsilon)K(\xi-\varepsilon)_{\eta=\Pi}\, d\varepsilon = \theta_1' - \theta_1$$

(15-8)

The solution of this integral equation enables the temperature distribution $f(\xi)$ at the beginning of a period for the state of equilibrium to be determined and it completely corresponds to the set of simultaneous linear Eqs. (15-4). It follows from Eq. (15-5) that it also agrees with the integral Eq. (14-4) found in Sec. 14-1. As indicated above, because of the way Eqs. (15-7) and (15-8) have been developed, *the heat pole method* can be considered to be *a method of approximation for solving the integral Eq.* (14-4) or (15-8). In practical calculations the heat pole method always achieves its goal more quickly and with greater ease than does the exact solution of Nusselt's integral equation which was discussed in Sec. 14-4.

On the other hand, however, the integral representations (15-7) and (15-8) can be used in order to *simplify the heat pole method* whereby a considerable increase in accuracy can be achieved with almost no difference in computational effort.

The total regenerator length Λ must again be imagined to be divided into N

equally long sections $\Delta\varepsilon$ so that we again have the sections of equal width $\Delta\varepsilon$ shown in Fig. 15-1. However, the average value of the temperature in these sections will not now be considered, instead the values will be used at the mesh points $\varepsilon = 0$, $\varepsilon = \Delta\varepsilon$, $\varepsilon = 2\Delta\varepsilon$ etc. up to $\varepsilon = N \cdot \Delta\varepsilon = \Lambda$.

The initial temperature of the heat-storing mass is considered to have the values f_0, f_1, f_2, etc., up to f_N at these points. The point ξ where the final temperature $\Theta(\xi)$ is required, using Eq. (15-7), or where the initial temperature $f(\xi)$ is required, using Eq. (15-8), lies on the boundary between the nth and the $(n+1)$th section, so that $\xi = n \cdot \Delta\varepsilon$ and thus $\Theta(\xi) = \Theta_n$ and $f(\xi) = f_n$. The values of $K(\xi - \varepsilon)$ for $\xi - \varepsilon = 0$, $\xi - \varepsilon = \Delta\varepsilon$, $\xi - \varepsilon = 2\Delta\varepsilon$, etc., up to $\xi - \varepsilon = n\Delta\varepsilon$ are denoted by K_0, K_1, K_2, etc., up to K_n. As these values can be calculated accurately using Eq. (15-5) or (15-6), the value of the product $f(\varepsilon) \cdot K(\xi - \varepsilon)$ can be computed accurately for all mesh points, given the values of f_0, f_1, etc., up to f_N.

The fundamental concept behind the refinement of the heat pole method, which has yet to be described, consists, on the basis of these values, of *calculating as accurately as possible the integrals* in Eq. (15-7) or (15-8) *with the help of a method of approximation*. Simpson's rule[5] is suitable for this and is based on the idea of fitting a parabolic curve through each of three neighbouring points on the curve to be integrated. As a first approximation the curvature of the temperature distribution in this way is also taken into account and thus high accuracy is achieved with relatively few individual values. If the number n of sections up to ξ is an even number, it follows from Simpson's rule that Eq. (15-9) can be obtained directly:

$$\int_0^\xi f(\varepsilon)K(\xi-\varepsilon)\,d\varepsilon = \frac{\Delta\varepsilon}{3}[f_0 K_n + 4f_1 K_{n-1} + 2f_2 K_{n-2} + 4f_3 K_{n-3} + 2f_4 K_{n-4} + \cdots$$

$$+ 2f_{n-2}K_2 + 4f_{n-1}K_1 + f_n K_0] \quad (n \text{ even}) \tag{15-9}$$

If, on the other hand, n is an odd number, the following can first be written for the integral between $\varepsilon = 0$ and $\varepsilon = (n-1)\Delta\varepsilon = \xi - \Delta\varepsilon$:

$$\int_0^{\xi-\Delta\varepsilon} f(\varepsilon)K(\xi-\varepsilon)\,d\varepsilon = \frac{\Delta\varepsilon}{3}[f_0 K_n + 4f_1 K_{n-1} + 2f_2 K_{n-2} + \cdots$$

$$+ 2f_{n-3}K_3 + 4f_{n-2}K_2 + f_{n-1}K_1]$$

In order to determine the remaining integrals as accurately as possible, we imagine a quadratic function $y = a + b\varepsilon + c\varepsilon^2$ to be fitted through the points $\varepsilon = (n-2)\Delta\varepsilon = \xi - 2\Delta\varepsilon$ and $\varepsilon = n\Delta\varepsilon = \xi$ so that the values of y, at positions $(n-2)\Delta\varepsilon$, $(n-1)\Delta\varepsilon$ and $n\Delta\varepsilon$, are exactly equal to the values of $f(\varepsilon) \cdot K(\xi - \varepsilon)$ at these positions respectively. If y is then integrated between $\varepsilon = (n-1)\Delta\varepsilon$ and $\varepsilon = n\Delta\varepsilon$ the resultant value is set equal to the required remaining integrals and the following is obtained

$$\int_{\xi-\Delta\varepsilon}^\xi f(\varepsilon)K(\xi-\varepsilon)\,d\varepsilon = \frac{\Delta\varepsilon}{12}[-f_{n-2}K_2 + 8f_{n-1}K_1 + 5f_n K_0]$$

---

[5] See, for example, Hütte, Des Ingenieurs Taschenbuch, Volume 1, 28th edition, 1955, p. 213 or [Z301] p. 231 ff.

In this way Eq. (15-10) results for the total integral when the last two equations are added and n is odd:

$$\int_0^{\xi} f(\varepsilon)K(\xi - \varepsilon_4)\,d\varepsilon = \frac{\Delta\varepsilon}{3}[f_0K_n + 4f_1K_{n-1} + 2f_2K_{n-2} + 4f_3K_{n-3} + 2f_4K_{n-4} + \cdots$$

$$+ 2f_{n-3}K_3 + 3.75f_{n-2}K_2 + 3f_{n-1}K_1 + 1.25f_nK_0]\quad (n\text{ odd})\qquad (15\text{-}10)$$

Care should be taken, since $f_{n-2} = f_{-1}$ does not exist for $\xi = \Delta\varepsilon$, i.e., for $n = 1$. $f_{n-2} = f_{-1}$ is chosen so that its value lies on the quadratic curve which is fitted through f_0, f_1 and f_2. Thus if it is assumed that $f_{-1} = 3f_0 - 3f_1 + f_2$ Eqs. (15-9) and (15-10) take the particular form for $n = 1, 2$, etc., set out below.

$$\int_0^{\xi = \Delta\varepsilon} f(\varepsilon)K(\xi - \varepsilon)\,d\varepsilon = \frac{\Delta\varepsilon}{12}[f_0(8K_1 - 3K_2) + f_1(5K_0 + 3K_2) - f_2K_2]$$

$$\int_0^{\xi = 2\Delta\varepsilon} f(\varepsilon)K(\xi - \varepsilon)\,d\varepsilon = \frac{\Delta\varepsilon}{3}[f_0K_2 + 4f_1K_1 + f_2K_0]$$

$$\int_0^{\xi = 3\Delta\varepsilon} f(\varepsilon)K(\xi - \varepsilon)\,d\varepsilon = \frac{\Delta\varepsilon}{3}[f_0K_3 + 3.75f_1K_2 + 3f_2K_1 + 1.25f_3K_0]\qquad (15\text{-}11)$$

$$\int_0^{\xi = 4\Delta\varepsilon} f(\varepsilon)K(\xi - \varepsilon)\,d\varepsilon = \frac{\Delta\varepsilon}{3}[f_0K_4 + 4f_1K_3 + 2f_2K_2 + 4f_3K_1 + f_4K_0]$$

$$\int_0^{\xi = 5\Delta\varepsilon} f(\varepsilon)K(\xi - \varepsilon)\,d\varepsilon = \frac{\Delta\varepsilon}{3}[f_0K_5 + 4f_1K_4 + 2f_2K_3 + 3.75f_3K_2$$

$$+ 3f_4K_1 + 1.25f_5K_0]$$

If these expressions are inserted one after the other in place of the integral in Eq. (15-7), equations for determining the final temperatures Θ_0, Θ_1, etc., up to Θ_N will result at all the grid positions. Similarly, by insertion into Eq. (15-8), $N+1$ linear equations are obtained from which the unknown initial temperatures f_0, f_1, etc., up to f_N can be calculated for the state of equilibrium of the regenerator. In both cases care must be taken by noting that $f(\xi) = f_n$ or $f(\Lambda - \xi) = f_{N-n}$. The equations obtained correspond, in principle, to Eq. (15-1) or (15-4) which were found using the simple heat pole method. If the numerical values K_0, K_1, etc., have once been calculated, these equations can be formed just as easily as these previous equations and can also be solved in the same way.

Iliffe [I301], who, as mentioned, developed independently the same method as the author for solving* the integral Eq. (15-8), made the notable suggestion that the integral should be evaluated over the first interval between $\xi = 0$ and $\xi = \Delta\varepsilon$ using the value $f_{1/2}$ instead of f_{-1} in the middle of the interval. Using Simpson's rule he

---

* Editor's note: For $\xi = k\Delta\xi$, $k = 2, 4, 6, \ldots$ (even), Iliffe represented the integral in Eq. (15-8) using Simpson's Rule. For $k = 3$, he used the three eighths rule. For $k = 5, 7, 9, \ldots$ (odd), Iliffe represented that part of the integral between 0 and $3\Delta\xi$ by the three eighths rule approximation while he represented the remainder of the integral using Simpson's Rule.

obtained Eq. (15-12) in place of the first equation in (15-11)

$$\int_0^{\xi = \Delta\varepsilon} f(\varepsilon) \cdot K(\xi - \varepsilon) \, d\varepsilon = \frac{\Delta\varepsilon}{6} [f_0 K_1 + 4f_{1/2} \cdot K_{1/2} + f_1 K_0] \tag{15-12}$$

where $K_{1/2}$ can be calculated for $\xi - \varepsilon = 1/2$ using Eq. (15-5). When determining the state of equilibrium, in order not to introduce $f_{1/2}$ as an additional unknown, Iliffe expressed $f_{1/2}$ in terms of f_0, f_1, f_2 and f_3 using a third order power series which passes through four values. He thus obtained the relationship

$$f_{1/2} = \frac{1}{16} [5f_0 + 15f_1 - 5f_2 + f_3] \tag{15-13}$$

However, as can be seen from Figs. 18-1 to 18-13, at the state of equilibrium the temperature distribution of the heat-storing mass at the gas entrance is only slightly curved at the beginning of a period so that it is usually sufficient to start with a power series of the second order. Thus is obtained:

$$f_{1/2} = \frac{1}{8} [3f_0 + 6f_1 - f_2] \tag{15-14}$$

By inserting either the relationship (15-13) or (15-14) into Eq. (15-12) the value of the integral between $\xi = 0$ and $\xi = \Delta\varepsilon$ can be found.

Recently Willmott and Duggan [W308] carried out practical calculations using Nahavandi's and Weinstein's method, described in Sec. 14-2 and the refined heat pole method in the form suggested by Iliffe. They were thus able to verify the practicability of both methods in a region of up to $\Lambda = 10$ and $\Pi = 10$. On the other hand they also pointed out that in determining the initial values f_0, f_1, etc., up to f_N for the state of equilibrium the difficulty could arise that the definitive linear equations could be almost singular. Such difficulty is more to be expected with the refined heat pole method than with the method of Nahavandi and Weinstein. The reader is also referred to the latest paper regarding this matter by Willmott and Thomas [W309].

15-3 CALCULATION OF THE EFFICIENCY AND THE HEAT TRANSMISSION COEFFICIENT USING THE HEAT POLE METHOD

(a) *Calculation using the simple heat pole method*
Both the average values f_1, f_2, etc., up to f_N of the initial excess temperatures and the corresponding values g_1, g_2, etc., up to g_N at the end of the period (see Eq. (15-3)) can be calculated for the N sections for the state of equilibrium using the heat pole method. It is then possible to determine the efficiency of the regenerator very simply in the following way. If F is the heating surface area of the complete regenerator, the part of the regenerator which corresponds to one section has a

heating surface area F/N and a thermal capacity $F/N \cdot \rho c \delta/2$ where δ again denotes the thickness of an element of the heat-storing mass which is assumed to be a plate in shape. The nth section of the heat-storing mass thus releases the following quantity of heat during the cold period:

$$\Delta Q_{Per} = \frac{F \cdot \rho c \delta}{2N} (f_n - g_n)$$

The quantity of heat being transferred by the total regenerator during a period is thus equal to

$$Q_{Per} = \frac{F \cdot \rho c \delta}{2N} \cdot \sum_{n=1}^{N} (f_n - g_n) \tag{15-15}$$

In the ideal case where a complete exchange of heat takes place, when $CT \leqslant C'T'$ the gas in the regenerator would be heated from θ_1 to θ'_1. The quantity of heat which would then be transferred would be:

$$Q_{id} = CT(\theta'_1 - \theta_1)$$

Equation (15-16) thus results as the efficiency of the regenerator, upon utilizing Eqs. (13-28) and (13-29)

$$\eta_{Reg} = \frac{Q_{Per}}{Q_{id}} = \frac{\Lambda}{\Pi N} \sum_{n=1}^{N} \frac{f_n - g_n}{\theta'_1 - \theta_1} \tag{15-16}$$

This equation is valid not only for counterflow but also for parallel flow for $CT \leqslant C'T'$. When $CT > C'T',\cdot$ on the other hand, Eq. (15-16) represents the "efficiency function" as expressed by Eq. (5-68) which is to be found in Part Three of this book.

When $CT = C'T'$ Eq. (15-16) can be simplified, with the help of the reversal requirement (15-3), namely that g_n can be replaced by f_{N-n+1}. Thus Eq. (15-17) is obtained:

$$\eta_{Reg} = \frac{\Lambda}{\Pi} \left[\frac{2}{N} \sum_{n=1}^{N} \frac{f_n}{\theta'_1 - \theta_1} - 1 \right] \quad \text{(with } CT = C'T') \tag{15-17}$$

(b) *Calculation using the refined heat pole method*
Since $\Lambda/N \doteq \Delta\varepsilon = \Delta\xi$ represents the width of a heat pole, the equation formed for the efficiency (15-16) takes the form (15-18) in the limiting case of infinitely narrow heat poles:

$$\eta_{Reg} = \frac{1}{\Pi(\theta'_1 - \theta_1)} \int_0^\Lambda [f(\xi) - g(\xi)] \, d\xi \tag{15-18}$$

If the values of f_0, f_1, etc., up to f_N and g_0, g_1, etc., up to g_N have been calculated for the state of equilibrium at the boundaries of the sections, using the refined heat pole method, the value of the integral can be computed for example, using the Simpson rule. Efficiency is then determined using Eq. (15-18).

Calculation of the Heat Transmission Coefficient Using the Heat Pole Method with $CT = C'T'$

Once the efficiency η_{Reg} has been determined using Eqs. (15-16), (15-17) or (15-18), the ratio k/k_0 of the true heat transmission coefficient k to the heat transmission coefficient k_0, which corresponds to the zero eigen function, can be calculated immediately using Eq. (13-105). The validity of this emerges from the derivation of Eq. (13-105) for $CT = C'T'$, Sec. 13-8.

In the case where $CT \neq C'T'$ quite complex relationships occur for k/k_0; these were discussed and derived in the first edition of this book, see Appendix 5. However, as k/k_0 is only slightly dependent upon the ratio $CT : C'T'$ and, moreover, as the values determined for k/k_0 when $CT = C'T'$ can be applied quite accurately to the more general case where $CT \neq C'T'$, using the dimensionless groups Λ and Π via Eq. (11-5), these more general considerations will not be repeated here.

Heat Transmission Coefficient at a Particular Position ξ

Generally, only the average heat transmission coefficient k is required in order to calculate the exchange of heat in a regenerator. However, as has already been shown in Sec. 12-6, in special cases knowledge of the heat transmission coefficient k_ξ at a particular position in the regenerator is required.

For example, if one has calculated the temperature distribution $f(\xi)$ at the beginning and $g(\xi)$ at the end of the period for the state of equilibrium using one of the heat pole methods, an equation for the ratio k_ξ/k_0 can be obtained in the following way. The quantity of heat per period

$$dQ_{Per} = k_\xi(T+T')(\bar{\theta}' - \bar{\theta})\,df = \frac{\rho c \delta}{2}\,df(f(\xi) - g(\xi))$$

will be obtained for an infinitely short section of the regenerator at position ξ with the heating surface area df. Equation (15-19) first follows from this relationship:

$$k_\xi = \frac{\rho c \delta}{2(T+T')} \cdot \frac{f(\xi) - g(\xi)}{\bar{\theta}' - \bar{\theta}} \tag{15-19}$$

On the other hand, it follows from Eq. (13-5) that

$$\frac{1}{k_0} = (T+T')\left[\frac{1}{\bar{\alpha}T} + \frac{1}{\bar{\alpha}'T'}\right]$$

applies to the heat transmission coefficient k_0 corresponding to the zero eigen function. By multiplying together these two equations we obtain:

$$\frac{k_\xi}{k_0} = \frac{\rho c \delta}{2}\left[\frac{1}{\bar{\alpha}T} + \frac{1}{\bar{\alpha}'T'}\right] \cdot \frac{f(\xi) - g(\xi)}{\bar{\theta}' - \bar{\theta}} \tag{15-20}$$

Finally, with the help of Eq. (13-29) these equations can be brought into the following form:

$$\frac{k_\xi}{k_0} = \left(\frac{1}{\Pi} + \frac{1}{\Pi'}\right) \cdot \frac{f(\xi) - g(\xi)}{\bar{\theta}' - \bar{\theta}} \tag{15-21}$$

First the case of $CT = C'T'$ will be mentioned. For this case a relationship is obtained from Eq. (15-21) which is very useful if η_{Reg} or k/k_0 have been previously determined. When $CT = C'T'$, $\bar{\theta}' - \bar{\theta}$ has the same value at all positions in the regenerator. Equation (15-22) also applies to this case.

$$\bar{\theta}' - \bar{\theta} = \theta'_1 - \theta_2 = (\theta'_1 - \theta_1)(1 - \eta_{Reg}) \quad (CT = C'T') \tag{15-22}$$

Using this, k_ξ/k_0 can easily be calculated using Eq. (15-21), given η_{Reg}. Finally, if we express η_{Reg} in terms of k/k_0, with the help of Eq. (13-105) again, Eq. (15-23) will be obtained by inserting Eq. (15-22) into Eq. (15-21) with $\Pi = \Pi'$:

$$\frac{k_\xi}{k_0} = \frac{1}{\Pi}\left(2 + \Lambda\frac{k}{k_0}\right)\frac{f(\xi) - g(\xi)}{\theta'_1 - \theta_1} \quad (CT = C'T') \tag{15-23}$$

For practical calculations of k_ξ/k_0, this equation could be particularly suitable since k/k_0 can be read off from Fig. 11-13. Presumably Eq. (15-23) can also be used, and provide a good approximation for the case where $CT \neq C'T'$ as long as Λ and Π are calculated using Eq. (11-5) or (11-6).

It follows from Eq. (15-23), as also from previous equations, that k_ξ/k_0 is generally not only dependent upon ξ but also upon Λ. However, if extremely long regenerators are considered, where the zero eigen function alone applies to the middle, k_ξ/k_0 proves to be independent of Λ at least for the case where $CT = C'T'$.

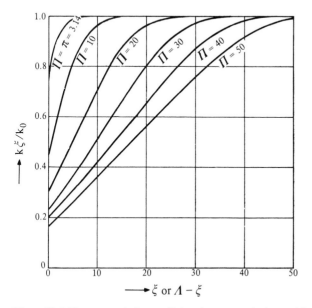

Figure 15-6 Heat transmission coefficient k_ξ at a particular position ξ or $\Lambda - \xi$ in a regenerator with various reduced period lengths Π. k_0 is the heat transmission coefficient corresponding to the zero eigen function.

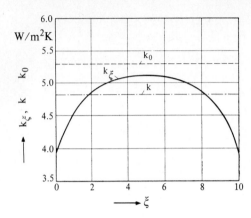

Figure 15-7 k_ξ as a function of ξ with $\Lambda = \Lambda' = 10$ and $\Pi = \Pi' = \pi$. k average heat transmission coefficient.

Here k_ξ/k_0 can be drawn as a function of ξ or $\Lambda - \xi$ as is shown in Fig. 15-6 on the basis of only slightly more precise calculations, which take advantage of Eq. (15-23).

Figure 15-7 illustrates an example of a short regenerator and shows how k_ξ is dependent upon ξ; as in Figs. 13-14 and 13-15, the values $\Lambda = \Lambda' = 10$ and $\Pi = \Pi' = \pi = 3.14$ are used, together with the data, $\bar{\alpha} = \bar{\alpha}' = 23.3 \, \text{W}/(\text{m}^2\text{K})$, $\rho c = 2070 \, \text{kJ}/(\text{m}^3 \, \text{K})$, $\lambda_s = 1.163 \, \text{W}/(\text{m K})$ and $\delta = 0.03 \, \text{m}$. It can be seen that in this case k_ξ does not quite reach the value k_0 in the centre of the regenerator. As stated previously this is due to the fact that the influence of the higher eigen functions (harmonic vibrations) does not completely fade away in the centre of the regenerator.

The following text is reproduced from the first edition.

In order to calculate $\dfrac{k_\xi}{k_0}$ for the case where $CT \neq C'T'$ we need to know the average temperature difference $\bar{\theta}' - \bar{\theta}$ between the two gases at position ξ. Further to determine this from the curve of $f(\xi)$ and $g(\xi)$ we have to write Eq. (685), by adding and subtracting θ'_1 and θ_1:

$$\bar{\theta}' - \bar{\theta} = (\theta'_1 - \theta_1) - (\bar{\theta} - \theta_1) - (\theta'_1 - \bar{\theta}'). \tag{685}$$

Furthermore, as a regenerator element with reduced length $d\varepsilon$ has the heating surface area $\dfrac{d\varepsilon}{\Lambda} F$, the quantity of heat

$$CT(\bar{\theta} - \theta_1) = \frac{\gamma c \delta F}{2\Lambda} \int_0^\xi (f(\varepsilon) - g(\varepsilon)) \, d\varepsilon = \frac{CT}{\Pi} \int_0^\xi (f(\varepsilon) - g(\varepsilon)) \, d\varepsilon$$

will be transferred per period in the section of the regenerator between 0 and ξ and

$$C'T'(\theta'_1 - \bar{\theta}') = \frac{\gamma c \delta F}{2\Lambda} \int_\xi^\Lambda (f(\varepsilon) - g(\varepsilon)) \, d\varepsilon = \frac{CT}{\Pi} \int_\xi^\Lambda (f(\varepsilon) - g(\varepsilon)) \, d\varepsilon$$

will be the quantity of heat transferred in the section between ξ and Λ. The last expression is obtained with the help of Eqs. (13-28) and (13-29). Finally, again taking Eqs. (13-28) and (13-29) into consideration, if $\bar{\theta} - \theta_1$ and $\theta'_1 - \bar{\theta}'$ are taken from the last two equations and inserted into Eq. (685) we get

$$\bar{\theta}' - \bar{\theta} = (\theta'_1 - \theta_1) - \frac{1}{\Pi} \int_0^{\xi} (f(\varepsilon) - g(\varepsilon))\,d\varepsilon - \frac{\Lambda'}{\Lambda\Pi'} \int_{\xi}^{\Lambda} (f(\varepsilon) - g(\varepsilon))\,d\varepsilon \quad (686)$$

When $CT = C'T'$ these equations are simplified to take the form of Eq. (687) because of $\Lambda/\Pi = \Lambda'/\Pi'$:

$$\bar{\theta}' - \bar{\theta} = (\theta'_1 - \theta_1) - \frac{1}{\Pi} \int_0^{\Lambda} (f(\varepsilon) - g(\varepsilon))\,d\varepsilon \quad (CT = C'T') \quad (687)$$

Using Eqs. (686) or (687) it is also possible to calculate the average temperature difference at an arbitrary position in the regenerator if the integral is evaluated approximately, for example using Simpson's rule. Equation (684) then gives the desired value of k_{ξ}/k_0.

STEP-BY-STEP METHOD FOR CALCULATING THE TEMPERATURE DISTRIBUTION IN REGENERATORS

16-1 THE UNDERLYING REASONS FOR USING THE STEP-BY-STEP PROCESS

The method of calculation mentioned in Sec. 14-6, which precisely determines all the details of the temperature differences in the cross-sections of the regenerator bricks, are most laborious and time consuming to apply.* The methods of approximation which have already been discussed, give results much more quickly, particularly the heat pole methods. Here the temperature differences mentioned above can be considered numerically by the introduction of the heat transfer coefficient $\bar{\alpha}$ which is related to the average brick temperature Θ_m. Even more simple are those step-by-step processes which are based on difference calculations and which afford a high degree of accuracy given a sufficient number of steps.

Prior to about 25 years ago it seemed desirable to develop diagrammatic methods using the difference method. More general and complicated expressions can be evaluated nowadays using a digital computer, within numerical calculations which have been developed using difference methods. Using these expressions the processes taking place at each step can also be calculated very accurately when the heat transfer coefficients and the physical properties are dependent upon the temperature, or the flowrate of the gases through the regenerator changes with time.

---

* Editor's note: On modern high speed digital computers, this may no longer be true but the overall philosophy presented here is quite correct.

However, in transforming to these more general expressions it must be remembered that when carrying out difference calculations using a tolerable number of steps, a high degree of accuracy can only be achieved if average values are introduced at each of the steps for all the different values which occur in the differential equations. Thus the difference coefficients are so formed that they represent a good average value of the differential coefficients.

In most of the step-by-step processes that follow, for the sake of simplicity, calculations will be undertaken using the average brick temperature $\Theta = \Theta_m$ and the heat transfer coefficient $\bar{\alpha}$ relating to it. Thus, given a large enough number of steps, the distribution of the average brick temperature can be computed with as high a degree of accuracy as with the heat pole method. However, it must be emphasized that the heat pole method will only correctly reproduce the curve of the time average gas temperature. It has already been shown, in Sec. 13-2, p. 352, that given knowledge of Θ_m the distribution of the gas temperature, including its rapid changes after reversal, can quite accurately be determined.

Finally, complex step-by-step processes will be mentioned which take into account the temperature changes within the brick cross-section.

Initial Considerations

In all step-by-step methods one must start with a known, or assumed, temperature distribution in the heat-storing mass. This initial temperature curve will be denoted by $\Theta = \theta_1 + f(\xi)$ as in the heat pole method, so that $f(\xi)$ denotes the excess temperature relative to the constant temperature θ_1 of the entry gas.

At the position of gas entry $\xi = 0$ or $f = 0$, the heat-storing mass will initially have the excess temperature $f(0)$. The chronological variation of the temperature of the heat-storing mass at this position is calculated using Eq. (16-1), which takes into account Eq. (13-27), and the fact that $\theta_1 = \text{const}$:

$$\Theta = \theta_1 + f(0) \cdot \exp(-\eta) = \theta_1 + f(0) \cdot \exp\left(-\frac{\bar{\alpha}\,df}{dC_s}t\right) \qquad (16\text{-}1)$$

Thus, at each arbitrary time η or t the calculation of the temperature of the heat-storing mass can start from a known value of Θ at $\xi = 0$ or $f = 0$.

Determination of the State of Equilibrium

The state of equilibrium can be determined in the following way with step-by-step methods. One starts with an appropriately estimated arbitrary temperature distribution in the heat-storing mass which is chosen so as to approach, as closely as possible, the temperature distribution at equilibrium. Then, using the step-by-step method, one performs the calculations over a large number of successive periods. The temperature distribution obtained at the end of each period serves as the initial temperature distribution at the beginning of the next period. One continues in this way until the temperature distribution at the end of a full cycle is no longer noticeably different. This process can be shortened in that, as soon as calculations have been performed over a few periods, the final temperature distribution after each full cycle is adjusted so that the state of equilibrium is approached more rapidly. Reference points for this adjustment are obtained from the manner in which the final temperature distribution has changed from one cycle to another as obtained from the calculation of previous cycles. A process of this type for accelerating calculations has been published by Willmott and Kulakowski [W310].

In the case where $\Lambda = \Lambda'$ and $\Pi = \Pi'$ this adjustment can be applied after each warm and cold period once the operation has begun. This is achieved by comparing the final temperature distribution obtained by calculation after each period with the temperature curve $g(\xi)$ which can be expected from the reversal requirement (15-2) using the previous initial temperature distribution $f(\xi)$. If no other reference points for the adjustment are available it is recommended to form the average of the calculated final curve and $g(\xi)$ and then to begin the calculation of the new period with this average curve. It will be seen later that a more favourable adjustment can be made. Thus it is possible, for example, to form the adjusted temperature curve at 2/3 from the calculated final temperature curve and at 1/3 from $g(\xi)$.

Estimation of the Temperature Distribution for the State of Equilibrium

In order to estimate the temperature distribution for the state of equilibrium before beginning a step-by-step calculation it is possible to commence with the zero eigen function in the following way. Using the value of k/k_0, which can be read off from Fig. 11-13, first a heating surface area $F_0 = k/k_o F$ is obtained from the true heating surface area F of the regenerator. It follows from Eqs. (13-56) to (13-59) or (13-62), (13-63) and (13-64) that the zero eigen function can be used for time $t = 0$ in the F_0 region. Here $t = 0$ denotes the middle of the warm or cold period. The constants D and B in these equations are so determined that $\theta = \theta_1$ when $f = 0$ (the start of the heating surface area F_0) and $\theta' = \theta'_1$ when $f = F_0$. In the case where $CT = C'T'$ a straight line results for Θ, as is shown in Fig. 16-1. Finally the line curves at its ends, as shown in this figure, so that the final values Θ_1 and Θ'_1 can be retained.

As a first approximation the curve obtained in this way can be interpreted to be the temperature distribution in the heat-storing mass in the middle of a warm or cold period. One can then begin the step-by-step process, starting from this curve, by first calculating through the second half of a warm or cold period.

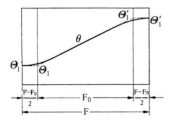

Figure 16-1 Approximate distribution of the temperature of the heat-storing mass in the middle of a warm or cold period when $CT = C'T'$.

Consideration of the Temperature Dependence of the Heat Transfer Coefficient and the Physical Properties

If the heat transfer coefficient and the physical properties are not dependent upon the temperature, it is recommended that calculations are carried out using Eqs. (13-24) and (13-25), even when using a dimensionless step-by-step process, that is with reduced independent variables ξ and η. In the following, the step-by-step methods will, in general, be presented so that they can be used without difficulty, even when the parameters mentioned vary with the temperature. In particular, the heat transfer coefficients are strongly dependent upon the gas temperature and are also a function of the temperature of the heat-storing mass in the case where radiative heat transfer takes place at higher temperatures. However since, in a particular case, the temperature differences between the gas and the heat-storing mass can be estimated initially, it is always possible to represent accurately the heat transfer coefficient as a function of the temperature of the heat-storing mass alone. Moreover, the temperature dependence of the heat transfer coefficients is moderated by the fact that they appear, as a rule, in the form $\bar{\alpha}/C$ and $\bar{\alpha}\,df/dC_s$, that is they are divided by values which themselves increase with the temperature. Thus it is always adequate to use the average values of these expressions at each calculation step, so that, in general, they alter only slightly from one step to another. Thus it is possible to consider variations with time, for example, which result from a chronological variation of the mass flow.

16-2 THE STEP-BY-STEP METHOD OF LAMBERTSON

Lambertson's method [L301] is particularly simple. A gas with the thermal capacity $C \cdot \Delta t$ flows, during the time interval Δt, through an element of the heat-storing mass, illustrated in Fig. 16-2, which has the surface area Δf and the thermal capacity ΔC_s. On average this gas enters the element of the heat-storing mass at the temperature θ_1 and leaves it after being heated by the average temperature θ_2: see Fig. 16-2. The quantity of heat required for this is extracted from the heat-storing element which is thus cooled from its average initial temperature Θ_1 to the average temperature Θ_2. For simplicity, Θ_1 is written in place of Θ_{m1} and Θ_2 is written in place of Θ_{m2}. The two initial temperatures θ_1 and Θ_1 are given and the final temperatures θ_2 and Θ_2 are thus sought.

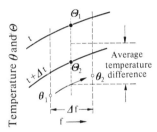

Figure 16-2 An explanation of the method of Lambertson.

During this process the gas takes up a quantity of heat

$$\Delta Q = C \, \Delta t (\theta_2 - \theta_1) \tag{16-2}$$

The same quantity of heat

$$\Delta Q = \Delta C_s (\Theta_1 - \Theta_2) \tag{16-3}$$

is extracted from the heat-storing mass. The transfer of heat is effected by the average temperature difference

$$\frac{\Theta_1 + \Theta_2}{2} - \frac{\theta_1 + \theta_2}{2}$$

which exists between the heat-storing mass and the gas. Thus there is a further relationship for the quantity of heat being transferred:

$$\Delta Q = \frac{\bar{\alpha}}{2} \cdot [(\Theta_1 + \Theta_2) - (\theta_1 + \theta_2)] \cdot \Delta f \, \Delta t \tag{16-4}$$

Since it follows from Eqs. (16-2), (16-3) and (16-4), that all the quantities ΔQ must be equal to each other, two equations are obtained which determine the two unknown final temperatures θ_2 and Θ_2. Solution of these two equations gives:

$$\theta_2 = \theta_1 + \frac{\Theta_1 - \theta_1}{\frac{1}{2}\left(1 + \frac{C \, \Delta t}{\Delta C_s}\right) + \frac{C}{\bar{\alpha} \, \Delta f}} \tag{16-5}$$

$$\Theta_2 = \Theta_1 - \frac{C \, \Delta t}{\Delta C_s} \cdot \frac{\Theta_1 - \theta_1}{\frac{1}{2}\left(1 + \frac{C \, \Delta t}{\Delta C_s}\right) + \frac{C}{\bar{\alpha} \, \Delta f}} \tag{16-6}$$

By applying Eqs. (16-5) and (16-6) step-by-step it is possible to calculate the total temperature distribution in the regenerator during a cold period and, then, also during a warm period. Only the temperature distribution in the heat-storing mass at any time is required, or must be assumed after estimation, and the temperature of the gas needs only to be known at its entry into the regenerator. For the time being it is appropriate here to consider calculations which are carried out at fixed times t and $t + \Delta t$ over the total length of the regenerator and only then to proceed to the next time interval.

Sandner [S301] has shown that it is possible to achieve a high degree of accuracy, using the method of Lambertson with a sufficiently large number of appropriately small steps.

16-3 TRANSFORMATION OF THE DIFFERENTIAL EQUATIONS INTO DIFFERENCE EQUATIONS AND THE NUMERICAL METHOD DERIVED FROM THIS

Other step-by-step procedures, in particular a numerical method described here and the Willmott method (Sec. 16-4), depend upon transforming the differential equations into difference equations and seeking their numerical solutions. As with the Lambertson method (Sec. 16-2), in order to simplify notation the average temperature of the heat-storing mass will be denoted by Θ instead of Θ_m. Thus the differential Eqs. (13-19) and (13-21) take the form

$$\left(\frac{\partial \theta}{\partial f}\right)_t = \frac{\bar{\alpha}}{C}(\Theta - \theta) \tag{16-7}$$

and

$$\left(\frac{\partial \Theta}{\partial t}\right)_f = \frac{\bar{\alpha}\, df}{dC_s}(\theta - \Theta) \tag{16-8}$$

If the "differentials" are replaced by finite differences these differential equations are transformed into the following difference equations:

$$\frac{\Delta \theta}{\Delta f} = \frac{\bar{\alpha}}{C}(\Theta^* - \theta^*) \tag{16-9}$$

and

$$\frac{\Delta \Theta}{\Delta t} = \frac{\bar{\alpha}\, df}{dC_s}(\theta^* - \Theta^*) \tag{16-10}$$

where Θ^* and θ^* denote the average values of Θ and θ over the interval Δf or Δt.

The appropriate transformation of the differential equation which only contains the temperature of the heat-storing mass will be explained in Sec. 16-5.

Graphical Process

H. Hausen [H305] has developed a graphical method for calculating θ and Θ from the difference Eqs. (16-9) and (16-10). Although graphical methods no longer play an important role nowadays, the graphical determination of θ, using Eq. (16-9), will be briefly explained, as this will serve to facilitate understanding of the computational process, which will be dealt with later.

The distribution of the temperature Θ of the heat-storing mass at the time in question t is known and plotted as a function of f, as shown by the solid line in Fig. 16-3. On the other hand, the gas temperature θ represented by the dashed line, is only given up to the position $f = f_1$, where it has the value θ_1. Its value at θ_2 at

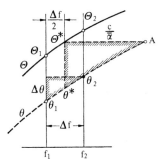

Figure 16-3 Graphical determination of the gas temperature θ for a given distribution of the temperature Θ of the heat-storing mass.

position $f_2 = f_1 + \Delta f$ is required. In order to find θ_2 one first determines the average value Θ^* of the temperature of the heat-storing mass by halving Δf. A horizontal line is drawn from the point which has been fixed in this way to point A lying at a distance $C/\bar{\alpha}$, and joining A with θ_1. This line crosses the perpendicular at position f_2 and gives the required temperature θ_2. That this construction satisfies Eq. (16-9) can be seen from the similarity of the two triangles whose edges are shaded. An appropriate process which corresponds to Fig. 16-3, and which satisfies Eq. (16-10) can be developed for the calculation of the temperature of the heat-storing mass. By applying first this process, in a suitable manner, and then the process mentioned previously, it is possible to draw the total temperature curve of the heat-storing mass and of the gas at an arbitrary time t, as long as the initial temperature of the heat-storing mass and the entry temperature of the gas are given. Moreover, the graphical process can be thoroughly examined, as described in [H305].

16-4 THE WILLMOTT METHOD

The numerical method of Willmott [W304] is based on the same ideas as the graphical process described above. In this Willmott method, the temperatures of the gas and the heat-storing mass, which are unknown, can be calculated without iteration, utilizing Eqs. (16-9) and (16-10). In the part of the process shown in Fig. 16-4, the brick and gas temperatures Θ_1, θ_1 and Θ_2, θ_2 are known at time t, whereas at time $t + \Delta t$ only the temperatures Θ_3 and θ_3 at position f_1 are known. The gas temperatures are not drawn in Fig. 16-4. Temperatures Θ_4 and θ_4 at position $f_2 = f_1 + \Delta f$ and at time $t + \Delta t$ are required.

It follows from Fig. 16-3 that

$$\Delta\theta = \theta_4 - \theta_3$$

$$\Theta^* = \frac{1}{2}(\Theta_3 + \Theta_4)$$

$$\theta^* = \frac{1}{2}(\theta_3 + \theta_4)$$

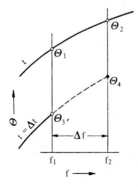

Figure 16-4 For the mathematical determination of the gas and heat-storing mass temperatures. Gas temperatures are not shown.

can be written for temperatures θ_3 and θ_4 at time $t + \Delta t$. Insertion into Eqs. (16-9) gives

$$\theta_4 = \theta_3 + \frac{\Theta_3 + \Theta_4 - 2\theta_3}{\dfrac{2C}{\bar{\alpha}\,\Delta f} + 1} \qquad (16\text{-}11)$$

Similarly, by assuming

$$\Delta\Theta = \Theta_4 - \Theta_2$$

$$\theta^* = \frac{1}{2}(\theta_2 + \theta_4)$$

$$\Theta^* = \frac{1}{2}(\Theta_2 + \Theta_4)$$

for position f_2 and the time interval Δt, Eq. (16-12) is obtained from Eq. (16-10):

$$\Theta_4 = \Theta_2 + \frac{\theta_2 + \theta_4 - 2\Theta_2}{\dfrac{2}{\bar{\alpha}\,\Delta t} \cdot \dfrac{dC_s}{df} + 1} \qquad (16\text{-}12)$$

According to Willmott, the two Eqs. (16-11) and (16-12) determine the unknown temperatures θ_4 and Θ_4. If θ_4 is eliminated from these the relationship[1] (16-13) results:

$$\Theta_4 = \frac{\left(\dfrac{2C}{\bar{\alpha}\,\Delta f} + 1\right)\left[\theta_2 + \left(\dfrac{2}{\bar{\alpha}\,\Delta t} \cdot \dfrac{dC_s}{df} - 1\right)\Theta_2\right] + \left(\dfrac{2C}{\bar{\alpha}\,\Delta f} - 1\right)\theta_3 + \Theta_3}{\left(\dfrac{2C}{\bar{\alpha}\,\Delta f} + 1\right)\left(\dfrac{2}{\bar{\alpha}\,\Delta t} \cdot \dfrac{dC_s}{df} + 1\right) - 1} \qquad (16\text{-}13)$$

If the brick temperature Θ_4 is obtained from this equation for the step in question, the gas temperature value θ_4 can at once be obtained from Eq. (16-11). The fact that both these two temperatures have to be determined at each step is a characteristic of this method.

Further Development of the Process[1]

A further particular of the Willmott method is that the two calculation steps using Eqs. (16-11) and (16-12) relate to the lower right-hand corner of the interval shown in Fig. 16-4. The mean value of the temperatures, used in deriving the equations relate to the corners labelled 2 and 3 in the figure. Thus, the effect of temperatures θ_1 and Θ_1 at point 1 of Fig. 16-4 is not taken into account. It is to be expected that the accuracy of the method can be increased if the best possible average values

---

[1] An as yet unpublished consideration by the author. (Essentially the same substitution for θ_4 in Eq. (16-12) is made in the original paper [W304]: Editor.)

within the complete interval surface can be formed using all available quantities. Average values, which use all four corner points of the interval are

$$\Theta^* = \frac{1}{4}(\Theta_1 + \Theta_2 + \Theta_3 + \Theta_4)$$

$$\theta^* = \frac{1}{4}(\theta_1 + \theta_2 + \theta_3 + \theta_4)$$

In addition, for the direction of the f-coordinate: one has

$$\Delta\theta = \frac{1}{2}[(\theta_2 + \theta_4) - (\theta_1 + \theta_3)]$$

and for the direction of the t-coordinate:

$$\Delta\Theta = \frac{1}{2}[(\Theta_3 + \Theta_4) - (\Theta_1 + \Theta_2)]$$

The following equations are obtained through insertion into Eqs. (16-9) and (16-10), using the abbreviations

$$\frac{\bar{\alpha}}{C}\Delta f = \Delta\xi \quad \text{and} \quad \frac{\bar{\alpha}\,df}{dC_s}\Delta t = \Delta\eta$$

(see Eqs. 13-24) and (13-25)

$$\theta_4 = \frac{\Delta\xi}{2+\Delta\xi}(\Theta_1 + \Theta_2 + \Theta_3 + \Theta_4) + \frac{2-\Delta\xi}{2+\Delta\xi}(\theta_1 + \theta_3) - \theta_2 \qquad (16\text{-}11a)$$

$$\Theta_4 = \frac{\Delta\eta}{2+\Delta\eta}(\theta_1 + \theta_2 + \theta_3 + \theta_4) + \frac{2-\Delta\eta}{2+\Delta\eta}(\Theta_1 + \Theta_2) - \Theta_3 \qquad (16\text{-}12a)$$

These two equations can also take the place of Eqs. (16-11) and (16-12). θ_4 and Θ_4 can also be determined purely mathematically from Eqs. (16-11) and (16-12). However θ_4 can also be eliminated from both equations and this gives

$$\Theta_4 = \frac{\dfrac{4}{2+\Delta\xi}\cdot\dfrac{\Delta\eta}{2+\Delta\eta}(\theta_1 + \theta_3) + \left[\dfrac{2-\Delta\eta}{2+\Delta\eta} + \dfrac{\Delta\xi}{2+\Delta\xi}\cdot\dfrac{\Delta\eta}{2+\Delta\eta}\right](\Theta_1 + \Theta_2)}{1 - \dfrac{\Delta\xi}{2+\Delta\xi}\cdot\dfrac{\Delta\eta}{2+\Delta\eta}} - \Theta_3$$

(16-13a)

Essentially simple equations of probably only slightly reduced accuracy are obtained on the basis of the consideration that

$$\Theta^* = \frac{1}{2}(\Theta_2 + \Theta_3)$$

and

$$\theta^* = \frac{1}{2}(\theta_2 + \theta_3)$$

already represent good average temperature values. If these expressions are put together with the expression obtained for $\Delta\theta$ and $\Delta\Theta$ into the differential Eqs. (16-9) and (16-10) one obtains

$$\theta_4 = \theta_1 - (1+\Delta\xi)\theta_2 + (1-\Delta\xi)\theta_3 + \Delta\xi(\Theta_2+\Theta_3) \qquad (16\text{-}11\text{b})$$

$$\Theta_4 = \Theta_1 + (1-\Delta\eta)\Theta_2 - (1+\Delta\eta)\Theta_3 + \Delta\eta(\theta_2+\theta_3) \qquad (16\text{-}12\text{b})$$

Because of their simplicity these equations have the advantage that they immediately give values for θ_4 and Θ_4 without the additional simultaneous equations having to be solved or one of the unknown quantities having to be eliminated. With the help of the method embodied in Eqs. (16-11) and (16-12) Willmott and Burns [W312] have recently examined the behavior of regenerators resulting from a step change in the entry temperature of one of the gases.

16-5 DETERMINATION OF THE AVERAGE TEMPERATURE ALONE OF THE HEAT-STORING MASS

It is often sufficient to determine only the temperature distribution in the heat-storing mass, particularly if the state of equilibrium is to be determined by calculations over a large number of successive periods. Hausen's method [H305], which is described below, is suitable for this purpose and from the author's experience, the computation associated with this method can be completed much more quickly than can those of the method shown in Fig. 16-3 or of the Willmott method. This is because it permits considerably larger steps, without reducing accuracy*; this will be shown at the end of this section. The gas temperature is then determined, using Fig. 16-3 or Eq. (16-11), only if it is needed for the final result.

The step-by-step process which will not be derived, starts with the differential Eq. (13-30) which contains only the temperature of the heat-storing mass. In a non-dimensionless form this equation, which is obtained from Eqs. (13-19) and (13-20), takes the following form:

$$\frac{\partial^2\Theta}{\partial f\,\partial t} + \bar{\alpha}\frac{df}{dC_s}\cdot\frac{\partial\Theta}{\partial f} + \frac{\bar{\alpha}}{C}\cdot\frac{\partial\Theta}{\partial t} = 0 \qquad (16\text{-}14)$$

Again $\Theta = \Theta_m$ denotes the brick temperature averaged over a cross-section.

As in Fig. 16-4, the temperatures Θ_1 and Θ_2 are assumed to be known at time t as is the temperature Θ_3 at time $t+\Delta t$ for the step under consideration. The temperature Θ_4 of the heat-storing mass at position f_2 at time $t+\Delta t$ is then required. In order to find Θ_4 we transform the differential Eq. (16-14) into a difference equation in which we set

$$\frac{\partial\Theta}{\partial f} = \frac{1}{2}\left(\frac{\Theta_2-\Theta_1}{\Delta f} + \frac{\Theta_4-\Theta_3}{\Delta f}\right) \qquad (16\text{-}15)$$

---

* Editor's note: The author has now revised his view of this matter. See the later discussion.

$$\frac{\partial \Theta}{\partial t} = \frac{1}{2}\left(\frac{\Theta_3 - \Theta_1}{\Delta t} + \frac{\Theta_4 - \Theta_2}{\Delta t}\right) \tag{16-16}$$

$$\frac{\partial^2 \Theta}{\partial f\, \partial t} = \frac{1}{\Delta t}\left(\frac{\Theta_4 - \Theta_3}{\Delta f} - \frac{\Theta_2 - \Theta_1}{\Delta f}\right) \tag{16-17}$$

Upon substituting these expressions into Eq. (16-14) we obtain

$$\Theta_4 = \Theta_1 + \frac{(\Theta_2 + \Theta_3 - 2\Theta_1) + \dfrac{\bar{\alpha}}{2}\left[\dfrac{\Delta f}{C} - \dfrac{df}{dC_s}\Delta t\right](\Theta_2 - \Theta_3)}{1 + \dfrac{\bar{\alpha}}{2}\left[\dfrac{\Delta f}{C} + \dfrac{df}{dC_s}\Delta t\right]} \tag{16-18}$$

From this equation is determined the required temperature Θ_4. The step lengths Δf and Δt in each stage of the calculation can be selected arbitrarily. However it is recommended that the values of Δf and Δt are held to be the same in each step. It is only in the region of the entrance of the gas at $\xi = 0$ to about $\xi = \dfrac{\bar{\alpha}}{C}f = 5$ that it is necessary to halve the step length as it is in this region that the temperature curves are most steeply bent.

Equation (16-18) can also be used, and provides a high degree of accuracy when $\bar{\alpha}/C$ and $\bar{\alpha}\,df/dC_s$ are functions of temperature and, in certain circumstances, also of time. Here, at each step, suitable average values are selected which are then held constant within the step but which, however, vary from one step to the next. In general the values of $\bar{\alpha}/C$ and $\bar{\alpha}\,df/dC_s$ permit this approach when the temperature Θ_1 (see Fig. 16-4) varies only slightly between Θ_1 and Θ_4. In case of doubt better average values are obtained which are based on the temperature $1/2(\Theta_2 + \Theta_3)$.

In the case where it is possible to perform calculations using constant heat transfer coefficients and physical properties with the dimensionless independent variables ξ and η, Eq. (16-18) can be reduced into the form of Eq. (16-20), taking into account Eqs. (13-24) and (13-25) together with relationships (16-19).

$$\frac{\bar{\alpha}}{C}\Delta f = \Delta \xi \qquad \text{and} \qquad \frac{\bar{\alpha}\,df}{dC_s}\Delta t = \Delta \eta \tag{16-19}$$

$$\Theta_4 = \Theta_1 + \frac{2(\Theta_2 + \Theta_3 - 2\Theta_1) + (\Delta\xi - \Delta\eta)(\Theta_2 - \Theta_3)}{2 + \Delta\xi + \Delta\eta} \tag{16-20}$$

The more general process, which is based on Eq. (16-18) can be simplified even further if it is assumed that

$$\Delta t = \frac{dC_s}{C\,df}\cdot \Delta f \tag{16-21}$$

Equation (16-18) is then transformed into

$$\Theta_4 = \Theta_1 + \frac{\Theta_2 + \Theta_3 - 2\Theta_1}{1 + \dfrac{\bar{\alpha}}{C}\Delta f} \tag{16-22}$$

This equation can not only be implemented computationally but also by means of a *graphical method*. This is described in Sec. 19-1 utilizing Fig. 19-2 for the special case of a wet regenerator. Equation (16-22) can be replicated simply by assuming that $\varepsilon = 1$ and $d\phi''/d\Theta = 0$ and making note of the importance of $\Delta\xi$ indicated by Eq. (13-24).

The dimensionless form of Eq. (16-21) is simply $\Delta\xi = \Delta\eta$. This restriction was introduced by the author [H305] in 1931 in order to develop the graphical method mentioned which is represented by Fig. 19-2. Ignoring this restriction immediately leads to Eq. (16-18) or (16-20) [H310]. Saunders and Smoleniec [S302], also Allen* [A301] developed a relationship which is essentially identical to Eq. (16-20).

Accuracy of the Step-by-Step Methods Set Out in Secs. 16-4 and 16-5†

For a long time it has not been clear whether the Willmott method, (Sec. 16-4) or the Hausen method (Sec. 16-5) provides the greater accuracy, assuming both methods to be operating with the same number of steps. The author's experience of similar numerical methods over almost 50 years had led him to the conclusion that the method described in Sec. 16-5 is the better one. However, research carried out recently has led him to conclude that both methods are equally accurate. His considerations will be described in detail in a separate publication. Here the basic ideas and conclusions will be described.

A step-by-step method is considered to be a good one if a parabolic curve, that is a curve which can be represented by a function of the second degree, can be replicated by the particular numerical method with only very few errors. In this way one proceeds from the most general solution of the second degree of the differential Eqs. (13-26) and (13-27), or even (13-30). The general expression of the second degree

$$\Theta = c + a\xi + b\eta + d \cdot \xi^2 + e \cdot \xi\eta + f \cdot \eta^2$$

leads to the following most general solution of the second degree (16-23) following insertion into Eq. (13-30) and with Eq. (13-27)

$$\left.\begin{array}{l} \Theta = c + a\xi + b\eta + \dfrac{a+b}{2}(\xi - \eta)^2 \\[3mm] \theta = (c+b) - b\xi + (a+2b)\eta + \dfrac{a+b}{2}(\xi - \eta)^2 \end{array}\right\} \qquad (16\text{-}23)$$

It will now be assumed that the curves in Fig. 16-4, and thus also the positions of the temperatures Θ_1, Θ_2, and Θ_3, correspond to solution (16-23). If one assumes

---

* Editor's note: Allen also indicated possible means of the acceleration of regenerator calculations.

† Author's note: This next section represents an alteration to the 1976 German text, which has been inserted at the request of the author.

$\Theta_1 = c, \xi = \Delta\xi, \eta = \Delta\eta$, the following are obtained from the Eq. (16-23):

$$\Theta_1 = c$$

$$\Theta_2 = c + a\,\Delta\xi + \frac{a+b}{2}(\Delta\xi)^2$$

$$\Theta_3 = c + b\,\Delta\eta + \frac{a+b}{2}(\Delta\eta)^2$$

Similar relationships result from the second Eq. (16-23) for the gas temperatures θ_1, θ_2, and θ_3.

We will try to ascertain which value of Θ_4 results from the given values of Θ_1, Θ_2, Θ_3 and $\theta_1, \theta_2, \theta_3$ using the two step-by-step methods described in Secs. 16-4 and 16-5 and how much this value deviates from the precise value which comes from the precise solution of the second degree (16-23).

Very complex but clear derivations can be obtained if the considerations are extended to arbitrary values of $\Delta\xi$ and $\Delta\eta$. On the other hand very simple relationships result when $\Delta\xi = \Delta\eta$. This usually proves to be exactly or approximately correct when the step-by-step methods are used. Thus

$$\Theta_4 = \Theta_1 + (a+b)\,\Delta\xi$$

is obtained for the Willmott method, using Eq. (16-13), and, likewise, the same expression is obtained for the Hausen method, using Eq. (16-22) together with Eqs. (13-24) and (13-25).

The same expression is reached for the solution (16-23) of the second degree, if $\xi = \eta = \Delta\xi = \Delta\eta$ is inserted into the solution. Thus the two processes of Willmott and Hausen replicate precisely the solution of the second degree of the differential equations when they have steps $\Delta\xi = \Delta\eta$ of an arbitrary size. Therefore, from this point of view, they should be considered to be equally precise.*

By far the largest number of regenerator calculations involve a temperature curve which within a step length, corresponds very approximately to the solution of the second degree discussed above. Only in the region of that end of the regenerator where the gas enters can this not be expected. Here there are not only relatively large temperature differences between the gas and the heat-storing mass during the first moments of a period but S-shaped temperature curves also appear which cannot be represented by a solution of the second degree.

A case where these differences appear much greater than for the state of equilibrium in a regenerator, is that of the first cooling of a heat-storing mass with an initially constant temperature, in which a gas enters with a constant lower temperature. The spatial and chronological temperature changes that take place here are shown in Figs. 13-5 and 13-6. This case is particularly suitable for testing the step-by-step processes described in Secs. 16-4 and 16-5 under the assumption that large deviations from the solution of the second degree appear. Such

---

* Editor's note: The same conclusion can be reached by noting that in the case where $\Delta\xi = \Delta\eta$, both the Hausen and Willmott methods have a truncation error associated with them of order $(\Delta\xi)^3$.

examinations were carried out so that, with the two methods, the temperature changes were calculated with steps of sizes $\Delta\xi = \Delta\eta = 0.5$, $= 1$ and $= 2$. As will be shown in detail in a separate publication, the Willmott method was superior in the first spatial and chronological steps up to about ξ or $\eta = 4$. With this method it was possible to achieve the same accuracy with larger steps, for example with $\Delta\xi = \Delta\eta = 1$ or $\Delta\xi = \Delta\eta = 2$, as was achieved with steps of less than half the size, using the Hausen method.* Thus, all the early steps, for example from $\xi = 0$ to $\xi = 2$ and $\eta = 0$ to $\eta = 2$ are critical. If, using the Willmott approach, one performs calculations starting with steps of size $\Delta\xi = \Delta\eta = 2$ but using the Hausen method, up to $\xi = 2$ and $\eta = 2$ one does the same calculation with steps of size $\Delta\xi = \Delta\eta = 1$ and from then onwards with $\Delta\xi = \Delta\eta = 2$, the same degree of accuracy will be achieved. The old rule of Hausen is thus confirmed that, when using his method, calculations should be carried out with steps of $\Delta\xi = \Delta\eta \leqslant 1$ up to about $\xi = 5$ and from then on with considerably larger steps, for example with $\Delta\xi = \Delta\eta = 3$.

That the Willmott method appears to be more suitable in this connection lies in the fact that only *one* temperature, Θ_2 in Fig. 16-4, is required at initial time step that is at $\eta = 0$. This takes account of the value of the high initial temperature of the heat-storing mass. On the other hand in the Hausen approach two temperatures, Θ_1 and Θ_2 are required in order to integrate from $\eta = 0$ to $\eta = \Delta\eta$. It follows that, in the second case, the result is displaced in the direction of higher initial temperature. Moreover, the Willmott method can be based upon the gas temperature distribution

$$\theta = \theta_0 + (\Theta_0 - \theta_0)(1 - e^{-\xi})$$

determined at the outset using Eq. (13-26), where θ_0 denotes the entry temperature of the gas and Θ_0 denotes the constant initial temperature of the heat-storing mass. Moreover, when carrying out the step-by-step procedure it is significant that, when $\xi = 0$ it follows from Eq. (13-27) that the temperature of the heat-storing mass at time η is equal to

$$\Theta = \theta_0 + (\Theta_0 - \theta_0)e^{-\eta}$$

If, as in most cases, we wish to determine the state of equilibrium of the regenerator, the differences between the results of using both methods are considerably less than indicated in the extreme example described above. As can be seen, for example, in Fig. 13-15, top, in the state of equilibrium when $\xi = 0$, the temperature difference between the gas and the heat-storing mass is much smaller than the highest temperature difference in the system, namely the difference between the temperatures of the two entry gases. Moreover the temperature of the heat-storing mass when $\eta = 0$, i.e., at the beginning of the cold period, is not constant but clearly increases from $\xi = 0$ onwards with increasing ξ. Both influences lessen the errors which are expected from both methods and thus also lessen the differences between these two procedures. Moreover, Hausen's Eq. (16-22) is simpler than Willmott's Eqs. (16-11)

---

* Editor's note: This can be anticipated from the fact that higher order derivatives are embodied in the truncation error associated with the Willmott method than are in the Hausen method.

and (16-13).* Since the gas temperatures do not have to be calculated, only slightly more than a third of the computational effort required by the Willmott process is necessary for the Hausen method.

16-6 STEP-BY-STEP METHOD PROVIDING A MORE PRECISE CALCULATION OF THE BRICK TEMPERATURE

Willmott [W307] and Schellmann [S309], and later Manzique and Cardenas [M302] have independently developed a method which not only allows the gas temperature and the average brick temperature to be calculated but also, at the same time, permits calculation of the spatial temperature distribution within a brick cross-section. In a way different to Figs. 11-6, 12-4, 12-6, 12-11 and 12-12 which show these temperature distributions for the zero eigen function, Willmott and Schellmann's methods are applied right up to the ends of the regenerator where the higher eigen functions also have a strong influence. The chronological variation of the gas temperature can thus be correctly obtained at all times during a period.

The heat transfer coefficients α and α' and the physical properties will be assumed to be independent of temperature and the gas flowrate will be considered to be constant so that, apart from the temperatures themselves, calculations can be made with dimensionless quantities. It follows from Eq. (13-24) that the reduced longitudinal coordinate ξ in the direction of flow and the reduced time η can be used as dimensionless variables but embodying the true heat transfer coefficient α rather than $\bar{\alpha}$. The heat-storing mass is considered to be constructed from parallel, smooth plates of thickness δ. In a cross-section perpendicular to the direction of flow z will be the distance of a position within the brick from its surface. The dimensionless distance ζ will be introduced for this using the equation below.

$$\zeta = \sqrt{\frac{\alpha}{a} \cdot \frac{df}{dC_s}} : z = \sqrt{\frac{2\alpha}{\lambda_s \delta}} \, z \qquad (16\text{-}24)$$

where a is the thermal diffusivity and λ_s is the thermal conductivity of the material of which the heat-storing mass is constructed. Using these dimensionless quantities one obtains the differential equations and the boundary conditions in the following form.

The differential equation for unsteady state thermal conduction in the brick

$$\frac{\partial \Theta}{\partial t} = a \frac{\partial^2 \Theta}{\partial z^2} \qquad (16\text{-}25)$$

is transformed:

$$\frac{\partial \Theta}{\partial \eta} = \frac{\partial^2 \Theta}{\partial \zeta^2} \qquad (16\text{-}26)$$

---

* Editor's note: The Editor feels in duty bound not to discuss this point in this book!

In the middle of the brick $z = \delta/2$ and on the basis of symmetry

$$\frac{\partial \Theta}{\partial \zeta} = 0 \qquad \text{when} \qquad \zeta = \sqrt{\frac{\alpha \delta}{2\lambda_s}} \left(z = \frac{\delta}{2} \right) \tag{16-27}$$

At the brick surface $\zeta = 0$, at temperature Θ_0, the boundary condition

$$\frac{\partial \Theta}{\partial z} = \frac{\alpha}{\lambda_s} (\Theta_0 - \theta) \qquad \text{with } z = 0 \tag{16-28}$$

will be satisfied which, with dimensionless variables, gives*

$$\frac{\partial \Theta}{\partial \zeta} = \sqrt{\frac{\alpha \delta}{2\lambda_s}} (\Theta_0 - \theta) \qquad \text{with } \zeta = 0 \tag{16-29}$$

The changes in gas temperature θ in the direction of flow are represented by the equation:

$$\frac{\partial \theta}{\partial \xi} = \Theta_0 - \theta \tag{16-30}$$

For the sake of simplicity it will be assumed that the gases enter the regenerator in the warm and cold periods, at the following temperatures:

$$\theta(\xi = 0, \eta) = +1 \qquad \text{for} \qquad 0 \leqslant \eta \leqslant \Pi \tag{16-31}$$

$$\theta'(\xi' = 0, \eta') = -1 \qquad \text{for} \qquad 0 \leqslant \eta' \leqslant \Pi' \tag{16-32}$$

Thus the dashed values always relate to the cold periods.

The sizes of the steps which are to be chosen will be denoted by $\Delta\zeta$, $\Delta\xi$ and $\Delta\eta$ so that a specific position in the three dimensional space of the spatial and chronological events is determined by $\zeta = i\,\Delta\zeta$, $\xi = j\,\Delta\xi$ and $\eta = k\,\Delta\eta$. This position will simply be denoted by i, j, k. At time $\eta = k \cdot \Delta\eta$ all the temperatures of the heat-storing mass and the gas are known and at time $\eta + \Delta\eta = (k+1)\,\Delta\eta$ they are only known as far as the regenerator cross-section $\xi = j \cdot \Delta\xi$. All the temperatures in the cross-section $j+1$ at time $k+1$ are required to be calculated. Thus, apart from just one exception, since all temperatures are related to the cross-section $j+1$, the index $j+1$ will be neglected.

In order to transform the differential Eq. (16-26) for thermal conduction into a difference equation, we assume

$$\left(\frac{\partial \Theta}{\partial \eta} \right)_{i,k+1/2} = \frac{\Delta\Theta}{\Delta\eta} = \frac{1}{\Delta\eta} (\Theta_{i,k+1} - \Theta_{i,k})$$

$$\left(\frac{\partial \Theta}{\partial \zeta} \right)_{i+1/2,k} = \frac{1}{\Delta\zeta} (\Theta_{i+1,k} - \Theta_{i,k})$$

$$\left(\frac{\partial \Theta}{\partial \zeta} \right)_{i-1/2,k} = \frac{1}{\Delta\zeta} (\Theta_{i,k} - \Theta_{i-1,k}) \tag{16-33}$$

---

* Editor's note: The approach described here preserves the familiar dimensionless variable η. The otherwise familiar Biot modulus usually associated with this boundary condition is retained but appears within the square root sign. Other authors sometimes use the dimensionless parameters $\eta = 4at/\delta^2$, $\zeta = z/\zeta$ with the same representation for ξ used here. The same Eqs. (16-26) and (16-30) emerge but the square root sign disappears in Eq. (16-29).

and thus

$$\left(\frac{\partial^2 \Theta}{\partial \zeta^2}\right)_{i,k} = \frac{1}{\Delta \zeta^2}(\Theta_{i+1,k} + \Theta_{i-1,k} - 2\Theta_{i,k})$$

A corresponding expression is obtained for $k+1$. Taking the average of both these we yield

$$\left(\frac{\partial^2 \Theta}{\partial \zeta^2}\right)_{i,k+1/2} = \frac{1}{2\Delta \zeta^2}[\Theta_{i+1,k+1} + \Theta_{i-1,k+1} - 2\Theta_{i,k+1} + \Theta_{i+1,k} + \Theta_{i-1,k} - 2\Theta_{i,k}]$$

$$(16\text{-}34)$$

It follows from the differential Eq. (16-26) that this is equal to the expression (16-33) for $\left(\dfrac{\partial \Theta}{\partial \eta}\right)_{i,k+1/2}$. Thus, with $p = \dfrac{\Delta \eta}{\Delta \zeta^2}$ we obtain

$$p \cdot \Theta_{i+1,k+1} - (2p+2)\Theta_{i,k+1} + p \cdot \Theta_{i-1,k+1}$$
$$+ p \cdot \Theta_{i+1,k} - (2p-2)\Theta_{i,k} + p \cdot \Theta_{i-1,k} = 0 \qquad (16\text{-}35)$$

Willmott and Schellmann derived this equation independently of one another and indicated that it originated in 1947 in the work of Crank and Nicholson [C305]. In addition, Willmott mentioned the findings of Mitchell and Pearce [M303], according to which they were able to achieve a high degree of accuracy, by making only a very slight alteration to the last Eq. (16-35). Indeed, the same numerical example shows that Crank and Nicholson's method is also very accurate and extends to all practical cases.*

The relationships shown below are obtained by replacing Eqs. (16-29) and (16-30), which were given in differential form, by corresponding difference equations.

Between the centre of the brick $z = \delta/2$ and the surface there are considered to be m steps of size $\Delta \zeta$. Equation (16-27) can thus be replaced by

$$\Theta_{m+1,k+1} = \Theta_{m-1,k+1} \qquad (16\text{-}36)$$

Equation (16-29) takes the form

$$\frac{\Theta_{1,k+1} - \Theta_{-1,k+1}}{2\Delta \zeta} = \sqrt{\frac{\alpha}{\lambda_s} \cdot \frac{\delta}{2}} \cdot (\Theta_{0,k+1} - \theta_{k+1}) \qquad (16\text{-}37)$$

and Eq. (16-30) with which, as previously mentioned, the position $\xi = j \cdot \Delta \xi$ is also involved, transforms into

$$\frac{\theta_{j+1,k+1} - \theta_{j,k+1}}{\Delta \xi} = \frac{1}{2}[(\Theta_0)_{j+1,k+1} + (\Theta_0)_{j,k+1} - (\theta_{j+1,k+1} - \theta_{j,k+1})] \quad (16\text{-}38)$$

The linear Eqs. (16-35) to (16-38) determine the $m+3$ unknown temperatures of the brick from $\Theta_{-1,k+1}$ to $\Theta_{m+1,k+1}$ at position $\xi(j+1)\Delta \xi$ at time $\eta = (k+1)\Delta \eta$

---

* Editor's note: The most important feature of both the Crank Nicolson scheme and that of Mitchel and Pearce is that the time step size is not restricted by stability problems.

and the unknown gas temperature $\theta_{j+1,k+1}$: thus, in total, they determine $m+4$ unknowns. The number of equations is just as large. This is because the Eqs. (16-35) can be written for all values from $i=0$ to $i=m$ so that $m+1$ equations are obtained. Thus we get the three additional Eqs. (16-36), (16-37) and (16-38).

These simultaneous linear equations can be solved using known methods. Indeed, Willmott has shown how these equations can be suitably modified in advance.

If we wish to compare the results obtained in this way with those obtained more simply by performing the calculations with $\bar{\alpha}$ and $\bar{\alpha}'$, we must take note that, in the method discussed so far with $f=F$ and $t=T$, the expressions $\dfrac{\alpha}{C}F$ and $\dfrac{2\alpha}{\rho c\,\delta}\cdot T$ are obtained and these are not equal to $\Lambda=\dfrac{\bar{\alpha}}{C}F$ and $\Pi=\dfrac{2\bar{\alpha}}{\rho c\,\delta}\cdot T$ as in Eqs. (13-28) and (13-29). In order to obtain the reduced length Λ and the reduced period length Π, the expressions which are obtained initially must first be multiplied by $\bar{\alpha}/\alpha$.

Comparative calculations, carried out by Schellmann, show that the simpler method described in Sec. 16-5 yields the variations of the average brick temperature as a function of ξ and η with a high degree of accuracy. In Sec. 13-2 it has already been shown, how, on the basis of these sorts of results, the chronological variation of the gas temperature can also be determined reasonably accurately.

16-7 REGENERATORS WITH VARIABLE FLOWRATES

During the normal operation of regenerators the gas, within each period, leaves the regenerator at a temperature which varies with time. However, for many purposes, blast furnaces for example, a largely constant temperature is required for the air being heated in the regenerator or for the other gas.

Frequently, in recent installations, the fluctuations in the temperature of the hot blast entering the blast furnace have been reduced by about half by means of the so-called "staggered" parallel scheme using two pairs of hot blast stoves. Here the switching of the two pairs of regenerators is staggered by half a blast period and the air leaving them is mixed before introduction into the blast furnace.

In another process, which is normal nowadays, the requirement for a constant temperature is satisfied by mixing together, in varying proportions the hot air coming out of a hot blast stove with cold air.* By suitably regulating the quantities of air flowing through the hot blast stove at a particular moment and the air which is to be mixed with it, it is possible to ensure that at any given instant constant flowrates and constant temperatures are supplied to the blast furnace. Below it will be shown how the processes in a regenerator can be calculated using this type of variable flow rate.

---

* Editor's note: Called *by-pass main operation* in English scientific literature.

In the following section it will first be shown how to determine the chronological variations involved. It will be assumed that the chronological change in the quantity of air flowing through the regenerator is known. Using the step-by-step process which has already been described in Sec. 16-5, which uses the heat transfer coefficient $\bar{\alpha}$ which relates to the average brick temperature, one is able to take into account the change in the flow rate by carrying out the calculations using an average value of the flowrate at each step which is varied from one time interval to another. Thus, one must insert the values of $\bar{\alpha}$ and C in Eq. (16-28) which corresponds to the current fluid velocity.

This same measure is also possible when using the Willmott method described in Sec. 16-4. However, Willmott [W305, 306] has developed a specific method for this purpose which emanates from the basic ideas built into his process which is mentioned in Sec. 16-4. He does not perform the calculations using the average values of $\bar{\alpha}$ and C but uses precise values of these quantities at points 2, 3 and 4 shown in Fig. 16-4. They are thus determined by the fluid velocities and physical properties which relate to these points. If the suffixes 2, 3 and 4 are applied to the quantities associated with these points Eqs. (13-19) and (13-21) can be transformed into the following difference equations:

$$\theta_4 = \theta_3 + \frac{\Delta f}{2}\left[\left(\frac{\bar{\alpha}}{C}\right)_3 (\Theta_3 - \theta_3) + \left(\frac{\bar{\alpha}}{C}\right)_4 (\Theta_4 - \theta_4)\right] \tag{16-39}$$

$$\Theta_4 = \Theta_2 + \frac{\Delta t}{2}\left[\left(\frac{\bar{\alpha}\,df}{dC_s}\right)_2 (\theta_2 - \Theta_2) + \left(\frac{\bar{\alpha}\,df}{dC_s}\right)_4 (\theta_4 - \Theta_4)\right] \tag{16-40}$$

With the abbreviations

$$\begin{array}{ll}
\frac{\Delta f}{2}\cdot\left(\frac{\bar{\alpha}}{C}\right)_3 = A_3; & \frac{\Delta f}{2}\cdot\left(\frac{\bar{\alpha}}{C}\right)_4 = A_4 \\[2mm]
\frac{\Delta t}{2}\cdot\left(\frac{\bar{\alpha}\,df}{dC_s}\right)_2 = B_2; & \frac{\Delta t}{2}\cdot\left(\frac{\bar{\alpha}\,df}{dC_s}\right)_4 = B_4
\end{array} \right\} \tag{16-41}$$

Eqs. (16-39) and (16-40) are transformed into

$$\theta_4 = \theta_3 + A_3(\Theta_3 - \theta_3) + A_4(\Theta_4 - \theta_4) \tag{16-42}$$

$$\Theta_4 = \Theta_2 + B_2(\theta_2 - \Theta_2) + B_4(\theta_4 - \Theta_4) \tag{16-43}$$

Solving Eq. (16-42) using θ_4 and Eq. (16-43) using Θ_4 yields

$$\theta_4 = \frac{(1 - A_3)\theta_3 + A_3\Theta_3 + A_4\Theta_4}{1 + A_4} \tag{16-44}$$

$$\Theta_4 = \frac{(1 - B_2)\Theta_2 + B_2\theta_2 + B_4\theta_4}{1 + B_4} \tag{16-45}$$

Finally, insertion of (16-44) into Eq. (16-45) gives

$$\Theta_4 = \frac{(1 + A_4)[(1 - B_2)\Theta_2 + B_2\theta_2] + B_4[(1 - A_3)\theta_3 + A_3\Theta_3]}{1 + A_4 + B_4} \tag{16-46}$$

The quantities A_3 and B_2 can be calculated from Eqs. (16-41) onwards. In order to determine A_4 and B_4, on the other hand, Θ_4 and θ_4 must first be evaluated, for example by setting these temperatures equal to Θ_1 and θ_1 or, better still, equal to $1/2(\Theta_2 + \Theta_3)$ and $1/2(\theta_2 + \theta_3)$. The approximation $A_4 = 1/2(A_2 + A_3)$ and $B_4 + 1/2(B_2 + B_3)$ could also be used. Thus Θ_4 is determined by Eq. (16-46) and θ_4 is determined by Eq. (16-40) for the step under consideration where the values of A_4 and B_4 do not have to be corrected iteratively.

In place of an iteration process Willmott [W306], who has derived these equations in a complicated form, has suggested the following approach. Using a more simple step-by-step method the calculations are carried out through a series of periods, first of all, for a constant flowrate which corresponds, as closely as possible, to the average value of the variable flowrates which occur in practice. For this starting process the calculation continues until the strictly periodic state of equilibrium is reached sufficiently closely. The temperatures determined in this way can serve as a starting point for the calculation of the variable flowrate using Eqs. (16-46) and (16-44). The values of A_3, A_4, B_2 and B_4 are fixed from this temperature distribution for each step which can be used for calculating the first of the periods which now follow. For each further period there results a different temperature distribution with new values of A_2 to B_4 which are based upon the previous period. After a sufficient number of periods have been calculated one finally reaches the required state of equilibrium with the prescribed variable flowrate.

Determination of the Chronological Dependence of the Variable Flowrate

In almost all the practical applications of the method just described there exists a further difficulty that it is not known how the flowrate varies with time. Rather it is required that the air leaving the regenerator at a variable temperature θ_2 should have a quantity of cold air added to it in such a way that at each moment a constant flowrate $\dot{m}^*$ at a constant temperature θ^* is delivered. As a rule the air, at flowrate $\dot{m}$, which is to be heated in the regenerator and the air, at rate $\dot{m}^* - \dot{m}$, which is to be added to it, originate from the same air stream and therefore start off from the same initial temperature θ_1. Thus this requirement for temperature independent thermal capacity of the air can be expressed through the equation

$$\dot{m}(\theta_2 - \theta_1) = \dot{m}^*(\theta^* - \theta_1) \tag{16-47}$$

The regenerator calculations are carried out in such a way that Eq. (16-47) is always satisfied. Moreover, at the end of the period, θ_2 should still be somewhat higher than θ^*. In order to fix the end value of θ_2 one can, for example, prescribe that at the end of the period, cold air amounting to 5 per cent of the total flow $\dot{m}^*$ is added so that $(\dot{m})_{end} = 0.95\dot{m}^*$ and thus $0.95(\theta_2 - \theta_1) = \theta^* - \theta_1$.

Willmott [W306] suggests that one can proceed in the following way in order to determine the chronological variation of the flowrate. After calculating through several periods using a more simple method for constant flowrates, one begins the more precise calculation with an assumption about the chronological variation of the flowrate, for example, that during the cold period $\dot{m}$ increases linearly with time from $0.75\,\dot{m}^*$ to $0.95\,\dot{m}^*$. Calculation through one period then gives a specific chronological variation in the exit temperature θ_2. However, using Eq. (16-47) a somewhat different chronological variation of the flowrate $\dot{m}$ can then be determined. With this new variation in flowrate one calculates through a new period. In this way, proceeding from cycle to cycle, after each period one can obtain a precise curve for the variation not only of the temperature but also of the flowrate. The calculation is terminated as soon as the differences in the temperature variations at the end of two successive cycles have become so small that they can be ignored.

16-8 SIMPLER APPROACH TO THE DETERMINATION OF THE TIME DEPENDENCE OF A VARIABLE FLOWRATE

The chronological variation of the flowrates in the state of equilibrium of the regenerator can be determined by means of the method described in Sec. 16-7. However, Hausen [H311] has shown that the chronological change in the flowrate and also in the exit temperature θ_2 can be determined more simply, more quickly and with greater accuracy by suitably transforming the change in the exit temperature found for a constant flowrate. An error calculation which fits in with the discussion which follows will confirm this. However, if sometimes accuracy cannot be achieved, the result of the transformation will at least afford a good starting point for a more precise calculation using the process described in Sec. 16-7.

The type of transformation which is to be described starts with a relationship, which is approximately accurate, between the chronological changes in the flowrate and the corresponding changes in the heat transfer coefficient. Again $\bar{\alpha}$ will denote the heat transfer coefficient related to the average brick temperature Θ and $\dot{m} = dm/dt$ will be the momentary flowrate in the regenerator. In addition, at a specific time, $(\Theta - \theta)_m$ will denote the average value of $\Theta - \theta$ along the total length of the regenerator. Then the quantity of heat transferred in the regenerator in a very short time will be

$$dQ = \bar{\alpha}F(\Theta - \theta)_M\,dt = \dot{m}c_p(\theta_2 - \theta_1)\,dt \qquad (16\text{-}48)$$

It can be seen from this equation that the ratio of the quantity of heat being transferred dQ to the quantity of gas $\dot{m} \cdot dt$ passing through the regenerator at a given temperature is given by the ratio $\bar{\alpha}/\dot{m}$.

The amount by which $\dot{m}$ itself may vary can be seen from a report by Kessels [K302] which deals with measurements in a large number of hot blast stoves, and also in example calculations carried out by Willmott [W305, 306]. It follows from these that the deviations in the flow of air which are to be expected at the beginning and the end of the blast period amount, at most, to ± 25 per cent of their average value $\dot{M}$ and usually lie below 10 per cent. The heat transfer coefficient is about 0.7th to 0.8th power of the fluid velocity and thus is also proportional to the flowrate. If calculations are carried out using the exponent 0.75, $\bar{\alpha}/\dot{m}$ will be proportional to $(\dot{m})^{-0.25}$. If the flowrate varies by 25 per cent of its average value,

$\bar{\alpha}/\dot{m}$ will thus vary by 6–7 per cent. A deviation in the flowrate of 10 per cent will effect a change in $\bar{\alpha}/\dot{m}$ of 2.5 per cent.

It is clear from this that no large errors will occur if $\bar{\alpha}/\dot{m}$ is considered to be constant. Care must be taken that the errors arising during the middle and largest part of the period under consideration are smaller than at the beginning and end of the period and, have opposite signs in the second half of the period to those in the first half. Strictly speaking these observations are valid for the true heat transfer coefficient α, since $\bar{\alpha}$ is rather less dependent upon $\dot{m}$.

The assumption that $\bar{\alpha}/\dot{m}$ = const leads to the following consideration. If the flowrate corresponds to less than the chronological mean within the period, then when $\bar{\alpha}/\dot{m}$ = const, the heat transfer coefficient will also be proportionally smaller. The ratio of the quantity of heat transferred to the flowrate thus remains unchanged. However the process requires more time as gas flowrate and rate of heat transfer are slowed down. Nevertheless the same amount of heat transferred and thus also the same change in θ_2 corresponds to the partial flowrate which passes through the regenerator. θ_2 thus becomes a clear function of the quantity m, which flows through the regenerator from the beginning of the period up to the point in time being considered. This function is independent of whether or not the flowrate varies with time. If this function is known for a constant flowrate it can also be transformed, unchanged, to the case of a chronologically variable flowrate.

Transfer from Constant Flowrates to Chronologically Varying Flowrates

The chronological change in θ_2 will be known for the state of equilibrium for a *constant* flowrate, which is equal to the chronological, average value of $\dot{M}$ of the variable flowrate. The dependence upon time t, which is thus determined, is replicated by means of an equation of the form

$$\theta_2 - \theta_1 = f(t) \tag{16-49}$$

If, like the quantity of fluid m, t is calculated from the beginning of the period, with a constant flowrate,

$$t = m/\dot{M} \tag{16-50}$$

can be set in each time point, as in this case $\dot{M}$ is equal to the flowrate of the fluid. Thus, it follows from Eq. (16-49) that

$$\theta_2 - \theta_1 = f(m/\dot{M}) \tag{16-51}$$

It will be seen from the considerations above that this equation is also valid for a time varying flowrate, although in this case Eqs. (16-49) and (16-50) would not be satisfied.

It has been discussed above that after leaving the regenerator, the variable flowrate $\dot{m}$ will be mixed with a stream of cold air at temperature θ_1 in such a way that a constant flowrate $\dot{m}^*$ at a prescribed constant temperature θ^* results. The equation representing this has already been established:

$$\dot{m}(\theta_2 - \theta_1) = \dot{m}^*(\theta^* - \theta_1) \tag{16-52}$$

The equation

$$\dot{m} \cdot f(m/\dot{M}) = \dot{m}^*(\theta^* - \theta_1)$$

follows by inserting (16-51) into (16-52). Upon noting that $\dot{m} = dm/dt$ and by integrating from 0 to m or 0 to t we obtain

$$\int_0^m f\left(\frac{m}{\dot{M}}\right) dm = \dot{m}^*(\theta^* - \theta_1) \cdot t$$

or

$$t = \frac{\dot{M}}{\dot{m}^*(\theta^* - \theta_1)} \int_0^m f\left(\frac{m}{\dot{M}}\right) d\left(\frac{m}{\dot{M}}\right) \tag{16-53}$$

It follows from this equation that $\theta_2 - \theta_1$ and $\dot{m}$ can be calculated as a function of t by assuming values of $m/\dot{M}$. Then $(\theta_2 - \theta_1)$ can be calculated from Eq. (16-51), $\dot{m}$ from Eq. (16-52) and t from Eq. (16-53). As the following estimation of error shows, such a calculation allows the required chronological change $\dot{m}$ in the flowrate to be determined, with sufficient accuracy, certainly for practical cases.

If the thermal capacity of the gas is dependent upon temperature the calculation can, in essence, be carried out in the same manner by inserting the corresponding value of the enthalpy in place of the temperatures θ_1, θ_2 and θ^*.

Estimation of the Errors Which Frequently Occur in the Transformation

An arbitrary time point will be considered for which, as in the transformation described, $\dot{m}$, θ_2 and, based on the assumption that $\bar{\alpha}/\dot{m} = \text{const}$, $\bar{\alpha}$ will be known. The corresponding temperatures Θ_1 and Θ_2 of the heat-storing mass are determined at this point in time by a previous calculation using a constant flowrate and by the transformation. The errors in the determination of $\dot{m}$ and θ_2 can then be estimated in the following way.

For the sake of simplicity, the average temperature difference $(\Theta - \theta)_M$ is set equal to the arithmetic mean of the temperature differences at both ends of the regenerator in Eq. (16-48). This equation will then be transformed:

$$\bar{\alpha}F[(\Theta_2 - \theta_2) + (\Theta_1 - \theta_1)] = 2\dot{m}c_p(\theta_2 - \theta_1) \tag{16-54}$$

After $\dot{m}$ has been obtained by means of the transformation, a more precise value of $\bar{\alpha}$ can be calculated, which will be denoted by $\bar{\alpha} + \Delta\bar{\alpha}$, by using the relevant heat transfer equation for the corresponding revised fluid velocity. A revised exit temperature thus results: $\theta_2 + \Delta\theta_2$. In order to determine this approximately, $\bar{\alpha} + \Delta\bar{\alpha}$ is inserted into Eq. (16-54) in place of $\bar{\alpha}$ and $\theta_2 + \Delta\theta_2$ is inserted into the same equation in place of θ_2. Finally, upon eliminating the expression $F/2\dot{m}c_p$ from the new equation and the original Eq. (16-54) we yield

$$\Delta\theta_2 = \frac{\dfrac{\Delta\bar{\alpha}}{\bar{\alpha}}(\Theta_2 + \Theta_1 - \theta_2 - \theta_1)}{\dfrac{\Delta\bar{\alpha}}{\bar{\alpha}} + \dfrac{\Theta_2 + \Theta_1 - 2\theta_1}{\theta_2 - \theta_1}} \tag{16-55}$$

It will be observed from this equation that $\Delta\theta_2$ can be calculated for various positions in time. In this way an idea can be obtained as to the errors which have arisen during the transformation. If necessary, θ_2 can also be corrected by adding $\Delta\theta_2$. With these revised values Eq. (16-52) also yields a better value for $\dot{m}$.

If, as an example of the use of Eq. (16-55), it is assumed that $\Theta_2 = 980°$, $\theta_2 = 900°$, $\Theta_1 = 110°$ and $\theta_1 = 50°$ at the beginning of the blast period and the unfavourable case mentioned above is used, namely that $\Delta\bar{a}$ is 7 per cent of $\bar{a}$, it follows from Eq. (16-55) that $\Delta\theta_2 = 7.9°$. On the other hand, it will be seen from Eq. (16-52) that $\dot{m}$ only decreases by 1 per cent. On average the errors are considerably smaller, for the reasons mentioned. It can be recognized from this that in all practical cases the chronological variation of $\dot{m}$ can be obtained very precisely through the simple transformation described.

Equation (16-55) can scarcely be made invalid by the fact that in Eq. (16-54) the arithmetic mean is used instead of the logarithmic mean. This is because in the second application of Eq. (16-54) almost the same error occurs so that this error can be pretty well cancelled out. It is to be noted that θ_2, Θ_1 and Θ_2 also change slightly but that, according to Eq. (16-55), these changes will have hardly any influence on the value of $\Delta\theta_2$.

16-9 EVALUATION OF THE CHRONOLOGICAL VARIATION IN THE EXIT TEMPERATURE θ_2 AND IN THE FLOWRATE $\dot{m}$ WHEN ONLY THE AVERAGE VALUE OF THE EXIT TEMPERATURE IS KNOWN

The method of transformation described for the determination of the time variation of the flowrate assumes that the state of equilibrium of the regenerator with a constant flowrate has been determined previously. This preliminary calculation is time consuming since for example, a large number of cycles must be calculated when a step-by-step process is used. For this reason it is often desirable as in [H311] and in what follows, to be able to simply and quickly evaluate the time variation of θ_2 for a constant flowrate.

In Sec. 11-4 was shown how the heat transmission coefficient k of a regenerator, and thus also its efficiency η_R, can be calculated in a simple manner. Because of the way efficiency is defined, such a calculation also allows the chronological average value $\bar{\theta}_2$ of the exit temperature of the gas to be determined. With this the chronological variation of all temperatures at the warm end of the regenerator can be simultaneously evaluated in the following way. Figure 16-5 shows the basic curves for these temperatures. The reduced time η or η', defined in Eq. (13-24) is the abscissa with η' shown from the right for the cold period and with η from the left for the warm period; indeed, if it is necessary, the two period lengths Π and Π' can also be represented by lines of the same length each drawn on a different scale. θ_1' is the constant entry temperature and Θ' is the brick temperature during the warm period while Θ and θ_2 denote the temperatures in the cold period.

First the ratio σ of the temperature differences $\theta_1' - \Theta_E$ and $\theta_1' - \Theta_A$ will be

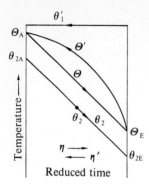

Figure 16-5 Evaluation of the chronological variation of temperature at the warm end of a regenerator when, apart from the entry temperature θ'_1 of the warm gas, only the average exit temperature $\bar{\theta}_2$ of the other gas is known.

determined. It will be seen in Fig. 16-5 that Θ_A and Θ_E represent the brick temperatures at the moment of reversal, that is, at the beginning and end of the cold period. When $\theta'_1 = $ const the solution of the differential Eq. (13-27) for the warm period takes the form

$$\theta'_1 - \Theta' = (\theta'_1 - \Theta_E)\exp(-\eta')$$

Equation (16-56) thus follows for the end of the warm period when $\eta' = \Pi'$:

$$\sigma = \frac{\theta'_1 - \Theta_E}{\theta'_1 - \Theta_A} = \exp(+\Pi') \tag{16-56}$$

Accordingly, σ can be assumed to be known in the following considerations.

Results of earlier calculations (see, for example, Fig. 13-14) show that, under the normal conditions which prevail in hot blast stoves and other regenerators, when there is a constant flowrate the temperatures Θ and θ_2 at the warm end of the regenerator vary almost linearly during the cold period. Thus in Fig. 16-5 these are represented by slightly curved lines. For the sake of simplicity any non-linearity will be neglected in the equation below

$$\theta'_1 - \theta_2 = (\theta'_1 - \theta_{2A})\left(1 + \beta \cdot \frac{\eta}{\Pi}\right) \tag{16-57}$$

where β is a constant which is still to be determined.

Equation (16-58) is obtained from Eq. (13-27) for the corresponding chronological variation of the temperature Θ of the heat-storing mass:

$$\theta'_1 - \Theta = (\theta'_1 - \theta_{2A})\left(1 + \beta \frac{\eta - 1}{\Pi}\right) \tag{16-58}$$

With $\eta = 0$ an expression for $\theta'_1 - \Theta_A$ results from this equation and with $\eta = \Pi$ it gives an expression for $\theta'_1 - \Theta_E$. If these two expressions are inserted into Eq. (16-56) and solved for β the Eq. (16-59) is then found:

$$\beta = \frac{\sigma - 1}{1 + \dfrac{\sigma - 1}{\Pi}} \tag{16-59}$$

The calculation of the variation of the exit temperature θ_2 using Eq. (16-57) now only requires knowledge of the initial temperature θ_{2A}. This is determined by assuming that the average exit temperature $\bar{\theta}_2$ is known. Because θ_2 varies linearly, $\bar{\theta}_2$ is equal to the value of θ_2 in the middle of the period, that is when $\eta = \Pi/2$. It thus follows from Eq. (16-57) that

$$\theta'_1 - \theta_{2A} = \frac{\theta'_1 - \bar{\theta}_2}{1 + \dfrac{\beta}{2}} \tag{16-60}$$

With the values of β and θ_{2A} obtained from Eqs. (16-59) and (16-60), the chronological variation of the exit temperature θ_2 of the air heated in the regenerator can be determined for the case where the flowrate does not vary with time. With the help of the transformation mentioned in Sec. 16-8 one can immediately determine the variation of θ_2 when there is a time-varying flowrate and thus the change in the flowrate itself can be found.

In the original paper [H311] a calculation was carried out which was based on these ideas; this calculation took into consideration the non-linearity of the chronological variation of θ_2. In this way it was also shown how one could approximately determine the extent of the non-linearity on the basis of both theoretical considerations and experience.

SEVENTEEN

SIMPLIFICATION OF THE CALCULATION OF REGENERATORS WITH THE HELP OF THE ZERO EIGEN FUNCTION

17-1 CALCULATION OF THE STATE OF EQUILIBRIUM OF LONG REGENERATORS USING THE ZERO EIGEN FUNCTION[1]

The calculation of the state of equilibrium of very long regenerators can be simplified by expressing the temperature distribution in the middle parts of such regenerators in terms of the zero eigen function. In this case it is only necessary to use one of the often laborious methods for the ends of the regenerator in those regions where the temperature curve no longer corresponds to the zero eigen function alone. Thus the calculation is first carried out for a short regenerator in which the middle region, to where the zero eigen function alone applies, is only small. From the results obtained in the short regenerator the temperature distribution in the longer regenerator can then be determined with the help of the zero eigen function; this will be shown in the following.

First consider the case where $\Lambda = \Lambda'$ and $\Pi = \Pi'$ where also $CT = C'T'$. Assuming that this is so, Fig. 17-1 shows the distribution of the temperature in the heat-storing mass in a short regenerator of length Λ_1. This distribution is calculated for the same period Π for which the temperature distribution in the longer regenerator is required. It follows from Fig. 17-1 that in the middle of the regenerator the curve is only straight for a short distance. As a rule this is achieved

---

[1] The author's method was published in the first edition of this book. Its usefulness was confirmed by Willmott and Thomas [W309] who were the first to examine the method briefly.

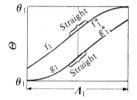

Figure 17-1 Short regenerator with $C = C'$ and $T = T'$.

reasonably precisely when Λ_1 is chosen so as to be three to four times as large as Π. In the left-hand half of the short regenerator f_1 represents the curve of the excess temperature at the beginning of the period relative to the entry temperature θ_1 of the gas, and g_1 is the excess temperature at the end of the period. Correspondingly, in the right-hand half of the regenerator the excess temperatures will be denoted by f^* or g^*. Figure 17-2 shows the required temperature distributions f_2 and g_2 in a long regenerator of length Λ_2. These distributions can be determined basically in such a way that in the outer sections of the length $\Lambda_1/2$ the temperature distribution found for the short regenerator is multiplied by a factor A, which still has to be determined, and in the middle parts the linear temperature distribution represented by the zero eigen function is inserted in a suitable manner.

It follows that (17-1) is set between the left-hand end of the long regenerator and the position $\xi = \Lambda_1/2$:

$$\left.\begin{array}{c} f_2 = Af_1 \\ g_2 = Ag_1 \end{array}\right\} \tag{17-1}$$

On the other hand, it will be noted from Eq. (13-60) that the zero eigen function gives Eq. (17-2) for the middle parts of the regenerator

$$\Theta = \frac{\theta_1' - \theta_1}{2} + B\left(\xi - \eta - \frac{\Lambda_2 - \Pi}{2}\right) \tag{17-2}$$

where B is an arbitrary constant. The constant D in Eq. (13-60) is fixed in such a way that, in the middle of the long regenerator where $\xi = \Lambda_2/2$ and in the middle of the period being considered when $\eta = \Pi/2$, Θ lies exactly in the middle between the two entry temperatures θ_1 and θ_1'. A and B in Eqs. (17-1) and (17-2) are determined by the requirement that, according to Eq. (17-1), the excess temperatures f_2 and g_2 at position $\xi = \Lambda_1/2$ must be equal to the values of Θ_0, which is calculated using Eq. (17-2) for $\eta = 0$ and $\eta = \Pi$. After solving the resultant Eq. (17-3) is obtained

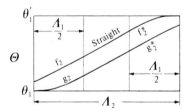

Figure 17-2 Long regenerator with $C = C'$ and $T = T'$. A straight section is inserted into the middle of the temperature distribution in Fig. 17-1.

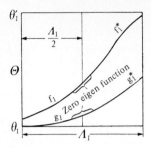

Figure 17-3 Short regenerator with $CT > C'T'$.

since $f_1 + g_1 = \theta'_1 - \theta_1$ when $\xi = \Lambda_1/2$:

$$
\left.
\begin{aligned}
A &= \frac{\Pi(\theta'_1 - \theta_1)}{\Pi(\theta'_1 - \theta_1) + (\Lambda_2 - \Lambda_1)(f_1 - g_1)} \\[2mm]
B &= \frac{(f_1 - g_1)(\theta'_1 - \theta_1)}{\Pi(\theta'_1 - \theta_1) + (\Lambda_2 - \Lambda_1)(f_1 - g_1)}
\end{aligned}
\right\}
\tag{17-3}
$$

where the values at position $\xi = \Lambda_1/2$ are to be inserted for f_1 and g_1. Thus with A and B found using Eq. (17-3), the temperature curve at the left-hand end of the regenerator and in the middle region can easily be determined. The temperature curve at the right-hand end of the regenerator is likewise obtained by multiplying the corresponding temperatures in the right-hand half of the shorter regenerator by A. Indeed, one must here calculate the excess temperatures relative to the temperature θ'_1 of the gas entering from the right. Thus, for the right-hand end of the regenerator one obtains

$$
\left.
\begin{aligned}
(\theta'_1 - \theta_1) - f_2^* &= A[(\theta'_1 - \theta_1) - f_1^*] \\
(\theta'_1 - \theta_1) - g_2^* &= A[(\theta'_1 - \theta_1) - g_1^*]
\end{aligned}
\right\}
\tag{17-4}
$$

The same process can also be used when the reduced regenerator dimensions Λ and Λ' and Π and Π' are different. However, the calculation is more laborious as the advantage of the point symmetrical temperature curve, as in Figs. 17-1 and 17-2 disappears and, one thus has to calculate four constants. Figure 17-3 again illustrates the small regenerator of length Λ_1 in which only a short section of the temperature curve in the middle of the regenerator is represented by the zero eigen function alone. On the other hand in Fig. 17-4, which represents the required

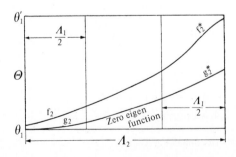

Figure 17-4 Long regenerator with $CT > C'T'$. The zero eigen function is introduced into the middle of a section of the temperature distribution shown in Fig. 17-3.

temperature curve in the longer regenerator, the zero eigen function is applicable, at least within the positions between $\xi = \Lambda_1/2$ and $\xi = \Lambda_2 - \Lambda_1/2$. As above we set the following in the outer sections of the longer regenerator, using two constants A and A', which have still to be determined.

$$f_2 = Af_1 \mid (\theta'_1 - \theta_1) - f_2^* = A'[(\theta'_1 - \theta_1) - f_1^*]$$
$$g_2 = Ag_1 \mid (\theta'_1 - \theta_1) - g_2^* = A'[(\theta'_1 - \theta_1) - g_1^*]$$

The zero eigen function which is applicable in the middle parts of the regenerator takes the form of Eq. (17-5), using Eq. (13-44), with two arbitrary constants B and D:

$$\Theta = D + B \cdot \exp\left(b\xi - \frac{b}{1+b} \eta \right) \tag{17-5}$$

It follows from Eq. (13-49) that b is written as an abbreviation namely

$$b = \left(\frac{\Lambda'_2}{\Lambda_2} - \frac{\Pi'}{\Pi} \right) \frac{\Pi}{\Pi + \Pi'}$$

It follows from Eq. (17-5) that the value of $\Theta - \theta_1$ at positions $\xi = \Lambda_1/2$ and $\xi = \Lambda_2 - \Lambda_1/2$ when $\eta = 0$ and $\eta = \Pi$ must be equal to the values of f_2, g_2 or f_2^*, g_2^*. Using the previous equations, four other equations result for the determination of A, A', B and D.

The Efficiency of the Longer Regenerator

If, in the case where $CT = C'T'$, the efficiency $(\eta_{Reg})_{\Lambda_1}$ of the shorter regenerator is known and if A has been calculated using Eq. (17-3) the efficiency $(\eta_{Reg})_{\Lambda_2}$ of the longer regenerator can also be easily determined. The warmer gas leaves the left-hand end of the short regenerator with a time mean temperature $(\theta'_2)_{\Lambda_1}$ and leaves at the left-hand end of the longer regenerator with the temperature $(\theta'_2)_{\Lambda_2}$. It will be seen from Eq. (13-94) that Eqs. (17-6) and (17-7) represent the efficiency:

$$(\eta_{Reg})_{\Lambda_1} = 1 - \frac{(\bar{\theta}'_2)_{\Lambda_1} - \theta_1}{\theta'_1 - \theta_1} \tag{17-6}$$

$$(\eta_{Reg})_{\Lambda_2} = 1 - \frac{(\bar{\theta}'_2)_{\Lambda_2} - \theta_1}{\theta'_1 - \theta_1} \tag{17-7}$$

Since all the excess temperatures in the left-hand section of the regenerator decrease in proportion to A, relative to θ_1, during the transformation from a shorter to a longer regenerator, this must also apply to the excess temperature of the warm exit gas. Thus Eq. (17-8) must also apply:

$$(\bar{\theta}'_2)_{\Lambda_2} - \theta_1 = A[(\bar{\theta}'_2)_{\Lambda_1} - \theta_1] \tag{17-8}$$

Upon inserting this into Eq. (17-7) and eliminating the temperature ratio by utilizing Eq. (17-6) the equation below is obtained.

$$1 - (\eta_{Reg})_{\Lambda_2} = A[1 - (\eta_{Reg})_{\Lambda_1}] \tag{17-9}$$

k/k_0 can also be determined for the longer regenerator using Eq. (13-105) using the value of $(\eta_{Reg})_{\Lambda_2}$ determined in the way described above.

General Method of Calculating the Efficiency from the Zero Eigen Function

Finally, it can be shown that, in general, when the zero eigen function alone determines the temperature distribution in the inner sections of a regenerator for the case where $CT = C'T'$, the efficiency of the heat transfer can be calculated very easily. It follows from Eq. (13-63) and (13-64) that the average temperature difference between the two gases is equal to

$$\bar{\theta}' - \bar{\theta} = B\left(1 + \frac{\bar{\alpha}}{\bar{\alpha}'} \cdot \frac{C'}{C}\right) \tag{17-10}$$

where the minus sign in front of the B has been omitted for the sake of simplicity.

It will be seen from the heat balance Eq. (11-2) that the average temperature difference, when $CT = C'T'$, has the same value at each end of the regenerator. Hence each of the gases undergoes the time mean temperature change

$$\theta'_1 - \theta_1 - B\left(1 + \frac{\bar{\alpha}}{\bar{\alpha}'} \cdot \frac{C'}{C}\right)$$

On the other hand, in an ideal regenerator the temperature change would be equal to $\theta'_1 - \theta_1$. Further since the quantities of heat transferred per period would be proportional to these changes in temperature of the gases, Eq. (17-11) emerges for the efficiency

$$\eta_{Reg} = 1 - \frac{B}{\theta'_1 - \theta_1}\left(1 + \frac{\bar{\alpha}}{\bar{\alpha}'} \cdot \frac{C'}{C}\right) \tag{17-11}$$

This equation is very simple and indeed, with very precise efficiency calculations, it is as good as the relationships developed earlier such as, for example, Eq. (15-18).

17-2 ADDITIONAL IDEAS FOR SIMPLIFYING THE CALCULATION WHEN THE PERIOD LENGTH IS LONG

In the first edition of this book, pp. 393 to 396 (Appendix 4), an extensive discussion was presented of the possibility of simplifying the calculation for the state of equilibrium for the case where the period length is very long. This involved the use of the rule of the *parallel displacement of the temperature curves*; this method will be described below. In the first edition this rule was combined with the zero eigen function and with the refined heat pole method (see Sec. 15-2). Furthermore it was shown how it is possible to develop the temperature curve in a short regenerator by making use of the results of the calculations for longer regenerators; this was described in detail. As the calculation processes developed for this purpose are today only of a slight interest they will not be reproduced here.

However two features of this method, which have proved useful, will be described briefly; these are the rule of parallel displacement and a further use of the zero eigen function for the case where period lengths are very long.

In most regenerators the rule of parallel displacement is approximately valid from about $\xi = 5$ to the end of the regenerator where the gas leaves, in the case where the reduced period length Π is greater than about 10. It follows from this rule that in the regions where the temperature distribution is not curved, or is only very slightly so, the temperature distribution at time η can be obtained precisely, or certainly approximately, from the temperature curve at time $\eta = 0$ such that each point on the curve is displaced by the amount $\Delta\xi = \eta$ parallel to the ξ-axis in the direction of gas flow. For the zero eigen function this result is exact, as can be deduced from previous considerations: see Fig. 11-10, top and Fig. 13-8. In the following development it will be shown however, that at other positions in the regenerator, where the temperature distribution is only slightly non-linear, this rule of parallel displacement can be applied to yield very approximate results. In the case of slight non-linearity the very small value of the differential quotient $\dfrac{\partial^2\Theta}{\partial\xi\partial\eta}$ in Eq. (13-30) can be neglected so that Eq. (13-30) takes the form

$$\frac{\partial\Theta}{\partial\xi} = -\frac{\partial\Theta}{\partial\eta}$$

For $\Theta = \text{const}$

$$d\Theta = \frac{\partial\Theta}{\partial\xi}d\xi + \frac{\partial\Theta}{\partial\eta}d\eta = \frac{\partial\Theta}{\partial\xi}(d\xi - d\eta) = 0$$

is obtained and thus

$$d(\xi - \eta) = 0 \quad \text{or} \quad \xi = \text{const} + \eta$$

Experience has shown that this rule of parallel displacement presented here is all the more precise the larger the value Π and Λ.

In long regenerators, if the period lengths are very long, for example, $\Pi > 50$, the application of the zero eigen function can be somewhat wider for the state of equilibrium. Again for the case where $\Lambda = \Lambda'$ and $\Pi = \Pi'$, the spatial and chronological temperature variations are known for most parts of the regenerator interior. It is known from experience that when the length of period is very long the temperature distribution at the end of the period forms an almost straight line right to the end of the regenerator where the gas leaves. This can be explained in the following way. The deviation from the linear temperature distribution in the region of the exit of the regenerator is represented by the balancing terms of the higher eigen functions (see Sec. 13-6 and Fig. 11-10). However, as Eqs. (13-82) to (13-85) show, the balancing terms decrease in magnitude almost proportionally with $\exp(-\eta/2)$, as a function of time. Thus, towards the end of the period, they are smaller, the longer the period length and indeed they can be neglected beyond a certain size of period length.

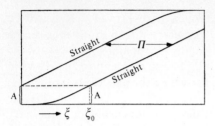

Figure 17-5 Regenerator with a very long period length Π.

However, as the temperature distribution of the heat-storing mass at the end of one period is the same as that at the beginning of the next period the temperature curve up to the exit of the regenerator must form a straight line. In this way a temperature distribution as shown in Fig. 17-5 is obtained.

EIGHTEEN

FURTHER DEVELOPMENTS

18-1 RESULTS OF CALCULATIONS USING THE APPROXIMATE METHODS

Figures 13-5 and 13-7, as well as Figs. 13-17 to 13-20, which serve to explain the exact theories, were partly calculated using one of the step-by-step processes (Sec. 16-5) or using the heat pole method (Secs. 15-1 and 15-2). Further results from these approximate methods for counterflow regenerators will be discussed below.

Figures 18-1 and 18-2 relate to the state of equilibrium when $\Lambda = \Lambda' = 10$ and $\Lambda = \Lambda' = 40$ for various values of the reduced period length $\Pi = \Pi'$. The curves replicate the temperature distribution in the heat-storing mass at the beginning and end of the cold period. Figure 18-3 refers to the temperature distribution for the state of equilibrium when the warm and cold periods are of unequal length; here $\Lambda = \Lambda' = 10$ and the length of the cold period is assumed to be $\Pi = 5$ and that of the warm period to be $\Pi = \pi = 3.1416$.

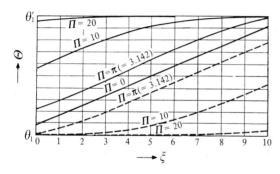

Figure 18-1 Temperature distribution for the state of equilibrium of a counter-flow regenerator when $\Lambda = \Lambda' = 10$ and $\Pi = \Pi'$.

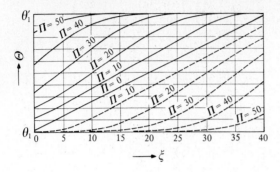

Figure 18-2 Temperature distribution for the state of equilibrium of a regenerator when $\Lambda = \Lambda' = 40$ and $\Pi = \Pi'$.

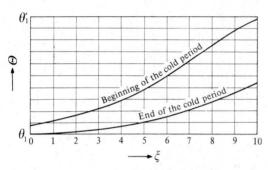

Figure 18-3 State of equilibrium of a regenerator when $\Lambda = \Lambda' = 10$, $\Pi = 5$ and $\Pi' = \pi$.

In the figures which follow is obtained an idea of the behaviour of regenerators used in low temperature technology, in which the reduced regenerator length is always very large. The water vapour content of the air, which has considerable influence and which is discussed in the next chapter, is not taken into account in these basic calculations.

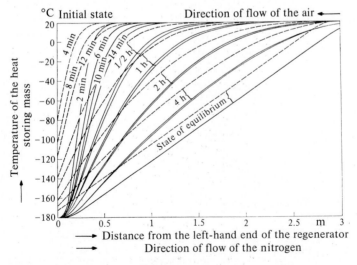

Figure 18-4 The starting up of a low temperature regenerator with regular reversals every two minutes.

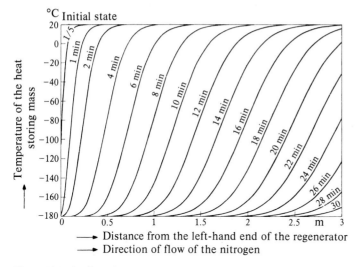

Figure 18-5 Cooling of the heat-storing mass of a regenerator without reversal.

The figures are therefore only strictly applicable to the case of completely dry air and this rarely occurs in practice. In Figs. 18-4, 18-5, 18-6 and 18-7 it is assumed that $\Lambda = \Lambda' = 120$ and $\Pi = \Pi' = 10$. From when the regenerator is started up equal quantities of air and nitrogen flow alternately through the regenerator which is reversed every two minutes. The entry temperature of the air is equal to 20°C and that of the nitrogen is equal to $-180°$C. Initially the heat-storing mass has a uniform temperature of 20°C. Figure 18-4 shows how the heat-storing mass is

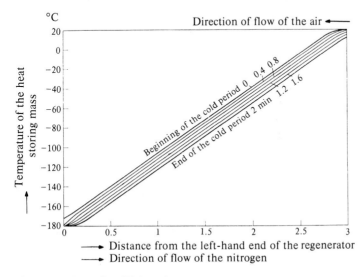

Figure 18-6 State of equilibrium of a regenerator at low temperatures without the precipitation of water vapour and carbon dioxide. (Compare with Figs. 19-3 to 19-8.)

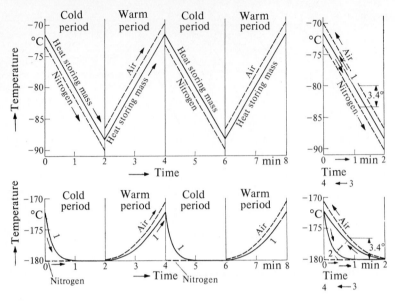

Key

1. Heat storing mass

2. Nitrogen

Figure 18-7 Chronological variation of temperature for the state of equilibrium in the middle and at the cold end of a regenerator.

gradually cooled by the successive reversals. Even after four hours the state of equilibrium has not been achieved. On the other hand, however, from the same initial conditions, if cold nitrogen is allowed to flow constantly through the regenerator, without any reversals, the heat-storing mass is cooled down, as has already been essentially displayed in Fig. 13-6, and as is now more accurately shown in Fig. 18-5.

The temperature distribution presented in Fig. 18-3, which is achieved at equilibrium following regular reversals, is shown in Fig. 18-6 as a function of the longitudinal coordinate. In Fig. 18-7 variation with time of the temperature at the middle of the regenerator and at the cold end of the regenerator is displayed.

Finally, Fig. 18-8 shows how the temperature distribution changes for the state of equilibrium if the thermal capacity of the air is chosen so as to be 5 per cent greater than that of nitrogen. In Figs. 18-6 and 18-8 the temperatures are calculated using the heat pole method only in the region of the ends of the regenerator. The large middle section is extrapolated using the zero eigen function as described in Sec. 17-1.

As a contrast to the regenerators used in low temperature technology, Fig. 18-9 shows the state of equilibrium of an unusually short regenerator with $\Lambda = \Lambda' = 4$ operating with $\Pi = \Pi' = 10$. The two left-hand diagrams illustrate the temperature distribution in the heat-storing mass and in the gas during the cold period at

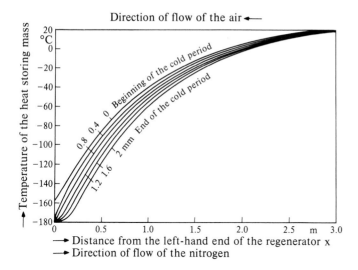

Figure 18-8 State of equilibrium of a regenerator where the gases flowing through the regenerator have unequal thermal capacities.

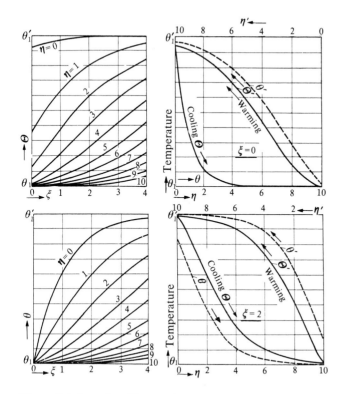

Figure 18-9 State of equilibrium of a very short regenerator: $\Lambda = \Lambda' = 4$ and $\Pi = \Pi' = 10$.

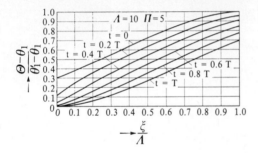

Figure 18-10 Temperature distribution where $\Lambda = 10$ and $\Pi = 5$.

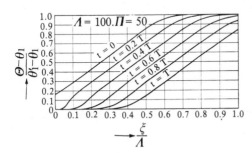

Figure 18-11 Temperature distribution where $\Lambda = 100$ and $\Pi = 50$.

various times η as a function of the longitudinal coordinate ξ. In this case the straight temperature curve corresponding to the zero eigen function, is completely suppressed by the strong influence of the higher order eigen functions which also have a considerable influence in the middle of the regenerator.

This can also be seen from the large hysteresis loops which are formed for the average brick temperatures Θ and Θ'. One (top right) is drawn for the cold end of the regenerator, the other (bottom right) is drawn for the middle of the regenerator. As a consequence of the heavy predominance here of the higher order eigen functions, the value of the ratio k/k_0 corresponding to Fig. 11-13 only amounts to 0.31.

Finally, how does the temperature distribution change if, with otherwise unaltered ratios, only the heat transfer coefficients α and α' or $\bar{\alpha}$ and $\bar{\alpha}'$ increase? Here we need to assume that $\Lambda = \Lambda'$, $\Pi = \Pi'$ and thus $\bar{\alpha} = \bar{\alpha}'$. It follows, from Eqs. (13-28) and (13-29), that the ratio $\Lambda : \Pi$ remains unaltered and that Λ and Π are

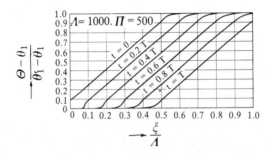

Figure 18-12 Temperature distribution where $\Lambda = 1000$ and $\Pi = 500$.

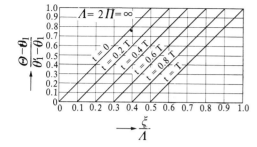

Figure 18-13 Temperature distribution where $\Lambda = 2\Pi = \infty$, that is with an infinitely large heat-transfer coefficient or an infinitely large heating surface area.

always proportional to the value of α. Figure 18-10 presents the temperature distribution when $\Lambda = 10$, $\Pi = 5$; Fig. 18-11 corresponds to the case where $\Lambda = 100$, $\Pi = 50$ and Fig. 18-12 to that where $\Lambda = 1000$ and $\Pi = 500$. Thus the heat transfer coefficient is 100 times greater in the last case than it is in the first. It follows from Fig. 18-10 that with small heat transfer coefficients the temperature distribution is curved uniformly throughout. With very large heat transfer coefficients, on the other hand, the temperature curves consist of the parallel displacements of a straight line. At the ends of this straight line bending occurs more sharply the larger the value of $\bar{\alpha}$. With an infinitely large heat transfer coefficient parallel-displaced straight lines are all that remain between θ_1 and θ_1' (Fig. 18-13).

18-2 THE USE OF THE ZERO EIGEN FUNCTION AS THE BASIS OF A METHOD OF MEASURING THE TRANSFER OF HEAT IN REGENERATOR HEAT-STORING MASSES

In order to measure the transfer of heat between the gases flowing through a regenerator and the heat-storing mass, the author of this book has suggested starting from the simple zero eigen function. As has been often mentioned, this function varies linearly, both spatially and chronologically, when $\Lambda = \Lambda'$ and $\Pi = \Pi'$. Based on this concept Glaser [G301] has developed a method of measurement which is suitable, in particular, where the heat-storing mass is constructed of metallic elements. These elements conduct the heat very well at right angles to the direction of flow while thermal conduction in the direction of flow is inhibited by the numerous divisions between the elements. Glaser chose the dimensions of the heat-storing mass of a pair of small experimental regenerators and the length of the periods in such a way that the required linear temperature curve appears in the central parts of the regenerators. In both periods the same gas, air in practice, is passed in equal quantities through the regenerators in opposite directions. As Glaser recognized, it is then sufficient to observe the distribution of the gas temperatures using a thermometer with a fast response, for example a resistance thermometer constructed of very thin platinum wire, at a particular position in the middle of the regenerator. As this variation in temperature is completely linear with time, the average temperature difference $\Delta\bar{\theta}$ between the warmer and colder gas can be extracted directly from the measurements, as can the chronological variation of the gas temperature in the warm and cold periods. This variation in gas temperature is equal to the change $\Theta_2 - \Theta_1$ in the temperature of the heat-storing mass during one period.

The heat transfer coefficients can then be obtained from these measured values in the following way. The quantity of heat that is transferred through the surface area ΔF of a part of the heat-storing mass with the thermal capacity ΔC_s during a period of duration T when $\alpha = \alpha'$, is equal to

$$\Delta C_s(\Theta_2 - \Theta_1) = \alpha \, \Delta F(\theta - \Theta) \cdot T = \alpha \, \Delta F \frac{\Delta \bar{\theta}}{2} \cdot T$$

It follows from this that the required heat transfer coefficient works out to be:

$$\alpha = \frac{2}{T} \cdot \frac{\Delta C_s}{\Delta F} \cdot \frac{\Theta_2 - \Theta_1}{\Delta \bar{\theta}} \qquad (18\text{-}1)$$

Glaser first used this method for measuring the transfer of heat in the storage elements constructed of corrugated metal strips as suggested by Fränkl [F301] for use in low temperature technology (see Fig. 2-22). That he, in fact, observed a completely linear temperature variation with time can be seen from Fig. 12-8 on p. 331.

In the same way Glaser also investigated the transfer of heat in pipes which were helically coiled in several layers and sited in a regenerator. Thus he realized practically cross flow in tube assemblies. The results of these, and other measurements by Glaser, were mentioned in Sec. 2-11.

Langhans [L42] has extended this process so that it can be applied to thicker storage elements with low thermal conductivity. The dimensions of the heat-storing mass and the length of the periods must be so selected that not only the zero eigen function prevails in the inner parts of the regenerator but also so that, in spite of the rapid temperature variations which take place after a reversal, at least part of the curves which are observed, which represent the chronological variation of the gas temperatures, remains linear. Figure 18-14 then results and this corresponds to Fig. 11-7. In his method of calculation, Langhans extended the linear section of the curve to the beginning of the period. As in Eq. (18-1) a heat transfer coefficient

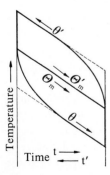

Figure 18-14 For the measurement of the transfer of heat in a heat-storing mass of low thermal conductivity.

is obtained from the completely linear variation in temperature produced in this way; this heat transfer coefficient does not represent α but a heat transfer coefficient which is related to the average brick temperature Θ_m. The true heat transfer coefficient α is obtained from this by using Eq. 13-3. The value of $\Phi = 1/6$ can be used because the rapid temperature variations at the beginning of the periods do not have to be taken into account within the calculation.

While investigating shaft-like heat-storing masses, Yasicizade [Y41] observed rapid temperature fluctuations which overlapped the smooth curves, such as those shown in Fig. 18-14. In these and similar cases he was able to increase the accuracy by taking measurements in two positions in the regenerator. In order to be able to evaluate the transfer of heat in heat-storing masses constructed of square bricks found in various normal, practical arrangements, Langhans and Yasicizade constructed geometrically scaled down models made of laminated paper to fill the experimental regenerators. Using the similarity principle, the experimental results thus obtained can be applied, in dimensionless form, to large scale models.

NINETEEN

WET REGENERATORS OPERATING AT LOW TEMPERATURES

19-1 NUMERICAL AND GRAPHICAL METHODS FOR DETERMINING THE TEMPERATURE DISTRIBUTION[1]

As has already been mentioned several times, in low temperature technology, regenerators are preferable to counterflow heat exchangers as the compressed air, or other cooling gases, does not have to be freed from carbon dioxide and water vapour before entering a regenerator. Rather, these components of the air, which is compressed to about 2 to 6 bar absolute, are deposited in the heat-storing mass by condensation or freezing during the warm period. Then, during the cold period, they are removed by the cold gas flowing back through the regenerator at approximately atmospheric pressure, by evaporation or sublimation. Because of the development of absorbant drying processes, this phenomenon does not have the significance it had formerly.* However, it is still an important principle in the operation of regenerators. Thus, its influence on the temperature distribution will be discussed as these same considerations are also of importance in the calculation of reversing exchangers (see Fig. 5-5).

Although the quantities of water vapour and carbon dioxide deposited during a period are only very small, they nevertheless exert a considerable influence on the temperature distribution in the regenerator, as will be shown in greater detail in Sec. 19-4 with the help of Figs. 19-3 to 19-7. More than anything else, this influence can be attributed to the fact that, as a result of the much larger volume of the cooled

---

[1] From Hausen [H306].

* Editor's note: However such considerations have recently become important in the development of rotary regenerators for air conditioning, especially in humid climates.

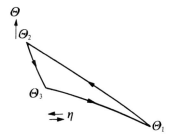

Figure 19-1 Chronological temperature variations of the heat-storing mass in the wet regenerator.

gas, the water and the carbon dioxide evaporate or sublimate far more rapidly than they were deposited during the warm period. Thus the heat-storing mass, which must supply the necessary heat for the evaporation or sublimation, is rapidly cooled at the beginning of the cold period. As a consequence, chronological hysteresis loops appear at the far end of the regenerator from the entrance of the warm gas; such a loop is shown in Fig. 19-1.

The chronological variation of the temperature Θ of the heat-storing mass is drawn in this diagram, as a function of time for a particular position in the regenerator. The time coordinates are drawn in the opposite directions for the two periods. In the warm period the heat-storing mass is heated from Θ_1 to Θ_2. In the "wet" part of the cold period it is first rapidly cooled from Θ_2 to Θ_3 and then during the "dry" part of the cold period it is slowly cooled from Θ_3 to Θ_1. The hysteresis loop formed in this way reduces the temperature differences between the gases and the heat-storing mass which are thus available for the exchange of heat and thus, where condensation and evaporation take place, considerably reduces the temperature drop in the longitudinal direction of the regenerator.

In order to be able to examine these temperature changes, a numerical and graphical method will be discussed, which are both an extension of the method for dry regenerators described in Sec. 16-5. The discussion will be carried out using, by way of example, a water vapour/air mixture; this can then be extended, without alteration, to any gases which contain carbon dioxide or other condensing components. In this discussion it will be assumed that a heat-storing mass of high thermal conductivity is placed across the direction of flow.

In order to simplify matters, calculations will be carried out using the dimensionless longitudinal and chronological coordinates ξ and η defined in Eq. (13-24); this assumes heat transfer coefficients and physical properties which are not dependent upon temperature. However, these limitations can be avoided by introducing the variables f and t from Eq. (13-24) into the resultant relationships. The temperature dependence of these values can thus be considered by calculating each of the values at each step with other average values.

The Differential Equations for the Wet Regenerator

In the regenerators used in low temperature technology, the layer of water or ice deposited upon the heat-storing mass is rarely thicker than about 1/100 mm. As a consequence the temperature of this layer can approximately be assumed to be equal to the temperature Θ of the heat-storing mass; the thermal capacity of this layer relative to that of the heat-storing mass, can be neglected. Similarly, again for simplicity, the thermal capacity of the water vapour contained in the gas can be omitted from the considerations. Following the nomenclature used earlier we have:

x the true water vapour content of the gas in kg per kg gas,

x'' the water vapour content that the gas had when it was saturated at the temperature Θ,

r the evaporation enthalpy (or sublimation enthalpy) of the water in kJ/kg,

σ the evaporation coefficient in $kg/(m^2\,s)$,

t the time since the last reversal in s,

m_w the quantity of water evaporating in time t per unit surface area of the heat-storing mass in kg/m^2.

Assuming the ideal gas law, Eq. (19-1) is applied to water vapour content x:

$$x = \frac{M}{M_G} \cdot \frac{p}{P-p} \tag{19-1}$$

where M is the molar mass and p the partial pressure of the water vapour. M_G is the molar mass of the gas and P is the total pressure.

The differential equation for the temperature distribution in a wet regenerator, the heat-storing mass of which, for example, is composed of sheet metal of uniform thickness, is obtained as follows.

As, in the cold period, the heat given off by the heat-storing mass is needed for the evaporation of the water as well as for increasing the temperature of the gas, Eq. (19-2) results for a section of the regenerator with heat-storing mass surface area df and the thermal capacity dC_s for time dt:

$$dC_s \left(\frac{\partial \Theta}{\partial t}\right)_f \cdot dt + r\left(\frac{\partial m_w}{\partial t}\right)_f \cdot df \cdot dt + C\,dt\left(\frac{\partial \theta}{\partial f}\right)_t df = 0 \tag{19-2}$$

Furthermore, for the exchange of heat between the gas and the heat-storing mass the equation takes the form[2]

$$C\,dt\left(\frac{\partial \theta}{\partial f}\right)_t df = \alpha\,df(\Theta - \theta)\,dt \tag{19-3}$$

and for the evaporation of the water (see, for example [B309])

$$\left(\frac{\partial m_w}{\partial t}\right)_f df \cdot dt = \sigma\,df(x'' - x)\,dt \tag{19-4}$$

The quantity of water vapour in the gas increases by as much as the water in the heat-storing mass evaporates. Thus Eq. (19-5) is obtained if one considers that C/c_p

---

[2] In Eq. (19-3) it is assumed that the evaporation of the water does not require an additional heating surface area, although heat for this is obtained from the heat-storing mass. Thus $\alpha = \infty$ is assumed for this partial process. This corresponds approximately, to the fact that the heat transfer coefficient is considerably higher in condensation than it is in the emission of heat to a gas.

represents the rate of flow gas through the regenerator

$$\frac{C}{c_p} dt \left(\frac{\partial x}{\partial f}\right)_f df = \left(\frac{\partial m_w}{\partial t}\right)_f df \cdot dt \tag{19-5}$$

Equation (19-6) follows from Eq. (19-3)

$$\left(\frac{\partial \theta}{\partial f}\right)_t = \frac{\alpha}{C} (\Theta - \theta) \tag{19-6}$$

and from Eqs. (19-4) and (19-5) yields

$$\left(\frac{\partial x}{\partial f}\right)_t = \frac{\sigma c_p}{C} (x'' - x) \tag{19-7}$$

Utilizing Eqs. (19-3) and (19-4), Eq. (19-8) results from Eq. (19-2)

$$\frac{dC_s}{df} \left(\frac{\partial \Theta}{\partial t}\right)_f + \alpha(\Theta - \theta) + \sigma \cdot r(x'' - x) = 0 \tag{19-8}$$

Furthermore, with a proportionality factor ε is obtained

$$\sigma = \frac{1}{\varepsilon} \frac{\alpha}{c_p} \tag{19-9}$$

The "reduced" water vapour content is introduced at this point:

$$\phi = \frac{r}{c_p} x \tag{19-10}$$

Thus, as in Eq. (13-22) or (13-24) in which $\bar{\alpha} = \alpha$, Eqs. (19-6), (19-7) and (19-8) take the form

$$\left(\frac{\partial \theta}{\partial \xi}\right)_\eta = \Theta - \theta \tag{19-11}$$

$$\left(\frac{\partial \phi}{\partial \xi}\right)_\eta = \frac{1}{\varepsilon} (\phi'' - \phi) \tag{19-12}$$

$$\left(\frac{\partial \Theta}{\partial \eta}\right)_\xi = -\left[(\Theta - \theta) + \frac{1}{\varepsilon} (\phi'' - \phi)\right] \tag{19-13}$$

These three differential equations, with the boundary condition specified below, allow the temperature distribution in the wet regenerator to be determined.

In order to obtain a differential equation in Θ alone, θ and ϕ are eliminated from Eqs. (19-11) to (19-13). By using the relationship following from these equations

$$\left(\frac{\partial(\theta + \phi)}{\partial \xi}\right)_\eta = -\left(\frac{\partial \Theta}{\partial \eta}\right)_\xi$$

and by differentiating Eq. (19-13) with respect to ξ we obtain

$$\varepsilon \frac{\partial^2 \Theta}{\partial \xi \partial \eta} = -\left(\frac{\partial \Theta}{\partial \eta}\right)_\xi + (\varepsilon - 1)\left(\frac{\partial \theta}{\partial \xi}\right)_\eta - \frac{d(\varepsilon \Theta + \phi'')}{d\Theta}\left(\frac{\partial \Theta}{\partial \xi}\right)_\eta \tag{19-14}$$

If one is limited to cases where the heat transfer coefficients α are large and thus there are small differences between Θ and θ or the value of $\varepsilon - 1$ is small, then

$$(\varepsilon - 1)\left(\frac{\partial \theta}{\partial \xi}\right)_\eta = (\varepsilon - 1)\left(\frac{\partial \Theta}{\partial \xi}\right)_\eta$$

can be inserted into Eq. (19-14) without introducing much error.[3] Thus Eq. (19-14) transforms into

$$\varepsilon \frac{\partial^2 \Theta}{\partial \xi\, \partial \eta} + \left(\frac{\partial \Theta}{\partial \eta}\right)_\xi + \left(1 + \frac{d\phi''}{d\Theta}\right)\left(\frac{\partial \Theta}{\partial \xi}\right)_\eta = 0 \qquad (19\text{-}15)$$

This equation has Θ only as an unknown. This is because $d\phi''/d\Theta$ can be calculated from the vapour pressure curve for water as a function of Θ on the basis of Eqs. (19-10) and (19-1). In so doing it is considered that ϕ'' and x'' correspond to the state of saturation at the temperature Θ.

The coefficient $\varepsilon = \alpha/\sigma c_p$ in Eq. (19-9), which is included in Eq. (19-15), represents *the ratio of the speed of the heat transfer to the speed of the mass transfer*. The value of ε is a function both of the condition of flow and the materials which take part in the mass transfer. If it is assumed that, in the transfer of heat and mass, only the resistance of a laminar boundary layer has to be overcome, it is clear that ε can be set equal to the ratio a/D of the thermal diffusivity $a = \lambda/\rho c_p$ to the diffusion coefficient D. a/D has the following values for the following pairs of materials:

for diffusion of water vapour in air	$a/D = 0.853$
for diffusion of carbon dioxide in air	$a/D = 1.32$
for diffusion of benzene in air	$a/D = 2.41$
for diffusion of water vapour in hydrogen	$a/D = 1.79$
for diffusion of carbon dioxide in hydrogen	$a/D = 2.28$
for diffusion of benzene in hydrogen	$a/D = 4.42$

These values of $\varepsilon = a/D$ are mainly applicable to laminar flow. With increasing turbulence, where mass transfer does not take place solely by diffusion but also, to a large extent by convection, ε supposedly approaches the value 1. The limiting case $\varepsilon = 1$ would be achieved with pure turbulence; it corresponds to Lewis's law [L303] according to which $\alpha = \sigma c_p$. In practice, calculations can be carried out for turbulent flow, using an average value between the values given for a/D and $\varepsilon = 1$; see, for example, [H309].

Numerical Step-by-Step Procedure

Equation (19-15) can be solved approximately by means of the step-by-step method described below which provides good accuracy. At time η the temperature is prescribed completely, or at least as far as the position $\xi + \Delta\xi$, while at time $\eta + \Delta\eta$, on the other hand, it is only prescribed as far as position ξ. This means, as in Fig.

---

[3] If it is desired to eliminate θ from Eqs. (19-11) to (19-13) in place of Eq. (19-15) a differential equation of the third order would result, which would be very unsuitable in the development of a step-by-step process.

16-4, that at the three positions the temperatures Θ_1, Θ_2 and Θ_3 are known. The temperature Θ_4 at position $\xi + \Delta\xi$ at time $\eta + \Delta\eta$ is required. The differential quotients in Eq. (19-15) are replaced by the expressions (16-15), (16-16) and (16-17). Thus we obtain

$$\Theta_4 = \Theta_1 + \frac{2\varepsilon[\Theta_2 + \Theta_3 - 2\Theta_1] + \left[\Delta\xi - \left(1 + \dfrac{d\phi''}{d\Theta}\right)\Delta\eta\right](\Theta_2 - \Theta_3)}{2\varepsilon + \Delta\xi + \left(1 + \dfrac{d\phi''}{d\Theta}\right)\Delta\eta} \qquad (19\text{-}16)$$

It follows from this equation that the temperature distribution Θ in the heat-storing mass can be calculated step-by-step in the wet part of the cold period and in the warm period. In doing this an average value of $d\phi''/d\Theta$ must be introduced at each step. It is advisable to select the value of $d\phi''/d\Theta$ using Θ_1 or $1/2(\Theta_2 + \Theta_3)$ in order to do this.

Graphical Step-by-Step Process

In place of Eq. (19-16) a graphical step-by-step method can be developed in which

$$\Delta\xi = \left(1 + \frac{d\phi''}{d\Theta}\right)\Delta\eta \qquad (19\text{-}17)$$

is set arbitrarily. If it is again sufficiently accurate to consider $d\phi''/d\Theta$ to be constant

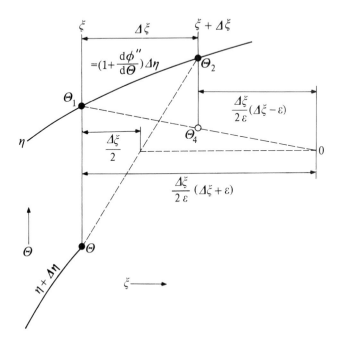

Figure 19-2 Graphical step-by-step method for determining the temperature of the heat-storing mass of a wet regenerator.

in each step, Eq. (19-15) can be transformed into the difference equation

$$\Theta_4 - \frac{1}{2}(\Theta_2 + \Theta_3) = \frac{\Delta\xi - \varepsilon}{\Delta\xi + \varepsilon}\left[\Theta_1 - \frac{1}{2}(\Theta_2 + \Theta_3)\right] \qquad (19\text{-}18)$$

Equation (19-18) is represented by means of the graphical process shown in Fig. 19-2. The points Θ_2 and Θ_3 are joined by a straight line. The point where this line is intersected by the perpendicular $\xi + 1/2\Delta\xi$ determines the average value $1/2(\Theta_2 + \Theta_3)$.

Proceeding along the horizontal line determined in this way and moving from the point ξ to the right by a distance $\Delta\xi/2\varepsilon(\Delta\xi + \varepsilon)$, the end point 0 is fixed. Finally if this end point 0 is joined to Θ_1, the connecting line intersects the perpendicular $\xi + \Delta\xi$ at the required value of Θ_4. It follows from the similarity of the two right-angled triangles, whose right-angles are formed by the horizontal line mentioned and the perpendiculars at ξ and $\xi + \Delta\xi$, and whose other angles are formed at Θ_1 or Θ_4 and the common point 0, that the process described satisfies Eq. (19-18).

As has already been mentioned in Sec. 16-5, the same process can also be applied to dry regenerators if it is assumed that $\Delta\xi = \Delta\eta$ and $\varepsilon = 1$. This also applies to the dry part of the cold period in a wet regenerator. When applied to the wet part of the cold period and to the warm period, the ratio of $\Delta\xi$ and $\Delta\eta$ varies from step to step. It is advisable to keep $\Delta\eta$ constant. $\Delta\eta$ was chosen in the range 1 to 5 when the curvature of the temperature distribution for the numerical example mentioned in Sec. 19-4, was calculated using this graphical method. Moreover, for the sake of simplicity, it was assumed that $\varepsilon = 1$ since it is to be expected that the values of ε are likely to be in the range between 0.853 and 1 in water vapour/air mixture when turbulent conditions at least partly obtain.

Calculation of θ and ϕ

The variation of the gas temperature θ, as well as the water-vapour content expressed by ϕ in Eq. (19-10), can also be determined at each Θ-curve obtained for $\eta = \text{const.}$ As Eq. (19-11) is identical to Eq. (13-26) for dry regenerators, the step-by-step method shown in Fig. 16-3, or a corresponding numerical method, is applicable for θ in an unchanged form.

Numerical or graphical methods can be used to find ϕ a function of ξ.

Because of the theoretical similarity between Eqs. (19-12) and (16-7) a suitable step-by-step procedure can easily be developed which corresponds to Fig. 16-3 or Eq. (16-11). However, as can be seen in the following, the calculation of ϕ can usually be neglected.

19-2 LENGTH OF TIME REQUIRED FOR THE RE-EVAPORATION AND SUBLIMATION OF THE WATER AND ICE DEPOSITED ON THE HEAT-STORING MASS[4]

In the following discussion, it will be assumed throughout that the amounts of ice and water deposited during the warm period have been completely evaporated or

---

[4] Procedure published by the author in the 1st edition of this book.

sublimated by the end of the cold period. The length of time necessary for this complete evaporation or sublimation determines the time Π^* of the wet part of the cold period (see Fig. 19-1).

The value of Π^* for each position ξ in the regenerator comes, basically, from the requirement that as much water is evaporated and sublimated at each position in the regenerator during the cold period as has been deposited during the previous warm period. This also applies when the state of equilibrium has not yet been reached or when the period lengths Π and Π' are of different duration. It will be shown how this requirement, which is called the "non-accumulation requirement"* can be expressed.

At a particular position ξ of the regenerator a quantity of water evaporates or sublimates per unit surface area in a very small time interval dt of the cold period, namely:

$$dm_w = \frac{C}{c_p} - dt\left(\frac{\partial x}{\partial f}\right)_\eta \tag{19-19}$$

where C/c_p denotes the flow rate of gas in kg/s, passing through the regenerator in the cold period and x is again the water vapour content of the gas. If Eq. (19-19) is integrated with respect to time for the damp part of the cold period, taking into account, Eqs. (13-22) and (19-10), the total quantity of water evaporating or sublimating per unit surface area is:

$$m_w = \frac{C}{c_p} \int_{\Theta_2}^{\Theta_3} \left(\frac{\partial x}{\partial f}\right)_\eta dt = \frac{1}{r}\frac{dC_S}{df} \int_{\Theta_2}^{\Theta_3} \left(\frac{\partial \phi}{\partial \xi}\right)_\eta d\eta \tag{19-20}$$

Equation (19-21) is obtained in a similar way

$$m_w' = \frac{1}{r}\frac{dC_S}{df} \int_{\Theta_1}^{\Theta_2} \left(\frac{\partial \phi}{\partial \xi}\right)_\eta d\eta \tag{19-21}$$

Like θ, ϕ decreases with ξ, that is in the direction of flow, during the warm period. Thus $\partial \phi / \partial \xi$ becomes negative and, it follows from Eq. (19-21) that so also does m_w'. This shows that the quantity of water m_w' does not evaporate or sublimate but is deposited. The non-accumulation requirement, namely that the quantity of evaporating or sublimating water is equal to the quantity of water deposited, thus gives $m_w' = -m_w$. From this, and following from Eqs. (19-20) and (19-21) the following equation is yielded:

$$\int_{\Theta_2}^{\Theta_3} \left(\frac{\partial \phi}{\partial \xi}\right)_\eta d\eta + \int_{\Theta_1}^{\Theta_2} \left(\frac{\partial \phi}{\partial \xi}\right)_\eta d\eta = 0 \tag{19-22}$$

This non-accumulation requirement can be used, without alteration, in the calculation of the temperature distribution in a wet regenerator, in order to determine the duration of the wet part of the cold period for a particular position in the regenerator. ϕ could first be calculated for various values of η in the way

---

* Editor's note: The literal translation of the German "Niederschlagsbedingung" is "deposit requirement". The underlying problem is that solid material should not accumulate on the heating surface.

described at the end of Sec. 19-1 as a function of ξ. Then $(\partial\phi/\partial\xi)_\eta$ could be determined for each position ξ and finally the integral in Eq. (19-22) could be evaluated using a method of approximation. It would be more difficult to determine the value of the integral analytically.

Such procedures would, moreover, be very laborious. Equation (19-22) is thus transformed by considering that, according to Eqs. (19-12) and (19-13)

$$\left(\frac{\partial\phi}{\partial\xi}\right)_\eta = -\left(\frac{\partial\Theta}{\partial\eta}\right)_\xi - (\Theta - \theta)$$

Then Eq. (19-23) is obtained as representing the non-accumulation requirement:[5]

$$\Theta_3 - \Theta_1 + \int_{\Theta_2}^{\Theta_3} (\Theta - \theta)\, d\eta = \int_{\Theta_1}^{\Theta_2} (\theta - \Theta)\, d\eta \qquad (19\text{-}23)$$

Equation (19-23) determines the temperature Θ_3 at the end of the wet part of the cold period at each position in the regenerator and also establishes the duration of this wet part if Θ_1 and Θ_2 are known from the previous warm period. Thus Θ_3 and Π^* can be determined in such a way that, knowing Θ, the value of θ can also be calculated, for example by using Eq. (16-11). First the value of the integral for the warm period must be determined using a method of approximation. The calculation of Θ as well as θ proceeds in the η direction for the cold period until the value of Θ_3 satisfies Eq. (19-23). The time Π^* for the wet part of the cold period generally varies from one position to another.

R. Schlatterer[5] has indicated that the non-accumulation requirement (19-23) can be evaluated in the following way for $\xi > 5$. Using the abbreviations

$$\text{for the warm period} \qquad 1 + \frac{d\phi''}{d\Theta} = w(\Theta) \qquad (19\text{-}24)$$

$$\text{for the cold period} \qquad 1 + \frac{d\phi''}{d\Theta} = k(\Theta) \qquad (19\text{-}25)$$

and employing with some intermediate calculations omitted here, Eq. (19-23) can be transformed into

$$\Theta_3 - \Theta_1 + \int_{\Theta_3}^{\Theta_2} \frac{d\Theta}{k(\Theta)_{\xi-\varepsilon}} = \int_{\Theta_1}^{\Theta_2} \frac{d\Theta}{w(\Theta)_{\xi-\varepsilon}} \qquad \text{(with } \xi > 5\text{)} \qquad (19\text{-}26)$$

It follows that in the integrals of this equation, $K(\Theta)$ and $w(\Theta)$ must be determined for the value of Θ at position $\xi - \varepsilon$, while $d\Theta$ and the limits Θ_1, Θ_2 and Θ_3 are related to the position ξ itself. The integrals can easily be worked out if Eqs. (19-27) and (19-28) are established fairly precisely:

$$\int_{\Theta_3}^{\Theta_2} \frac{d\Theta}{k(\Theta)_{\xi-\varepsilon}} = \frac{(\Theta_2 - \Theta_3)_\xi}{(\Theta_2 - \Theta_3)_{\xi-\varepsilon}} \cdot \int_{\Theta_3(\xi-\varepsilon)}^{\Theta_2(\xi-\varepsilon)} \frac{d\Theta}{k(\Theta)} \qquad (19\text{-}27)$$

---

[5] This method, which comes from the first edition of this book, was developed by R. Schlatterer, then in Höllkrigsreuth near Munich, using preliminary work undertaken by the author.

and

$$\int_{\Theta_1}^{\Theta_2} \frac{d\Theta}{w(\Theta)_{\xi-\varepsilon}} = \frac{(\Theta_2 - \Theta_1)_\xi}{(\Theta_2 - \Theta_1)_{\xi-\varepsilon}} \cdot \int_{\Theta_1(\xi-\varepsilon)}^{\Theta_2(\xi-\varepsilon)} \frac{d\Theta}{w(\Theta)} \qquad (19\text{-}28)$$

The limits of the newly obtained integrals are determined by the values of Θ_1, Θ_2, and Θ_3 at position $\xi - \varepsilon$, which also lie in the direction of flow at a distance ε from position ξ.

The non-accumulation requirement (19-26) can easily be worked out with the help of Eqs. (19-27) and (19-28). Then the integrals

$$\int_0^\Theta \frac{d\Theta}{K(\Theta)} \quad \text{and} \quad \int_0^\Theta \frac{d\Theta}{w(\Theta)}$$

which are determined using the vapour pressure curve and the instantaneous gas pressure, can be calculated from the beginning by fairly accurate integration and can then be drawn as a function of Θ. The integrals in Eqs. (19-27) and (19-28) can then easily be determined from the curves obtained, as the difference of two values each of which can be read off from these curves. In a corresponding manner, the integrals can also be calculated directly.

19-3 APPROXIMATE DETERMINATION OF THE STATE OF EQUILIBRIUM OF WET REGENERATORS[5]

The state of equilibrium can basically be calculated in such a way that starting with an estimated temperature distribution in the heat-storing mass a sufficient number of subsequent warm and cold periods can be calculated using one of the step-by-step methods described and the non-accumulation requirement, until there is no longer any marked change in the temperature distribution from one period to another. So that the number of periods to be calculated does not become too large, it is desirable that the temperature distribution is estimated at the outset. This is made possible by a relationship, according to which, when $CT = C'T'$, the chronological average value $\bar{\Theta}'$ of the temperature of the heat-storing mass in the warm period can be calculated as a function of ξ' as a first approximation.

First, Eq. (760) on page 434 of the first edition of this book shows, quite accurately, how the chronological average value $\bar{\vartheta}'$ of the gas temperature is dependent upon ξ'. This approximate equation, the derivation of which will not be repeated here, takes the form

$$\xi' = \text{const} - \left(1 + \frac{\Pi'}{\Pi}\right) \frac{\bar{\vartheta}'}{\Delta\bar{\theta}} - \frac{\Pi'}{2\Delta\bar{\theta}} \cdot \frac{P' - P}{P'} \phi'' \quad \text{(when } CT = C'T') \quad (19\text{-}29)$$

where the dashed quantities ξ', Π' and P' relate to the warm period. P' is the pressure in the warm period, P is the pressure in the cold period. $\Delta\bar{\theta}$ is the uniform, chronological, average value of the temperature difference between the two gases.

---

[5] This method, which comes from the first edition of this book, was developed by R. Schlatterer, then in Höllkrigsreuth near Munich, using preliminary work undertaken by the author.

$\bar{\theta}'$ can be set equal to $\bar{\Theta}'$ in Eq. (19-29) because, as can be seen from Fig. 19-7, right, only very small differences between the two temperatures can be expected. Further, if this relationship is applied to the entry temperature $\bar{\theta}' = \theta'_1$ of the warmer gas when $\xi' = 0(\xi = \Lambda)$, the following Eq. (19-30) is obtained by subtraction:

$$\xi' = \left(1 + \frac{\Pi'}{\Pi}\right)\frac{\theta'_1 - \bar{\Theta}'}{\Delta\bar{\theta}} + \frac{\Pi'}{2\Delta\bar{\theta}} \cdot \frac{P' - P}{P'}(\phi''(\theta'_1) - \phi''(\bar{\Theta}')) \tag{19-30}$$

Within the framework of the inaccuracy of this approximate equation it is possible to conceive of the average temperature of the heat-storing mass $\bar{\Theta}'$ determined by this equation as being also the temperature in the middle of the warm period and to use this as a starting point for the more accurate calculation using Eq. (19-16).

However, in general, it is preferable to use the beginning of the warm period as the starting point for the more accurate calculation, because above all, at the end of the warm period, values of the deposited water can be obtained which themselves determine the duration of the wet part of the cold period.

From Eq. (19-30) an estimated curve can be obtained for the temperature of the heat-storing mass at the *beginning of the warm period* by altering ξ' at each value of $\Theta = \text{const}$, by an amount $\Delta\xi'$ which corresponds to the half period length $\Delta\eta' = -\Pi'/2$. It follows from Eq. (19-17) that the necessary displacement in ξ' is equal to

$$\Delta\xi' = -\left(1 + \frac{d\phi''}{d\Theta'}\right)\frac{\Pi'}{2} \tag{19-31}$$

Thus, from Eq. (19-30) is obtained

$$\begin{aligned}\xi' = \xi'_A = &\left(1 + \frac{\Pi'}{\Pi}\right)\frac{\theta'_1 - \Theta'_A}{\Delta\bar{\theta}} \\ &+ \frac{\Pi'}{2}\left[\frac{1}{\Delta\bar{\theta}} \cdot \frac{P' - P}{P'}(\phi''(\theta'_1) - \phi''(\Theta'_A)) - \left(1 + \frac{d\phi''}{d\Theta'}\right)_A\right]\end{aligned} \tag{19-32}$$

where the suffix A denotes the value at the beginning of the warm period. It thus gives a fairly close approximation for the temperature distribution $\Theta' = \Theta'_A$ of the heat-storing mass at the beginning of the warm period.

19-4 RESULTS OF THE CALCULATIONS

The temperature distribution calculated using the method described here will be explained by way of example from low temperature technology.[6] Again only the condensation of water vapour, and not that of carbon dioxide or other components, will be considered. The dimensionless regenerator length will be $\Lambda = \Lambda' = 250$ and the dimensionless period length will be $\Pi = \Pi' = 50$. These values of Λ and Π correspond, for example, to the case where each one of the two

[6] From H. Hausen [H306].

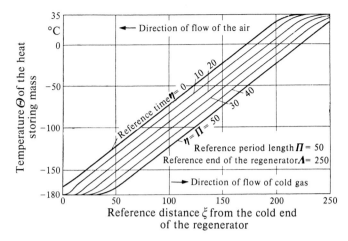

Figure 19-3 Temperature distribution in a regenerator supplied with dry gas. Cold period in the state of equilibrium.

regenerators working together has a heat-storing mass with a thermal capacity of 420 kJ/K and a heat-transferring surface of 1200 m², where the heat transfer coefficient is $\alpha = 58$ W/(m² K), where the flowrates are such that about 1000 kg of air and a quantity of nitrogen of the same thermal capacity ($C = C' = $ const) would flow through the regenerator in one hour. Reversals take place every five minutes. The air enters the regenerator saturated with water at a temperature of $+35°C$; then nitrogen enters the regenerator completely dry, at a temperature of $-180°C$. The air pressure will be assumed to be 5 bar absolute, the nitrogen pressure will be assumed to be 1.1 bar absolute.

Figure 19-3 basically shows the temperature distribution during the cold period when only dry gas is flowing through the regenerator. The individual curves

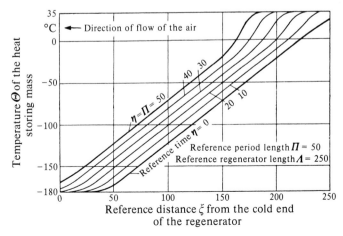

Figure 19-4 Change over of the regenerator from dry to damp air. First wet, warm period. Pressure in warm period 5 bar, in the cold period 1.1 bar.

represent the temperature distribution at various times η. The temperature of the heat-storing mass oscillates up and down with an amplitude of 50°C where the temperature curve is linear.

If the regenerator is then supplied with air which is saturated with water vapour at 5 bar absolute and is at 35°C, the temperature of the heat-storing mass changes suddenly as shown in Fig. 19-4. The figure illustrates the first warm period with wet operation. Here the air enters the regenerator from the right and is cooled by the condensation of the water vapour. The latent heat which is thus released has the effect that the heat-storing mass is heated much more quickly in the region where the water vapour condenses than where only dry air is cooled. This can be seen by the bending of the curves in the top of the figure.

However this temperature distribution cannot be maintained because in so doing, the requirement in Eq. (11-2) would not be satisfied, namely that in the state of equilibrium the average temperature difference $\Delta\bar{\theta}$ should be the same everywhere because $\Lambda = \Lambda'$, $\Pi = \Pi'$ and thus $CT = C'T'$. Figures 19-5 and 19-6 illustrate the state of equilibrium in wet operation. In the cold part of the regenerator (left), where the water deposit is not appreciable, the temperature distribution is still linear but is considerably steeper than in Figs. 19-3 and 19-4. Here the amplitude of the temperature oscillations has increased from 50° to 80°. On the other hand, the region where the water is deposited (right) is very much drawn out. In this region the amplitude of the temperature oscillation, which was very large in Fig. 19-4 has decreased sharply. The right-hand part of Fig. 19-5 shows that the disturbance to the linear temperature distribution, which is caused by the constant entry temperature of the air, is transmitted much further into the interior of the regenerator than in the case of dry operation. In Fig. 19-6, which illustrates the temperature distribution in the cold period, the dashed line indicates the temperature (Θ_3) at each position in the regenerator at which the water ceases to be

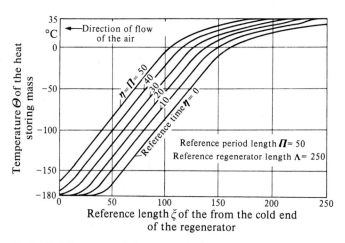

Figure 19-5 Regenerator being operated with wet, saturated air at 35°C. Warm period in the state of equilibrium.

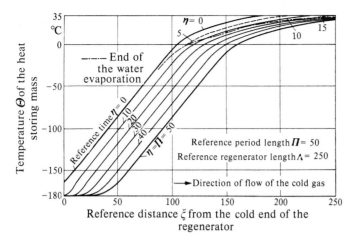

Figure 19-6 Regenerator being operated with wet, saturated air at 35°C. Cold period in the state of equilibrium.

deposited. In the perpendicular direction from the solid curves, particularly between $\xi = 120$ and $\xi = 180$, it can be easily seen that the temperature of the heat-storing mass decreases much more quickly in the wet part of the cold period than in

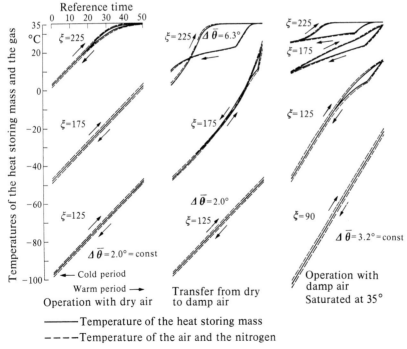

Figure 19-6 Regenerator being operated with wet, saturated air at 35°C. Cold period in the state of equilibrium.

the dry part of this same period. The evaporation of the water is complete after the first fifth of the cold period. Below $\xi = 120$ the evaporation or sublimation proceeds much more quickly. Above $\xi = 180$ on the other hand, the evaporation time is somewhat delayed and thus finishes later. Indeed, above $\xi = 180$ it does not start immediately at the beginning of the cold period. This is because right at the beginning of this period nitrogen flows past the heat-storing mass; this nitrogen has just been heated to the highest temperature, namely 35° and is thus already saturated with water vapour.

The processes just described can be followed more precisely using Fig. 19-7 which shows the chronological temperature variations at various positions in the regenerator. The reduced time η, which runs in opposite directions for the warm and cold periods, is shown horizontally; the temperature Θ of the heat storing mass (solid line), and the gas θ (dashed lines) are shown vertically. The known chronological temperature changes in the dry regenerator are shown first on the left-hand side of the figure. Thus the average temperature difference between the two gases is $\Delta\bar{\theta} = 2°$ at all positions in the regenerator. The chronological temperature variations in the first warm and cold periods directly after the change over from dry to wet operation are shown in the centre of Fig. 19-7 (see Fig. 19-4). The right-hand side of Fig. 19-7 shows the chronological temperature variations which occur at the state of equilibrium. $\Delta\bar{\theta}$ also has the same value throughout the period in this case but is about 60 per cent larger than in a dry regenerator. A considerable capacity for adaptation by the regenerator is expressed in the temperature distribution described. If the temperature distribution in the longitudinal direction of the regenerator remained essentially the same for wet air as it does for dry operation (see Figs. 19-3 and 19-4), the gas, which is to be heated, would leave the regenerator relatively very cold as a result of the temperature oscillation which is increased in amplitude by the condensation and evaporation and the large hysteresis loop thus created. The exchange loss thus becomes unacceptably large. On the other hand, the regenerator automatically has more heat-storing mass available, because of the bending of the temperature curve as shown in Figs. 19-5 and 19-6, which can be used in addition for condensation and evaporation. Conversely, the exchange of heat and cold is increased by the steep rise in temperature in the cold parts of the regenerator. This all has the effect that the strain on the regenerator which is increased by the wetness of the air, is uniformly distributed over its entire length and thus the quality of exchange is affected only slightly over the whole operation.

Figure 19-8 shows the temperature distribution, for this same example, taking into account, in addition, the influence of the deposit and sublimation of the carbon dioxide. Here the non-linearity of temperature distribution is again weak at very low temperatures but the distribution curves off, as previously, in the higher temperature region under the influence of the water vapour content in the air.

Finally, the following should be pointed out. In the examples discussed, it is assumed that the thermal capacity of the gases flowing in the warm and cold periods are the same. In practice however, the thermal capacity of the nitrogen is 2 to 3 per cent greater than that of air which flows through regenerators when, for

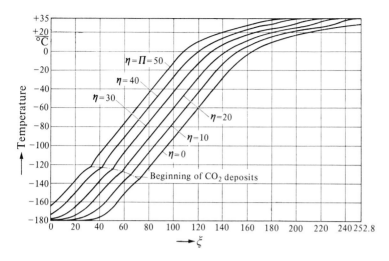

Figure 19-8 Wet regenerator with a carbon dioxide deposit, warm period.

example, they are used in oxygen production. Thus, in the coldest part of the regenerators, very small temperature differences occur, which favour the sublimation of the carbon dioxide. This ensures that the heat-storing mass will be completely free of carbon dioxide towards the end of the period. Then, however, the temperature difference at the warm end of the regenerator becomes larger. This can be avoided if surplus cold gas is added to the gas stream but is then removed again in the middle of the regenerator. Thus the temperature differences are increased in the middle of the regenerator although these differences are smaller at the warm and cold ends. This idea emerged during the design of the first low temperature regenerators but was not then put into practice. It was later developed in America and became known as the "unbalance system" [L304]. The arrangement basically corresponds to the "reversing exchanger" illustrated in Fig. 5-5. Here about 11 per cent of the air in a particular cross-section cooled to the lowest temperature in the exchanger, is fed back again into the middle of the exchanger.

19-5 LITERATURE RELATING TO PART FOUR: REGENERATORS

See Foreword, Sec. 1-5.

A

301. Ackermann, G.: Die Theorie der Wärmeaustauscher mit Wärmespeicherung. Z. angew. Math. Mech. 11 (1931) 192.
302. Allen, D. N. de G.: The Calculation of the Efficiency of Heat Regenerators. Quarterly J. Mech. Appl. Math. 5 (1952) 455–461.
303. Altenkirch, E.: Schnelläufige Regeneratoren. Wissenschaftliche Berichte, Series 4 Maschinenbau, H. 1. Berlin: Verlag Technik 1952.
304. Anzelius, A.: Über Erwärmung vermittels durchströmender Medien. Z. angew. Math. Mech. 6 (1926) 291.

B

301. Bartsch, A.: Regeneratoren der Tieftemperaturtechnik. Berlin: Verlag Technik 1962.
302. Becker, R.: Anwendung von Regeneratoren in der Tieftemperaturtechnik, besonders zur Anreicherung von Äthylen aus Koksofengas. Linde-terichte aus Wissensch. und Technik (1958) 3, 25f.
303. Becker, R.: Regeneratorverfahren zur Koksgaszerlegung. Kältetech. 16 (1964) 239–241.
304. Binder, L.: Äußere Wärmeleitung und Erwärmung elektrischer Maschinen. Diss. München 1911.
305. Binder, L.: Über Wärmeübergang auf ruhige und bewegte Luft sowie Lüftung und Kühlung elektrischer Maschinen. Halle: W. Knapp 1912.
306. Blass, E.: Die Eignung von Drahtgeweben als Wärmespeichermassen. Brennstoff-Wärme-Kraft 16 (1964) 6, 267–272.
307. Bock, H.: Regeneratoren und Sorptiv-Regeneratoren. Forschg. Ing. Wes. 28 (1962) 37–46.
308. Boestad, G.: Die Wärmeübertragung im Ljungström-Luftvorwärmer. Feuerungstechnik 26 (1938) 9, 282–286.
309. Bošnjaković, Fr.: Technische Thermodynamik, 2nd part Dresden und Leipzig: Th. Steinkopff, 4th ed. 1965, p. 22, Equations (53) and (54).
310. Buker, J. C.; Simcic, N. F.: Blast-Furnace Stove Analysis and Control. Instrum. Soc. Amer. Trans. 2 (1962) 2, 160–167.

C

301. Chase, C. A. jr.; Gidaspow, D.; Peck, R. E.: A Regenerator-Prediction of Nusselt Numbers. Int. J. Heat Mass Transfer 12 (1969) 6, 727–736.
302. Chase, C. A.; Gidaspow, D.; Peck, R. E.: Transient Heat and Mass Transfer in an Adiabatic Regenerator—a Green's Matrix Representation. Int. J. Heat Mass Transfer 13 (1970) May 817–834.
303. Coppage, J. E.; London, A. L.: The Periodic-Flow Regenerator. A Summary of Design Theory. Trans. Am. Soc. Mech. Eng. 75 (1953) 5, 779–787.
304. Courant, R.: Hilpert, D.: Methoden der mathematischen Physik. Vol. I. Berlin: Springer 1924, p. 32ff.
305. Crank, J.; Nicolson, P.: A Practical Method for Numerical Evaluation of Solutions of Partial Differential Equations of the Heat Conduction Type. Proc. Camb. Phil. Soc. 34 (1947) 50–76.
306. Creswick, F. A.: A Digital Computer Solution of the Equations for Transient Heating of a Porous Solid including Effects of Longitudinal Conduction. Ind. Math. 8 (1957) 61.

E

301. Edwards, J. V.; Evans, R.; Probert, S. D.: Computation of Transient Temperatures in Regenerators. Int. J. Heat Mass Transfer 14 (1971) Aug., 1175–11202.

F

301. Fränkl, M.: DRP 490878 and 492431.

G

301. Glaser, H.: Der Wärmeübergang in Regeneratoren. Z. VDI Beiheft "Verfahrenstechnik" (1938) 4, 112–125.
302. Glaser, H.: Der Regenerator als Hochleistungs-Wärmeaustauscher. Chem. Ing. Tech. 26 (1954) 1, 39–41.
303. Glaser, H.: Regeneratoren mit bewegter Speichermasse. Forschg. Ing.-Wes. (1951) 1, 9–15.

H

301. Handley, D.; Heggs, P. J.: The Effect of Thermal Conductivity of the Packing Material on Transient Heat Transfer in a Fixed Bed. Int. J. Heat Mass Transfer 12 (1969) 549–570.

302. Harper, D. B.; Rohsenow, W. M.: Effect of Rotary-Regenerator Performance on Gas-Turbine-Plant Performance. Trans. ASME 75 (1953) 759–765.
303. Hausen, H.: Über die Theorie des Wärmeaustausches in Regeneratoren, Habilitationsschrift vom 21. 2. 1927 (Prof. Nußelt was the first to read), published in Z. angew. Math. Mech. 9 (1929) 173–200; see also Hausen, H.: Wärmeaustausch in Regeneratoren. Z. VDI 73 (1929) 431–433.
304. Hausen, H.: Über den Wärmeaustausch in Regeneratoren. Techn. Mech. Thermodyn. 1 (1930) 219 and 250.
305. Hausen, H.: Näherungsverfahren zur Berechnung des Wärmeaustausches in Regeneratoren. Z. angew. Math. Mech. 11 (1931) 105–114.
306. Hausen, H.: Feuchtigkeitsablagerung in Regeneratoren. Z. VDI Beiheft "Verfahrenstechnik" 1937, No. 2, 62–67.
307. Hausen, H.: Berechnung der Steintemperatur in Winderhitzern. Arch. Eisenhüttenw. 12 (1938/39) 473–480.
308. Hausen, H.: Vervollständigte Berechnung des Wärmeaustausches in Regeneratoren. VDI-Beiheft "Verfahrenstechnik" (1942) 2, 31–43.
309. Hausen, H.: Einfluß des Lewisschen Koeffizienten auf das Ausfrieren von Dämpfen aus Gas-Dampf-Gemischen. Angew. Chemie (B), 20 (1948) 177–182.
310. Hausen, H.: Berechnung der Wärmeübertragung in Regeneratoren bei temperaturabhängigen Stoffwerten und Wärmeübergangszahlen. Int. J. Heat Mass Transfer 7 (1964) 112–123.
311. Hausen, H.: Berechnung der Wärmeübertragung in Regeneratoren bei zeitlich veränderlichem Mengenstrom. Int. J. Heat Mass Transfer 13 (1970) 1753–1766.
312. Hausen, H.: Berechnung von Regeneratoren nach der Gaußschen Integrationsmethode. Int. J. Mass Heat Transfer 17 (1974) 1111–1113.
313. Heiligenstaedt, W.: Regeneratoren, Rekuperatoren, Winderhitzer. Leipzig: Spamer 1931, p. 169.
314. Heiligenstaedt, W.: Die Berechnung von Wärmespeichern. Arch. Eisenhüttenw. 2 (1928/1929) 217–222; see also Heiligenstaedt, W.: Wärmetechnische Rechnungen für Industrieöfen. 2nd ed. Düsseldorf: Stahleisen 1941.
315. Herzog, E.: Der heutige Stand unserer Kenntnisse vom Siemens-Martin-Ofen. p. 12f. Bericht No. 120 d. Ver. deutsch. Eisenhüttenl. Düsseldorf: Stahleisen 1927 (most important older Literature on Regenerators).
316. Hlinka, J. W.; Puhr, F. S.; Paschkis, V.: AISE Manual for Thermal Design of Regenerators. Iron Steel Engr. (1961) 59.
317. Hochgesand, C. P.: Das Linde-Fränkl-Verfahren zur Zerlegung von Gasgemischen. Mitt. Forschanst. d. GHH-Konzerns 4 (1935) 1, 14–23.
318. Hofmann, E. E.; Kappelmayer, A.: Strömungstechnische Modellversuche an einem Winderhitzer mit außenstehendem Brennschacht und Vergleich der Ergebnisse mit Betriebsmessungen. Arch, Eisenhüttenw. 40 (1969) 4, 311–322.
319. "Hütte", 28th ed., vol. I, 1955, p. 107.

I

301. Iliffe, C. E.: Thermal Analysis of the Contra-Flow Regenerative Heat Exchanger. Proc. Inst. Mech. Engrs. 159 (1948) 363–371.

J

301. Jenne, O.; Schwarz v. Bergkampf, E.: Eine einfache Berechnung von Regeneratoren. Gaswärme 3 (1954) 240–246.
302. Johnson, J. E.: Regenerative Heat Exchangers for Gas-Turbines. Technical Report No. 2630 of the Aeronautical Research Council (1948) 1025–1094.

K

301. Kardas, A.: On a Problem in the Theory of the Unidirectional Regenerator. Int. J. Heat Mass Transfer 9 (1966) 567.

302. Kessels, K.: Ergebnisse der Untersuchung von Hochofenwiderhitzern. Stahl und Eisen 75 (1955) 958–1119.
303. Knapp, H.: Umschalt-Wärmeaustauscher und Regeneratoren in Luftzerlegungsanlagen. Chem. Ing. Tech. 39 (1967) 5/6, 390–394.

L

301. Lambertson, T. J.: Performance Factors of a Periodic-Flow Heat Exchanger. Trans. Am. Soc. Mech. Eng. A 80 (1958) 586–592.
302. Larsen, F. W.: Rapid Calculation of Temperature in a Regenerative Heat Exchanger Having Arbitrary Initial Solid and Entering Fluid Temperatures. Int. J. Heat Mass Transfer 10 (1967) 149–168.
303. Lewis, W. K.: The Evaporation of a Liquid into a Gas. Mech. Engineering 44 (1922) 445. See also Merkel, Fr.: Verdunstungskühlung, VDI-Forschungsheft 275 (1925).
304. Lobo, W. E.; Skaperdas, G. T.: Chem. Eng. Progress 43 (1947) 69.
305. Loske, K.; Wiezorke, B.: Ergebnisse bei der Anwendung eines mathematischen Winderhitzer-Modells. Arch. Eisenhüttenw. 39 (1968) 4, 265–272.
306. Lowan, A. N.: On the Problem of Heat Recuperator. Phil. Mag. (7), Vol. 17 (1934) pp. 914–933 (Phys. Ber. 15, 2 (1934) 1289).
307. Lund, G.; Dodge, B. F.: Fränkl-Regenerator Packings. Heat Transfer and Pressure Drop Characteristics. Ind. Eng. Chem. 40 (1948) 6, 1019–1032.

M

301. MacLaine-Cross, I. L.; Banks, P. J.: Coupled Heat and Mass Transfer in Regenerators — Prediction Using an Analogy with Heat Transfer. Int. J. Heat Mass Transfer 15 (1972) 1225–1242.
302. Manrique, J. A.; Cardenas, R. S.: Digital Simulation of a Regenerator. 5. Int. Heat Transfer Conference Tokyo, Spet. 1974. Preprints vol. V, pp. 190–194.
303. Mitchell, A. R.; Pearce, R. P.: High Accuracy Difference Formulae for the Numerical Solution of the Heat Conduction Equation. Computer J. 5 (1962) 142–146.
304. Modest, M. F.; Tien, C. L.: Thermal Analysis of Cyclic Cryogenic Regenerators (takes into account the influences which are usually neglected such as the internal energy exchange of the fluid as a result of the periodically changing pressure and the longitudinal thermal conduction of the heat-storing mass). Int. J. Heat Mass Transfer 17 (1974) 37–49.

N

301. Nahavandi, A. H.; Weinstein, A. S.: A Solution to the Periodic Flow Regenerative Heat Exchanger Problem. Appl. Sci. Res. A, 10 (1961) 335–348.
302. Naumann, G.: Versuche an einem Winderhitzer. Mitt. Wärmestelle Düsseldorf VDEh No. 82 (1926).
303. Nußelt, W.: Die Theorie des Winderhitzers. Z. VDI 71 (1927) 85.
304. Nußelt, W.: Der Beharrungszustand im Winderhitzer. Z. VDI 72 (1928) 1052–1054.

P

301. Peiser, A. M.; Lehner, J.: Design Charts for Symmetric Regenerators. Ind. Eng. Chem. 45 (1953) 10, 2166–2170.
302. Price, C. B. A.: Heat and Momentum Transfer in Thermal Regenerators. University of London. Ph.D. Thesis 1964.

R

301. Razelos, P.; Lazaridis, A.: A Lumped Heat Transfer Coefficient for Periodically Heated Hollow Cylinders. Int. J. Heat Mass Transfer 10 (1967) 1373–1387.
302. Ridgion, J. M.; Willmott, A. J.; Thewlis, J. H.: An Analogue Computer Simulation of a Cowper Stove. Computer Jl. 7 (1964) 188–196.

303. Riedel, H.: Über regenerative Wärmeübertrager mit poröser Speichermasse. Forschg. Ing. Wes. 31 (1965) 6, 187–191.

304. Rummel, K.: Die Berechnung der Wärmespeicher auf Grund der Wärmedurchgangszahl. Stahl und Eisen 48 (1928) 1712–1715.

305. Rummel, K.; Schack, A.: Die Berechnung von Regeneratoren. Stahl und Eisen 49 (1929) 1300; Arch. Eisenhüttenwes. 2 (1928/29) 473.

306. Rummel, K.: Die Berechnung der Wärmespeicher. Arch. Eisenhüttenw. 4 (1930/31) 367.

S

301. Sandner, H.: Beitrag zur linearen Theorie des Regenerators. Diss. Techn. Univ. München 1971.

302. Saunders, O. A.; Smoleniec, S.: Heat Regenerators. VII. Int. Congr. Appl. Mech. 3 (1948), 91–105.

303. Schack, A.: Praktische Berechnung zeitlich veränderlicher Wärmeströmungen. Arch. Eisenhüttenw. 1 (1927/28) 357.

304. Schack, A.: Über den Einfluß des Staubbelages auf den Wirkungsgrad von Regeneratoren. Z. techn. Physik 9 (1928) 390–398; Arch. Eisenhüttenw. 2 (1928/29) 287–292.

305. Schack, A.: Die zeitliche Temperaturänderung im Regenerator. Arch. Eisenhüttenw. 2 (1928/29) 481.

306. Schack, A.: Die Gas- und Luftvorwärmung durch Stahlrekuperatoren in der Großindustrie. Z. kompr. und flüssige Gase 36 (1941) 101.

307. Schack, A.: Die Berechnung der Regeneratoren. Arch. Eisenhüttenw. 17 (1943/44) 101–118.

308. Schefels, G.: Reibungsverluste in gemauerten engen Kanälen und ihre Bedeutung für die Zusammengänge zwischen Wärmeübergang und Druckverlust in Winderhitzern. Arch. Eisenhüttenw. 6 (1933) 477–486.

309. Schellmann, E.: Näherungsverfahren zur Berechnung der Wärmeübertragung in Regeneratoren unter Berücksichtigung der Wärmeverluste. Chem. Ing. Tech. 42 (1970) 22, 1358–1363.

310. Schlatterer, R.: see p. 475.

311. Schmeidler, W.: Mathematische Theorie der Wärmespeicherung. Z. angew. Math. Mech. 8 (1928) 385–393.

312. Schmidt, E.: Über die Anwendung der Differenzenrechnung auf technische Anheiz- und Abkühlungsprobleme. Beiträge zur technischen Mechanik und technischen Physik. Festschrift August Föppl zum 70. Geburtstag. Berlin: Springer 1924, pp. 179–189.

313. Schmidt, E.: Das Differenzenverfahren zur Lösung von Differentialgleichungen der nicht stationären Wärmeleitung. Diffusion und Impulsausbreitung. Forschung Ing. Wes. 13 (1942) 177–185.

314. Schofield, J.; Butterfield, P.; Young, P. A.: Hot Blast Stoves. J. Iron Steel Inst. 199 (1961) 229–240.

315. Schultz, B. H.: Regenerators with Longitudinal Heat Conduction. IME-ASME General Discussion on Heat Transfer, London (1951) Sept. 440–443.

316. Schultz, B. H.: Approximative Formulae in the Theory of the Thermal Regenerators. Appl. Sc. Research (The Hague) Ser. A: Mechanics 3 (1953) 165–173.

317. Schumacher, K.: Großversuche an einer zu Studienzwecken gebauten Regenerativkammer. Arch. Eisenhüttenw. 4 (1930/31) 63–74.

318. Schumann, T. E. W.: Heat Transfer: A Liquid Flowing through a Porous Prism. J. Franklin Inst. 208 (1929) 405.

319. Stuke, B.: Berechnung des Wärmeaustausches in Regeneratoren mit zylindrischem oder kugelförmigem Füllmaterial. Angew. Chemie B 20 (1948) 262–268.

T

301. Tipler, W.: A Simple Theory of the Heat Regenerator. Shell Technical Report ICT 14 (1947); Proc. 7th Int. Congr. Appl. Mech. Section III, p. 196, London (1948).

302. Traustel. S.: Auslegung von Regeneratoren. Brennstoff-Wärme-Kraft 24 (1972) 14–16.

W

301. Wagner, C.: Über einen einfachen Sonderfall zur Berechnung der Temperaturverteilung in Wärmespeichern beim Wärmeaustausch mit strömenden Gasen. Ingenieurarchiv 14 (1943) 136–140.

302. Weishaupt, J.: Messungen an Regeneratoren von Groß-Sauerstoff-Anlagen. Kältetechnik 5 (1953) 4, 99–103.
303. Willmott, A. J.; Voice, E. W.: Development of Theoretical Methods for Calculating the Thermal Performance of the Hot Blast Stove. Troisièmes Journées Internationales de Sidérurgie, Luxemburg (1962) 473–482.
304. Willmott, A. J.: Digital Computer Simulation of a Thermal Regenerator. Int. J. Heat Mass Transfer 7 (1964) 1291–1302.
305. Willmott, A. J.: Operation of Cowper-Stoves under Conditions of Variable Flow. J. Iron Steel Inst. 206 (1968) 33–38.
306. Willmott, A. J.: Simulation of a Thermal Regenerator under Conditions of Variable Mass Flow. Int. J. Heat Mass Transfer 11 (1968) 1105–1116.
307. Willmott, A. J.: The Regenerative Heat Exchanger Computer Representation. Int. J. Heat Mass Transfer 12 (1969) 997–1014.
308. Willmott, A. J.; Duggan, R. C.: Refined Closed Methods for the Contra-Flow Thermal Regenerator Problem. Int. J. Heat Mass Transfer 23 (1980), 655–662.
309. Willmott, A. J.; Thomas, R. J.: Analysis of the Long Contra-Flow Regenerative Heat Exchanger. J. Inst. Maths. Applics 14 (1974) 267–280.
310. Willmott, A. J.; Kulakowski, B.: Numerical Acceleration of Thermal Regenerator Simulations. Int. J. Numerical Methods in Engineering, 11, (1977), 533–551.
311. Willmott, A. J.; Hinchcliffe, C.: The Effect of Gas Heat Storage upon the Performance of the Thermal Regenerators. Int. J. Heat Mass Transfer 19 (1976) 821–826.
312. Willmott, A. J.; Burns, A.: Transient Response of Periodic Flow Regenerators. Int. J. Heat Mass Transfer, 20, (1977), 753–761.

Z

301. Zurmühl, R.: Praktische Mathematik. Berlin, Heidelberg, New York: Springer 1965.

These tables relate the Equation, Figure and Table numbers in the German second edition to the numbers used in this English-language edition.

Equations

German	English	German	English	German	English
1	2-1	24	2-24	47	2-47
2	2-2	25	2-25	48	2-48
3	2-3	26	2-26	49	2-49
4	2-4	27	2-27	50	2-50
5	2-5	28	2-28	51	2-51
6	2-6	29	2-29	52	2-52
7	2-7	30	2-30	53	2-53
8	2-8	31	2-31	54	2-54
9	2-9	32	2-32	55	2-55
10	2-10	33	2-33	56	2-56
11	2-11	34	2-34	57	2-57
12	2-12	35	2-35	58	2-58
13	2-13	36	2-36	59	2-59
14	2-14	37	2-37	60	2-60
15	2-15	38	2-38	61	2-61
16	2-16	39	2-39	62	2-62
17	2-17	40	2-40	63	2-63
18	2-18	41	2-41	64	2-64
19	2-19	42	2-42	65	2-65
20	2-20	43	2-43	66	2-66
21	2-21	44	2-44	67	2-67
22	2-22	45	2-45	68	2-68
23	2-23	46	2-46	69	2-69

German	English	German	English	German	English
70	2-70	122	4-16	174	5-49
71	2-71	123	4-17	175	5-50
72	2-72	124	4-18	176	5-51
73	2-73	125	4-19	177	5-52
74	2-74	126	5-1	178	5-53
75	2-75	127	5-2	179	5-54
76	2-76	128	5-3	180	5-55
77	2-77	129	5-4	181	5-56
78	2-78	130	5-5	182	5-57
79	2-79	131	5-6	183	5-58
80	3-1	132	5-7	184	5-59
81	3-2	133	5-8	185	5-60
82	3-3	134	5-9	186	5-61
83	3-4	135	5-10	187	5-62
84	3-5	136	5-11	188	5-63
85	3-6	137	5-12	189	5-64
86	3-7	138	5-13	190	5-65
87	3-8	139	5-14	191	5-66
88	3-9	140	5-15	192	5-67
89	3-10	141	5-16	193	5-68
90	3-11	142	5-17	194	5-69
91	3-12	143	5-18	195	5-70
92	3-13	144	5-19	196	5-71
93	3-14	145	5-20	197	5-72
94	3-15	146	5-21	178	5-73
95	3-16	147	5-22	199	5-74
96	3-17	148	5-23	200	5-75
97	3-18	149	5-24	201	5-76
98	3-19	150	5-25	202	5-77
99	3-20	151	5-26	203	5-78
100	3-21	152	5-27	204	5-79
101	3-22	153	5-28	205	5-80
102	3-23	154	5-29	206	5-81
103	3-24	155	5-30	207	5-82
104	3-25	156	5-31	208	5-83
105	3-26	157	5-32	209	5-84
106	3-27	158	5-33	210	5-85
107	4-1	159	5-34	211	5-86
108	4-2	160	5-35	212	5-87
109	4-3	161	5-36	213	5-88
110	4-4	162	5-37	214	5-89
111	4-5	163	5-38	215	5-90
112	4-6	164	5-39	216	5-91
113	4-7	165	5-40	217	5-92
114	4-8	166	5-41	218	5-93
115	4-9	167	5-42	219	5-94
116	4-10	168	5-43	220	5-95
117	4-11	169	5-44	221	5-96
118	4-12	170	5-45	222	5-97
119	4-13	171	5-46	223	5-98
120	4-14	172	5-47	224	6-1
121	4-15	173	5-48	225	6-2

German	English	German	English	German	English
226	6-3	278	7-39	330	8-22
227	6-4	279	7-40	331	8-23
228	6-5	280	7-41	332	8-24
229	6-6	281	7-42	333	8-25
230	6-7	282	7-43	334	8-26
231	6-8	283	7-44	335	8-27
232	6-9	284	7-45	336	8-28
233	6-10	285	7-46	337	8-29
234	6-11	286	7-47	338	8-30
235	6-12	287	7-48	339	8-31
236	6-13	288	7-49	340	8-32
237	6-14	289	7-50	341	8-33
238	6-15	290	7-51	342	8-34
239	6-16	291	7-52	343	8-35
240	7-1	292	7-53	344	8-36
241	7-2	293	7-54	345	8-37
242	7-3	294	7-55	346	8-38
243	7-4	295	7-56	347	8-39
244	7-5	296	7-57	348	8-40
245	7-6	297	7-58	349	8-41
246	7-7	298	7-59	350	8-42
247	7-8	299	7-60	351	8-43
248	7-9	300	7-61	352	8-44
249	7-10	301	7-62	353	8-45
250	7-11	302	7-63	354	8-46
251	7-12	303	7-64	355	8-47
252	7-13	304	7-65	356	8-48
253	7-14	305	7-66	357	8-49
254	7-15	306	7-67	358	8-50
255	7-16	307	7-68	359	8-51
256	7-17	308	7-69	360	8-52
257	7-18	309	8-1	361	8-53
258	7-19	310	8-2	362	8-54
259	7-20	311	8-3	363	8-55
260	7-21	312	8-4	364	8-56
261	7-22	313	8-5	365	8-57
262	7-23	314	8-6	366	8-58
263	7-24	315	8-7	367	8-59
264	7-25	316	8-8	368	8-60
265	7-26	317	8-9	369	8-61
266	7-27	318	8-10	370	8-62
267	7-28	319	8-11	371	8-63
268	7-29	320	8-12	372	8-64
269	7-30	321	8-13	373	8-65
270	7-31	322	8-14	374	8-66
271	7-32	323	8-15	375	8-67
272	7-33	324	8-16	376	8-68
273	7-34	325	8-17	377	8-69
274	7-35	326	8-18	378	8-70
275	7-36	327	8-19	379	8-71
276	7-37	328	8-20	380	8-72
277	7-38	329	8-21	381	8-73

German	English	German	English	German	English
382	8-74	434	9-44	486	12-19
383	8-75	435	9-45	487	12-20
384	8-76	436	9-46	488	12-21
385	8-77	437	9-47	489	12-22
386	8-78	438	9-48	490	12-23
387	8-79	439	9-49	491	12-24
388	8-80	440	9-50	492	12-25
389	8-81	441	9-51	493	12-26
390	8-82	442	9-52	494	12-27
391	9-1	443	9-53	495	12-28
392	9-2	444	9-54	496	12-29
393	9-3	445	9-55	497	12-30
394	9-4	446	9-56	498	12-31
395	9-5	447	9-57	499	12-32
396	9-6	448	10-1	500	12-33
397	9-7	449	10-2	501	12-34
398	9-8	450	10-3	502	12-35
399	9-9	451	10-4	503	12-36
400	9-10	452	10-5	504	12-37
401	9-11	453	10-6	505	12-38
402	9-12	454	11-1	506	12-39
403	9-13	455	11-2	507	12-40
404	9-14	456	11-3	508	12-41
405	9-15	457	11-4	509	12-42
406	9-16	458	11-5	510	12-43
407	9-17	459	11-6	511	12-44
408	9-18	460	11-7	512	12-45
409	9-19	461	11-8	513	12-46
410	9-20	462	11-9	514	12-47
411	9-21	463	11-10	515	12-48
412	9-22	464	11-11	516	12-49
413	9-23	465	11-12	517	12-50
414	9-24	466	11-13	518	12-51
415	9-25	467	11-14	519	13-1
416	9-26	468	11-15	520	13-2
417	9-27	469	12-1	521	13-3
418	9-28	470	12-2	522	13-4
419	9-29	471	12-3	523	13-5
420	9-30	472	12-4	524	13-6
421	9-31	473	12-5	525	13-7
422	9-32	474	12-6	526	13-8
423	9-33	475	12-7	527	13-9
424	9-34	476	12-8	528	13-10
425	9-35	477	12-9	529	13-11
426	9-36	478	12-10	530	13-12
427	9-37	479	12-11	531	13-13
428	9-38	480	12-12	532	13-14
429	9-39	481	12-13	533	13-15
430	9-40	482	12-15	534	13-16
431	9-41	483	12-16	535	13-17
432	9-42	484	12-17	536	13-18
433	9-43	485	12-18	537	13-19

German	English	German	English	German	English
538	13-20	590	13-72	642	14-7
539	13-21	591	13-73	643	14-8
540	13-22	592	13-74	644	14-9
541	13-23	593	13-75	645	14-10
542	13-24	594	13-76	646	14-11
543	13-25	595	13-77	647	14-12
544	13-26	596	13-78	648	14-13
545	13-27	597	13-79	649	14-14
546	13-28	598	13-80	650	14-15
547	13-29	599	13-81	651	14-16
598	13-30	600	13-82	652	14-17
549	13-31	601	13-83	653	14-18
550	13-32	602	13-84	654	14-19
551	13-33	603	13-85	655	14-20
552	13-34	604	13-86	656	14-21
553	13-35	605	13-87	657	14-22
554	13-36	606	13-88	658	14-23
555	13-37	607	13-89	659	14-24
556	13-38	608	13-90	660	14-25
557	13-39	609	13-91	661	14-26
558	13-40	610	13-92	662	14-27
559	13-41	611	13-93	663	14-28
560	13-42	612	13-94	664	14-29
561	13-43	613	13-95	665	14-30
562	13-44	614	13-96	666	14-31
563	13-45	615	13-97	667	14-32
564	13-46	616	13-98	668	14-33
565	13-47	617	13-99	669	14-34
566	13-48	618	13-100	670	14-35
567	13-49	619	13-101	671	14-36
568	13-50	620	13-102	672	14-37
569	13-51	621	13-103	673	14-38
570	13-52	622	13-104	674	15-1
571	13-53	623	13-105	675	15-2
572	13-54	624	13-106	676	15-3
573	13-55	625	13-107	677	15-4
574	13-56	626	13-108	678	15-5
575	13-57	627	13-109	679	15-6
576	13-58	628	13-110	680	15-7
577	13-59	629	13-111	681	15-8
578	13-60	630	13-112	682	15-9
579	13-61	631	13-113	683	15-10
580	13-62	632	13-114	684	15-11
581	13-63	633	13-115	685	15-12
582	13-64	634	13-116	686	15-13
583	13-65	—	—	687	15-14
548	13-66	636	14-1	688	15-15
585	13-67	637	14-2	689	15-16
586	13-68	638	14-3	690	15-17
587	13-69	639	14-4	691	15-18
588	13-70	640	14-5	692	15-19
589	13-71	641	14-6	693	15-20

German	English	German	English	German	English
694	15-21	730	16-34	766	17-10
695	15-22	731	16-35	767	17-11
696	15-23	732	16-36	768	18-1
697	16-1	733	16-37	769	19-1
698	16-2	734	16-38	770	19-2
699	16-3	735	16-39	771	19-3
700	16-4	736	16-40	772	19-4
701	16-5	737	16-41	773	19-5
702	16-6	738	16-42	774	19-6
703	16-7	739	16-43	775	19-7
704	16-8	740	16-44	776	19-8
705	16-9	741	16-45	777	19-9
706	16-10	742	16-46	778	19-10
707	16-11	743	16-47	779	19-11
708	16-12	744	16-48	780	19-12
709	16-13	745	16-49	781	19-13
710	16-14	746	16-50	782	19-14
711	16-15	747	16-51	783	19-15
712	16-16	748	16-52	784	19-16
713	16-17	749	16-53	785	19-17
714	16-18	750	16-54	786	19-18
715	16-19	751	16-55	787	19-19
716	16-20	752	16-56	788	19-20
717	16-21	753	16-57	789	19-21
718	16-22	754	16-58	790	19-22
719	16-23	755	16-59	791	19-23
720	16-24	756	16-60	792	19-24
721	16-25	757	17-1	793	19-25
722	16-26	758	17-2	794	19-26
723	16-27	759	17-3	795	19-27
724	16-28	760	17-4	796	19-28
725	16-29	761	17-5	797	19-29
726	16-30	762	17-6	798	19-30
727	16-31	763	14-7	799	19-31
728	16-32	764	17-8	800	19-32
729	16-33	765	17-9		

NB. Equation No. 635 is missing in the original German text.

Figures

German	English	German	English	German	English
1	2-1	8	2-8	15	2-15
2	2-2	9	2-9	16	2-16
3	2-3	10	2-10	17	2-17
4	2-4	11	2-11	18	2-18
5	2-5	12	2-12	19	2-19
6	2-6	13	2-13	20	2-20
7	2-7	14	2-14	21	2-21

German	English	German	English	German	English
22	2-22	74	5-24	126	11-4
23	2-23	75	6-1	127	11-5
24	2-24	76	7-1	128	11-6
25	2-25	77	7-2	129	11-7
26	2-26	78	7-3	130	11-8
27	2-27	79	7-4	131	11-9
28	2-28	80	7-5	132	11-10
29	2-29	81	7-6	133	11-11
30	2-30	82	7-7	134	11-12
31	2-31	83	8-1	135	11-13
32	2-32	84	8-2	136	11-14
33	2-33	85	8-3	137	11-15
34	2-34	86	8-4	138	12-1
35	2-35	87	8-5	139	12-2
36	2-36	88	8-6	140	12-3
37	3-1	89	8-7	141	12-4
38	3-2	90	8-8	142	12-5
39	3-3	91	8-9	143	12-6
40	3-4	92	8-10	144	12-7
41	3-5	93	8-11	145	12-8
42	3-6	94	8-12	146	12-9
43	3-7	95	8-13	147	12-10
44	3-8	96	8-14	148	12-11
45	3-9	97	8-15	149	12-12
46	4-1	98	8-16	150	12-13
47	4-2	99	8-17	151	12-14
48	4-3	100	8-18	152	13-1
49	4-4	101	8-19	153	13-2
50	4-5	102	8-20	154	13-3
51	5-1	103	8-21	155	13-4
52	5-2	104	8-22	156	13-5
53	5-3	105	8-23	157	13-6
54	5-4	106	8-24	158	13-7
55	5-5	107	8-25	159	13-8
56	5-6	108	8-26	160	13-9
57	5-7	109	8-27	161	13-10
58	5-8	110	8-28	162	13-11
59	5-9	111	9-1	163	13-12
60	5-10	112	9-2	164	13-13
61	5-11	113	9-3	165	13-14
62	5-12	114	9-4	166	13-15
63	5-13	115	9-5	167	13-16
64	5-14	116	9-6	168	13-17
65	5-15	117	9-7	169	13-18
66	5-16	118	9-8	170	13-19
67	5-17	119	9-9	171	13-20
68	5-18	120	9-10	172	14-1
69	5-19	121	9-11	173	14-2
70	5-20	122	10-1	174	14-3
71	5-21	123	11-1	175	14-4
72	5-22	124	11-2	176	14-5
73	5-23	125	11-3	177	15-1

German	English	German	English	German	English
178	15-2	191	17-3	204	18-11
179	15-3	192	17-4	205	18-12
180	15-4	193	17-5	206	18-13
181	15-5	194	18-1	207	18-14
182	15-6	195	18-2	208	19-1
183	15-7	196	18-3	209	19-2
184	16-1	197	18-4	210	19-3
185	16-2	198	18-5	211	19-4
186	16-3	199	18-6	212	19-5
187	16-4	200	18-7	213	19-6
188	16-5	201	18-8	214	19-7
189	17-1	202	18-9	215	19-8
190	17-2	203	18-10		

Tables

German	English	German	English	German	English
1	2-1	6	2-6	10	7-3
2	2-2	7	3-1	11	7-4
3	2-3	8	7-1	12	7-5
4	2-4	9	7-2	13	9-1
5	2-5				

Editor's note on Appendices 2, 3, 4 and 5

The Appendices represent English translations of parts of the 1st edition (1950), which were not included in the 2nd edition (1976) but to which reference is made in the 2nd edition. The Appendices are included, with the author's agreement, to eliminate unnecessary gaps in the English text of the 2nd edition, gaps caused by the fact that the 1st edition was never published in English and, in any case, has been out of print for over twenty years.

No attempt has been made to change the nomenclature or the equation numbers used in the 1st edition, except where reference is made to equations outside the body of the Appendix concerned. In these cases, the equation number (of the form (7-23)) used in this English edition is employed.

In the German text, the author refers the reader to the appropriate material in the 1st edition. In this English translation the reader is referred, instead, to the appropriate Appendix provided here.

TWO

APPROXIMATION PROCESS FOR CALCULATING THE TEMPERATURE CURVE WITH VARIABLE SPECIFIC HEATS

When the thermal capacities C, C' and C'' vary with temperature or even when k' and k'', and thus K' and K'', are not constant, as in Eq. (9-43), and as long as these variations are only slight, it is often possible to retain the above mentioned equations by substituting suitable average values for the quantities mentioned. When there are large variations it is often possible to consider the thermal capacities to be constant, at least piecewise. Then it would be appropriate to assume that the temperature differences at the ends of the recuperator be given or estimated and to carry out each section of the calculation step-by-step using Eqs. (9-50) to (9-52).

The following step-by-step process allows a more precise consideration of all variations in the thermal capacities. If we introduce the quantities of gas G, G' and G'' per unit time and the thermal contents i, i' and i'' per unit quantity of the respective gases, in place of Eqs. (9-37), (9-38) and (9-39) we can also write

$$d\dot{q} = -G\,di = -(G'\,di' + G''\,di'') \tag{408}$$

$$d\dot{q}' = -G'\,di' = k'\frac{F'}{F}\,df\,(\theta - \theta') \tag{409}$$

$$d\dot{q}'' = -G''\,di'' = k''\frac{F''}{F}\,df\,(\theta - \theta'') \tag{410}$$

If the differentials in these are replaced by finite differences and the temperatures are replaced by their average values θ_m, θ'_m and θ''_m in the interval Δf, we obtain the difference equations

$$\Delta i = \frac{G'}{G}\,\Delta i' + \frac{G''}{G}\,\Delta i'' \tag{411}$$

$$\Delta i' = \Delta f \cdot \frac{k'F'}{G'F}\,(\theta_m - \theta'_m) \tag{412}$$

$$\Delta i'' = \Delta f \cdot \frac{k''F''}{G''F}\,(\theta_m - \theta''_m) \tag{413}$$

θ, θ' and θ'' as a function of f up to the beginning of the interval being considered are assumed to have already been determined. Up to now these temperatures, or the temperature differences $\theta - \theta'$ and $\theta - \theta''$ can be drawn dependent upon f (see Fig. 123). By extrapolating the curves thus obtained up to the middle of the arbitrarily assumed new interval Δf we can determine $\theta_m - \theta'_m$ and $\theta_m - \theta''_m$. Thus we obtain $\Delta i'$ and $\Delta i''$ using Eqs. (412) and (413) and Δi using Eq. (411), whereby the values of i, i' and i'' are determined at the end of the interval. Finally we can read off the temperatures θ, θ' and θ'' at the end of the interval from the i, T-diagrams for the material in question, in which the thermal content is shown to be dependent upon temperature. We can then test whether $\theta_m - \theta'_m$ and $\theta_m - \theta''_m$ have been correctly determined using the extrapolation mentioned above. Otherwise the calculation must be repeated using improved values for these temperature differences.

When $\theta - \theta'$ or $\theta - \theta''$ vary rapidly with f, to achieve an adequate level of accuracy, we must ensure that none of these temperature differences in the interval Δf being considered decrease by more than half or are more than doubled; if this happens there is the danger that the temperature differences determined at position

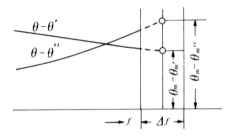

Figure 123 Determination of the temperature distribution during the exchange of heat between three materials.

$f + \Delta f/2$ in Fig. 123 will deviate too much from the standard average values for heat transfer. This condition can always be satisfied by selecting a sufficiently small Δf.

However, the calculation can be carried out with considerably larger values of Δf if, in place of $\theta_m - \theta'_m$ and $\theta_m - \theta''_m$, the corresponding *logarithmic* average values of the temperature differences in the interval in question are introduced. In order to do this we must first estimate the values of θ, θ' and θ'' at the end of the interval and we must then calculate the logarithmic average values using Eq. (5-56). The values of Δi, $\Delta i'$ and $\Delta i''$ which result from Eqs. (412), (413) and (411) then determine new values for θ, θ' and θ''. As a rule the calculation must be carried out several times until the assumed and calculated temperatures at the end of the interval are in agreement. Nevertheless, this process is much quicker than the calculation with small steps as described previously.

THREE

SECTION 65. THE FUNDAMENTAL OSCILLATION OF THE REGENERATOR GIVEN UNEQUAL THERMAL CAPACITIES OF BOTH GASES PER PERIOD ($CT \neq C'T'$)

As has already been mentioned on several occasions, the relationships developed so far for k_0, in particular Eqs. (12-40) and (12-49), can be used without noticeable errors when $CT \neq C'T'$. In order to demonstrate this, and to obtain an estimation of possible deviations, the zero eigen function will be accurately developed in the following for the general case $CT \neq C'T'$. As hitherto, C and C' and α and α' will be considered to be independent of temperature.

We will again proceed with the assumption that, according to the zero eigen function, the temperatures in the longitudinal direction of the regenerator will curve in the same way as under the same conditions in a working recuperator. In a recuperator with unequal values for C and C', according to Eq. (5-43) (in Sec. 5-6) and also following from the considerations in Sec. 5-7, not only the gas temperatures but also the air temperatures (e.g., at one of the wall surfaces) are determined using equations of the form

$$t = D + A \cdot e^{bf} \tag{485}$$

where A, b and D are constants and f is again the heating surface area serving as the longitudinal coordinate. For the zero eigen function of the regenerator we now need to reproduce also the temperature of the heat storing mass in the longitudinal direction of the regenerator, given fixed values of y and τ according to Eq. (485). However, in order to express the dependence of y and τ, A should be understood as a function of y and τ. The temperature t of the heat storing mass on the other hand, if the heat storing mass is assumed to be plate shaped, can be represented by the following solution of the differential Eq. (12-1):

$$t = D + \sum_{n=0}^{\infty} B_n e^{-\beta_n^2 a \tau} \cos \beta_n \left(y - \frac{\delta}{2} \right) \tag{486}$$

where D and β_n again denote arbitrarily selected constants while the coefficient B_n, on the other hand, is dependent upon f. Equation (486) transforms into the form of Eq. (485) if we set

$$B_n = \text{constant} \cdot e^{bf} \tag{487}$$

whereby

$$\frac{1}{B_n} \frac{d B_n}{df} = b \tag{488}$$

As is shown in Sec. 13-5 (see Eq. (13-56)), Eq. (489) provides a very good approximation for b:

$$b = \frac{1}{C'T'} - \frac{1}{CT}(T + T')k_0 \tag{489}$$

As k_0 can only be determined more precisely using the following equations, when using Eq. (489) it is necessary to insert the value for k_0 obtained from Eq. (12-40).

The values of β_n should now be so determined that Eq. (486) also satisfies the differential Eqs. (12-3) and (12-4). If we develop expressions for t_m and t_0, as in Sec. 12-5, and thus obtain an expression for θ, Eq. (12-3) can be satisfied taking Eq. (488) into consideration, by setting Eq. (490) for each β_n:

$$\beta_n \frac{\delta}{2} \tan \beta_n \frac{\delta}{2} = \frac{\dfrac{\alpha}{\lambda B} \dfrac{\delta}{2}}{\dfrac{\alpha}{bC} + 1} \tag{490}$$

The positive real solutions of these equations determine the required values for β_n.

Thus, for the temperature t' in the cold period, we require the solution

$$t' = D' + B'_0 e^{\beta_0^2 a\tau'} \cosh \beta'_0 \left(y - \frac{\delta}{2} \right) + \sum_1^{\infty} B_{n'} e^{-\beta_{n'}^2 a\tau'} \cos \beta_{n'} \left(y - \frac{\delta}{2} \right) \tag{491}$$

If we consider that, during the cold period, f is to be calculated in the opposite direction to the warm period and thus, according to Eq. (488)

$$\frac{1}{B_{n'}} \frac{d B_{n'}}{df} = -b \tag{492}$$

then the defining Eqs. (493) and (494) are obtained for

$$\beta'_0 \frac{\delta}{2} \cdot \tanh \beta'_0 \frac{\delta}{2} = \frac{\dfrac{\alpha'}{\lambda B} \dfrac{\delta}{2}}{\dfrac{\alpha'}{bC'} - 1} \qquad \text{for} \quad n' = 0 \tag{493}$$

$$\beta_{n'}\frac{\delta}{2}\cdot\tan\beta_{n'}\frac{\delta}{2}=\frac{\dfrac{\alpha'}{\lambda B}\dfrac{\delta}{2}}{\dfrac{\bar{\alpha}}{bC'}-1}\qquad\text{for}\quad n'\geqslant 1 \tag{494}$$

As will be shown in more detail, one of the values of B_n and $B_{n'}$, for example B_0, can be arbitrarily assumed for a specific brick cross-section. The remainder of these values will be determined from the reversal requirement. The basic idea here is the same as in Sec. 12-5, however it is more complicated as β_n and $\beta_{n'}$ are different. In order to satisfy the requirement that the brick temperature t_a at the beginning of the warm period and the temperature $t_{e'}$ at the end of the cold period are the same, only the values for B_n for $n\geqslant 1$ will be treated as unknown and these values will be determined so that t_a and $t_{e'}$, in the sense of the method of least squares, will be in as close as possible agreement throughout the brick. Thus if $\Delta = t_a - t_{e'}$

$$M = \int_0^\delta \Delta^2\,dy$$

should be a minimum. The requirement for this is $\partial M/\partial B_n = 0$ from which we get

$$\int_0^\delta \Delta\frac{\partial\Delta}{\partial B_n}\,dy = 0 \tag{495}$$

By calculating $\Delta = t_a - t_{e'}$ from Eqs. (486) and (491) and inserting it into Eq. (495), an equation of condition is obtained in which the values $B_{n'}$ also appear. In order to find other equations we carry out the same consideration for the end of the warm period and the beginning of the cold period whereby the values of $B_{n'}$, including B_0' are treated as unknowns. In addition, as the reversal condition is also fully satisfied in the neighbouring cross-section,

$$\left(\frac{\partial t}{\partial f}\right)_{\tau y} = -\left(\frac{\partial t'}{\partial f}\right)_{\tau y}$$

must also be valid at the moment of reversal. Here the minus sign again shows that f is measured in opposite directions in both periods. Equation (496) follows from this requirement when t and t' from Eqs. (486) and (491) are inserted and Eqs. (488) and (492) are taken into account:

$$D = D' \tag{496}$$

Thus, in this way, we finally obtain the following relationships for B_n and $B_{n'}$:

$$B_n\left[\frac{\sin\beta_n\delta}{\beta_n\delta}+1\right]=2\left[\frac{\dfrac{\alpha}{\lambda B}\dfrac{\delta}{2}}{\dfrac{\alpha}{bC}+1}+\frac{\dfrac{\alpha'}{\lambda B}\dfrac{\delta}{2}}{\dfrac{\alpha'}{bC'}-1}\right]$$

$$\times\left[B_0'e^{\beta_0'^2 aT'}\frac{\cos\beta_n\dfrac{\delta}{2}\cdot\cosh\beta_0'\dfrac{\delta}{2}}{(\beta_n^2+\beta_0'^2)\dfrac{\delta^2}{4}}+\sum_{n'=1}^{\infty}B_{n'}e^{-\beta_{n'}^2 aT'}\frac{\cos\beta_n\dfrac{\delta}{2}\cdot\cos\beta_{n'}\dfrac{\delta}{2}}{(\beta_n^2-\beta_{n'}^2)\dfrac{\delta^2}{4}}\right] \tag{497}$$

$$B_0' \left[\frac{\sin \beta_0' \delta}{\beta_0' \delta} + 1 \right]$$

$$= 2 \left[\frac{\dfrac{\alpha}{\lambda B} \dfrac{\delta}{2}}{\dfrac{\alpha}{bC} + 1} + \frac{\dfrac{\alpha'}{\lambda B} \dfrac{\delta}{2}}{\dfrac{\alpha'}{bC'} - 1} \right] \cdot \sum_{n=0}^{\infty} B_n e^{-\beta_n^2 aT} \cdot \frac{\cos \beta_n \dfrac{\delta}{2} \cdot \cosh \beta_0' \dfrac{\delta}{2}}{(\beta_n^2 + \beta_0'^2) \dfrac{\delta^2}{4}} \quad (498)$$

$$B_{n'} \left[\frac{\sin \beta_{n'} \delta}{\beta_{n'} \delta} + 1 \right]$$

$$= 2 \left[\frac{\dfrac{\alpha}{\lambda B} \dfrac{\delta}{2}}{\dfrac{\alpha}{bC} + 1} + \frac{\dfrac{\alpha'}{\lambda B} \dfrac{\delta}{2}}{\dfrac{\alpha'}{bC'} - 1} \right] \cdot \sum_{n=0}^{\infty} B_n \cdot e^{-\beta_n^2 aT} \frac{\cos \beta_n \dfrac{\delta}{2} \cdot \cos \beta_{n'} \dfrac{\delta}{2}}{(\beta_n^2 - \beta_{n'}^2) \dfrac{\delta^2}{4}}$$

$$\text{(for } n' \geq 1\text{)} \quad (499)$$

All values of B_n and $B_{n'}$ can be calculated from these equations when, as mentioned, B_0 is assumed arbitrarily. The solution of the equations appears to be complicated by the appearance of the summation terms. However, it is usually sufficient to consider only the terms with B_0 and B_0' in the very rapidly converging series in these equations. In this way, at least, it is possible to obtain a first approximation. Then if necessary, it is possible to repeat the calculation more accurately by inserting the values of B_n and $B_{n'}$, obtained for $n \geq 1$ and $n' \geq 1$ in the first calculation, into the sums, in place of the first neglected terms.

The arbitrary value B_0 can, for example, be determined so that, in the considered cross-section, a largest chronological fluctuation $(t_m)_e - (t_m)_a$ of the average temperature t_m of the heat storing mass is prescribed. As the highest terms in Eq. (486) contribute little to t_m, we can set, as an approximation:

$$(t_m)_e - (t_m)_a = -B_0 \left(1 - e^{-\beta_0^2 aT}\right) \frac{\sin \beta_0 \dfrac{\delta}{2}}{\beta_0 \dfrac{\delta}{2}} \quad \text{(approximation)} \quad (500)$$

whereby B_0 can easily be determined.

The values of all B_n and $B_{n'}$, which are valid for a specific cross-section, can, according to Eq. (487), be transferred to other arbitrary cross-sections which are at a distance of $\pm \Delta f$ by multiplying the value by $e^{\pm b \Delta f}$.

Simplified Determination of β_n, $\beta_{n'}$, B_n and $B_{n'}$

The somewhat laborious method of determining β_n, $\beta_{n'}$, B_n and $B_{n'}$ can be simplified in the following way. If the abbreviations

$$\frac{\dfrac{\alpha}{\lambda B} \dfrac{\delta}{2}}{\dfrac{\alpha}{bC} + 1} = \varepsilon \quad \text{and} \quad \frac{\dfrac{\alpha'}{\lambda B} \dfrac{\delta}{2}}{\dfrac{\alpha'}{bC'} - 1} = \varepsilon' \quad (501)$$

are introduced into Eqs. (490), (493) and (494), Eqs. (502) and (503) will be obtained by expansion from ε or ε', neglecting the higher terms:

$$\beta_0 \frac{\delta}{2} = \sqrt{\varepsilon}\left(1 - \frac{\varepsilon}{6} + \frac{11\varepsilon^2}{360}\right); \qquad \beta_0' \frac{\delta}{2} = \sqrt{\varepsilon'}\left(1 + \frac{\varepsilon'}{6} - \frac{11\varepsilon'^2}{360}\right)$$

$$\beta_n \frac{\delta}{2} = n\pi + \frac{\varepsilon}{n\pi} - \frac{\varepsilon^2}{(n\pi)^3} \quad \text{(for } n \geq 1) \quad (502)$$

$$\beta_{n'} \frac{\delta}{2} = n'\pi - \frac{\varepsilon'}{n'\pi} - \frac{\varepsilon'^2}{(n'\pi)^3} \quad \text{(for } n' \geq 1) \quad (503)$$

These equations are about 1 percent accurate up to $\varepsilon = 1$ and $\varepsilon' = 1$. Higher values of ε and ε' occur only seldom even in unfavourable cases. However, as long as we are only dealing with only small values of ε and ε', the defining Eqs. (497) to (499) for B_n and $B_{n'}$ can be brought into more simplified form by using Eqs. (502) and (503) and neglecting terms containing ε^2 and ε'^2:

$$B_0'\left(1 + \frac{\varepsilon'}{3}\right) = B_0 e^{-\beta_0^2 aT}\left(1 - \frac{\varepsilon - \varepsilon'}{2}\right) + (\varepsilon + \varepsilon')\left(1 + \frac{\varepsilon'}{2}\right)\sum_{n=1}^{\infty} B_n e^{-\beta_n^2 aT} \frac{(-1)^n}{(n\pi)^2} \quad (504)$$

and for $n \geq 1$ or $n' \geq 1$

$$B_n\left(1 + \frac{\varepsilon}{(n\pi)^2}\right) = B_0' e^{+\beta_0'^2 aT'} \frac{(-1)^n(\varepsilon + \varepsilon')(2 + \varepsilon')}{(n\pi)^2 + 2\varepsilon + \varepsilon'}$$

$$+ \sum_{n'=1}^{\infty} B_{n'} e^{-\beta_{n'}^2 aT'} \frac{(-1)^{n+n'} 2(\varepsilon + \varepsilon')}{(n^2 - n'^2)\pi^2 + 2(\varepsilon + \varepsilon')} \quad (505)$$

$$B_{n'}\left(1 - \frac{\varepsilon'}{(n'\pi)^2}\right) = -B_0 e^{-\beta_0^2 aT} \frac{(-1)^{n'}(\varepsilon + \varepsilon')(2 - \varepsilon)}{(n'\pi)^2 - \varepsilon - 2\varepsilon'}$$

$$+ \sum_{n=1}^{\infty} B_n e^{-\beta_n^2 aT} \frac{(-1)^{n+n'} 2(\varepsilon + \varepsilon')}{(n^2 - n'^2) + 2(\varepsilon + \varepsilon')} \quad (506)$$

Furthermore, in the above equations, given large values of n and n', the terms for $n \neq n'$ can often be neglected so that, in place of Eqs. (505) and (506), Eqs. (507) and (508) can be used to give a good approximation:

$$B_n - B_{n'} e^{-\beta_{n'}^2 aT'} = \frac{(\varepsilon + \varepsilon')(2 + \varepsilon')(-1)^n}{(n\pi)^2 + (\varepsilon' + 2\varepsilon)} B_0' e^{\beta_0'^2 aT'} \quad (507)$$

$$B_{n'} - B_n \cdot e^{-\beta_n^2 aT} = -\frac{(\varepsilon + \varepsilon')(2 - \varepsilon)(-1)^{n'}}{(n'\pi)^2 - (\varepsilon + 2\varepsilon')} B_0 e^{-\beta_0^2 aT} \quad (508)$$

A further simplification is usually possible by neglecting the second term on the left hand side of Eqs. (507) and (508) and by neglecting $2\varepsilon + \varepsilon'$ in return for $(n\pi)^2$, $\varepsilon + 2\varepsilon'$ in return for $(n'\pi)^2$, and ε and ε' in return for 2 on the right hand side. The values of B_n thus obtained are always sufficiently accurate for evaluating the summation in Eq. (504). Thus, if we insert these simplified expressions for B_n into the summation mentioned, Eq. (509) is obtained from Eq. (504):

$$B_0' = B_0 \frac{\left(1 - \dfrac{\varepsilon - \varepsilon'}{2}\right) e^{-\beta_0^2 aT}}{\left(1 + \dfrac{\varepsilon'}{3}\right) - (2 + \varepsilon')(\varepsilon + \varepsilon')^2 e^{\beta_0^2 aT'} \displaystyle\sum_{n=1}^{\infty} \frac{1}{(n\pi)^4} e^{-\beta_0^2 aT}} \tag{509}$$

Often the summation in this equation is so small that it can be neglected. The desired values of B_n and $B_{n'}$ can, in most cases, be found sufficiently accurately using Eqs. (509), (507) and (508).

Result of the Calculation

Having ascertained all the values of B_n and $B_{n'}$ the temperature curve in a brick cross-section can now be calculated accurately, even when $CT \neq C'T'$, using Eqs. (486) and (491). Figure 159 illustrates the result of such a calculation for $C = 2C'$ and $T = T'$, whereby $CT = 2C'T'$ also. In other respects the case is the same as in Fig. 12-4. Assuming

$$\alpha = \alpha' = 20 \frac{\text{kcal}}{\text{m}^2 \text{h}°\text{C}}$$

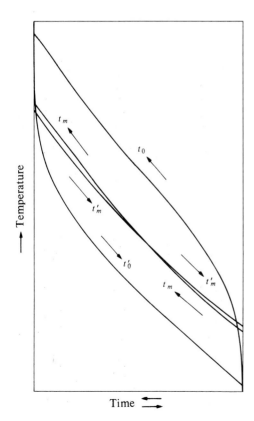

Time

Figure 159 Chronological temperature distribution in a brick when $CT = 2C'T'$.

and using the values of δ, λ etc. obtained from Fig. 12-4, according to Eq. (12-40) $k_0 = 4.04$ and according to Eq. (489) $bC = 8.08$ and $bC' = 4.04$; and B_n and $B_{n'}$ can be calculated from the accurate Eqs. (497) to (499), although Eqs. (507) to (509) give slightly different values. Figure 159 illustrates the curve of the surface area temperature t_0 or t'_0 and the curve of the average brick temperature t_m or $t_{m'}$ dependent upon time where, as earlier, time in the cold period and time in the warm period are drawn in the opposite directions. Compared with Fig. 12-4 the temperature distribution, on the whole, curves slightly towards the bottom. In Fig. 159 the values of t_m and $t_{m'}$ do not coincide but form a very small hysteresis loop whereby the curves intersect. The difference between t_m and $t_{m'}$ is so small that the figure would have to be enlarged fifty-fold in order for this difference to be clearly observed. Moreover, the temperature difference $t_m - t_{m'}$ changes its plus or minus sign during the intersection so that it increases considerably over the time average.

According to Figs. 159 and 12-4 the differences in the temperature curve when $CT = 2C'T'$ and when $CT = C'T'$ are numerically only small. In particular it is possible to show, by calculation, that the chronological average values of the temperature differences $\theta - t_0$, $t_0 - t_m$ etc. are almost exactly the same in Figs. 12-4 and 159. As these temperature differences are a determining factor for heat exchange, the heat transmission coefficient k_0 in the case handled in Fig. 159 is not noticeably different from the value calculated earlier from Eqs. (417) and (475) for $CT = C'T'$. The exact equation for k_0 when $CT \neq C'T'$, which is mentioned below, will confirm this.

Calculation of the Heat Transmission Coefficient k_0

If we differentiate expressions for θ, θ' and t_0 from Eqs. (486) and (491) and calculate the chronological average values $\bar{\theta} - \bar{\theta'}$ and $\bar{\theta} - \bar{t_0}$ by integrating over τ, insertion into Eq. (12-34) for the heat transmission coefficient k_0 when $CT \neq C'T'$ will give

$$\frac{1}{k_0} = \frac{T + T'}{T} \frac{\alpha + bC}{\alpha bC}$$

$$\times \left[\frac{\alpha'}{\alpha' - bC'} \cdot \frac{B'_0 \dfrac{e^{\beta_0'^2 aT'} - 1}{\beta_0'^2 aT'} \cos \beta'_0 \dfrac{\delta}{2} + \sum_{n'=1}^{\infty} B_{n'} \dfrac{1 - e^{-\beta_n'^2 aT'}}{\beta_n'^2 aT'} \cos \beta_{n'} \dfrac{\delta}{2}}{\sum_{n=0}^{\infty} B_0 \dfrac{1 - e^{-\beta_n^2 aT}}{\beta_n^2 aT} \cos \beta_n \dfrac{\delta}{2}} - \frac{\alpha}{\alpha + bC} \right]$$

$$(510)$$

In order to simplify the evaluation of this equation it is recommended that Eq. (511) can be used for $n \geq 1$ and $n' \geq 1$ up to $\varepsilon = 1$ and $\varepsilon' = 1$ and this gives an accuracy of about 1 percent:

$$\cos \beta_n \frac{\delta}{2} = (-1)^n \left[1 - \frac{\varepsilon^2}{2(n\pi)^2} \right] \quad \text{and} \quad \cos \beta_{n'} \frac{\delta}{2} = (-1)^{n'} \left[1 - \frac{\varepsilon'^2}{2(n'\pi)^2} \right] \quad (511)$$

In the case shown in Fig. 159, Eq. (510) gives the value $k_0 = 4.042$, i.e., within the accuracy of calculation, the same value as that obtained using Eqs. (417) or (475) for $CT = C'T'$ which give $k_0 = 4.04$. If we consider that $CT = 2C'T'$ denotes an unusually large difference between the thermal capacities of both gases per period, we can conclude from this result that Eqs. (11-4) and (12-40) are equally accurate in almost all practical cases.

Finally it must be pointed out that the precise value for k_0 obtained from Eq. (510) satisfies Eq. (489) approximately but not exactly. The strong relationship between k_0 and the assumed values of C, C', T, T' and b is given by the process described for determining β_n, $\beta_{n'}$, B_n and $B_{n'}$ and by Eq. (510).

FOUR

SIMPLIFIED CALCULATION OF THE STATE OF EQUILIBRIUM USING THE REFINED HEAT POLE METHOD[1]

The calculation of the state of equilibrium can be considerably simplified if the following two facts are observed:

1. Only the zero eigen function is valid in the central section of a long regenerator.
2. In the regions of the regenerator in which the temperature distribution is not curved, or is only very slightly so, the temperature curve at time η proceeds precisely, or very approximately from the temperature curve at time $\eta = 0$ so that, each point on the curve parallel to the ξ-axis is displaced in the direction of flow by $\Delta\xi = \eta$. For the zero eigen function, this follows directly from Eq. (13-60) when $CT = C'T'$ as $\xi - \eta = $ constant results for $t = $ constant.

For this zero eigen function the temperature distribution is a straight line. However, the following consideration shows that this principle is very close to parallel displacement in the parts of the regenerator where the temperature distribution is very slightly curved. With very slight curves the very small value of the differential coefficient $\partial^2 t/\partial\xi\partial\eta$ in Eq. (13-30) can be neglected whereupon Eq. (13-30) transforms into

$$\frac{\partial t}{\partial \xi} = -\frac{\partial t}{\partial \eta}$$

Thus for $t = $ constant we obtain

$$dt = \frac{\partial t}{\partial \xi}d\xi + \frac{\partial t}{\partial \eta}d\eta = \frac{\partial t}{\partial \xi}(d\xi - d\eta) = 0$$

---

[1] Author's method, not yet published.

and thus

$$d(\xi - \eta) = 0 \quad \text{or} \quad \xi - \eta = \text{constant}$$

as in the zero eigen function mentioned. The principle of parallel displacement thus proved is all the more empirically valid the larger are Π and Λ.

In the considerations which follow we will limit ourselves, primarily, to the case where $CT = C'T'$. Furthermore it will be assumed that $\Lambda \geqq 4\Pi$ so that only the zero eigen function prevails in the centre of the regenerator. Here Π will be so large, i.e., about >10, that the principle of parallel displacement will be sufficiently accurately valid for all temperatures which are larger than or the same as f_0 (see Fig. 187). Here f_0, f_1, f_2 etc. will again denote the temperature at the beginning of the cold period and g_0, g_1, g_2 will denote the temperature at the end of the cold period. The calculation is made more simple if $\Delta\varepsilon$ is chosen so that the reduced period length Π is equal to a whole multiple of $\Delta\varepsilon$. By way of example $\Pi = 3\Delta\varepsilon$ will be assumed. Then Eq. (668) may be applied following the principle of parallel displacement, whereby the end temperature proceeds from the initial temperature by horizontal displacement by $\eta = \Pi$ (see Fig. 187):

$$g_3 = f_0; \; g_4 = f_1; \; g_5 = f_2; \; g_6 = f_3 \; \text{etc.} \tag{668}$$

As, in the shortest of the hypothetical regenerators, f_6 and g_6 represent the temperatures in the centre of the regenerator, f_6 and $g_6 = f_3$ must satisfy the zero eigen function. This must also apply to the temperatures f_4 and f_5 which lie between f_3 and f_6.

According to Eq. (13-60) the zero eigen function has the form

$$t = D + B(\xi - \eta)$$

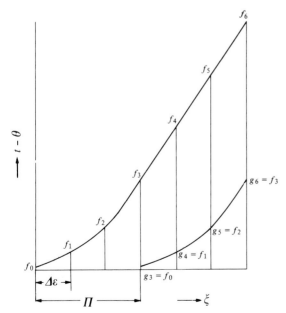

Figure 187 Temperature curve at the state of equilibrium at the beginning and end of the cold period (left-hand end of the regenerator).

where $+B$ has been written in place of $-B$ in order to remove the negative values of B. Time η will have its origin at the beginning of the period, the longitudinal coordinate will start from the left-hand end of the regenerator, beginning with 0. A relationship can immediately be formed between the two, primarily arbitrary, constants B and D on the basis of the following considerations. As can be seen, for example, from the top part of Fig. 11-10, in the centre of the regenerator ($\xi = \Lambda/2$) and in the middle of the period being considered ($\eta = \Pi/2$), the temperature of the heat storing mass must lie precisely in the middle between the two entry temperatures θ_1 and θ'_1. Thus according to the zero eigen function

$$D + B\left(\frac{\Lambda}{2} - \frac{\Pi}{2}\right) = \frac{\theta_1 + \theta'_1}{2}$$

If we take D from this equation and insert it into the previous equation, using the zero eigen function we obtain Eq. (669) for the excess temperature $u_0 = t - \theta_1$:

$$u_0 = t - \theta_1 = \frac{\theta'_1 - \theta_1}{2} + B\left(\xi - \eta - \frac{\Lambda - \Pi}{2}\right) \tag{669}$$

With $\eta = 0$ we thus obtain Eq. (670) for f_3, f_4, f_5 and f_6 by setting $\xi = 3\Delta\varepsilon = \Pi$, $\xi = \frac{4}{3}\Pi, \frac{5}{3}\Pi$ and 2Π in series (see Fig. 187):

$$f_3 = \frac{\theta'_1 - \theta_1}{2} - \left(\frac{\Lambda}{2} - \frac{3\Pi}{2}\right)B \quad f_4 = \frac{\theta'_1 - \theta_1}{2} - \left(\frac{\Lambda}{2} - \frac{11\Pi}{6}\right)B$$

$$\tag{670}$$

$$f_5 = \frac{\theta'_1 - \theta_1}{2} - \left(\frac{\Lambda}{2} - \frac{13\Pi}{6}\right)B \quad f_6 = \frac{\theta'_1 - \theta_1}{2} - \left(\frac{\Lambda}{2} - \frac{5\Pi}{2}\right)B$$

Equation (671) follows if we apply the refined heat pole method using Eqs. (663) and (667) taking note that $\Delta\varepsilon = \Pi/3$ and taking Eq. (668) into consideration:

$$g_3 = f_0 = f_3 \cdot e^{-\Pi} + \frac{\Pi}{9}[f_0 K_3 + 3.75 f_1 K_2 + 3 f_2 K_1 + 1.25 f_3 K_0]$$

$$g_4 = f_1 = f_4 \cdot e^{-\Pi} + \frac{\Pi}{9}[f_0 K_4 + 4 f_1 K_3 + 2 f_2 K_2 + 4 f_3 K_1 + f_4 K_0]$$

$$\left.\begin{array}{l} \\ \\ \\ \\ \end{array}\right\}\tag{671}$$

$$g_5 = f_2 = f_5 \cdot e^{-\Pi} + \frac{\Pi}{9}[f_0 K_5 + 4 f_1 K_4 + 2 f_2 K_3 + 3.75 f_3 K_2 + 3 f_4 K_1 + 1.25 f_5 K_0]$$

$$g_6 = f_3 = f_6 \cdot e^{-\Pi} + \frac{\Pi}{9}[f_0 K_6 + 4 f_1 K_5 + 2 f_2 K_4 + 4 f_3 K_3 + 2 f_4 K_2 + 4 f_5 K_1 + f_6 K_0]$$

Finally if f_3, f_4, f_5 and f_6 are taken from Eq. (670) and inserted into Eq. (671) four equations will be obtained, the solution of which allows the four remaining unknowns f_0, f_1, f_2 and B to be calculated. Knowing the value of B, f_3, f_4, f_5 and f_6 can then be calculated using Eq. (670) and thus, using Eq. (668), g_3, g_4, g_5 and g_6 can also be determined. By means of the refined heat pole method (Eqs. (15-7) and (15-11)), g_0, g_1 and g_2 can be found from f_0, f_1 and f_2. Thus, it follows from the reversal requirement, Eq. (15-2), that the values of f can also be determined for the right-hand end of the regenerator so long as the zero eigen function from Eq. (669) is valid for the central parts of the regenerator. In this way the temperature distribution for the state of equilibrium can be fully determined at the beginning and, using the principle of parallel displacement, also at the end of the period.

The larger the value of Π, the more simple becomes the process described. With increasing Π first f_2 and then f_1 and finally f_0, approach the values derived from the zero eigen function. Then first the last, then the next to the last and, given very large values of Π, the second of the Eqs. (671), can be omitted. Then, in Eq. (670), the equation for f_6 will be replaced by an equation for f_2 etc.

If $CT \neq C'T'$ almost the same considerations can be applied but using the zero eigen function from Eq. (13-49). In this case the principle of parallel displacement will be applied using

$$\Delta \xi = \frac{\Pi + \Pi'}{\Pi} \cdot \frac{\Lambda}{\Lambda + \Lambda'} \eta$$

However the zero eigen function cannot be determined with so clear a connection between B and D as above. If we apply all the considerations made hitherto to the left-hand end of the regenerator, we will find, at first, only one relationship between B and D. In order to obtain a second equation for B and D the same considerations must be applied as for the right-hand end of the regenerator, taking the warm period as the basis for these considerations.

CALCULATION OF k/k_0 USING THE HEAT POLE METHOD IN GENERAL CASES

Even when $CT \neq C'T'$ the efficiency factor can be calculated first and then k/k_0 can be determined from it. However, from a practical point of view it is more appropriate to find an expression for the ratio k/k_0 straight away, as mentioned in Sec. 13-8; this has the advantage of being only slightly dependent upon $CT:C'T'$.

When $CT \neq C'T'$, k must be related to the true average temperature difference $\Delta\theta_M$ between the two gases, calculated as for recuperators, and k_0 must be related to the average temperature difference $\Delta\theta_{M_0}$, when only the zero eigen function is applicable. Again θ is the gas temperature during the cold period, θ' is the gas temperature in the warm period and $\bar{\theta}_2$ and $\bar{\theta}'_2$ are the true average exit temperatures of both gases. If, on the other hand, the temperature distribution corresponds to the zero eigen function, the exit temperatures in the cold period will be denoted by $\bar{\theta}_{2,0}$. Then Eq. (678) represents the quantity of heat Q_{Per} transferred during one period:

$$Q_{Per} = k(T+T')\Delta\theta_M F = CT(\bar{\theta}_2 - \theta_1) = C'T'(\theta'_1 - \bar{\theta}'_2) \qquad (678)$$

and Eq. (679) is the corresponding quantity of heat Q_0 which is transferrred relative to the zero eigen function:

$$Q_0 = k_0(T+T')\Delta\theta_{M_0} F = CT(\bar{\theta}_{2,0} - \theta_1) \qquad (679)$$

From these two equations we then obtain

$$\frac{k}{k_0} = \frac{\bar{\theta}_2 - \theta_1}{\Delta\theta_M} \cdot \frac{\Delta\theta_{M_0}}{\bar{\theta}_{2,0} - \theta_1} \qquad (680)$$

Furthermore, as in Eq. (5-56)

$$\Delta \theta_M = \frac{(\theta_1' - \bar{\theta}_2) - (\bar{\theta}_2 - \theta_1)}{\ln \dfrac{\theta_1' - \bar{\theta}_2}{\bar{\theta}_2 - \theta_1}}$$

Equation (681) is obtained again taking Eq. (678) into consideration

$$\frac{\Delta \theta_M}{\bar{\theta}_2 - \theta_1} = \frac{\dfrac{CT}{C'T'} - 1}{\ln \dfrac{(\theta_1' - \theta_1) - \dfrac{Q_{Per}}{CT}}{(\theta_1' - \theta_1) - \dfrac{Q_{Per}}{C'T'}}} \tag{681}$$

Now if, for example, the excess temperatures f_n and g_n were determined for all sectors at the beginning and end of the cold period using the heat pole method, we would obtain the quantity of heat Q_{Per} transferred during the cold period using Eq. (672). With this, Eq. (681) is transformed into the following equation, taking Eqs. (13-28) and (13-29) into consideration:

$$\frac{\bar{\theta}_2 - \theta_1}{\Delta \theta_M} = \frac{\Lambda \Pi'}{\Lambda' \Pi - \Lambda \Pi'} \ln \frac{(\theta_1' - \theta_1) - \dfrac{\Lambda}{\Pi} \cdot \dfrac{1}{N} \sum_{n=1}^{N} (f_n - g_n)}{(\theta_1' - \theta_1) - \dfrac{\Lambda'}{\Pi'} \cdot \dfrac{1}{N} \sum_{n=1}^{N} (f_n - g_n)}$$

A similar consideration can be employed for the zero eigen function by using Eqs. (13-49), (13-51) and (13-52). According to these equations the ratio of the temperature change $\theta_2 - \theta_1$ between $\xi = 0$ and $\xi = \Lambda$ to the average value of $\theta' - \theta$ is the same at all times η. Thus it is sufficient to calculate this ratio for $\eta = 0$. Thus we obtain

$$\frac{\Delta \theta_{M_0}}{\bar{\theta}_{2,0} - \theta_1} = \frac{\Pi + \Pi'}{\Lambda \Pi'} \tag{682}$$

Finally, if we insert the last two expressions in Eq. (680) we get

$$\frac{k}{k_0} = \frac{\Pi + \Pi'}{\Lambda' \Pi - \Lambda \Pi'} \ln \frac{(\theta_1' - \theta_1) - \dfrac{\Lambda}{\Pi} \cdot \dfrac{1}{N} \sum_{n=1}^{N} (f_n - g_n)}{(\theta_1' - \theta_1) - \dfrac{\Lambda'}{\Pi'} \cdot \dfrac{1}{N} \sum_{n=1}^{N} (f_n - g_n)} \tag{683}$$

Thus we have found the required equation for k/k_0.

For $\Lambda = \Lambda' = 10$, $\Pi = 5$ and $\Pi' = \pi = 3.1416$ and using, for example, Eq. (683), $k/k_0 = 0.885$ while if we use Fig. 11-13 for $\Lambda = 10$ and the reciprocal average value

$$\Pi_M = 2 : \left(\frac{1}{\Pi} + \frac{1}{\Pi'} \right) = 3.86$$

the ratio k/k_0 is read off as 0.88. The difference lies in the accuracy of Fig. 11-13. Thus is proved the statement made previously (see Secs. 11-4 and 13-8) that k/k_0 can always be read off from Fig. 11-13 with a sufficient degree of accuracy, even when $CT \neq C'T'$, as long as suitable average values are used for Λ and Λ' and Π and Π', e.g., from Eqs. (11-5) or (11-6).

Using Eq. (683) k/k_0 can also be calculated, with the help of the refined heat pole method described in Sec. 15-2; this is achieved by replacing the summation

$$\frac{1}{N} \sum_{n=1}^{N} (f_n - g_n)$$

by the integral

$$\frac{1}{\Lambda} \int_0^n [f(\xi) - g(\xi)]\, d\xi$$

and evaluating this employing the Simpson rule, using Eq. (676).

INDEX